Cytogenetic Assays of Environmental Mutagens

One of the most pressing problems in reversing our environmental deterioration has been the identification of physical and chemical agents capable of inducing genetic toxicity. The cytogenetic test system is one of the most useful short-term systems for screening out these mutagenic agents. In addition to the classic chromosome aberrations, other endpoints have been developed for these assays, such as sister chromatid exchanges, unscheduled DNA synthesis, and DNA content measurements. This volume presents a limited spectrum of materials and methods in use for those who wish to embark on a mutagen testing program or for those who have a test program but wish to learn other systems.

Cytogenetic Assays of Environmental Mutagens

Edited by
T. C. HSU

ALLANHELD, OSMUN Publishers

ALLANHELD, OSMUN & CO. PUBLISHERS, INC.
(A Division of Littlefield, Adams and Co.)

Published in the United States of America in 1982
by Allanheld, Osmun & Co. Publishers, Inc.
81 Adams Drive, Totowa, NJ 07512

Copyright © 1982 by Allanheld, Osmun & Co.

All rights reserved. No part of this publication may be reproduced, stored in a retrieval system, or transmitted in any form or by any means, electronic, mechanical, photocopying, recording, or otherwise, without the prior permission of the publisher.

Chapters 11 and 14 were produced under contract to the U.S. Government and the publisher acknowledges the U.S. Government's right to retain a nonexclusive, royalty-free license in and to any copyright covering these chapters

Library of Congress Cataloging in Publication Data
Main entry under title:

Cytogenetic assays of environmental mutagens.

   Includes index.
   1. Mutagenicity testing. 2. Cytogenetics. 3. Chromosomes—Examination. 4. Environmentally induced diseases. I. Hsu, T. C. [DNLM: 1. Chromosome aberrations. 2. Environmental pollutants. 3. Cytogenetics. 4. Mutagens. QH462.A1 C997]
QH465.AlC97   575.2'2'028   79-88262
ISBN 0-916672-56-5

Printed in the United States of America.

Contents

Preface		vii
Acknowledgments		ix
1	Introduction *T. C. Hsu*	1
2	DNA Repair *Robert B. Painter*	11
3	Sister Chromatid Exchange Analysis: Methodology, Applications, and Interpretation *S. A. Latt, R. R. Schreck, K. S. Lovejoy, and C. F. Shuler*	29
4	Root Tips of *Vicia faba* as a Material for Studying the Induction of Chromosomal Aberrations and Sister Chromatid Exchanges *B. A. Kihlman*	81
5	Insect Cells for Testing Clastogenic Agents *Mary Esther Gaulden and Jan C. Liang*	107
6	Detection of Sister Chromatid Exchanges In Vivo Using Avian Embryos *Stephen E. Bloom*	137
7	The Use of Cytogenetics to Study Genotoxic Agents in Fishes *A. D. Kligerman*	161
8	Human Peripheral Blood Lymphocyte Cultures: An In Vitro Assay for the Cytogenetic Effects of Environmental Mutagens *K. E. Buckton and H. J. Evans*	183
9	Assays for Chromosome Aberrations Using Mammalian Cells in Culture *William Au and T. C. Hsu*	203
10	The Micronucleus Test: An In Vivo Bone Marrow Method *W. Schmid*	221
11	The Use of the Host-Mediated Assay for Cytogenetic Studies *R. J. Preston and J. G. Brewen*	231

12	Measurement of DNA Repair Synthesis in Cultured Human Fibroblasts as a Short-term Bioassay for Chemical Carcinogens and Carcinogenic Mixtures *R.H.C. San and H. F. Stich*	237
13	Male Germ Cell Cytogenetics *I.-D. Adler*	249
14	Cytogenetic Analysis of Mammalian Oocytes in Mutagenicity Studies *J. G. Brewen and R. J. Preston*	277
15	The Heritable Translocation Test in Mice *B. M. Cattanach*	289
16	Application of Flow Cytometry to Cytogenetic Testing of Environmental Mutagens *Larry L. Deaven*	325
17	Premature Chromosome Condensation for the Detection of Mutagenic Activity *Walter N. Hittelman*	353
18	The Epidemiological Approach: Chromosome Aberrations in Persons Exposed to Chemical Mutagens *Erich R. E. Gebhart*	385
19	Short-term Cytogenetic Tests in Modern Society *Sheldon Wolff*	409
Name Index		413
Subject Index		423
About the Editor		431

Preface

Many factors have contributed to our environmental deterioration. One of the most pressing problems in reversing this trend is to identify physical and chemical agents capable of inducing genetic toxicity. Without data carefully collected from a variety of test systems, regulatory actions may be either inadequate or excessive. Carcinogenesis tests are reliable, but tedious and costly. Therefore, short-term test systems to screen out the most potent mutagens are highly desirable. The cytogenetic test system is one of the most useful short-term systems.

Cytogeneticists have employed a variety of test materials and test methods to measure the effects of mutagens. In addition to the classic chromosome aberrations, orther endpoints (such as sister chromatid exchanges, unscheduled DNA synthesis, and DNA content measurements) have been developed for assays. The chapters of this monograph offer a limited spectrum of materials and methods in use for those who wish to embark on a mutagen testing program or for those who have a test program but wish to learn other systems. The variety of materials being used is extremely important, not only because they can complement one another, but also because they may be suitable for different laboratory conditions. For example, in our laboratory we use mammalian cells in culture as our principal test material. But we have no facilities to work on plant or grasshopper systems. Conversely, a laboratory without cell culture facilities may easily use the plant, chick embryo, or insect protocols.

I would like to thank the scientists who took pains to write their chapters for this monograph. I would also like to express my appreciation to Mr. Matthew Held of Allanheld, Osmun and Co., Publishers, who initially contacted me to see if I would be willing to organize such a book.

T. C. Hsu

Contributors and Their Affiliations

Adler, Dr. I.-D.
Department of Genetics
Gesellschaft f. Strahlen
    Umweltforschung
D-0842 Neuherberg
West Germany

Bloom, Dr. Stephen
Department of Poultry Science
Cornell University
Ithaca, New York 14853

Buckton, Drs. K. E. and *H. J. Evans*
MRC Clinical & Population
    Cytogenetics Unit
Western General Hospital
Edinburgh EH4 2XU
Scotland

Cattanach, Dr. Bruce M.
MRC Radiobiology Unit
Harwell, Didcot, Oxon OX11 ORD
England

Deaven, Dr. Larry L.
EV-32, Mail Stop E-201, Germantown
Office of Environment
U.S. Department of Energy
Washington, D.C. 20545

Gaulden, Drs. Mary Esther and *Jan C. Liang*
Department of Radiology
The Univ. of Texas Health Science
    Center
5323 Harry Hines Blvd.
Dallas, Texas 75235

Hittelman, Dr. Walter
Dept. of Developmental Therapeutics
The University of Texas System Cancer
    Center
M.D. Anderson Hospital and Tumor
    Institute
Texas Medical Center
Houston, Texas 77030

Kihlman, Prof. B. A.
Dept. of General Genetics
University of Uppsala
S750 07
Uppsala 7, Sweden

Kligerman, Dr. A. D.
CIIT
P.O. Box 12137
Research Triangle Park, N.C. 27709

Latt, Dr. Samuel A.
Division of Genetics
Children's Hospital Medical Center
Department of Pediatrics
Harvard Medical School
Boston, Massachusetts 02115

Painter, Dr. Robert B.
Laboratory of Radiobiology
University of California
San Francisco, California 94143

Preston, Drs. R. J. and *J. G. Brewen*
Biology Division
Oak Ridge National Laboratory
P.O. Box Y
Oak Ridge, Tennessee 37830

Medical Affairs
Allied Chemical Corporation
Morristown, New Jersey 07960

San, Drs. R. H. C. and *H. F. Stich*
Environmental Carcinogenesis Unit
British Columbia Cancer Research
    Center
601 West 10th Ave.
Vancouver, B.C., Canada

Schmid, Dr. Werner
Institute of Medical Genetics
University of Zurich
Zurich, Switzerland

Wolff, Dr. Sheldon
Laboratory of Radiobiology
University of California
San Francisco Medical Center
San Francisco, California 94143

Acknowledgments

The work in Chapter 2 was supported by the U.S. Department of Energy. Chapter 5 is dedicated to Dr. J. Gordon Carlson, who introduced the authors directly (Dr. Gaulden) and indirectly (Dr. Liang) to insect chromosomes. The research for Chapter 12 was supported by the National Cancer Institute of Canada. The experimental work for Chapter 13 was performed under contract No. 136-77-ENV D of the Environmental Research Programme of the European Communities. The work reported in the tables of Chapter 13 was supported by the EEC contract No. 136-77-1 ENV D. The research for Chapter 19 was supported by the U.S. Department of Energy.

1
Introduction

by T. C. HSU*

Soon after the discoveries by H. J. Muller and E. Altenberg that X-ray and ultraviolet light, respectively, can induce mutations, studies on radiation-induced chromosome damage began. In the 1930's and 1940's, a good deal of information was gathered relating to radiation-induced chromosome aberrations, their mechanisms, behavior in cell division, and fate in subsequent cell generations, both in mitosis and in meiosis. Primarily, insect (*Drosophila*, grasshoppers) and plant (maize, *Allium, Tradescantia,* etc.) systems were used to obtain pertinent information and conclusions.

Study of the effects of chemicals on chromosomes and mitotic apparatus started when colchicine was discovered to arrest dividing cells at metaphase. Albert Levan and other cytologists made some significant contributions in this area. However, research activities relating to chemical effects on chromosomes were mainly limited to academic exercises until the recent decade when concern over our deteriorating environment became acute and demand for effective short-term assay systems to screen mutagens and carcinogens became strong. Numerous test systems, from microorganisms to laboratory animals, and numerous endpoints, from point mutations to tumor induction, have been developed. Cytogenetic tests have become one of the tiers in the overall test program.

Chromosome damage constitutes a set of efficient, reliable and economical criteria to measure genetic toxicity. There are two major types of chromosome abnormalities, both of which cause genetic disturbance of the cell: (1) chromosome breakage and (2) changes in chromosome number. Agents that can cause chromosome breakage are referred to as clastogens, whereas agents that can cause numerical chromosome changes are referred to as mitotic poisons. Clastogens induce damage directly in the genetic apparatus, whereas mitotic poisons induce a doubling or an imbalance of the genome. Both types occur spontaneously, but many agents

*Department of Cell Biology, The University of Texas System Cancer Center, M. D. Anderson Hospital and Tumor Institute, Houston, Texas 77030.

2 Introduction

(physical, chemical, and biological) can significantly increase their frequencies. In somatic cells, an increase in chromosome abnormalities may increase the chance of developing neoplasia; in the germline cells, an increase in chromosome abnormalities may lead to a higher frequency of spontaneous abortions, birth defects, and heritable chromosomal rearrangements. Thus, it becomes increasingly evident that cytogenetic effects of environmental agents should be carefully studied and, more importantly, effectively screened.

Regardless of the materials and the endpoints used, there are basically only two types of cytogenetic test systems: the classic chromosome breakage measurement and the more recent sister chromatid exchange (SCE) measurement. Some investigators hesitate to employ chromosome breakage as a short-term test system because they consider that the cytological features of chromosome aberrations are too complex to learn and perhaps too subjective to read. Therefore, the SCE test, with its more clear-cut endpoints which can be easily counted as a quantitative approach, has gained popularity. However, there are several reasons for advocating that the rate of chromosome breakage should be employed in mutagen test programs, at least in conjunction with SCE counts:

1. The mechanism of chromosome breakage is better understood than that of SCE.
2. The confidence level of chromosome breakage is somewhat higher than that of SCE. For example, the X-ray, one of the earliest carcinogens and mutagens discovered, causes a linear dose response in the frequency of chromosome breakage but causes a limited increase in the SCE rate. Similar discrepancies were found in bleomycin, neocarzinostatin, etc.
3. The procedure for SCE staining, expecially for cells in vivo, is more cumbersome than that for chromosome aberrations. Conventional Giemsa staining suffices for the latter.

The comparison above does not suggest that SCE is not a good test system. It merely emphasizes that the classic chromosome breakage tests should be and can be effectively utilized as an assay procedure, and that SCE can be applied as an additional criterion.

As just mentioned, a number of laboratories engaged in mutagen testing are reluctant to use chromosome aberrations (breakage and rearrangements) as criteria of measuring genetic toxicity of an agent. They think that chromosome aberrations are too difficult to learn, too complex to decipher, have too many categories to score, and are hence not efficient for quantitation. Cytogeneticists may have been guilty of insisting on classifying and recording all types of aberrations and demanding complicated calculations. In conducting in-depth research on the effects of an agent, this approach is meritorious; but when using chromosome aberrations as criteria for an assay system, some modifications in concept as well as in procedure should be made to achieve efficiency and economy without severely sacrificing information. Thus, many experiments designed to find out details of the mechanism of drug action, relation to the cell cycle, the fate of aberrations, localization of breaks, and other details are not necessary for drug assays. For a drug assay, two main questions are asked: (1) whether an agent can cause chromosome damage; and (2) if so, how much. If these two goals are reached, some details can be omitted or can be left to others to do comprehensive research. Therefore, the methodology and the protocol should be geared toward these two objectives. Workers should possess some background knowledge in this field and should be trained to do competent work, but the training is not as time-consuming and as difficult as people are led to believe. To recognize the various aberrations is not an insurmountable job, especially when one understands the basic lesions and the subsequent changes after the lesions are induced.

There is really only one basic lesion, a break in a chromatin fiber. Unfortunately, cytologists must examine the lesions when the cells come to metaphase, so that many events can take place between the time at which a break is induced and the ensuing metaphase (M_1). But the seemingly confusing and complicated aberration patterns are not frighteningly difficult to recognize and to quantitate if one realizes (1) it takes two breaks, and *only* two breaks, to form an exchange; and (2) a chromosome in interphase may contain one chromatid (G_1 and early S) or two chromatids (late S or G_2). The common aberrations can be classified briefly and diagrammatically represented as follows.

Chromatid-type Aberrations (late S or G_2)

Let us first examine the lesions occurring in interphase when the fiber is decondensed and no repair (restitution) takes place; then, a chromatid break will manifest at metaphase.

Interphase   Metaphase

Multiple chromatid breaks should be counted individually regardless of whether they occur in one chromosome or several chromosomes:

4 breaks

or

4 breaks

Multiple breaks may restitute in a variety of ways, but there are only three principal types:

Interchromosomal Exchanges

In the classic literature, Type A is known as symmetrical exchange and Type B as asymmetrical exchange. A symmetrical exchange results in two regular chromosomes with a translocated segment, whereas an asymmetrical exchange results in a dicentric and an acentric fragment. During anaphase, chromosomes with symmetrical exchange will divide normally, but those with asymmetrical exchange will display a chromatin bridge and a fragment. In cytogenetic assays, this distinction is not necessary; each means two breaks and a restitution.

4 Introduction

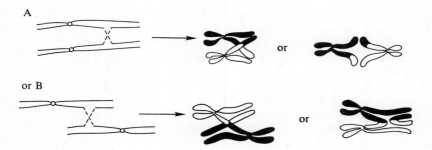

Some recent cytogeneticists use the terms "quadriradial" to describe chromatid exchange between two metacentric chromosomes and "triradial" to describe exchange between a metacentric and an acrocentric. Such terms are not useful in a genetic toxicology program.

Since exchanges are individual events, multiple exchanges, whether they involve two chromosomes or many chromosomes, may be counted separately and converted into number of breaks:

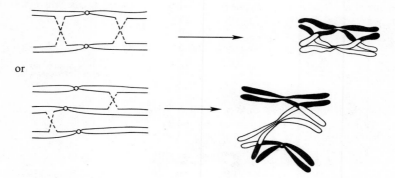

In actual recording into the raw data book, both types illustrated above can be recorded as two exchanges, hence four breaks. A metaphase with five exchange points are so recorded irrespective of the number of chromosomes involved. The number of breaks is then 10.

Intrachromosomal Exchanges

An interchromosomal exchange leads to a translocation and an intrachromosomal exchange leads to an inversion. But those are genetic consequences. In mutagen assay work, the purpose is to find out whether an agent can cause chromosome damage and how potent it is, not to find the consequence of the damage. Therefore, an intrachromosomal exchange should also be recorded just as an exchange, hence two breaks.

Interstitial Deletion

This is not a common aberration, but it can be observed occasionally. Two breaks in one chromatid restitute, leaving the middle segment free:

At metaphase the deleted segment is usually associated with its sister segment which forms a hairpin-like bulge. An interstitial deletion should again be considered as two breaks.

So the recording of aberrations is not complicated as long as each aberration is entered into the data book individually without worrying about which chromosomes are involved.

Chromatid Gap

Chromatid gap is the type of chromatid lesion that is very similar to a chromatid break but the attenuated segment is shorter than chromatid breaks. Some cytogeneticists believe that gaps and breaks are not the same. Therefore, they consider that gaps should be disregarded or scored separately from breaks. An arbitrary method for distinguishing a gap from a break is that when the length of an attenuated region is shorter than the diameter of the chromatid, it is diagnosed as a gap. If the length of the attenuated region is equal to or longer than the diameter of the chromatid, it is called a break. This definition is not only arbitrary but also has problems. What stage of metaphases should one read? The diameter of the chromatid at prometaphase is much smaller than that at full metaphase, especially when a mitotic arrestant is used to accumulate metaphases. Therefore, at full metaphase, there will be more gaps than breaks and, at prometaphase, it will be the reverse.

Many times a chromatid gap or sometimes even a chromatid break (according to the definition just described) will exhibit a thin string running in the achromatic area connecting the two chromatic ends. This string may be so thin that it is barely perceptible with an oil-immersion lens or it may be nearly half the width of the chromatid. Again, this causes problems in deciding whether it should be considered as a break or a gap.

One of the considerations for those who attempted to define chromatid breaks and chromatid gaps (Chatham Bars Conference, 1971) was that many observers may confuse secondary constrictions as gaps. Normally, secondary constrictions represent the nucleolus organizer regions. Each species has its own characteristic number and locations of secondary constrictions. In the human karyotype, for example, there may be ten such secondary constrictions located in the short arms of the D- and G-group chromosomes. In the laboratory mouse, the secondary constrictions are located in the paracentric regions of five pairs of chromosomes, but not all of them are detectable in every metaphase. The investigators working on the chromosomes of a particular species should be acquainted with the locations of the nucleolus organizers. Under the influence of certain mutagens, e.g., actinomycin D, the secondary constrictions may be stretched to give an impression of a break. In the Chinese hamster karyotype, the X chromosome frequently exhibits a constriction in its long arm. This constriction is not a nucleolus organizer nor a lesion.

In flame-dried preparations, the C-band areas of human chromosomes, particularly Nos. 1, 9, and 16, may be weakly stained as if they were lesions. Again, investigators should be prepared to recognize such areas. But once they are cognizant of these facts, we believe that to distinguish chromatid gaps and chromatid breaks is unnecessary, at least in mutagen assays.

Comings (1974) offered an interesting hypothesis regarding the nature of the gap. In essence, he thinks gaps and breaks represent the same basic lesions but differ in the time at which the lesion was induced. The basic lesion is a break on a chromatin fiber. At metaphase, the fiber runs up and down the chromatid after condensation so that it appears that a chromatid contains a bundle of fibers across its diameter. In interphase, especially in the S phase, the chromatin fiber may be fully extended. The fiber begins to fold at late S and G_2 stages. When the chromatin is fully extended, and a break occurs at that time, subsequent condensation of the chromosomes will result in a chromatid break. However, if the chromosome is already partially condensed (e.g., G_2), each chromatid has many folding fibers running longitudinally. Although a break at one location severs the fiber, the chromatid is still held together by other fibers, giving a gappy appearance. Thus, the earlier the occurrence of the fiber break, the longer will be the achromatic segment; and the difference between a gap and a break may be only the final morphological expression, not their cause.

Whatever the real cause of chromatid gaps, the arbitrary definition produces more harm than good when chromosome aberrations are used to assay mutagens. The insistence of some cytogeneticists on differentiating chromatid breaks and gaps is one of the reasons many laboratories refrain from using chromosome aberrations for cytogenetic testing. A decision must be made every time such chromatid lesions are observed (and they are most common).

Under the influence of mutagens, the frequencies of both gaps and breaks increase. So a gap is also a chromosome lesion. It may be convenient, in routine analyses of mutagen effects, to consider both chromatid gaps and chromatid breaks as chromatid lesions without distinguishing them. However, for experienced cytogeneticists, gaps and breaks may be recorded separately in the data sheets as they see them. In this connection, another phenomenon should be mentioned: the induction of banding appearance by some chemicals, such as actinomycin D. The lightly stained bands are not breaks or gaps.

Chromosome-type Aberrations (G_1 or early S)

Superficially, chromosome-type aberrations are simpler to analyze than chromatid-type aberrations. Each chromosome has one chromatid at the time a lesion occurs (G_1 and early S); therefore, rearrangements are not as complicated. A single break without restitution will yield an acentric fragment:

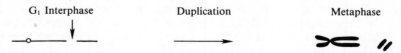

G_1 Interphase   Duplication   Metaphase

Thus, at M_1, a fragment represents one break.

Rearrangements are of two major types, again interchromosomal and intrachromosomal.

A. Interchromosomal

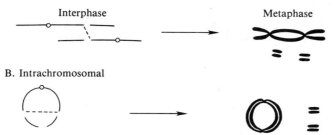

B. Intrachromosomal

To determine whether an agent is a clastogen, chromosome-type aberrations are as good indicators as chromatid-type aberrations. Dicentrics and rings are very conspicuous objects. However, if the question is how potent is this agent?, then one needs quantitative data. And here chromosome-type aberrations become troublesome. Let us return to interchromosomal rearrangements.

First, an exchange between two G_1 chromosomes does not always result in a dicentric. A symmetrical exchange will result in two normal-looking chromosomes:

Without banding, they do not suggest that anything has happened. But in routine screening programs, banding chromosomes to see if such exchanges have occurred is not practical. Using chromatid-type aberrations for recording, one can easily identify both types.

If the exchange is asymmetrical, then a dicentric will be formed. In the ensuing metaphase, one sees a dicentric and two acentric fragments. The two fragments and the dicentric are the results of two breaks, not four. Therefore, if a dicentric is accompanied by two fragments, the two fragments should be disregarded in final computation. However, the two fragments may also restitute into one longer acentric fragment:

If these are the only aberrations in this metaphase, at least one can surmise what has happened with some degree of confidence. But suppose another chromosome has a simple break in addition to the dicentric formation. Now we get two fragments:

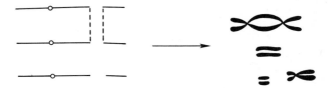

8 Introduction

Unless banding is done, there is no way to tell whether two or three breaks have been induced. The same dilemma will occur in cells with ring formation.

Interstitial deletion with restitution gives a similar problem:

Microscopically, the fragment is no different from a simple break. Yet it represents two breaks instead of one. These are the reasons that using chromosome-type aberrations alone for scoring is not desirable. Furthermore, a drug entering a cell in G_1 phase will stay with the DNA throughout the cell cycle or even in the next cycle. Therefore, the cells may simultaneously exhibit both chromosome- and chromatid-type aberrations. Quantitation will, therefore, be more complex.

The brief discussion just presented emphasizes several points:

1. The purpose of cytogenetic assays of environmental mutagens is different from that of in-depth research on cytogenetic effects of individual drugs. Therefore, protocols and methods for data collection should also be different.

2. For obtaining better quantitative data, chromatid aberrations seem to be more suitable than chromosome aberrations. Thus, a short-term incubation to cover only a portion of the S phase and the G_2 phase, for both in vitro and in vivo materials, is more desirable than long-term incubations. Longer exposure may be used as confirmation and for slow-acting drugs.

3. Studies on chromosome aberrations are essential. The SCE test should not be used alone to measure genetic toxicity of mutagens.

A few words should be said concerning competence of investigators. To do a competent job in any profession, some basic knowledge and technical training are absolutely necessary. Cytogenetics is no exception. We have known cases of industry employing biochemists or bacteriologists to be in charge of a cytogenetic toxicology program. Not only is it a waste of their respective talents and trainings, but also it contributes to inaccurate data and confused interpretation. A cytogenetic test program does not necessarily require expert cytogeneticists to do all the microscopic work. But those engaged in cytogenetic assays must have an adequate basic knowledge and technical competence. Even cytogeneticists who have worked previously on other problems should take some time to adjust to this specific program.

Another important factor is competent technical personnel. A certain period of training is absolutely necessary for each technician or assistant. If the turnover rate is high, the job cannot be done well. Stability is the key. But even with competent technical personnel, at least one investigator with research experience should be placed in this program.

There is an urgent need to train a number of investigators and technicians to work on genetic toxicology with special reference to cytogenetic toxicology. An intensive course (with laboratory exercise) of three weeks should be a good starter. This course should cover the basic DNA and chromatin structure, chromosome organization, repair mechanism, cell cycle, and the causes, expressions and consequences of chromosome damage, SCE, and related subjects. The trainees need to know something about banding, but they do not have to become experts in chromosome identification (training clinical cytogeneticists is the opposite). Perhaps the government or industrial agencies should seriously consider sponsoring such training programs — it would be money well spent.

An alternative is to urge those engaged in cytogenetic toxicology to learn and to participate in research in laboratories working in such problems. Since this is not a didactic course, it may require a longer time, say six months. The experience will be invaluable for future work.

Literature Cited

Chatham Workshop Conference on Karyological Monitoring of Normal Cell Populations. 1971. Intern. Assoc. of Biol. Standardization.

Comings, D. E. 1974. What is a chromatid break? *In* German, J., ed. *Chromosomes and cancer*. New York: John Wiley & Sons; 95–133.

2

DNA Repair

by ROBERT B. PAINTER*

Abstract

Ionizing radiation causes base damage, single strand breaks, double strand breaks and complex lesions that include some or all of these lesions. Repair of single strand breaks and base damage is rapid and almost complete within an hour after irradiation. Double strand breaks are repaired more slowly but, in surviving cells, all double strand breaks are probably repaired. Ultraviolet light induces base damage, mostly in the form of pyrimidine dimers, DNA-DNA crosslinks, and DNA-protein crosslinks. In human cells, dimers are repaired over a time period of a few days by a process known as nucleotide excision repair, which is described. DNA-DNA crosslinks and DNA-protein crosslinks are slowly repaired but the mechanisms are unknown.

After treatment of cells with chemical agents, most lesions are in the form of adducts in which the agent is bound to DNA by covalent bonds, although breaks in DNA are formed directly by a few chemicals. There is a spectrum of addition sites for each chemical agent. The kinetics of repair varies considerably for each adduct site and also from cell type to cell type. A process called base excision repair, which is also described, is responsible for repair of many simple adducts, such as alkyl groups. Nucleotide excision repair probably plays an important role in the repair of bulky addition products such as those formed by benzo(a)pyrene or aflatoxin B_1. Methods for measuring various kinds of DNA repair are briefly described. The possible roles for DNA repair in the formation of chromosome aberrations and sister chromatid exchanges are briefly discussed.

Introduction

There has been tremendous interest in DNA repair in the past several years, and much of it has come about because of the relationships between DNA damage and mutation and possible cancer. As the epidemiological and laboratory evidence has grown for DNA damage playing a major role in carcinogenesis, more and more em-

*Laboratory of Radiobiology, University of California, San Francisco, California 94143.

phasis has been placed on how cellular DNA repair systems reverse and modify this damage.

"Repair" is a word that has been heavily used in the literature in toxicology, radiobiology, and genetics for many years. It was not until relatively recently, however, that it became evident that many of the phenomenological manifestations of repair at the organismal or cellular level originate at the level of DNA. The first evidence for a DNA repair system came when Rupert (1960) showed that ultraviolet (UV) light-induced damage to *Hemophilus* DNA could be reversed if the DNA was incubated with yeast extract in the presence of visible light. This in vitro reactivation of transforming DNA proved that previous demonstrations of photoreactivation at the cellular level were mediated through DNA.

The first demonstrations of biochemical removal of damage in DNA came simultaneously in reports by Setlow and Carrier (1964) and Boyce and Howard-Flanders (1964) of the specific excision of UV light-induced thymine dimers from the cellular DNA of *Escherichia coli*. These demonstrations were the beginning of the era of DNA repair. In the same year, Rasmussen and Painter (1964) reported evidence for DNA repair in mammalian cells. They found that human cells that had been irradiated with UV light exhibited a phenomenon later called unscheduled DNA synthesis (Djordjevic and Tolmach 1967). This was detected autoradiographically as the simultaneous participation of all cells in an irradiated asynchronous culture in low levels of DNA synthesis. In unirradiated cultures, only those 30% or so of cells that are in the normal process of replicating their genome (in the S phase) synthesize DNA at any instant. Later work showed that unscheduled DNA synthesis was actually a manifestation of repair synthesis, i.e., the insertion of new bases into DNA after damaged ones had been excised (Painter and Cleaver 1969).

The biological importance of eukaryotic DNA repair was first demonstrated by Cleaver (1968), who showed that individuals with the autosomal genetic disease, xeroderma pigmentosum, were incapable of carrying out unscheduled DNA synthesis. Later work proved that these patients (whose clinical symptoms always include multiple skin cancers) do not excise pyrimidine dimers from their DNA (Cleaver and Trosko 1970). This established the first convincing correlation between defective DNA repair and carcinogenesis.

Since 1968, when Roberts et al. showed repair replication in the DNA of human cells after exposure to nitrogen mustard, there has been a host of papers demonstrating various kinds of DNA repair induced by chemical damage to DNA. An excellent review of this literature was recently published (Roberts 1978). It is now well established that most agents known to cause cancer do so by mechanisms involving DNA damage. Outstanding exceptions are asbestos and some of the hormones that seem to act by other mechanisms. But the great majority of chemicals that cause cancer are also known to be mutagens (McCann and Ames 1976). Thus, many short-term tests have been developed to detect mutagenic carcinogens; it is believed that many or all of the tests covered in this book reflect the consequences of DNA damage or repair at the chromosomal level.

The first part of this paper treats *DNA lesions induced by mutagens and carcinogens*.

Lesions Caused by Radiation

Much of the early work on cytogenetics used radiation as an agent to damage chromosomes. For this reason, and because of the use of ionizing radiation for cancer therapy and the impact of the atomic age, DNA damage and repair following exposure to radiation has been more heavily studied than DNA damage and repair after chemical exposure. Moreover, the biochemical processes following damage induced by ionizing radiation and by UV light form two extreme classes of repair, which serve as useful models for the study of DNA repair of damage induced by chemicals. For this reason, the damage and repair of DNA after exposure to radiation will be discussed first.

Lesions Induced by Ionizing Radiation

Single-strand breaks (and alkali labile bonds). The best-studied lesions induced by ionizing radiation are single-strand breaks (ssb). These are actually a class of lesions because any damage that results in a disruption in the continuity in one of the two strands of the double helical DNA molecule is called an ssb. Single-strand breaks can occur at any one of four points in the sugar-phosphate backbone (Fig. 2.1). It has

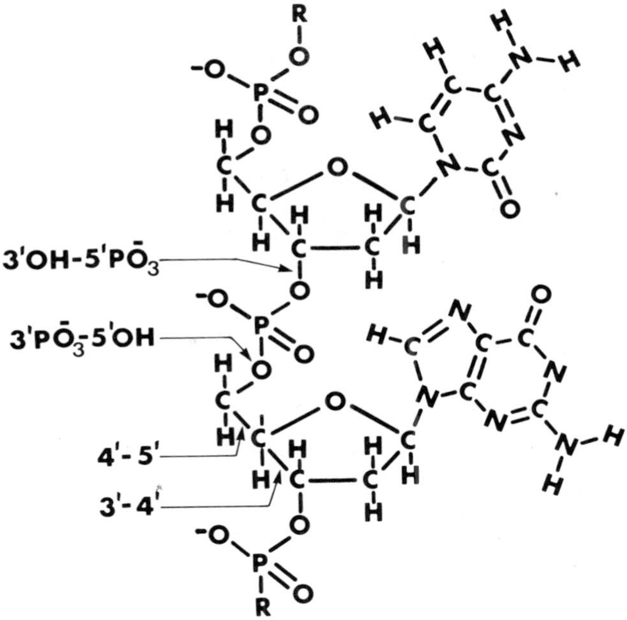

Figure 2.1 Sites in DNA where breaks in DNA can result in single-strand breaks. Note that damage in the bases or disruptions in the 2'-3' or 1'-2' bonds will not cause single-strand breaks.

only been possible to obtain a small amount of information on the relative abundance of these kinds of breaks. When thymus cells are heavily irradiated with ionizing radiation, about two-thirds of induced ssb carried a 3' hydroxyl group (Lennartz et al. 1975a,b). One would expect, then, that an equal percentage would carry 5' phosphoryl groups, but this is not the case. Only about one-fourth of the ssb carry the 5' phosphoryl end group (Coquerelle et al. 1973). These data indicate that the majority of breaks induced by ionizing radiation do not simply disrupt the bond between two atoms; more extensive damage occurs that results in loss of more than one atom at the site of the damage.

The study of ssb has been complicated by the fact that certain kinds of damage which are not actually single-strand breaks are converted to ssb upon exposure to alkali. This is important because the favored method for study of ssb has been the alkaline sucrose gradient method first introduced by McGrath and Williams (1966). For this technique, cells are lysed at pH 12 or higher to free the DNA from other cellular material and to separate the helix into single strands. Regions in DNA where purines or pyrimidines have been removed from the sugar phosphate backbone (spontaneously or by action of a damaging agent, directly or enzymatically), i.e., apyrimidinic or apurinic sites, are very sensitive to alkaline conditions and will be converted into ssb. These and other more poorly characterized lesions then will be registered along with those breaks produced directly by the damaging agent. It appears that about 20-30% of ionizing radiation-induced ssb registered by the alkaline sucrose technique are actually due to alkali-sensitive sites (Lennartz et al. 1975a,b).

Double-strand breaks. To be classed as a double-strand break (dsb) there must be a discontinuity in both strands, at least within a few nucleotides of each other. Dsb are induced in cells much more rarely than ssb; about 1/10 to 1/70 of ssb are also dsb (Corry and Cole 1968; Lehmann and Ormerod 1970). After low-LET irradiation, they are formed linearly with dose in cells (Coquerelle et al. 1973) and therefore are rarely due to two separate photons. Apparently, the ionizations along a track of even sparsely ionizing radiation are sometimes close enough to cause damage simultaneously in both strands of the double helix. Methods for study of dsb are still not very accurate and little is known about their chemical characteristics.

Base damage. Radiation studies with DNA in solution indicate that damage to the bases occurs as often, or more often, than does damage causing ssb. However, technical problems have hindered the study of intracellular base damage caused by ionizing radiation; the only information available is on thymine (Cerutti 1974). The most common effect of ionizing radiation on thymine is to induce an unsaturation of the ring, resulting in glycols and similar substances (Fig. 2.2). The in vitro radiation studies make it highly probable that other bases are also damaged, but the relative levels of damage in adenine, guanine, and cytosine compared to thymine are not known.

Complex lesions. When an agent damages DNA, it is naive to imagine each lesion as an entity, completely separate from every other chemically defined lesion. It is more likely that there are regions where similar or differing kinds of damage overlap each other. For instance, if a directly induced ssb is near a region where a purine has been removed from the complementary strand, the probability of a dsb being formed

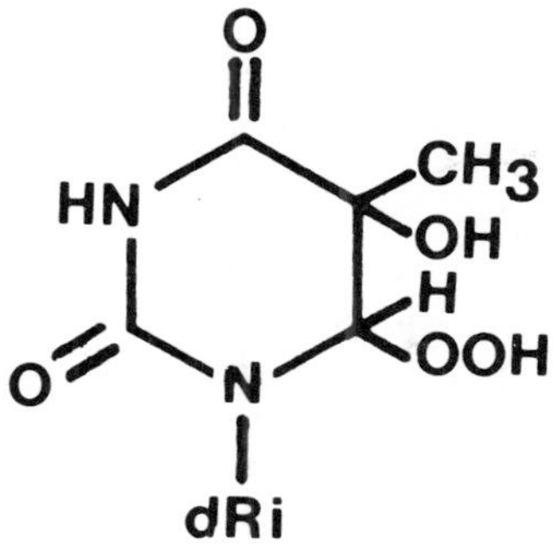

Figure 2.2 One of the products of the type, 5-hydroxy-6-hydroperoxydihydrothymine, formed by the action of ionizing radiation on intracellular DNA thymine.

depends very much on whether the ssb is repaired before the apurinic site is converted, chemically or enzymatically, to an ssb. Similarly, ssb and base damage or base loss may occur very closely to each other on the same strand, thus producing a substrate that may not be recognizable by enzymes normally involved in the repair of either of the lesions alone. Such possible complications of changes in DNA structure in time and space make the use of the term "DNA lesion" somewhat ambiguous.

Lesions Induced by Ultraviolet Radiation

Pyrimidine dimers. Because of the phenomenon of long wavelength photoreactivation of UV light-induced killing in bacteria, many early studies on DNA damage and repair utilized UV light. In 1960, Beukers and Berends showed that thymine dimers were a major product found after UV-irradiation of DNA. Later work showed that dimers could be formed from any two pyrimidines adjacent to each other on the same strand of DNA. Thus DNA pyrimidine dimers can be one of three types: thymine-thymine (TT), thymine-cytosine (TC), and cytosine-cytosine (CC). However, in the DNA of *E. coli*, which has an AT/GC ratio of 0.5, i.e., equal frequency in each pyrimidine, TT is the most abundantly formed and CC the least abundantly formed by UV (Setlow and Carrier 1966). Dimers are extremely stable structures and can be isolated from DNA after treatment with strong acids and high temperatures. This has made their study much more convenient than other less stable photoproducts induced by UV.

Pyrimidine dimers are formed in monolayers of mammalian cells at a frequency of about 5×10^{-5} per J/m^2 at 260 nm light (Regan et al. 1968). About 90% of mammalian cells in culture survive an exposure of 1 J/m^2; this means that each cell is capable of coping with more than 100,000 dimers in its genetic material.

Other base damage. When DNA or free bases are irradiated in vitro, it is well known that several photoproducts other than pyrimidine dimers, notably hydrates, are formed (Smith 1964). These products are unstable and are probably converted back to the original base form during isolation of DNA from cells; therefore, there has been almost no work done on them in cells. Dihydrothymine of the kind formed after X-irradiation is also formed by irradiation of cells with 313 nm light and to a lesser extent after 260 nm light (Hariharan and Cerutti 1977), but no other base damage has been studied in eukaryotic cells after UV-irradiation.

DNA-protein crosslinks. There is no doubt that chemically stable crosslinks between DNA and protein are formed by UV-irradiation of eukaryotic cells (Habazin and Han 1970), but very little is known about the properties of these lesions. Model systems (Smith 1962; Varghese 1974) suggest that cysteine-thymine or glutathione-thymine may be active species in this class, but in no case have the crosslinks formed in cells been characterized. Because there is considerable circumstantial evidence suggesting that these may cause very profound effects, more study should be given to this class of photoproduct.

Lesions Caused by Chemical Agents

Alkylating Agents

This class of DNA-damaging agent includes such chemicals as the alkylmethane sulfonates, the nitrogen mustards, and the nitrosamines. The reactions of many of these agents with various sites in DNA have been extensively studied (for review, see Singer 1975) and classified on the basis of relative affinity for different nucleophilic centers (Ehrenberg 1971). Reaction can occur with essentially all the nitrogens and oxygens in the purines and pyrimidines and with the phosphodiester groups (see review by Singer 1979).

Alkylated bases. Work up until 1976 showed that reaction of alkylating agents could occur at the N-1, N-3, and N-7 positions in adenine; N-3, N-7, and O-6 in guanine; N-1 and N-3 in cytosine; and N-3 and O-4 in thymine. Singer (1976) then showed that all the oxygens in nucleic acids could react, so that for DNA the O-2 in thymine and in cytosine was added to this list (Fig. 2.3). Indeed, the oxygens are the main targets in DNA bases for ethylnitrosourea, a very powerful mutagen, and are in general attacked by ethylating agents more than by methylating agents (although the alkyl sulfates react almost exclusively with base nitrogens [Singer 1979]).

Phosphotriesters. The alkylation reaction also occurs in DNA at phosphorous atoms to convert the normally occurring phosphodiesters to the phosphotriesters (Fig. 2.3); after treatment with ethylnitrosourea or methylnitrosourea, about two-thirds of the total alkylations are of this type (Sun and Singer 1975). There is no evidence that the exocyclic amino groups are attacked in vivo. The deoxyribose in DNA apparently does not react, although the ribose of RNA can be alkylated at the 2 position (Fig. 2.3).

DNA crosslinks. Certain alkylating agents, e.g., some of the sulfur and nitrogen mustards, are difunctional and have two groups that can be active in alkylation reactions. When these agents are incubated with cells, a certain fraction of total reaction, depending on the compound, will form crosslinks, where both functional groups react with a nucleophilic center in the DNA. The crosslinks can be intramolecular (within the same molecule), intrastrand (between different molecules on the same strand), or interstrand (between molecules in the two strands). Interstrand crosslinks

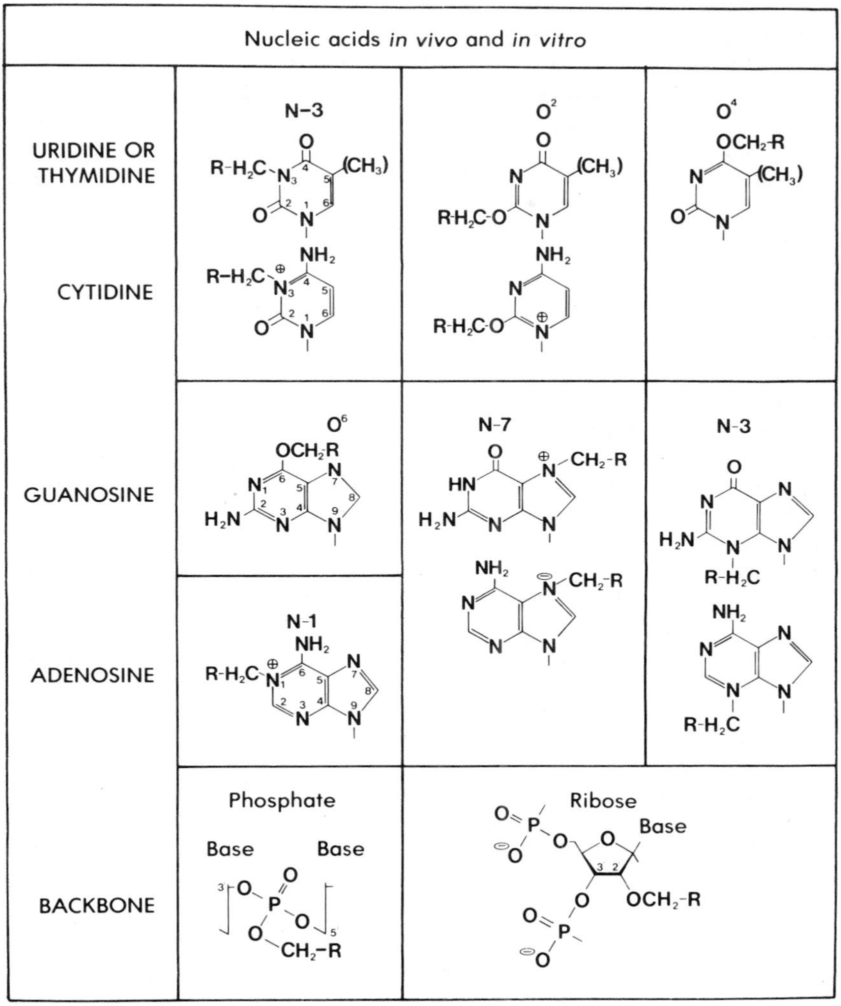

Figure 2.3 Products of reactions of DNA (and the ribose of RNA) with alkylating agents. (Adapted from Singer 1979, with permission.)

will form a chemically stable bridge between the strands and, if not removed, act as a lesion that is a severe, probably insurmountable, block to DNA duplication. About one in 24 alkylations caused by nitrogen mustard results in an interstrand crosslink (Kohn et al. 1966).

Single-strand breaks. The alkylation reactions lead to some products that are themselves chemically unstable; e.g., in the 3- and 7-alkyl purine nucleosides, depurination can occur spontaneously. During breakdown of these substances, single-strand breaks can be produced, or the new lesion can be unstable in alkali so that when the DNA is analyzed on alkaline sucrose gradients, it is counted as a single-strand break. Phosphotriesters in DNA also lead to strand breaks upon exposure to alkali, and these are probably responsible for the majority of single-strand breaks measured in alkaline sucrose gradients after treatment of cells with many of the alkylating agents.

True single-strand breaks may also be formed by alkylating agents, either by direct reaction or via radicals formed during metabolism of the agent by the cell. Cerutti (1978) especially has championed the so-called indirect action of chemicals where, in parallel with ionizing radiation, damage to DNA resulting in strand breaks and base damage is caused by the very reactive OH radical.

Other Agents

The mechanism of action of very large molecules that act as carcinogens differs from that of simple alkylating agents. The best example is benzo(a)pyrene. This molecule is converted via a complicated series of microsomally located metabolic steps to a reactive species, 7α, 8β-dihydroxy-9β, 10β-epoxy-7,8,9,10-tetra-hydro-benzo(a) pyrene, the so-called diol epoxide I. This product then reacts through the epoxide directly and preferentially with the exocyclic N-2 of guanine to form the DNA adduct. Similarly complicated metabolic reactions convert other polycyclic hydrocarbons, such as dimethylbenzanthracene (DMBA), to diol epoxides which bind to DNA, although for DMBA it is not clear that these are the only products to react with DNA (for review, see Roberts 1978).

The aromatic amine, acetyl-2-aminofluorene (AAF), is metabolically converted to N-acetoxy-AAF, which then reacts with guanine in DNA to form adducts. It appears that about 80% of the adducts occur at the C-8 group and the other at the exocyclic 2-amino group (Kriek et al. 1967; Kriek and Westra 1973). It is interesting that the attachment of the large acetoxy AAF at the C-8 of guanine is believed to change the conformation of the DNA, resulting in the "base displacement" model in which the modified guanine is displaced from its normal coplanar relation with adjacent bases, and is replaced by the fluorene group of the acetoxy AAF (Grunberger and Weinstein 1976).

A particularly interesting chemical agent is bleomycin, which acts in many ways like X-rays. Indeed, not only does it directly produce single-strand breaks, but it produces double-strand breaks in a ratio to single-strand breaks very similar to that produced by X-rays, i.e., about one in 10 (Povirk et al. 1977). This agent has been used for chemotherapy in cancer and is an interesting link between chemical and physical DNA-damaging agents.

Another large class of chemicals that act on DNA are the intercalating agents. A good example of these is actinomycin D. These agents may or may not react to form covalent bonds to DNA, but their main action is to form stable noncovalent bonds between the two strands of DNA, i.e., intercalate. Thus, even in the absence of true chemical changes in the DNA molecule itself, these agents can change the secondary, tertiary, or higher-ordered structure of DNA and affect profoundly those processes, such as replication and transcription, that depend on normal conformation of the DNA molecule.

The next section of this paper treats the *repair of DNA*.

Single-strand Break Repair

More studies on the repair of single-strand breaks have been reported than on any other DNA lesions. This came about primarily because of the introduction by McGrath and Williams in 1966 of the alkaline sucrose gradient technique for studying large DNA molecules. The breakthrough here was the introduction of lysis of cells directly on top of the preformed gradient, so that subsequent handling of the freed DNA molecules was avoided. Previous methods involving pipetting, etc., of the sample had always introduced shear that broke the very long native DNA molecules. This technique was first adapted to mammalian cells by Lett et al. (1967).

When mammalian cells are irradiated with X-rays, about 10 single-strand breaks per cell are induced per rad (Lett and Sun 1970). Even after extremely high doses (tens of thousands of rads), over 95 per cent of these are repaired with a half-time on the order of 10 minutes (Koch and Painter 1975). This is rather remarkable because, at these doses, this very active repair process is occurring in cells that will soon die. There is dispute about whether all single-strand breaks are repaired, especially after doses (about 100 rads) where survival is high. None of the techniques so far developed are sensitive enough to detect one to 10 unrepaired breaks per cell. Nevertheless, the very rapid and nearly complete repair of these lesions after X-irradiation is believed by many to indicate that single-strand breaks are not important in cell killing. Because the fidelity of this repair process is still completely unknown, caution should be used in this interpretation.

The single-strand breaks measured after treatment of cells with chemical agents are, as mentioned earlier, due to many different kinds of original lesions, including phosphotriesters and depurinated or depyrimidinated sites. It is not surprising then that reports of "single-strand break" repair after exposure of cells or whole animals to chemicals indicate very heterogeneous repair times, from a few minutes (Cox et al. 1974) to several days (Cox et al. 1973; Damjanov et al. 1973). The latter may be due to repair of phosphotriesters, which are apparently very stable in living cells (Bodell et al. 1979).

Double-strand Break Repair

Relatively little is known about double-strand break repair in eukaryotes because the technology has not advanced enough to make meaningful measurements after doses of DNA-damaging agents that allow reasonable survival. X-radiation–induced double-strand breaks are repaired, as measured after very high doses (50,000 rads) by neutral sucrose gradients (Corry and Cole 1973; Sawada and Okada 1972). The problem, however, is that in these studies repair was never complete and lower doses could not be used (probably because of artifacts caused by tertiary or higher orders of DNA structure). Therefore, the completeness of double-strand break repair at moderate levels of damage is unknown.

It can be argued that after doses of X-radiation that average one lethal hit per cell, double-strand breaks must be repaired. The yield of double-strand breaks in mammalian cells has been reported to be from 1/7 to 1/50 of the total breaks measured in alkaline sucrose gradients (Corry and Cole 1968; Lehmann and Ormerod 1970). Taking the lower of these, there will still be an average of 20 double-strand breaks per cell at 100 rads, a dose at which 50% or more of mammalian cells will survive. Poisson statistics will permit only a vanishingly small percentage of cells ($\ll 0.1\%$) to escape with zero double-strand breaks. Therefore, those cells that survive must either repair about 20 double-strand breaks or continue to grow indefinitely with them in their DNA. Replication of DNA past double-strand breaks is almost impossible to conceive of, and since there is good evidence that they are repairable, the favored hypothesis must then be that essentially all double-strand breaks are repaired after exposure of mammalian cells to low doses of ionizing radiation.

Base Damage Repair

There are at least two kinds of excision repair of base damage. The first is the kind that removes pyrimidine dimers from DNA. In this repair system, nucleotide repair, the first step (other than recognition of damage, which is a very important step in all repair schemes) is the production of a "nick" in the DNA near the lesion. This is the incision step of nucleotide excision and is performed by an endonuclease. For dimer excision, it is very likely that this endonuclease is specific for dimers alone because the two dimer-recognizing endonucleases that have been purified (from *Micrococcus luteus* and from phage T4-infected *E. coli*) do not seem to recognize lesions other than dimers (Patrick and Harm 1973; Friedberg 1972).

Incision is followed by excision of the damage and adjoining undamaged nucleotides. After UV-irradiation, 30 to 100 nucleotides per dimer are removed (Edenberg and Hanawalt 1972; Regan and Setlow 1974); for many chemicals a similar process occurs (Regan and Setlow 1974). Following or accompanying the excision step is the insertion or repair replication step, in which the "long patch" is filled in with exogenous nucleotides. This step uses a polymerase that may be the same one used for normal DNA replication. The final step is the sealing of the newly inserted nucleotides to the extant strand, catalyzed by the enzyme ligase.

In contrast to the nucleotide excision system, base excision begins with the simple splitting of the base from the deoxyribose, catalyzed by a glycosylase (Friedberg et al. 1978), leaving an AP (apurinic or apyrimidinic) site. After this, one of two pathways may be used. In one, an AP-specific endonuclease nicks the DNA backbone (Lindahl and Andersson 1972) and the deoxyribose lacking a base is removed. Apparently, only a relatively few (0–5) adjacent nucleotides are also excised, so that this is "short-patch" repair. Presumably, repair replication and ligation occur subsequently in a manner similar to that for nucleotide excision repair.

The second pathway that may be used after the action of the glycosylase is one that has only recently been reported and presumably is the simplest of all—an undamaged base is simply added to the site vacated by the damaged base. This "insertase" activity has been reported to be present in extracts of both mammalian (Deutsch and Linn 1979) and bacterial cells (Livneh et al. 1979), but doubts about its reality have been raised on theoretical thermodynamic grounds. A summary of the pathways for excision repair is presented in Figure 2.4. In bacteria, three different DNA lesions are known to be repaired by a base excision pathway: 3-methyladenine (Riazuddin and Lindahl 1978), hypoxanthine (Karran and Lindahl 1978) and uracil (Lindahl et al. 1977). 3-Methyladenine is formed by methylation of adenine, while the other two are formed by deamination, which probably occurs spontaneously in cellular DNA.

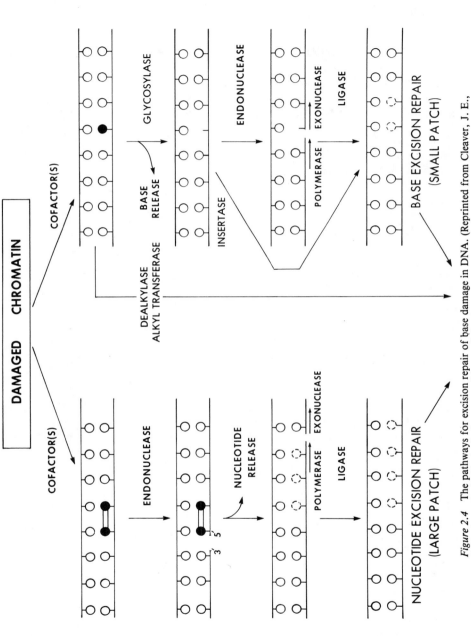

Figure 2.4 The pathways for excision repair of base damage in DNA. (Reprinted from Cleaver, J. E., J. Environ. Path. Toxicol., in press, with permission.)

Excision repair can be detected by several techniques. The most satisfactory method is to measure chemically the removal of lesions from intact intracellular DNA, but this is often difficult to accomplish, especially at doses where survival remains high. Radioactive compounds are required and, after their attachment to DNA, the separation techniques must show a significant loss of radiolabel from the relatively "hot" DNA. For this reason, it is often easier to use the next step of excision repair, insertion or repair replication, to detect repair.

The original technique developed by Pettijohn and Hanawalt (1964) uses equilibrium density gradients to separate DNA made by semiconservative synthesis from that made by repair synthesis (Fig. 2.5). This technique, which requires some changes in protocol for work with mammalian cells (Painter and Cleaver 1967), is actually the only one that proves that DNA synthesis stimulated by DNA-damaging agents is due to nonconservative (repair) replication.

The most popular and, in many cases, the easiest methods for measuring excision repair synthesis are those associated with unscheduled DNA synthesis (Djordjevic and Tolmach 1967). Autoradiography, the first method used to detect unscheduled DNA synthesis in mammalian cells (Rasmussen and Painter 1964), was described earlier. Another method is to suppress normal DNA synthesis by using stationary cells in G_1 phase and/or by incubation with hydroxyurea, and to measure the increase in total uptake of ^{3}H-thymidine caused by stimulation of excision repair synthesis after treatment of mammalian cells with DNA-damaging agents.

Several other methods for measuring excision repair have been reviewed by Cleaver (1977).

Evidence for repair of damage by simply removing the alkyl group or transferring it to another place has recently been reported. Karran et al. (1979) have shown that the methyl group added to the O-6 of guanine formed by treatment of so-called adapted strains of *E. coli* (Samson and Cairns 1977) with N-methyl-N'-nitro-N-nitrosoguanidine (MNNG) is enzymatically transferred to another as yet unidentified site in DNA. Presumably, this process removes the alkyl group from a "dangerous" site, namely, the O-6 of guanine, which is known to miscode during in vitro DNA replication (Gerchman and Ludlum 1973), to a place where its effects are relatively innocuous. A similar process may occur in mammalian cells; Pegg (1978) has reported evidence for removal of the alkyl groups from the O-6 position of guanine in rat liver DNA.

Repair of DNA-DNA Crosslinks

Interstrand DNA crosslinks are repaired in mammalian cells and bacteria (Fujiwara et al. 1977; Cole et al 1976). The exact details of this repair are not known. In bacteria, the genes that are responsible for recombination at the single-strand level (rec genes) are required for this process (Cole et al. 1976), but in mammalian cells there has not been any convincing evidence for single-strand recombinational events and therefore it seems unlikely that this process is involved in crosslink repair. Models involving sequential excision repair with intervening spurts of DNA synthesis can be generated for crosslink repair, but experimental evidence for them is lacking. Humans with the genetic disease, Fanconi's anemia, are deficient in the repair of DNA crosslinks. Conflicting reports on the ability of cells from patients with this disease to perform excision repair make any model for crosslink repair in normal cells impossible at the present time.

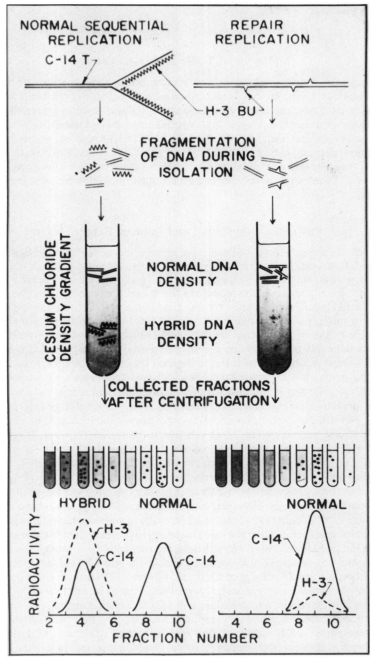

Figure 2.5 Rationale and procedure for distinguishing repair replication from semiconservative replication. Cells were prelabeled with ¹⁴C-thymidine (C-14T) to distinguish parental DNA from nascent DNA, which was labeled with ³H-bromodeoxyuridine (H-3BU). (Reprinted from Hanawalt, P. C., in *Genetics and Neoplasia*. Baltimore: Williams and Wilkins; 528–47, with permission.)

Repair of DNA-protein Crosslinks

Several drugs as well as UV light form covalent bonds between DNA and protein. In studies with UV-irradiation of mammalian cells, these lesions persist for long periods of time (Habazin and Han 1970). However, the doses used in these studies were relatively large and it is possible that crosslinks are repaired in cells exposed to doses that allow greater cell survival. In cells treated with the chemical carcinogen, nitrogen mustard, some but not all protein crosslinks are repaired (Ewig and Kohn 1977). Because many carcinogens and mutagens react strongly with proteins as well as DNA, these lesions and the mechanism of their repair deserve more concentrated study.

Postreplication Repair and Induced Repair Systems

In 1968, Rupp and Howard-Flanders presented evidence for a postreplication repair process in *E. coli*. This repair occurs in newly created ("nascent") DNA strands made after the insult with DNA-damaging agents and is concerned with events at the growing fork when it arrives in the region of a lesion in the parental strand. The Rupp and Howard-Flanders model indicated that gaps in the nascent strands were caused by the growing point reinitiating downstream from the lesion after halting before it. Because strains of *E. coli* that are deficient in normal recombination (rec$^-$) are also deficient in this postreplication repair, Rupp et al. (1971) presented evidence for single-strand recombination as the mechanism for postreplication repair. In the succeeding years, much intense study has indicated that, although recombinational events are abundant in bacteria after UV-induced damage, the actual role of the rec gene in postreplication repair may be concerned with events other than recombination. Nevertheless, the Rupp and Howard-Flanders model and their evidence for single-strand recombination initiated a search for similar processes in mammalian cells. As of this writing, there is still no compelling evidence that single-strand recombination occurs in somatic mammalian cells.

Lehmann (1972) presented evidence for gaps across from dimers in mammalian cells, but these gaps were apparently later filled by *de novo* synthesis rather than by recombinational events. However, even the occurrence of gaps has been challenged (Painter 1974; Edenberg 1976) and the best evidence suggests that if these gaps do occur they are not across from the lesions (dimers) in parental strands (Meneghini 1976; Clarkson and Hewitt 1976). Nevertheless, the concept of gap production in nascent DNA and the postreplication repair mechanism is still adhered to by many in the field. Because our knowledge of the details of normal DNA replication, especially at the growing point, is very sparse, it is probably best to be cautious about accepting postreplication repair in mammalian cells as a real phenomenon.

In bacteria, certain repair processes are inducible, i.e., the repair system is turned on in response to DNA damage. There are a host of responses induced in *E. coli* after stimulation by DNA-damaging agents, including several manifestations of an error-prone (SOS) repair system (Witkin 1976). The biological consequences of these are several, including increased survival and increased mutagenesis. A good example is the so-called Weigle or W-reactivation of λ bacteriophage. This phenomenon, first observed many years ago (Weigle 1953), concerns the enhanced recovery and increased mutation rate measured when λ phage is plated on irradiated bacteria compared to unirradiated bacteria. Only in the past five years has it been shown that this is due to an induced error-prone repair system in the bacteria (Witkin 1974). There has been one report (DasGupta and Summers 1978) of mutagenic W-reactivation in mammalian cells.

The error-prone repair system is not the only induced repair system in bacteria. Samson and Cairns (1977) reported that after *E. coli* were grown in a low, relatively nontoxic concentration (1 μg/ml) of MNNG, they became much more resistant to the toxic and mutagenic effects of larger doses. It has now been shown (Karran et al. 1979) that this is due to the induction of an enzyme that rapidly transfers the methyl from the 0-6 of guanine to another site in DNA, as discussed earlier. There is very limited evidence that a similar process may occur in mammals (Pegg 1978).

The final section of this paper discusses some *possible roles for DNA repair in cytogenetics*.

There still is no absolute proof that DNA repair is directly involved in the production of chromosome breaks or sister chromatid exchanges (SCE). The general consensus, however, is that unrepaired double-strand breaks are the basis of chromosome breaks. Indeed, Bender et al. (1976) have proposed that DNA repair even converts single-strand breaks into double-strand breaks and therefore causes chromosome aberrations under certain conditions.

The role of repair in SCE production is unresolved. There is no good correlation of excision repair with SCEs (Wolff et al. 1974). Caffeine, which inhibits certain steps believed to be concerned with postreplication repair after UV-irradiation, has no effect on SCE yields, although it does increase the frequency of chromosome aberrations (Kihlman 1975; Palitti and Becchetti 1977). Goth-Goldstein (1977) has shown that the 0-6-alkylation product of guanine is poorly excised by cells from xeroderma pigmentosum (XP) (which also cannot excise thymine dimers). It is possible that these DNA lesions (guanine 0-6-alkyl) are responsible for increased yields of SCEs in XP cells as compared to normal cells after exposure to chemicals that induce the same measured amounts of excision repair in XP and normal cells. There is also evidence that SCE production is proportional to dimer yield (Reynolds et al. 1980) and that, in chicken embryo cells, photoreactivation, which photo-reverses dimers back to monomers, decreases the number of SCEs (Natarajan et al. 1980). Thus, repair probably can play a role by decreasing the damage in parental strands, but the actual production of SCEs is probably a process concerned with localized disruptions in DNA replication rather than a manifestation of a repair system.

Literature Cited

Beukers, R.; Berends, W. 1969. Biochim. Biophys. Acta 41: 550–51.
Bodell, W. J.; Singer, B.; Thomas, G. H.; Cleaver, J. E. 1979. Nucleic Acids Res. 6: 2819–29.
Boyce, R. P.; Howard-Flanders, P. 1964. Proc. Nat. Acad. Sci. USA 51: 293–300.
Cerutti, P. A. 1974. Naturwissenschaften 61: 51–59.
Cerutti, P. A. 1978 *In* Hanawalt, P. C.; Friedberg, E. C.; Fox, C. F. eds., DNA repair mechanisms. New York: Academic Press; 1–14.
Clarkson, J. M.; Hewitt, R. R. 1976. Biophys. J. 16: 1155–64.
Cleaver, J. E. 1968. Nature 218: 652–56.
Cleaver, J. E. 1977. *In* Kilbey, B. ed. Handbook of mutagenesis test procedures. Amsterdam: Elsevier/North-Holland; 19–48.
Cleaver, J. E.; Trosko, J. E. 1970. Photochem. Photobiol. 11: 547–50.

Cole, R. S.; Levitan, D.; Sinden, R. R. 1976. J. Mol. Biol. 103: 39–59.
Coquerelle, T.; Bopp, A.; Kessler, B.; Hagen, U. 1973. Int. J. Rad. Biol. 24: 397–404.
Corry, P. M.; Cole, A. 1968. Rad. Res. 36: 528–43.
Corry, P. M.; Cole, A. 1973. Nature New Biol. 245: 100–101.
Cox, R.; Damjanov, I.; Abanobi, S. E.; Sarma, D. S. R. 1973. Cancer Res. 33: 2114–21.
Cox, R.; Daoud, A. H.; Irving, C. C. 1974. Biochem. Pharmacol. 23: 3147–51.
Damjanov, I.; Cox, R.; Sarma, D. S. R.; Farber, E. 1973. Cancer Res. 33: 2122–28.
DasGupta, U. B.; Summers, W. C. 1978. Proc. Nat. Acad. Sci. USA 75: 2378–81.
Deutsch, W. A.; Linn, S. 1979. Proc. Nat. Acad. Sci. USA 76: 141–44.
Djordjevic, B.; Tolmach, L. J. 1967. Rad. Res. 32: 327–46.
Edenberg, H. J. 1976. Biophys. J. 16: 849–60.
Edenberg, H.; Hanawalt, P. 1972. Biochim. Biophys. Acta 272: 361–72.
Ehrenberg, L. 1971. *In* Hollaender, A. ed. Chemical mutagens: principles and methods of their detection. Vol. 2. New York: Plenum Press; 365.
Ewig, R. A. G.; Kohn, K. W. 1977. Cancer Res. 37: 2114–22.
Friedberg, E. C. 1972. Mutat. Res. 15: 113–23.
Friedberg, E. C.; Bonura, T.; Cone, R.; Simmons, R.; Anderson, C. 1978. *In* Hanawalt, P. C.; Friedberg, E. C.; Fox, C. F. eds. DNA repair mechanisms. New York: Academic Press; 163–73.
Fujiwara, Y.; Tatsumi, M.; Sasaki, M. S. 1977. J. Mol. Biol. 113: 635–49.
Gerchman, L. L.; Ludlum, D. B. 1973. Biochim. Biophys. Acta 308: 310–16.
Goth-Goldstein, R. 1977. Nature 267: 81–82.
Grunberger, D.; Weinstein, I. B. 1976. *In* Yuhas, J. M.; Tennant, R. W.; Regan, J. D. eds. Biology of radiation carcinogenesis. New York: Raven Press; 175–87.
Habazin, V.; Han, A. 1970. Int. J. Rad. Biol. 17: 569–75.
Hariharan, P. V.; Cerutti, P. A. 1977. Biochemistry 16: 2791–95.
Karran, P.; Lindahl, T. 1978. J. Biol. Chem. 253: 5877–79.
Karran, P.; Lindahl, T.; Griffin, B. 1979. Nature 280: 76–77.
Kihlman, B. A. 1975. Chromosoma 51: 11–18.
Koch, C. J.; Painter, R. B. 1975. Rad. Res. 64: 256–69.
Kohn, K. W.; Spears, C. L.; Doty, P. 1966. J. Mol. Biol. 19: 266–88.
Kriek, E.; Westra, J. G. 1973. Eur. Assoc. Cancer Res. 2nd, 1973 Abstract; 240.
Kriek, E.; Miller, J. A.; Juhl, U.; Miller, E. C. 1967. Biochemistry 6: 177–82.
Lehmann, A. R. 1972. J. Mol. Biol. 66: 319–37.
Lehmann, A. R.; Ormerod, M. G. 1970. Biochim. Biophys. Acta 217: 268–77.
Lennartz, M.; Coquerelle, T.; Hagen, U. 1975a. Int. J. Rad. Biol. 28: 181–85.
Lennartz, M.; Coquerelle, T.; Bopp, A.; Hagen, U. 1975b. Int. J. Rad. Biol. 27: 577–87.
Lett, J. T.; Sun, C. 1970. Rad. Res. 44: 771–87.
Lett, J. T.; Caldwell, I.; Dean, C. J.; Alexander, P. 1967. Nature 214: 790–92.
Lindahl, T.; Andersson, A. 1972. Biochemistry 11: 3618–23.
Lindahl, T.; Ljungquist, S.; Siegert, W.; Nyberg, B.; Sperens, B. 1977. J. Biol. Chem. 252: 3286–94.
Livneh, Z.; Elad, D.; Sperling, J. 1979. Proc. Nat. Acad. Sci. USA 76: 1089–93.
McCann, J.; Ames, B. N. 1976. Proc. Nat. Acad. Sci. USA 73: 950–54.
McGrath, R. A.; Williams, R. W. 1966. Nature 212: 534–35.
Meneghini, R. 1976. Biochim. Biophys. Acta 425: 419–27.
Natarajan, A. T.; van Zeeland, A. A.; Verdegaal-Immerzeel, E. A. M.; Filon, A. R. 1980. Mutat. Res. 69: 307–17.
Painter, R. B. 1974. Genetics 78: 139–48.
Painter, R. B.; Cleaver, J. E. 1967. Nature 216: 369–70.
Painter, R. B.; Cleaver, J. E. 1969. Rad. Res. 37: 451–66.
Palitti, F.; Becchetti, A. 1977. Mutat Res. 45: 157–59.
Patrick, M. H.; Harm, H. 1973. Photochem. Photobiol. 18: 371–86.
Pegg, A. E. 1978. Biochem. Biophys. Res. Commun. 84: 166–73.
Pettijohn, D.; Hanawalt, P. 1964. J. Mol. Biol. 9: 395–410.
Povirk, L. F.; Wübker, W.; Köhnlein, W.; Hutchinson, F. 1977. Nucleic Acids Res. 4: 3573–80.
Rasmussen, R. E.; Painter, R. B. 1964. Nature 203: 1360–62.

Regan, J. D.; Setlow, R. B. 1974. Cancer Res. 34: 3318-25.
Regan, J. D.; Trosko, J. E.; Carrier, W. L. 1968. Biophys. J. 8: 319-25.
Reynolds, R. J.; Natarajan, A. T.; Lohman, P. H. M. 1980. Mutat Res. 64: 353-56.
Riazuddin, S.; Lindahl, T. 1978. Biochemistry 17: 2110-18.
Roberts, J. J. 1978. Adv. Rad. Biol. 7: 211-436.
Roberts, J. J.; Crathorn, A. R.; Brent, T. P. 1968. Nature 218: 970-72.
Rupert, C. S. 1960. J. Gen. Physiol. 43: 573-95.
Rupp, W. D.; Howard-Flanders, P. 1968. J. Mol. Biol. 31: 291-304.
Rupp, W. D.; Wilde, C. E., III; Reno, D. L.; Howard-Flanders, P. 1971. J. Mol. Biol. 61: 25-44.
Samson, L.; Cairns, J. 1977. Nature 267: 281-83.
Sawada, S.; Okada, S. 1972. Int. J. Rad. Biol. 21: 599-602.
Setlow, R. B.; Carrier, W. L. 1964. Proc. Nat. Acad. Sci. USA 51: 226-31.
Setlow, R. B.; Carrier, W. L. 1966. J. Mol. Biol. 17: 237-54.
Singer, B. 1975. Prog. Nucleic Acid Res. Mol. Biol. 15: 219-84.
Singer, B. 1976. Nature 264: 333-339.
Singer, B. 1979. J. Nat. Cancer Inst. 62: 1329-39.
Smith, K. C. 1962. Biochem. Biophys. Res. Commun. 8: 157-63.
Smith, K. C. 1964. Photophysiology 2: 329-88.
Sun, L.; Singer, B. 1975. Biochemistry 14: 1795-1802.
Varghese, A. J. 1974. Photochem. Photobiol. 20: 339-43.
Weigle, J. J. 1953. Proc. Nat. Acad. Sci. USA 39: 628-36.
Witkin, E. M. 1974. Proc. Nat. Acad. Sci. USA 71: 1930-34.
Witkin, E. M. 1976. Bacteriol. Revs. 40: 869-907.
Wolff, S.; Bodycote, J.; Painter, R. B. 1974. Mutat. Res. 25: 73-81.

3

Sister Chromatid Exchange Analysis: Methodology, Applications, and Interpretation

by S. A. LATT, R. R. SCHRECK, K. S. LOVEDAY, C. F. SHULER*

Abstract

BrdU-dye techniques for sister chromatid exchange (SCE) analysis are being used extensively to characterize the impact of mutagens and carcinogens on chromosomes. A number of such agents, known to damage DNA, have been observed to cause significant increases in SCE frequencies, typically at doses below those necessary to induce an appreciable increase in chromosome aberrations. The validity of SCE analysis as a sensitive and convenient test for mutagen-carcinogens is currently being investigated both in vitro, with cultured cells, and in vivo, in different tissues of intact animals. In addition, the biological significance of SCE formation is being examined in studies comparing SCE induction with events such as mutagenesis at specific loci, and an increased amount of effort is being directed at characterizing chemical events associated with SCE formation. The present paper will review methodology used for SCE detection, some of the information derived from application of SCE methodology, and hypotheses about the mechanism and significance of SCE formation.

Introduction

Sister chromatid exchanges represent the interchange of DNA strands between replication products at apparently homologous loci. These exchanges, which are generally detected in cytological preparations of metaphase chromosomes, presumably involve DNA breakage and reunion, although the molecular basis of sister chromatid exchange formation, as well as the biological significance of exchanges, is not completely understood. In spite of these uncertainties, analysis of sister chromatid ex-

---

*Division of Genetics, Mental Retardation Center, Children's Hospital Medical Center and Department of Pediatrics, Harvard Medical School, Boston, MA. Literature review concluded December 1978; manuscript submitted January 29, 1979.

change formation in cytological systems has already provided information about chromosome structure and has been used to detect the effects of clastogens and to differentiate between chromosome fragility diseases.

Sister chromatid exchanges were first described by J. Herbert Taylor and associates, who utilized autoradiography to detect differentially labelled sister chromatids in cells which had undergone one cycle of ^{3}H-dT incorporation followed by a replication cycle in nonradioactive medium (229). Reciprocal alterations in labelling (SCEs) were detected along the chromatids of a number of metaphase chromosomes. Analysis of SCE formation in cytological chromosome preparations has been facilitated by recently developed BrdU-dye techniques for detecting DNA synthesis.

BrdU Methodology for SCE Detection

BrdU substitution into DNA quenches the fluorescence of certain bound dyes, such as 33258 Hoechst (122, 128, 129), acridine orange (45, 48, 97), and, at alkaline pH, 4′, 6-diamidinophenylindole (DAPI) (144). Light energy absorbed but not emitted by such dyes can also promote the selective degradation of BrdU-substituted DNA (71, 72, 135), leading to reduced staining by Giemsa (178). Additional effects of BrdU on chromatin structure permit detection by yet other Giemsa protocols (10, 31, 114, 191, 195, 224, 257), and immunological methods for BrdU detection have been introduced (73).

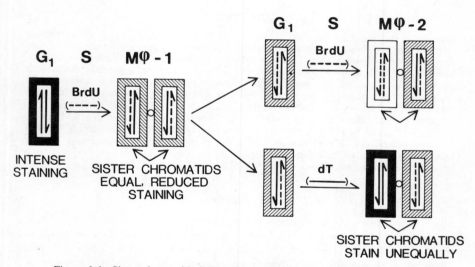

Figure 3.1 Sister chromatid differentiation by BrdU-dye techniques. Cells are allowed to incorporate BrdU for one cycle, followed by a second cycle of replication in which the presence of BrdU is optional. Sister chromatids in metaphase chromosomes from such second division cells will exhibit unequal fluorescence, if stained, e.g., with 33258 Hoechst, or unequal intensity following Giemsa staining, reflecting different numbers of BrdU-substituted polynucleotide chains. Solid, hatched, and open areas surrounding each rectangle represent intense, intermediate, and pale staining, respectively.

Sister chromatid differentiation, which is necessary for SCE detection, can be achieved in most chromosomal regions, by two related protocols. Both require one cycle of BrdU incorporation into chromosomal DNA; they differ in that only one involves the presence of BrdU during the second cycle (Fig. 3.1). In vitro studies typically utilize two cycles of BrdU incorporation, primarily to avoid the difficulty of changing cell culture medium to remove the BrdU. In contrast, in vivo studies usually involve BrdU incorporation for the first cycle only. BrdU is rapidly degraded in intact animals, and levels of BrdU drop rapidly as soon as external sources of BrdU are removed. A second cycle of BrdU incorporation has only a small effect on the baseline level of SCEs (156). In vitro cultures must be protected from light (e.g., ≤313nm) that can degrade BrdU-substituted DNA; such precautions do not appear to be necessary for most in vivo studies. For both protocols, cells are trapped at metaphase of the second cycle following initial exposure to BrdU, and cytological chromosome preparations are then made by standard techniques.

Exceptions to the above discussion are chromosomal regions containing DNA with a markedly unequal distribution of thymine on complementary polynucleotide strands. In these regions, such as mouse centromeres (145) and the distal part of the long arm of the human Y (134), sister chromatid differentiation and SCE detection can be achieved after one cycle of BrdU incorporation. After a second replication, SCEs formed during the first cycle appear as isolabelling, while second cycle SCEs appear as reciprocal interchanges (68). This latter situation is analogous to patterns observed in most chromosomal regions after three, rather than two, cycles of BrdU incorporation (232).

Staining protocols used for BrdU detection, originally described elsewhere (122, 125, 130, 178), are reviewed in some detail here. The basic steps required in one fluorescent protocol, as well as in related Giemsa methods, are summarized in Table 3.1. BrdU administration protocols can achieve greater than 80% substitution of BrdU for dT in one or both DNA strands (9, 130). BrdU detection under these circumstances presents little problem, since appreciable fluorescence quenching occurs even at one-third to one-half maximal BrdU substitution levels (124, 130). When sister chromatid differentiation is detected by 33258 Hoechst fluorescence, the contrast between chromatids can be optimized by mounting slides at a pH slightly above neutrality (e.g., pH 7.5) and an ionic strength of approximately 0.15 (129). Unfortunately, conditions promoting fluorescence contrast also lead to rapid fading of fluorescence. Conversely, use of a mounting medium at low pH or one containing glycerol stabilizes fluorescence but reduces BrdU-dependent contrast.

Of the many Giemsa protocols, one based on the procedure introduced by Perry and Wolff (178) is probably the most convenient. In this procedure, the dye (e.g., 33258 Hoechst) serves to photosensitize degradation of BrdU-substituted DNA. When bound to DNA, the dye exhibits very high absorbance in the near-UV region ($\varepsilon_{max} \sim 3 \times 10^4 M^{-1} cm^{-1}$), so that staining with 33258 Hoechst should increase the sensitivity of BrdU-substituted chromosomes to DNA breakage by light of wavelengths between 350–400 nm by several orders of magnitude (compared with the sensitivity of BrdU-substituted but unstained chromosomes). This estimate is corroborated by observations of the ability of 33258 Hoechst to sensitize BrdU-substituted cells to killing by light (217, 218). Effective photosensitization requires dye to be complexed with DNA under conditions (e.g., pH 7.5 with 33258 Hoechst) such that BrdU quenches fluorescence (72). That is, light energy absorbed but not

Table 3.1 BrdU Substitution and Cytological Procedures for SCE Analysis

A. BrdU incorporation protocols
   1. In vitro
      a. Human peripheral lymphocytes
         Blood is collected in preservative-free heparin, 10 units/ml, and allowed to settle 90 min. before removal of the buffy coat. Aliquots of plasma containing 5-8 x 10^6 leukocytes are added to 16 x 125 mm culture tubes together with the following:
         (1) Medium: Eagle's MEM with Earle's BSS and 2mM L-glutamine, 100 U/ml pennicillin, 100 µg/ml streptomycin, and 20% fetal bovine serum. Total volume used is 5 ml.
         (2) Phytohemagglutinin: either 0.2 ml of a crude saline extract of red kidney beans or 0.1 ml of Burroughs-Wellcome phytohemagglutinin (HA-15, reagent grade).
         (3) Nucleosides: we routinely employ BrdU, fluorodeoxyuridine (FdU), and uridine (U) at final concentrations of 10^{-5} M, 4×10^{-7} and 6×10^{-6} M. Other workers have omitted the FdU and U. Addition of deoxycytidine (dC) (10^{-4} M) improves cell growth but also results in higher baseline SCE frequencies. Addition of dC without adding FdU will ultimately lead to reduced BrdU substitution. It is convenient to add the BrdU together with the cells, although an 18-24 hR. delay in BrdU addition is possible.
         Cells are cultured at 37°C in the absence of light. Tubes are sealed tightly to prevent CO_2 escape and inclined at an angle of approximately 45°. Good yields of second division metaphases are generally obtained after 72-74 hr. total culture time. Retardation of cell growth by various clastogens may necessitate the use of a longer culture time before cell harvest.
         Agents to be tested for potential SCE induction are typically added at the start of culture or 24 hr. before harvest. The latter protocol, which exposes cells for only the second cycle of BrdU incorporation, reduces bias due to selection of those cells which are able to reach the second metaphase after addition of BrdU. All manipulations of cells after addition of BrdU to the cultures should be done in subdued light and preferably in the absence of light with wavelength less than 500 nm.
         Lymphocyte harvest consists of the following steps: (1) addition of 0.1 µg/ml colcemid for 2 h; (2) removal of medium and resuspension in 0.075 M KCl for 12 min.; and (3) fixation in at least three changes of a 3:1 mixture of methanol and acetic acid prior to slide preparation. Minor variations in mitotic arrest or hypotonic treatment might be appropriate in individual circumstances. For example, metaphase preparations of lymphoblasts appear to be improved by reducing the colcemid concentration to 0.023 µg/ml and the time in KCl to 8 min. At each step, the cells are centrifuged, all but about 0.2 ml liquid is removed, and the cell pellet is gently resuspended before adding the next solution. Careful cell resuspension appears to increase the probability of obtaining good metaphase spreads.
      b. Cultured cell lines
         Analogous procedures are employed when examining cells other than lymphocytes. The type of medium and container (e.g., petri dish vs. T-flask) varies with the cell type. BrdU is generally added a few hours following cell transfer and cell harvest is timed to occur at the second metaphase following BrdU addition. As with lymphocyte cultures, it is generally convenient to allow cells to incorporate BrdU for two replication cycles, although sister chromatid differentiation can also be effected if BrdU is present for the first but not the second cycle. Examples of specific conditions that have been employed for different cell types have been tabulated previously (131). Cell harvest typically begins with proteolytic release of cells (0.05% trypsin plus 0.02% EDTA, or 0.1% Pronase) followed by the steps described for lymphocytes.
   2. in vivo
      a. For bone marrow, spleen, or thymus cells, small rodents (e.g., mice, hamsters) are implanted with a 55 mg tablet of BrdU made on a Parr pellet press (Moline, Ill.) us-

Table 3.1 (Continued)

ing a 0.178 in. diameter punch and dye set. The animals are mildly etherized and a small incision is made in the abdominal skin. The tablet is then implanted subcutaneously and the wound sealed with autoclips (Fisher). Cells are collected 18-30 hours later (the time for obtaining second division metaphase cells varies, among other things, with drug treatment). Prior to cell collection mitotic arrest was achieved by injecting the animals with colcemid ($\sim$ 3 mg/kg). Cells are collected in hypotonic KCl (0.075 M) and after 12-20 min. transferred to 3:1 methanol: acetic acid for slide preparation.

b. To obtain cells from regenerating liver, the animals are subjected to partial hepatectomy; approximately 65% of the liver is removed. Since the amount of BrdU released by dissolving tablets appears to inhibit liver regeneration, BrdU is administered by injection. The animals receive 13 0.2 ml, half hourly intraperitoneal injections of 10^{-2} M BrdU in water, commencing 31 hours after the partial hepatectomy. Between 16 and 19 hours after the start of the BrdU series, the animals receive 2 i.p. injections of colcemid. The liver is removed and soaked in 0.5 mg/ml collagenase in a pH 7.6 buffer with 5mM $CaCl_2$, minced finely with scissors, and transferred to 0.075 M KCl (20 min) and then 3:1 methanol:acetic acid fixative, before slide preparation.

c. Spermatogonial cells are obtained from mice implanted with 55 mg BrdU tablets (Parr pellet press as in (a). The tablets are implanted subcutaneously through an abdominal skin incision and the incision closed with Autoclip sutures (Fisher). Spermatogonial cells were harvested 60 hours following tablet implantation and 4 hours after colcemid treatment (0.6 mg/kg). Testes were dissected and seminiferous tubules teased apart in 1% sodium citrate hypotonic solution. The tubules were transferred twice in 1% sodium citrate (12 min.) and placed into 3:1 methanol: acetic acid fixative. Spermatogonial cells were released from the tubules by gentle agitation in 60% acetic acid and the released cells dropped on warm slides (60°) for metaphase cell preparation.

d. Cheek pouch epithelial cells are obtained from Chinese hamsters (Dr. George Yerganian, Northeastern University, Boston, MA) implanted with a 55 mg BrdU tablet (Parr pellet press as in a). A one cm incision is made through the skin of the lower abdomen of an etherized animal. The subcutaneous space is opened by blunt dissection. The tablet is implanted subcutaneously, and the incision closed with Autoclip sutures (Fisher). The cheek pouch tissue was harvested 72 hours after tablet implantation. Mitotic arrest was achieved by 0.4 ml injection of 0.1% vinblastine sufate 2 hours prior to harvest. The cheek pouch tissue is everted and excised, and the entire tissue placed in hypotonic KCl (0.075M) for 40 minutes and then transferred to 3:1 methanol:acetic acid fixative. The cells are dispersed and slides prepared by mincing portions of the tissue in hot 60% acetic acid.

B. Slide Preparation

Cells are centrifuged and resuspended in fresh fixative. Clay Adams Goldseal slides (3 in. x 1 in. x 1 mm) are washed in chromic acid, rinsed with water, dipped in methanol and air dried. The cell suspension in fixative is applied dropwise to the slides. A covered plastic box (e.g., 12 in. x 8 in. x 4 in.) containing warm water can serve as a humidity chamber to control drying of material added to slides. Slides containing cells are placed on a rack inside this chamber and the top removed after a time (e.g., 15 sec.) determined to give satisfactory metaphase spreads. For example, if chromosome spreading (examined under a phase microscope) at low ambient relative humidity is found to be insufficient, the time before removal of the chamber top is increased. The chamber top can be used to fan the slides to effect even drying in a few seconds. Cells from bone marrow or regenerating liver are routinely dropped on wet slides (50% methanol) and passed quickly through a flame to facilitate chromosome spreading.

(continued on following page)

Table 3.1 (Continued)

C. Slide Staining
  1. 33258 Hoechst
      a. Slides are successively dipped in PBS (0.14 M NaCl, 0.004 M KCl, 0.01 M phosphate, pH 7.0) (5'), 0.5 µg/ml 33258 Hoechst in PBS (10'), PBS (1'), PBS (5') and H$_2$O (2-3 changes). Stock solutions of dye (50 µg/ml) in H$_2$O can be stored at 4° in the dark for a least two weeks. (Samples of 33258 Hoechst were originally obtained from Dr. H. Loewe, Hoechst AG, Frankfurt, Germany, although the dye can now be purchased from Calbiochem.)
      b. Microscopic observation of 33258 Hoechst fluorescence is guided by the position of the high wavelength absorption band of the dye-DNA complex (maximal near 350 nm and appreciable up to or slightly beyond 425 nm). While excitation under darkfield conditions is certainly possible, excitation with incident illumination, using a UG-1 (360 nm peak) bandpass and TK 400 (reflect ≤ 400 nm) dichroic mirror is especially convenient and effective. A 460 nm high wavelength pass filter in the observation pathway removes most unwanted exciting light, from the fluorescence, which peaks near 475 nm.
      For optimal quenching of dye fluorescence by incorporated BrdU, the slides are mounted in a buffer of moderate ionic strength at or slightly above neutrality (e.g., pH 7.5 McIlvaine's buffer). Primarily because of the specificity of the stain, under these conditions, for (A-T rich) DNA, and because the free dye has very weak fluorescence, slides observed as described have little background fluorescence. However, the fluorescence of 33258 Hoechst bound to BrdU-substituted chromatin fades rapidly, and photography requires speed. Using a microscope with incident illumination, acceptable photographs can be obtained in 5-10 seconds, using Tri-X film. Reduction of the mounting medium pH shifts the dye fluorescence color from blue towards green or yellow, and the fluorescence fades less rapidly, but specific quenching due to BrdU substitution is decreased.
  2. Giemsa staining
      a. Following fluorescence photomicrography of 33258 Hoechst-stained slides (mounted in pH 7 or pH 7.5 buffer), the slides can be washed, incubated in 2 x SSC at 65° for 15-30 minutes, and restained in Giemsa (e.g., 4% Gurr's R-66 in 5 mM, pH 6.8 phosphate buffer). Sister chromatid differentiation previously apparent only with fluorescence can then be observed with transmitted light, as initially described by Kim (110) and by Perry and Wolff (178). In this procedure, the illumination used to excite 33258 Hoechst fluorescence also presumably promotes selective destruction of the chromatin which has been highly substituted with BrdU.
      b. Direct photosensitization of chromosomes can be achieved (178) by mounting slides previously stained with 33258 Hoechst in a 10^{-4} M solution of the dye in one-third strength PBS (or a comparable buffer with a pH near 7). A coverslip is applied and the slides are then placed in a Petri dish and exposed to light for a period of time which depends on the illuminating conditions. For example, approximately 6-12 hours exposure is sufficient after positioning the slides 6 cm from a Sylvania 20 watt cool white bulb (131). The slides are then rinsed in H$_2$O, incubated at 65° C in 2 x SSC (0.30 M NaCl, 0.03 M Na citrate, pH 7) for 15', rinsed thoroughly with H$_2$O and stained with Giemsa as before. Contrast can be heightened by increasing the time during which slides are exposed to light. For photography, slides can be mounted in a standard embedding medium, or immersion oil can be applied directly to the slide without a coverslip. A 544 nm interference filter can be used to enhance contrast, and the optical image is recorded on High Contrast Copy film. Highly corrected microscope objectives can be employed.

emitted as fluorescence is responsible for DNA degradation. Experiments with appropriately labelled, sychronized cells (135) indicate that the procedure produces single-strand breaks in BrdU-substituted DNA (Fig. 3.2). Incubation of illuminated slides in warm (e.g., >65°) buffer promotes elution of single-stranded fragments. Importantly, the contrast in Giemsa-stained slides can be controlled by varying the time of illumination and hence DNA degradation (Fig. 3.3).

Excellent sister chromatid differentiation can be achieved with either fluorescence or fluorescence plus Giemsa protocols (Fig. 3.4). For routine studies, Giemsa staining has the advantage of producing permanent chromosome preparations, permitting repeated examination by several observers, and bright-field rather than fluorescence microscopy can be used. Moreover, Giemsa-stained slides lend themselves to automated analysis. Detection of an SCE, with essentially redundant reciprocal information signalling exchange on sister chromatids, is probably simpler than automated recognition of banded chromosomes. Automated detection of SCEs can now be done at nearly the speed of manual studies, albeit with somewhat lower accuracy (256), though this work is still at a fairly early stage. Some form of automation in SCE scoring may ultimately prove necessary, e.g., to screen hundreds of compounds or large numbers of individuals potentially exposed to clastogenic compounds.

Basic Information about SCEs

Newer techniques for sister chromatid differentiation have confirmed most of the conclusions about the overall features of SCE drawn from previous autoradiographic studies, e.g., that sister chromatid exchange is constrained by the polarity of the DNA helix (26, 228, 232), that segregation at mitosis of sister chromatids into pairs of homologues is random (42, 138), and that apposition of

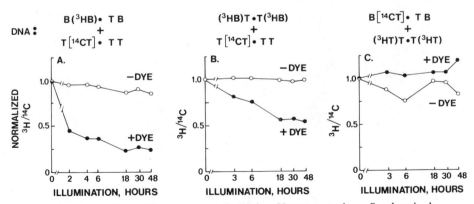

Figure 3.2 DNA elution during a BrdU-dye–Giemsa procedure. Synchronized CHO cells were cultured to produce DNA substituted as shown at the top of each frame. Mixtures of Colcemid-treated cells (average mitotic index approximately 40%) were applied to coverslips, mounted at pH 7 with or without prior staining with 33258 Hoechst, exposed 6 cm below a 20 watt cool white lamp for time periods indicated in the graphs, and subsequently incubated 15 min. in 2 X SSC at 65°. Relative elution of DNA species was estimated from the residual $^3H^{14}C$ ratio (135).

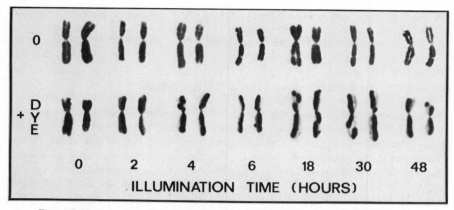

Figure 3.3 Induction of sister chromatid differentiation in CHO chromosomes. The chromosomes shown are from synchronized CHO cells, described in Figure 3.2, that were allowed to incorporate ^{3}H BrdU for one cycle followed by a cycle of nonradioactive BrdU. Slide treatment, including light exposure, was as described in the caption to Figure 3.2, with or without 33258 Hoechst staining prior to light exposure. Photosensitization by 33258 Hoechst enhanced subsequent differential staining with Giemsa, and this increased with illumination time, up to 4-6 hours.

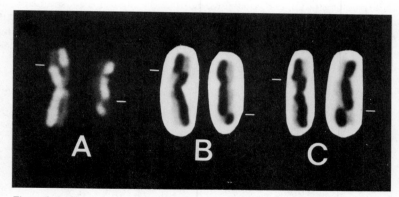

Figure 3.4 Sister chromatid exchanges. The chromosomes in this figure are from human lymphocytes which replicated twice in medium containing 10^{-5}M BrdU, 6 x 10^{-6} M U, and 4 x 10^{-7} M FdU. Those in (A) were stained with 33258 Hoechst and photographed under conditions described for fluorescence microscopy. Chromosomes in (B) were previously photographed to record fluorescence, as in (A), and then washed with H$_2$O, incubated 15 min. at 60-65° C in 2 X SSC, and stained with Giemsa. Chromosomes in (C) were exposed to fluorescent light while mounted in buffer containing 10^{-4} M 33258 Hoechst as described in Table 3.1, incubated in 2 X SSC and stained with Giemsa. Chromosomes shown were chosen to demonstrate relatively unambiguous sister chromatid exchanges (indicated by short, horizontal lines) (131).

newly synthesized polynucleotide chains is external to old chains with respect to centromeres (81, 125, 203, 242, 247).

The position of SCEs detected by fluorescence or Giemsa can be reasonably well localized relative to chormosome banding patterns, for example, in human chromosomes, in quinocrine (Q)-negative bands, or at the junctions of Q-positive and Q-negative regions (44, 123, 158). Similar studies detected a clustering of SCE at junctions between heterchromatic and euchromatic regions in muntjac (35), rat kangaroo (25), microtus, and hamster chromosomes (84). The significance of these "junctional" regions is as yet unknown. Reexamination of the position of SCEs in highly extended chromosomes, prepared e.g., as described by Yunis (254, 255), or by premature chromosome condensation (140), should help elucidate systematic characteristics of SCE localization.

The greater effective resolution of BrdU dye techniques has facilitated the detection of multiple, closely spaced SCEs (113, 124, 247). This capability has increased the accuracy and simplicity with which SCE induction by many clastogenic agents can be quantitated (96, 124, 176). BrdU itself, like ^{3}H-dT (27, 69, 150), induces SCEs (63, 97, 124, 154, 246, 253), and may be responsible for most of the baseline SCE's observed in the absence of additional clastogens. However, increments in SCEs can easily be scored and the extent of SCE induction (at least by mitomycin C) does not seem to be very sensitive to BrdU levels to which cells are exposed (92).

Induction of Sister Chromatid Exchange by Clastogens

Thus far, the most extensive use of SCE analysis has been to assess the impact of clastogens on chromosomes. Kato (96, 98) had originally employed autoradiography to demonstrate SCE induction by alkylating agents and proflavine. However, quantitation of high SCE frequencies was difficult with this method. BrdU-dye methodology was used to show that low doses of alkylating agents such as mitomycin C (Fig. 3.5) or nitrogen mustard induced large numbers of SCEs at concentrations well below those causing significant numbers of chromosome breaks (124). Numerous subsequent reports confirmed these observations and extended them to include other agents known to damage chromosomes either directly or after metabolic activation.

Dozens of mono- and bifunctional alkylating agents have been shown to induce SCEs (Table 3.2). Since many of the agents initially used to induce SCEs were also well-known mutagens and/or carcinogens, it was suggested that SCE analysis could be used as an assay for mutagens and carcinogens (176). Comparison of SCE induction results with mutagenesis, carcinogenesis, and unscheduled DNA synthesis data (91A, 147, 152) generally support this contention. Agents for which information on SCE induction is conflicting, or for which induction is at most minimal and examined in only one system, are tabulated separately (Table 3.3). Importantly, a number of agents, which are known to cause genetic damage, but which are relatively ineffective at inducing SCEs (X-irradiation, monomeric acrylamide, bleomycin), are able to induce chromosome breaks and/or rearrangements. The combination of SCEs *and* chromosome aberrations thus appears to give very few "false negatives" when examining mutagen-carcinogens. Diethylstilbesterol, the only compound listed in this table as definitely negative in SCE induction, but positive as a mutagen (91A) or unscheduled DNA syntheses inducer (152), may act on systems which are inoperative

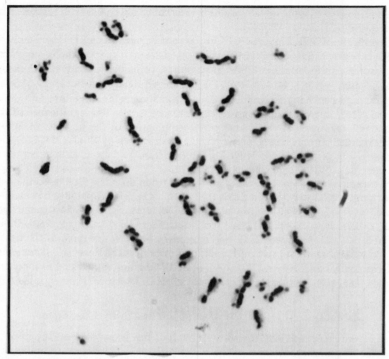

Figure 3.5 Induction of sister chromatid exchanges in a human peripheral lymphocyte by mitomycin C. Mitomycin C (0.075 µg/ml) was present during the third and final day of cell culture. Slides of metaphase chromosomes were stained with 33258 Hoechst, exposed to light and 2 X SSC, and then stained with Giemsa. More than 50 SCEs can be detected in this cell; untreated cells exhibit approximately 15 SCEs.

in many cultured cells; moreover, certain metabolites of dietlylstilbesterol can in fact induce SCEs (188A).

Direct contradictions are observed in only 2 out of the compounds, found positive for SCE (Table 3.2) but negative in some other system. Aniline (147) and diphenyl (91A) show no chromosome breakage or bacterial mutagenesis, but neither of these compounds is an especially potent SCE inducer. Moreover, there does not yet appear to be any convincing example thus far of an agent which is highly effective at inducing SCEs that is not also mutagenic or carcinogenic in at least some other system. Of equal importance, a large number of agents that are not thought to be mutagenic or carcinogenic appear incapable of inducing SCEs (Table 3.4).

A number of viruses have been observed to induce SCEs, although the extent of this induction is quite varied. For example, transformation by SV40 (although apparently not acute infection with SV40) more than triples SCE frequencies in some human diploid fibroblasts (162). In other SV40 transformed human cells, the SCE frequencies are only slightly elevated above those in comparable untransformed cells (250). Human lymphoblastoid cell lines, which are presumably transformed with Epstein-Barr virus (163), show either normal SCE levels (197) or a very slight SCE elevation (116). SCEs in mouse embryo fibroblasts are nearly doubled by infection with Rauscher leukemia virus, while SCEs in peripheral lymphocytes from vaccinated individuals or individuals with miscellaneous other virus infections (e.g., herpes simplex, cold/flu, or hepatitis), show at most a modest SCE elevation (114, 117). However, herpes simplex virus, administered at the end of S, does not appreciably increase SCEs in human diploid fibroblasts (103). This last result may be due, at least in part, to the general insensitivity of cells (127) to SCE induction at the end of S; exposure of cells to virus at the start of S might produce a different response.

Extension of SCE Studies from In Vitro to In Vivo Systems

Bloom and Hsu (23) described the formation of SCEs in ovo in chick embryos exposed to BrdU. The chick embryo system has excellent potential for examining tissue specific cytogenetic effects of mutagen-carcinogens during development (22). Subsequent reports described the induction by alkylating agents of SCE formation in marrow cells or spermatogonia of mice which received repeated doses of BrdU (6, 7, 239), and extension of in vivo SCE analysis to other rodent systems, as well as to the mudminnow (112), has been accomplished. The host-mediated (11, 141) aspects of in vivo systems, together with the obvious relevance of spermatogonial damage to germ cell formation, make this approach unique for studying environmental mutagenesis.

In contrast to combined in vivo–in vitro studies, in which a microsomal system capable of activating some agents is added directly to in vitro cultures (160, 216), or in which cultured cells are enclosed in porous chambers and implanted in animals (59, 85, 211), the in vivo systems permit examination of different processes in multiple tissues of a given organism. In vivo SCE analysis may prove particularly valuable, since some data (21) suggest that the array of products produced by in vivo versus in vitro activation of potential clastogens may be different. Also, in vitro "activating" conditions are capable of actually reducing the SCE inducibility of some agents, such as N-acetoxy-acetylaminofluorene (225), and in at least one instance (i.e., styrene) (47b), a microsomal activating system was effective only if accompanied by cyclohexene oxide, an epoxide hydratase inhibitor.

In our laboratory, SCE formation has been detected in a number of tissues, including mouse spermatogonia, bone marrow, thymus, and spleen cells (6, 7, 8, 9). Interestingly, spermatogonia have a lower baseline SCE level than the other tissues, and SCE induction by mitomycin C or cyclophosphamide is also lowest in spermatogonia. Recently, a system for detecting SCE induction in regenerating mouse liver has been developed (9, 200) (Fig. 3.6, 3.7). Since the liver contains the highest level of microsomal activating activity (24, 74, 80), chemical activation and SCE induction should be possible within the same cell. This system increases the sensitivity of detection of SCE induction by agents (e.g., acetylaminofluorene) which require activation, but which have thus far appeared to be relatively ineffective at SCE induction (200, 225, 226), perhaps because, once they are activated in the liver, they react without reaching more peripheral tissues. A system for SCE detection in regenerating liver has been developed (199A) to examine the relative sensitivity of mice with different genetically determined basal and inducible liver arylhydrocarbon hydroxlase activity levels to clastogens requiring metabolic activation by associate enzyme complexes.

A major methodological difficulty with in vivo studies has been the requirement

Table 3.2 Agents Capable of Inducing SCEs (Strongly Positive)

Agent	Cell type	Activation	Baseline SCE/cell	Induction SCE/cell	Conditions	Comments	Reference
acetaldehyde	human WBC	−	4.0	24.0	4.5×10^{-4} M		185
	CHO	−	4.7	28.4	2.3×10^{-4} M		167
N-acetylaminofluorene	CHO	+	11	12	10^{-4} M	S9	225
	CHO	+a	12	34	10^{-4} M	S9	226
	V-79	−	10.8	9.7	2×10^{-5} M		180
	mouse (CBA) marrow	+	4.2	10.7	225 µg/g (10^{-3} moles/kg)	in vivo	200
	marrow	+	5.9	22.7	225 µg/g	in vivo; partial hepatectomy	200
	regenerating liver	+	6.8	16.7	225 µg/g	in vivo; partial hepatectomy	200
N-acetoxyacetylaminofluorene	human fibroblasts	−	10	38	4×10^{-6} M		197
N-hydroxyacetylaminofluorene	V-79	−	10.8	32.4	2×10^{-6} M		180
	CHO	−	11	33	10^{-7} M	↓ with S9	225
adriamycin	V-79	−	10.8	41.3	2×10^{-5} M		180
	CHO	−	11	33	10^{-4} M	↓ with S9	225
	human WBC	−	4.8	24	2×10^{-7} M		161
	human WBC	+b	4.8	16	1.6×10^{-7} M		161
	human WBC	−	10	24	10^{-7} M		60
	CHO	−	12.2	72	3×10^{-7} M		176
	mouse marrow (AKR)	+	5.1	20	12 µg/g (2.2×10^{-5} moles/kg)	in vivo	115
aflatoxin B1	marrow (C57)	+	4.8	32	12 µg/g	in vivo	115
alkeran	CHO	+	11	37	10^{-4} M	↑ with S9	225
aminofluorene	human WBC	−	10.8	23.5	1.3×10^{-9} M		181
	CHO	−	11	18	10^{-4} M		225

Chemical	Cell type				Concentration	Notes	Ref
4-aminoquinoline-1-oxide	DON	−		3.5	24.1	5×10^{-4} M		3
aniline	DON	−		7.7	21.8	2×10^{-3} M		2
benzo(α)pyrene	human WBC	−	15	28		10^{-6} M		188
	human WBC	−	15	27		10^{-6} M		199
	human WBC	−	7	12.9		5×10^{-5} M		43
	Chinese hamster bone marrow	+	3	7		4×10^{-4} M (100 mg/kg)	in vivo	17
	DON	−	3.5	6.5		10^{-4} M		3
	CHO	+	11	26		10^{-4} M	S9	225
	V-79	+c	5.5	11.3		150 μg/g (6×10^{-4} moles/kg)		211
trans 4,5-dihydro-4,5-dihydroxybenz(α)pyrene	V-79	−	10.8	9.2		4×10^{-6} M		180
	V-79	+d	10.8	37.9		4×10^{-6} M		180
	CHO	−	4.4	9.8		3×10^{-5} M		168
trans 9,10-dihydro-9,10-dihydroxybenz(α)pyrene	CHO	−	4.1	6.7		3×10^{-5} M		168
trans 7,8-dihydro-7,8-dihydroxybenz(α)pyrene	CHO	−	3.4	7.9		3×10^{-5} M		168
betapropriolactone	CHO	−	4.7	23		3×10^{-5} M		168
	DON	−	7.7	32.7		10^{-4} M		2
	CHO	−	12.2	83.4		3×10^{-4} M		176
busulfan	human WBC	−	10.8	27.7		2×10^{-7} M		181
BrdU + light	human WBC	−	4.6	17		(3×10^{-6} M + 5×10^{3} ergs/mm²)	near-UV (unfiltered)	241
	CHO	−	1.6	10.2		(10^{-5} M + several light flashes)	light at end of S	91
	DON	−	4ᵉ	32ᵉ		(3×10^{-6} M + light)	20′, 20W bulb	99
	Vicia faba	−	22	65		(10^{-4} M + light)	30′, 40W near-UV	108
BrdU	human WBC	−		15		10^{-4} M	(4×10^{-7} M FdU)	122
	human WBC	−		37		4×10^{-4} M	(4×10^{-7} M FdU)	126

continued

Table 3.2 (Continued)

Agent	Cell type	Activation	Baseline SCE/cell	Induction SCE/cell	Conditions	Comments	Reference
BrdU (cont.)	human WBC	−		27	7×10^{-4} M		49
	human WBC	−		42	5×10^{-4} M		118
	human fibroblasts	−		5.6–5.8	10^{-5} M	one cycle	156
	human fibroblasts	−		6.5–7.8	10^{-5} M	two cycles	156
	CHO	−		16	2×10^{-5} M		127
	DON	−		5^e	10^{-4} M		106
	chick embryo	+		0.75^f	250 μg/gg (8×10^{-4} moles/kg)		23
butylbutanolnitrosamine	mudminnow	+		2.5	500 μg	in vivo	112
	Allium cepa	−		2.8	10^{-4} M	one cycle	201
	Allium cepa	−		5.5	10^{-4} M	two cycles	201
N-n-butylurea	DON	−	3.5	5.8	1.5×10^{-3} M		3
N-n-butyl-N-nitrosourea	DON	−	3.5	6.8	1×10^{-3} M		3
N-n-butyl-N-nitrosourethane	DON	−	3.5	21.9	1×10^{-3} M		3
chlorambucil	DON	−	8.8	18.4	10^{-4} M		2
	human WBC	−	5.1	56	10^{-5} M		212
	human WBC	−	10.8	33.4	3×10^{-7} M		181
chlorpropamide	V-79	−	8.8	15.4	3×10^{-3} M		29
cyclophosphamide	human WBC	−	19.6	22.9	2×10^{-3} M		7
	human WBC	+^{c}	2–5	20–35	30 μg/g (1.2×10^{-4} moles/kg)		85
	Chinese hamster fibroblasts	+	13.2	44	10^{-3} M	S9 (human)	231
	Chinese hamster fibroblasts	+	13.2	83.6	10^{-3} M	S9 (rat)	231
	mouse 3T3	−	23.1	34.0	2×10^{-3} M		7

Cell type				Dose	Conditions	Ref.
marrow (CBA)	+	7.4	22.4	5 μg/g (2 × 10⁻⁵ moles/kg)	in vivo	9
marrow (AKR)	+	5.1	50	5 μg/g	in vivo	115
marrow (C57)	+	4.8	90	5 μg/g	in vivo	115
marrow (NMRI)	+	3.7	13.7	10 μg/g (4 × 10⁻⁵ moles/kg)	in vivo	15
marrow (NMRI)	+	4	24	25 μg/g (10⁻⁴ moles/kg)	in vivo	239
marrow (CBA)	+	7.7	57.5	20 μg/g	in vivo	8
marrow (AKR)	+	1.4	17.7	25 μg/g	in vivo	184
thymus (CBA)	+	9.1	33.1	5 μg/g	in vivo	9
spleen (CBA)	+	6.8	25.1	5 μg/g	in vivo	9
spleen (CBA)	+	6.7	46.3	20 μg/g	in vivo	8
spermatogonia (CBA)	+	3.4	8.1	5 μg/g	in vivo	9
spermatogonia (CBA)	+	1.7	8.8	20 μg/g (8 × 10⁻⁵ moles/kg)	in vivo	7
regenerating liver (CBA)	+	5.4	22.8	10 μg/g	in vivo; partial hepatectomy	200
CHO	−	12.2	21.2	10⁻³ M		176
CHO	+	11	55	10⁻³ M	S9	216
V-79	+c	4	30	15 μg/g (6 × 10⁻⁵ moles/kg)		59
CHO	+	10	58	10⁻³ M	S9	47
CHO	+	15	120	8 × 10⁻⁵ M	S9-Aroclor induced	47
CHO	+	10	120	8 × 10⁻⁵ M	S9-phenobarbital induced	47
CHO	+	15	98	10 × 10⁻⁴ M	S9-3-methylcholanthrene induced	47
CHO	−	11	12	2 × 10⁻³ M	No S9	47
V-79	+	5.5	16	15 μg/g	implanted in mice	211
Chinese hamster cheek pouch	+	4.8	10.1	5 μg/g	in vivo	210
rabbit WBC	+b	5.7	20.7	35 μg/g (1.4 × 10⁻⁴ moles/kg)		215
chick embryo	+	1.2f	13.6f	50 μg/g (2 × 10⁻⁵ moles/kg)	in vivo	22

continued

Table 3.2 (Continued)

Agent	Cell type	Activation	Baseline SCE/cell	Induction SCE/cell	Conditions	Comments	Reference
deoxythymidine	Chinese hamster lung	−	10	50	10^{-2} M		237
	chick embryo	+	1.2[f]	7.5[f]	50 mg/gg (0.22 moles/kg)	in vivo	22
N-dibutylamine	DON	−	3.5	7.1	10^{-3} M		3
dibutylnitrosamine	DON	−	3.5	6.0	7×10^{-4} M		3
dibutylphthalate	DON	−	8.8	13.6	10^{-4} M		2
diethylnitrosamine	human WBC	−	7	83	5×10^{-5} M		43
	CHO	+	11	25	0.1 M	S9	160
	V-79	+c	5.5	8.8	600 μg/g (6×10^{-3} moles/kg)		211
	mouse (AKR) marrow	+	1.4	3.3	100 μg/g (10^{-3} moles/kg)	in vivo	184
	mouse (NMRI) marrow	+	3.7	4.2	200 μg/g (2×10^{-3} moles/kg)	in vivo	15
diepoxybutane	CHO	−	12.2	91	3×10^{-6} M		176
dimethylamine	DON	−	3.5	6.2	1.2×10^{-3} M		3
7,12-dimethylbenzanthracene	CHO	−	10.2	13.7	10^{-3} M		62
	CHO	+	10.5	17.1	10^{-3} M	S9	62
	Chinese hamster cheek pouch	+	6.7	11.5	0.5 mg in mineral oil	in vivo	210
	Chinese hamster bone marrow	+	3	11	4×10^{-4} M	in vivo	17
	DON	−	3.5	10.0	10^{-3} M		3
	V-79	+c	5.5	11.1	150 μg/g (6×10^{-4} moles/kg)		211
	rat gliosarcoma	−	15.5	26.9	10^{-4} M		62
	rat gliosarcoma	+	15.2	34.8	10^{-4} M	S9	62

compound	cell type					ref
dimethylnitrosamine	Chinese hamster bone marrow	+	3.3	4.5	$100\ \mu g/g$ $(1.4 \times 10^{-3}$ moles/kg)	in vivo	16
	human WBC	−		7.6	5×10^{-4} M		43
	CHO	+	7	100	4×10^{-2} M	S9	160
	DON	−	11	25.7	0.12M		3
	V-79	+c	3.5	9.1	$30\ \mu g/g$ $(4 \times 10^{-4}$ moles/kg)		211
			5.5				
	mouse (NMRI) marrow	+	3.7	10.6	$2\ \mu g/g$	in vivo	15
	Chinese hamster fibroblasts	+	13.2	48.4	40mM	+S9 (human)	231
	Chinese hamster fibroblasts	+	13.2	48.0	40mM	+S9 (rat)	231
dimethylphenyltriazine	mouse (NMRI) marrow	+	3.7	6.9	$6\ \mu g/g$ $(3 \times 10^{-5}$ moles/kg)	in vivo	15
diphenyl	DON	−	7.7	13.1	10^{-3} M		2
ethylmethane sulfonate	human WBC	−	7	17.1	5×10^{-4} M		43
	human WBC	−	15	35	2×10^{-3} M		138
	human WBC	−	10	27	10^{-3} M		60
	human fibroblasts	−	10	31	4×10^{-4} M		197
	CHO	−	12.2	103	3×10^{-3} M		176
	CHO	−	7.0	82	2×10^{-3} M		33
	CHO	+	11	26	5×10^{-3} M	S9	215
	DON	−	4e	42e	5×10^{-3} M		100
	BHK	−	21.1	88.4	4.2×10^{-3} M		95
	THK (clone A)	−	9.8	65.3	4.2×10^{-3} M		95
	THK (clone B)	−	21.5	80.7	4.2×10^{-3} M		95
	THK (clone E)	−	18.6	113.9	4.2×10^{-3} M		95
	THK (clone G)	−	9.4	41.5	1.7×10^{-3} M		95
	Vicia faba	−	20	85	4×10^{-2} M		109
	rabbit WBC	+b	5.7	13.6	$0.2\ mg/g$ $(1.6 \times 10^{-3}$ moles/kg)		215
	chick embryo	+	1.2f	8.6f	$3\ mg/g$ $(2.5 \times 10^{-2}$ moles/kg)	in vivo	22

continued

Table 3.2 (Continued)

Agent	Cell type	Activation	Baseline SCE/cell	Induction SCE/cell	Conditions	Comments	Reference
ethylnitrosourea	CHO	—	7.0	62	1.5×10^{-3} M		33
33258 Hoechst	CHO	—	12.2	67	10^{-5} M		176
	M3-1 Chinese hamster	—	4.4	7.0	3.4×10^{-6} M		213
8-methoxypsoralen	human WBC	—	12.1	34.2	5×10^{-7} M + 2.3×10^{5} ergs/mm^2 near-UV light		36
	human WBC	—	4.6	25	2.5×10^{-4} M + 5×10^{3} ergs/mm^2 near UV light		241
	human WBC	—	7	20	2×10^{-5} M + 4×10^{4} ergs/mm^2 near-UV light		159
	CHO	—	15.0	88.7	6×10^{-6} M + 1.7×10^{4} ergs/mm^2 near-UV light		127
4,5′,8-trimethylpsoralen	human WBC	—	11	34	5×10^{-6} M + 4×10^{4} ergs/mm^2 near-UV light		126
methylazoxymethanol acetate	human WBC	—	5.3	11.6	7.6×10^{-5} M		50
	DON	—	7.7	51.9	10^{-4} M		2
7-methylbenz(α)-anthracene	CHO	—	6.8	19.5	3×10^{-5} M		168

trans-5,6-dihydro-5,6-dihydroxy-7-methylbenz(α)-anthracene	CHO	–	7.9	13.2	3×10^{-5} M	168
trans-1,2-dihydro-1,2-dihydroxy-7-methylbenz(α)-anthracene	CHO	–	6.8	18.7	3×10^{-5} M	168
trans-8,9-dihydro-8,9-dihydroxy-7-methylbenz(α)-anthracene	CHO	–	8.4	21.7	3×10^{-5} M	168
trans-3,4-dihydro-3,4-dihydroxy-7-methylbenz(α)-anthracene	CHO	–	8.6	43.0	3×10^{-5} M	186
4-methyl-N'-nitro-N-nitrosoquanidine	CHO	–	12.2	106.2	10^{-6} M	176
	V-79	–	10.8	59.1	4×10^{-6} M	180
	mouse (NMRI) marrow	+	3.7	7.2	0.3 µg/g (2.3×10^{-6} moles/kg) in vivo	15
methylnitrosourea	Vicia faba	–	20	85	2×10^{-5} M	109
	DON	–	8.8	50.6	10^{-3} M	2
3-methylcholanthrene	V-79	+[c]	5.5	9.3	100 µg/g (3.7×10^{-4} moles/kg)	24
	V-79	–	10.8	8.9	3.7×10^{-6} M	180
	V-79	+[d]	10.8	38	3.7×10^{-6} M	180
	DON	–	3.5	4.4	10^{-4} M	3
	human WBC	–	7	15.6	3×10^{-5} M	43
methylmethane sulfonate	human WBC	–	7	34.7	10^{-4} M	43
	CHO	–	12.2	98	3×10^{-4} M	176
	Chinese hamster marrow	+	3.3	9.0	10 µg/g (10^{-4} moles/kg) in vivo	151

continued

Table 3.2 (Continued)

Agent	Cell type	Activation	Baseline SCE/cell	Induction SCE/cell	Conditions	Comments	Reference
methylmethane sulfonate (cont.)	rabbit WBC	+[b]	5.7	11.4	25 µg/g (2.5 × 10^{-4} moles/kg)		215
	mouse (AKE) marrow	+	1.4	9.8	100 µg/g (10^{-3} moles/kg)	in vivo	184
	Vicia faba	−	20	88	1.5 × 10^{-3} M		109
	chick embryo	+	1.2[f]	9.5[f]	1.5 mg/g (1.5 × 10^{-2} moles/kg)		22
mitomycin C	human WBC	−	12	120	9 × 10^{-7} M		124
	human WBC	−	15	29	9 × 10^{-8} M		137
	human WBC	−	10.8	48.6	1.2 × 10^{-7} M		181
	human WBC	−	11	47	4.5 × 10^{-7} M		92
	human WBC	−	10	92	3 × 10^{-7} M		60
	human lymphoblasts	−	10	39	4.5 × 10^{-7} M		197
	CHO	−	12.2	128	10^{-7} M		176
	CHO	−	7	77	7.5 × 10^{-8} M		33
	DON	−	7[e]	30[e]	2 × 10^{-6} M	autoradiography	98
	DON	−	2.4	28.6	10^{-6} M		105
	Vicia faba	−	20	70	2.2 × 10^{-6} M (.75 µg/ml)		109
	BHK	−	21.3	138.3	9 × 10^{-8} M		95
	THK (clone A)	−	7.9	79.5	9 × 10^{-8} M		95
	THK (clone B)	−	18.9	116.6	9 × 10^{-8} M		95
	THK (clone E)	−	21.4	178.1	9 × 10^{-8} M		95
	THK (clone G)	−	14.0	102.3	9 × 10^{-8} M		95
	Muntjac	−	8.0	35	3 × 10^{-6} M		88
	Muntjac	−	6	52	6 × 10^{-6} M		32
	mouse (AKR) marrow	+	5.1	50	5 µg/g (1.5 × 10^{-5} moles/kg)	in vivo	118
	mouse (C57) marrow	+	4.8	90	5 µg/g	in vivo	115

Compound	Cell type						
	mouse spermatogonia (CBA)	+		1.8	7.2	$0.5\ \mu g/g$ $(1.5 \times 10^{-6}$ moles/kg)	in vivo	6
	chick embryo	+		1.2^f	5.2^f	$1\ \mu g/gg$ $(3 \times 10^{-6}$ moles/kg)	in vivo	22
nitrogen mustard								
	human WBC	−		12	45	2×10^{-8} M		124
	CHO	−		12.2	109	3×10^{-6} M		176
2-nitro-p-phenylenediamine								
	CHO	−		11	45	5×10^{-4} M		177
4-nitro-o-phenylenediamine								
	CHO	−		11	53	10^{-3} M		177
4-nitroquinoline-1-oxide								
	CHO	−		12.2	57	10^{-6} M		176
	V-79	−		10.8	46.9	2.6×10^{-6} M		180
	DON	−		7^e	32^e	4×10^{-6} M		98
	DON	−		3.5	35.6	10^{-6} M	autoradiography	3
N-nitrosodiphenylamine								
	DON	−		7.7	13.8	2.5×10^{-5} M		2
procarbazine								
	Chinese hamster marrow	+		3.3	12	$200\ \mu g/g$ $(9 \times 10^{-4}$ moles/kg)	in vivo	16
	mouse (AKR) marrow	+		1.4	6.5	$0.2\ \mu g/g$ $(9 \times 10^{-7}$ moles/kg)	in vivo	184
proflavine								
	CHO	−		7.0	12	1.6×10^{-6} M		33
	V-79	−		10.8	16.5	3.3×10^{-6} M		180
	DON	−		7^e	21^e	4×10^{-6} M		98
propane sulfone								
	DON	−		7.7	21.0	10^{-3} M		2
1-(pyridyl)-3-3-dimethyltriazine								
	V-79	$+^c$		5.5	17.3	$100\ \mu g/g$ $(5.3 \times 10^{-4}$ moles/kg)		211
quinacrine mustard								
	human WBC	−		5.1	85	2×10^{-6} M		212
	CHO	−		12.2	121.1	10^{-6} M		176
	Vicia faba	−		20	105	2.5×10^{-6} M		109

continued

Table 3.2 (Continued)

Agent	Cell type	Activation	Baseline SCE/cell	Induction SCE/cell	Conditions	Comments	Reference
saccharin	human WBC	–	9.8	17.0	2.2×10^{-2} M		248
	CHO	–	8.8	12.0	5.5×10^{-2} M		248
	DON	–	7.7	15.2	10^{-3} M	fewer SCE at 5×10^{-2} M saccharin	2
sodium nitrite	DON	–	3.5	12.0	3×10^{-3} M		3
styrene	CHO	+	14.1	28.0	10^{-5} M	S9 + cyclohexene-oxide to inhibit epoxide hydrase	47a
styrene oxide	CHO	–	11.9	62	8×10^{-7} M	↓ with S9	47a
thiotepa	Vicia faba	–	20.6	76.0	2×10^{-4} M		108
triaziquone	human WBC	–	7.3	89.9	2.2×10^{-7} M		78
	human WBC	–	4.1	14.7	10^{-8} M		18
	human WBC	–	5.4	47	10^{-7} M		64
	human fibroblasts	–	9.3	79.6	2.2×10^{-9} M		78
	V-79	–	13.4	40.3	4.3×10^{-10} M		240
	mouse (NMRI) marrow	+	4	30	0.125 μg/g (5.5×10^{-7} moles/kg)	in vivo	241
	marrow	+	3.7	17	0.125 μg/g	in vivo	15
tris (2,3-dibromo-propyl) phosphate	V-79	–	5	25	2.9×10^{-5} M		59
	V-79	+c	3	16	0.5 mg/g (7.2×10^{-4} moles/kg)		59
tritiated deoxythymidine	rat kangaroo	–		9	18 C/mM	autoradiography	69

UV light (254 nm)	CHO	–	6.6	13.2	26 ergs/mm^2	autoradiography	250
	DON	–	7^e	38^e	80 ergs/mm^2	autoradiography; ↓ with caffeine	96
X-ray	V-79	–	13.4	50.2	50 ergs/mm^2		240
	human WBC		5.1	10	150 rads, G_1		212
	human WBC		5.2	15.5	200 rads, G_1	↓ by L-cysteine	4
	human WBC		10	10.9	200 rads, S		60
	CHO		12.2	27	200 rads, G_1		176
	CHO		12.2	35	200 rads, S		176
virus (SV-40)	human fibroblasts	–	7.5	8.3		T-antigen negative	162
	human fibroblasts	–	7.5	18.0		T-antigen positive	162

^{a}Extended S9 exposure conditions.
^{b}Treatment in vivo; cell culture in vitro.
^{c}Implanted in mice.
^{d}Syrian hamster feeder layer.
^{e}Assumes chromosome #1 = 10% of genome.
^{f}Macrochromosomes.
^{g}Assumes 0.1 gm embryo.

Table 3.3 Agents Exhibiting Mixed or at Most Weak SCE Induction Behavior

Agent	Cell type	Activation	Baseline SCE/Cell	Induction SCE/Cell	Conditions	Comments	Reference
acridine orange	V-79	−	10.8	16.9	3.3×10^{-6} M		180
		−	10.8	11.8	8.3×10^{-6} M		180
acrylamide	mouse (DDY) marrow	+	2.9	3.7	0.1 mg/g $(1.4 \times 10^{-3}$ moles/kg)	chromosome breakage	205
	spermatogonia	+	3.1	4.2	0.1 mg/g	chromosome breakage	205
anthracene	V-79	+c	5.5	4.3	150 μg/g $(5.6 \times 10^{-4}$ moles/kg)		211
	DON	−	3.5	5.4	1×10^{-4} M		3
	DON	−	3.5	4.8	1×10^{-4} M		3
arsenic	human WBC	+a	5.8	14.0			30
barbital	DON	−	7.7	9.7	8×10^{-3} M	inconsistent concentration dependence	2
bleomycin	human WBC	−	5.4	5.4	3.2×10^{-6} M	chromosome breakage	64
	CHO	−	12.2	23.6	3×10^{-6} M	chromosome breakage	176
	Vicia faba	−	20	21	2×10^{-6} M	chromosome breakage	109
butylbutanolamine	DON	−	3.5	5.0	3×10^{-4} M		3
butylhydroxyanisole	DON	−	7.7	11.0	10^{-4} M		2
caffeine	human WBC	−	4.6	5.5	10^{-3} M		241
		−	11	22	1.5×10^{-3} M	8-methoxypsoralen + light also potentiates alkylating agents	93
		−	25	32	10^{-3} M		241
		−	6.1	8.0	5×10^{-4} M		51
	human WBC	−	47	80	1.5×10^{-3} M	+ mitomycin C	93
	human fibroblasts	−	2.9	6.8	1.5×10^{-3} M	+ thiotepa ($\leq$10% increase with 5 alkylating agents)	193
	Vicia faba	−	76	77.3	5×10^{-4} M		108
							109

Compound	Cell type				Concentration	Notes	Ref
	V-79	–	13.4	14.6	10^{-3} M	+ triaziquon	240
		–	40.3	32.4	10^{-3} M	+ triaziquon	240
		–	40.3	27.9	2×10^{-3} M	+ UV light	240
		–	50.2	34	10^{-4} M		240
cytosine arabinoside	human WBC	–	10.8	14.3	8.2×10^{-6} M	marked intersample variation	181
deoxycytidine	human WBC	$+^a$	10.9	8	8.2×10^{-6} M		181
	human WBC	–	10	15	10^{-4} M (10^{-5} M BrdU + 4×10^{-7} M FdU)		137
	human	–	10.6	11	10^{-4} M (10^{-4} M BrdU but no FdU)		60
di-(2-ethylhexyl)-phthalate	DON	–	8.8	11.0	10^{-3} M		2
fluorescent brightener 24 (Kayaphor SN)	DON	–	7.7	11.3	10^{-4} M		2
fluorescent brightener 225 (Kayaphor LSK)	DON	–	7.7	10.6	10^{-4} M		2
maleic hydrazide	Vicia faba	–	20	98	5×10^{-5} M	no change in SCE between 10^{-5} M and 10^{-3} M ±S9	109
	CHO	–	12.2	15	10^{-3} M		176
	CHO		11	11	10^{-3} M		216
2-methyl-4-dimethylaminobenzene	DON	–	7.7	11.3	10^{-4} M		2
phenanthrene	Chinese hamster bone marrow		3	4	5.6×10^{-4} M	in vivo	17
	V-79	–	7.7	11.3	10^{-4} M		180
		$+^b$	11.8	11.1	5.6×10^{-5} M		180
	DON	–	8.8	10.6	10^{-3} M		2
potassium metabisulfite	DON	–	7.7	10.6	10^{-3} M		2
potassium sorbate	DON	–	7.7	12.4	2×10^{-2} M		2

continued

Table 3.3 (Continued)

Agent	Cell type	Activation	Baseline SCE/Cell	Induction SCE/Cell	Conditions	Comments	Reference
pyrene	V-79	+c − +b	5.5 10.8 10.8	4.7 12.7 16.6	7.5×10^{-4} M 5×10^{-5} M 5×10^{-5} M		180 211 211
pyridine	DON	−	7.7	10.9	5×10^{-3} M	inconsistent concentration dependence	2
sodium benzoate	DON	−	7.7	12.7	10^{-2} M		2
sunset yellow FCF (food yellow #5)	DON	−	7.7	10.5	2×10^{-3} M	highly toxic	2
4-0-tolylazo-0-toluidine	DON	−	7.7	10.0	10^{-5} M		2
urethane	DON	−	7.7	14.4	8×10^{-2} M		2
vincristine	human WBC human WBC	− −	10.8 11.7	25.5 3.7	6.1×10^{-8} M 3×10^{-6} M		181 220
virus (vaccinia)	human WBC		7.9	9.8		in vivo	113

aExposure in vivo; culture in vitro.
bSyrian hamster feeder layer.
cImplanted in mice.

Table 3.4 Agents Found Not to Induce SCE's[a]

Agent	Cell type	Activation	SCE Baseline	SCE Treated	Treatment Limit	Comments	Reference
acetone	DON	−	3.4	3.4	7×10^{-5} M		3
	chick embryo	+	1.2[b]	—	50 µl/g (7×10^{-4} moles/kg)		22, i
alcohols							
butanol	CHO	−	5.5	5.7	1.4×10^{-5} M		167
	chick embryo	+	1.2[b]	—	100 µl/g[c] (1.1×10^{-3} moles/kg)		22
ethanol	DON	−	3.4	3.7	8.7×10^{-5} M		3
	CHO	−	4.5	4.8	2.2×10^{-5} M		167
	chick embryo	+	1.2[b]	—	150 µl/g (2.6×10^{-3} moles/kg)	30% ethanol	22, i
methanol	CHO	−	4.5	4.2	3×10^{-5} M		167
propanol	CHO	−	5.5	4.7	1.7×10^{-5} M		167
aminopyrine[d]	DON	−	7.7	7.3	10^{-4} M		2
arochlor 1254[e]	V-79	+[f]	5.5	5.4	0.5 mg/g		211
bilirubin	human WBC	−	16.5	17.2	3.4×10^{-7} M		202
	human WBC	−	16.2	17.1	3.4×10^{-7} M + light	17 J/cm^2	202
	human WBC	+[g]	9.0	9.0	2.3×10^{-7} M + light		202
N-n-butylurethane	DON	−	8.8	7.6	10^{-3} M		2
dibutylhydroxytoluene	DON	−	7.7	7.8	10^{-3} M		2
ε-caprolactone	DON	−	7.7	7.8	10^{-3} M		2

continued

Table 3.4 (Continued)

Agent	Cell type	Activation	SCE Baseline	SCE Treated	Treatment Limit	Comments	Reference
cycloheximide	human fibroblasts	–	2.9	1.0	1.8×10^{-6} M		193
diethylstilbesterol	DON	–	7.7	7.8	10^{-4} M		2
dimethylsulfoxide	human WBC	–	15.5	16.5	1.3×10^{-2} M		202
	V-79	+f	5.5	5.0	not given		211
	mouse (NMRI) marrow	+	3.7	4.1	0.7 mg/g (9×10^{-4} moles/kg)		15
8-ethoxycaffeine	Vicia faba	–	21.4	20.5	10mM		107
ethylene glycol (50%)	chick embryo	+	1.2^b	—	100 $\mu l/g^c$ (1.75×10^{-3} moles/kg)		22, i
fluorescent brightener (#260)							
fluorescent light	DON	–	7.7	8.3	10^{-4} M	highly toxic	202
hydroxyurea	human WBC	–	15.5	16.2	17 J/cm^2		180
lead acetate	V-79	–	11.8	8.9	1.3×10^{-5} M		18
8-methoxypsoralen	human WBC	–	4.1	4.6	10^{-5} M		
	human WBC	–	12.1	12.1	5×10^{-7} M	no light	36
	human WBC	–	7	7	2×10^{-5} M	no light	159
	CHO	–	10.6	13.0	5×10^{-5} M	no light	127
methylene blue	V-79	–	11.8	10.4	10^{-4} M		180
N-methylurea	DON	–	7.7	7.2	10^{-3} M		2
near=UV light	human WBC	–	13.1	11.6	2.3×10^5 ergs/mm^2		36
	human WBC	–	7	8.5	1.5×10^5 ergs/mm^2	at most a 20% increase	159
	CHO	–	15.0	14.1	1.5×10^4 ergs/mm^2		127

Agent	System	Mutagenic			Dose	Comments	Ref.
ozone	human WBC	—	6.2	6.6	2 hr, 0.6 ppm	exposure in vivo; assay in vitro	148
penicillin G	chick embryo	+	1.2[b]	—	3 mg/g[c] (1.4×10^{-2} moles/kg)		22, i
perylene	V-79	—	11.8	10.2	4×10^{-5} M		180
	V-79	+[f]	5.5	4.2	150 μg/g (6×10^{-4} moles/kg)		211
quinoline	DON	—	7.7	6.9	10^{-3} M		2
S-9	mouse C3H	—	12.6	13.5	—		19
	Syrian hamster	—	6.6	6.8	—		19
salt solutions							
sodium acetate	human WBC	—	4.1	4.6	10^{-5} M		18
Hanks balanced salt solution	chick embryo	+	1.2[b]	—	1 ml/g[c]		22
0.3M NaCl 0.03M citrate	chick embryo	+	1.2[b]	—	200 μl/g[c]		22
0.2M phosphate–0.1M citrate	chick embryo	—	1.2[b]	—	200 μl/g[c]		22
sodium dehydroacetate	DON	—	7.7	8.4	10^{-3} M		2
streptomycin	chick embryo	+	1.2[b]	—	5 mg/g[c] (8.5×10^{-3} moles/kg)		22
tetracycline	chick embryo	+	1.2[b]	—	10 mg/g[c] (2×10^{-2} moles/kg)		i

[a] Negative results based only on a single test system, especially one that does not involve metabolic activation, should be viewed as tentative.
[b] Macrochromosomes only.
[c] Assumes 0.1 g embryo.
[d] This agent has been described as being mutagenic (118a).
[e] Arochlor 1254 is a potent inducer of mono-oxygenase activating enzymes (10), in addition to any direct genetic effect it might have.
[f] Implanted in mice.
[g] WBC from infants receiving phototherapy.
[h] Microsome-containing liver extract.
[i] S. E. Bloom, personal communication.

for multiple BrdU injections (6, 7, 240) or continuous BrdU infusion (173, 196, 198) because of rapid host metabolism of BrdU. The BrdU infusion method may prove especially valuable in studies in which sustained, known concentrations of clastogens must be administered to animals. We have introduced a simplified procedure, involving the use of BrdU in the form of a small tablet that can be implanted subcutaneously (8). Tablets can be prepared with a small, commercially available pill press (e.g., Parr Co., Moline, Ill.). Nearly 100% unifilar replacement of dT by BrdU during a single cycle can thus be effected, and tablets with different release kinetics have been prepared (9). The tablets will probably be more useful for large scale in vivo SCE studies in tissues such as bone marrow, the replication of which is apparently not seriously inhibited by the high BrdU levels provided by the tablets. However, relative to BrdU tablets, multiple BrdU injections give better results (e.g. a higher mitotic index) with regenerating liver cells, and result in lower baseline SCE levels (200).

In vivo SCE analysis has now been performed on cells from Chinese hamster cheek pouch mucosa (210) (Fig. 3.8). This tissue is accessible not only to systemic agents, but to topically applied agents, such as 7,12-dimethylbenzanthracene. In the latter situation, one cheek pouch can be exposed to clastogens, with the other serving as an internal control. This system should be especially useful for cytogenetic evaluation of putative topical carcinogens, and has recently been used to show SCE induction by a clastogen (8-methoxypsoralen) that is administered systemically, but activated topically (210).

A different type of "in vivo" SCE analysis involves the use of SCE frequencies to assess the cytogenetic impact of clastogenic agents administered to patients, usually in the course of chemotherapy (117, 181). Peripheral lymphocytes withdrawn from patients exposed to various drugs are cultured for two cycles in medium containing BrdU prior to SCE analysis. Nevstad (161) utilized this approach to detail the time course of SCE elevation due to adriamycin, a compound previously stated to induce SCE in patients (176). Perry (174) has continued this type of study. Widespread use of this procedure will require means to account for variations in the persistence of SCE elevation following treatment, as well as lymphocyte toxicity, which compromises the yield of analyzable metaphases. The information obtained may assist interpretation of emerging data on SCE elevations in humans exposed to other environmental conditions, e.g., due to occupational conditions (58) or to cigarette smoking habits (82, 119).

Sister Chromatid Differentiation in Meiotic Cells

In vivo administration of BrdU has permitted sister chromatid differentiation (SCD) in meiotic cells. Previous studies of SCD in meiosis had utilized autoradiography (172, 228) which afforded limited resolution. Initial success with BrdU was achieved in the X-Y bivalent of the mouse (7), in which SCE was detected. Meiotic interchange is not known to occur in the mouse X-Y pair, however, and only very limited sister chromatid differentiation was effected in autosomes, perhaps because of marked BrdU sensitivity. Allen et al. (9) have investigated meiosis in the Armenian hamster, an animal in which meiotic interchange presumably occurs in the X-Y bivalent (139), and have detected non-sister chromatid exchange, most likely due to meiotic recombination. BrdU-dye techniques have also been used to study meiotic interchange in locust chromosomes (230).

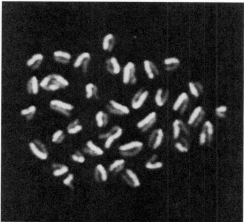

Figure 3.6 Sister chromatid exchanges in a regenerating mouse liver cell. A 6 week old male CBA mouse was subjected to partial (~65%) hepatectomy. Thirty-two hours later, the animal received a series of 13 half-hourly intraperitoneal injections of 10^{-2} M BrdU. Cell harvest 49 hours after partial hepatectomy was preceded (4 hours) by i.p. injection of 160 µg colcemid. Slides were stained with 33258 Hoechst, and SCEs detected by fluorescence microscopy. Three SCEs can be seen in this cell; controls on the average had five SCEs per cell (200).

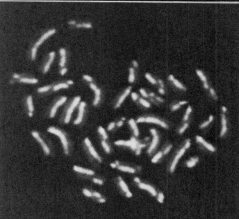

Figure 3.7 Induction of SCEs in regenerating mouse liver cells by cyclophosphamide. The experimental protocol was essentially that of Figure 3.4, except that cyclophosphamide (5 mg/kg) was injected intraperitoneally one hour after the final BrdU injection. This cell exhibits more than 32 SCEs; cells treated by this protocol had, on the average, 22 SCEs per cell (200).

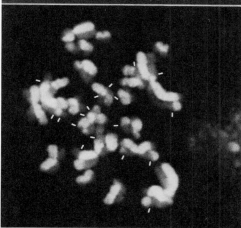

Figure 3.8 Sister chromatid exchange induction in a Chinese hamster cheek pouch cell. A 35 gm 1 year-old male Chinese hamster was implanted subcutaneously with a 55 mg BrdU tablet. Thirty hours after the implantation the left cheek pouch mucosa was painted with 0.1 ml of 0.5% 7,12-dimethyl-benzanthracene in mineral oil. The contralateral pouch was painted with mineral oil only. Forty-six hours after carcinogen exposure, and after a 4 hour treatment with 0.4 ml 0.1% vinblastine sulfate to achieve mitotic arrest, the cheek pouch tissue was excised and the cells prepared for microscopic examination. This cell from the DMBA treated cheek pouch displays 18 SCEs, indicated by white bars, while cells from the mineral oil-treated cheek pouch had a mean SCE frequency of 5.6.

Interpretation of SCE Induction Tests

A number of potentially confounding variables and other limitations must be kept in mind when interpreting the results of SCE tests. For example, exposure to BrdU must be high enough to permit good sister chromatid differentiation, but not so high that it produces a variable and unacceptably high background level of SCEs. Moderate (e.g., $< 10^{-4}$ M) doses of BrdU do not appear to alter the response of cultured cells to other clastogens, but much higher levels of BrdU induce a sharp increase in SCEs, out of proportion to additional BrdU incorporation (124, 126), which might compromise the sensitivity with which additional SCE induction can be detected. Also, while most early studies of SCEs were done with alkylating agents, chosen primarily to exemplify efficient SCE induction, it is desirable that future studies be capable of examining agents for which clastogenic activity is less certain. In these instances, at most a small increment in SCEs might be observed, and variables, such as effects due to the vehicles used to dissolve the agent, or the time required for metabolic activation, may become important.

A major problem in arriving at a decision about the clastogenicity of a new compound is the upper limit of the concentration to be tested before negative results are to be accepted. Typically, this upper limit will be a treatment level that is sufficiently toxic to cells that proliferation for one or two cycles, which is necessary for SCE detection, is inhibited. Such toxicity may become evident either in chromosome breakage or in alteration of specific cell kinetic parameters (and a reduced mitotic index). In any case, particularly in view of the approximately linear dependence of SCE induction on clastogen concentration observed by Carrano et al. (33) in CHO cells, it would seem meaningful ultimately to describe results in terms of the SCE increment per cell per concentration of specific agent. A minimal requirement would seem to be the acquisition of data in a range over which the SCE frequency showed progressive increase with increasing treatment levels. Since agents to be tested may perturb the cell cycle, the most accurate estimate of SCE induction would probably require several collections of metaphases, to include all cells exposed, as utilized by Carrano et al. (33).

Finally, the limitations of the test system employed must be considered. Most frequently, one wishes to know whether an unknown agent will cause genetic damage to a variety of human tissues. If this agent is active without metabolic modification, a human peripheral lymphocyte test system may be adequate, subject primarily to the possibility that different human tissues might have different repair capacities or drug metabolism rates. If metabolic activation of an agent is required, a rodent test system is most frequently used. However, DNA repair in rodents is known to differ from that in humans (235), and interpretation of results with rodent cells should consider this difference. If microsome preparations are used to activate the agent to be tested, differences between the modifications effected in vivo and those caused by isolated microsomes may prove important. Typically, most artifacts due to particular test systems will tend to produce false negative rather than false positive results. Because of the former possibilities, utilization of multiple test systems is probably advisable. However, comparison of test data on different substances would be facilitated if the plethora of test systems currently utilized (e.g., Tables 3.2, 3.3, 3.4) was reduced to a standard set, which was then applied to each compound. While it is difficult to specify the optimal composition of such a set precisely, it would probably include human peripheral lymphocytes, rodent (e.g., hamster) cells with/without S9 activating material, and an in vivo test (e.g., mouse bone marrow). Other tests utilizing, e.g., regenerating mouse liver, Chinese hamster cheek pouch cells, or chick embryos* would probably prove valuable in appropriate specific circumstances.

It is possible that "positive" results might depend on the use of an unrealistically high treatment dosage. This problem is inherent in many short-term tests, for which high-level short-term exposure is used to estimate the effect of low dose exposure over an interval of many years. Quantitative estimates of SCE induction efficiency per unit exposure will be important, both to characterize the potential hazard of an individual chemical and to estimate the possible additive effects of many agents, each present in low amounts. Introduction of quantitative, rather than *qualitative*, evaluations of chemicals may prove to be very important in large scale mutagen-carcinogen testing. SCE induction tests are very well suited for such a quantitative analysis.

Relationship of SCE Induction to DNA Damage, Repair, and Synthesis

A variety of chemical and physical agents, exhibiting diverse modes of interaction with DNA (Table 3.2) as well as transformation of cells by SV40 virus (162), are capable of inducing SCEs. Alkylating agents of many different types seem to be especially effective. SCEs can also be induced by irradiation of BrdU-substituted DNA (91, 99, 101), a treatment causing predominantly (though not exclusively) single-strand breaks (87). Only fragmentary information exists, however, about the quantitative relationship between the number and types of alkylation products or DNA strand interruptions, the efficiency of their repair, and the number of SCEs produced. Quantitation of DNA alkylation and removal can be accomplished by chemical analysis of reaction products or, if suitable isotopic derivatives can be obtained, by measurement of radioactivity in newly formed DNA adducts.

We have obtained evidence that SCE may account for only a small fraction of DNA damage by 8-methoxypsoralen plus near-UV light (36A). The combination of 8-methoxypsoralen plus 365 nm light, but not either agent alone, is effective in inducing SCE in human and CHO chromosomes (36, 126, 127, 131, 132, 133, 159, 241). The dependence of SCE on either light or 8-methoxypsoralen, keeping the other agent fixed, has been quantitated (126) and an assay for measuring the binding of tritiated 8-methoxypsoralen was developed, so that the ratio between these two quantities can be compared. Data thus far indicate that one SCE is induced (in the two cycles following DNA damage) per approximately 200 8-methoxypsoralen-DNA adducts (36A). This result is currently being analyzed into components due to mono- and bifunctional adducts.

We have thus far obtained both cytological and biochemical evidence for the persistence of alkylation by 8-methoxypsoralen during at least a few replication cycles.

---

*An extensive summary of the utility of the chick embryo system for clastogen testing is presented by Bloom in Chapter 6. A more detailed description of SCE test systems, analysis, and results, prepared after the present chapter was completed, will be published as part of the Gene-Tox program (Latt et. al, Mutat. Res. Reviews, in press).

The cytological data (127) consist of the observation of reciprocal interchanges of dark chromatids in third cycle metaphases, indicative of SCE formation after the second cycle (156); SCE formed during the first two cycles appear as isolated segments of darkly staining chromatids in third division metaphases. Similar data implicating SCE induction during the third cycle following DNA damage have now been described by Ishii and Bender in cells treated with mitomycin C (92). Thus, alkylation damage might underlie the observation of Stetka et al. (214) that repeated exposure of rabbits to mitomycin C ultimatley leads to persistently elevated SCE levels (in peripheral lymphocytes cultured in vitro).

Shafer (204) has recently postulated that SCE formation involves the bypass of DNA crosslinks during replication. This model is compatible with the observation that 8-methoxypsoralen adducts are slowly removed by cells. It will now be important to determine whether, as predicted by Shafer, those adducts remaining after replication are still in the form of crosslinks.

It is instructive to note that, since SCEs reflect less than 1% of DNA adducts, and that chromosome breaks are less than 1% as frequent as SCE (123, 124), chromosome breaks may detect $10^{-4} - 10^{-5}$ or less of the total DNA damage in a cell. The disparity between the numbers of DNA adducts, SCEs, and chromatid breaks might contribute to the multiplicity of results obtained by investigations comparing SCE-to-break ratios and the relative location of chromosome breaks and incomplete SCE following exposure of cells to different clastogens.

Consistent with an earlier suggestion by Heddle et al. (79), nearly half of the breaks in chromosomes in lymphocytes from Fanconi's anemia patients treated at the start of S with mitomycin C occur at incomplete SCE sites (137), as do 25–50% of the breaks induced at the end of S by UV-irradiation of BrdU-substituted Chinese hamster cells (101). Also, treatment of rat cells with dimethylbenzanthracene a few hours prior to harvest (i.e., at the end of S for the metaphases scored) gives a similar distribution of SCEs and breaks (236). In other systems, breaks occur in the absence of SCE (90, 194). While an explanation of these divergent observations is not apparent, there is ample room within the confines of observed stoichiometry for a given combination of damage and cell response to cause SCE and chromosome breaks by completely or largely divergent paths.

Sister chromatid exchange formation appears to be tightly coupled to DNA synthesis. Wolff et al. (249) demonstrated that UV-damaged rodent cells needed to pass through S phase for SCE induction to be detected. Variation in SCE inducibility within the S phase was investigated by Kato (99, 102), who used near-UV light to induce SCEs in unsynchronized, BrdU-substituted, Chinese hamster cells. The position of cells within S at the time of irradiation was estimated from the time between irradiation and metaphase collections, and by the extent of incorporation of a ^{3}H-dT pulse which was administered at the time of irradiation and then detected at metaphase. While SCE induced at the end of S were observed to occur preferentially in late replicating regions, the efficiency of SCE induction appeared to be maximal near mid-S, coinciding with the maximum in the rate of DNA synthesis.

Analysis of SCE induction by 8-methoxypsoralen plus light in synchronized cells (127, 132) led to a different conclusion, namely, that SCE induction was maximal at the start of S and decreased progressively throughout the S phase. The difference between this result and that of Kato (99) may be due to lack of cell synchrony in Kato's experiment or to a difference in the type of DNA damage effected.

Preliminary evidence for the latter possibility has recently been presented by Shafer (204). This possibility is especially easy to test, e.g., by treating synchronized cells with BrdU plus light. Loss of coherence in cell phasing during S would tend to broaden the SCE versus S phase traverse curve, especially for data attributed to the start of S. For data obtained at the end of S, there was better agreement between the two studies; a need for DNA synthesis in a given chromosome region, subsequent to DNA damage, appeared necessary for SCE induction. The molecular events accounting for the coupling between SCE induction and DNA synthesis, however, remain to be determined.

Biological Significance of SCE Formation

Implicit in many of the above studies is the assumption that SCE formation bears some relationship to DNA damage, repair, and mutagenesis. Certain evidence lends support to this idea. Carrano and associates (33) have observed an increase in mutations at the HGPRT (and APRT) (34) locus of the CHO cells in proportion to concentrations of ethylmethanesulfonate, ethylnitrosourea, mitomycin C and proflavine, in concentration ranges also causing a linear response in SCE induction. Relatively fewer mutations were observed with the bifunctional agent mitomycin C, or with its monofunctional decarbamyl analogue. Assuming the existence of 50,000 genes per cell, all with the same mutagenic susceptibility as HGPRT, Carrano et al. estimated that 0.01–1.0 mutation per SCE occurred during the first two S phases. It would seem desirable, though admittedly difficult, to develop a method of measuring SCE induction and mutagenesis in the same cells, to rule against the possibility that these two phenomena reflect disparate effects of alkylation in different members of a cell population.

Alkylation by psoralen derivatives plus light, a powerful inducer of SCE formation (36, 126, 127, 131, 132, 133, 159, 241), is known to stimulate DNA strand interchange in recombination proficient but not in recombination deficient (Rec A) bacteria (40). This observation prompted the suggestion (124) that SCE formation in metaphase chromosomes was somehow analogous to recombinational repair (189, 190) in bacteria. A feature complicating this analogy is the possible difference between DNA repair processes in bacterial and mammalian cells (143). Recombinational repair in bacteria may be errorprone, e.g., it has been reported that induction of mutations in bacteria by UV or psoralen plus light requires a functional Rec A system (89). The relationship of these observations to the error-prone SOS repair system (244) in bacteria remains to be determined. Of potential interest, in this regard, is the claim (111) that the tumor promoter TPA (12-0-tetradecanoyl-phorbol-13-acetate) can induce SCEs although work from our laboratory (145A), as well as that of Baker and associates (230A), was unable to confirm this observation. The relationship of these results to data (153, 236A) indicating that protease inhibitors (e.g., antipain and elastatinal) inhibit SCE induction, is not clear at present.

Another event, in addition to mutagenesis, paralleling SCE induction by clastogens is the release of SV40 virus from transformed cells. This has been demonstrated (95) in a number of different hamster kidney cell lines, using mitomycin C and EMS. A 10,000-fold greater concentration of EMS (relative to mitomycin C) was needed both for SCE induction and for virus induction.

Sister Chromatid Exchange Formation in Human Chromosome Fragility Diseases

Analysis of SCE formation has been used to differentiate between various inherited human diseases characterized by chromosome fragility and a predisposition for the development of neoplasia (65). These diseases, which include Bloom's syndrome, Fanconi's anemia, and ataxia telangiectasia, presumably involve defects in DNA repair. The diseases potentially constitute test systems, with specific DNA repair defects, for dissecting the SCE process, and cells from other diseases (e.g., xeroderma pigmentosum, see below) may permit extraordinary sensitive clastogen detection. All three conditions listed above are rare, but they follow an autosomal recessive inheritance mode, and the respective heterozygotes amount to 1–2% of the total population (222, 223). Since heterozygotes for some of these conditions also appear to be at an increased risk for certain forms of cancer, they make up several percent of all individuals with those neoplastic conditions.

Cells from patients with Fanconi's anemia have been shown to be highly susceptible to killing (55, 56) and to chromosome breakage (12, 192, 194) by bifunctional alkylating agents, and they appear to exhibit reduced ability to excise UV- (179) and gamma-irradiation products (183), and DNA crosslinks (57).

Lymphocytes from Fanconi's anemia patients, while exhibiting essentially normal SCE frequencies in the presence of BrdU, respond to mitomycin C treatment with a subnormal increase is sister chromatid exchange formation (137). This observation has now been confirmed in one other laboratory (quoted in reference 60) but not in others (see e.g., Latt, Ann Rev Genet, in press). The reduced stimulation of SCE formation by mitomycin C in Fanconi's anemia is associated with increased chromatid breakage. However, the relative contribution of mitomycin C monoadducts and crosslinks to the SCE and chromosome breakage results has not yet been determined. Interestingly, approximately half of the breaks induced in F.A. lymphocytes by mitomycin C occurred at sites of incomplete sister chromatid exchange formation (126, 137), compatible with the hypothesis that the break increment and at least some of the exchange deficit are causally related.

Our initial studies of lymphocytes from four patients with Fanconi's anemia have been repeated, with similar results, on two other patients with this disease. Fibroblasts from Fanconi's anemia patients show only a marginal deficit in SCE response, although chromosome breakage in the presence of mitomycin C is elevated, and the response in cells from different sources is heterogenous. The results can be interpreted to suggest that Fanconi's anemia cells are defective in a form of DNA repair.

We have not detected abnormalities in short term SCE induction in heterozygotes from patents with Fanconi's anemia. However, extended exposure of carriers to low levels of the potentially bifunctional alkylating agent diepoxybutane (13) does seem to elicit abnormally high chromosome breakage in both diseased and heterozygote cells. This latter observation may reflect accumulation over several cell cycles of incompletely repaired DNA damage.

In Bloom's syndrome, the baseline SCE frequency is greatly elevated (37). It is not yet apparent how this relates to retarded rate of DNA replication fork progression

(67, 75) or increased sensitivity to ultraviolet light (67) in these cells. Tice et al. (233) have observed an approximately 50% elevation in SCE frequencies in normal fibroblasts cocultivated with cells isolated from patients with Bloom's syndrome. One interpretation of these data is that a humoral factor is responsible for the SCE elevation in Bloom's syndrome. German et al. (66) reported that, in certain Bloom's syndrome patients, a subpopulation of lymphocytes does not exhibit elevated SCEs, perhaps suggesting that, if such a humoral factor exists, not all cells are equally susceptible. Van Buul et al. (238) were unable to repeat Tice's observations (233), but instead found, in related experiments, that cocultivation of Bloom's syndrome fibroblasts with Chinese hamster cells led to a modest (~30%) *reduction* in SCEs in the Bloom's syndrome cells (238). Since this effect required the simultaneous presence of both cell types, the action of a humoral factor was considered to be unlikely. With regard to the cocultivation results, it will be important to determine the extent to which the large number of hamster cells present depleted the BrdU content of the medium (which was initially 5×10^{-6} M, a relatively low value), to determine whether any of the change in SCE in the Bloom's syndrome cells reflected a lower effective BrdU concentration. Bryant et al (29A) have reported that fusion of Bloom's syndrome cells with normal cells leads to a normalization of the SCE frequency in the former.

Recently, Shiriashi and Sandberg (208) have shown that lymphocytes from a patient with Bloom's syndrome undergo a modest additional increase in SCEs upon exposure to mitomycin C. This increase may in part be limited by the high baseline level of SCEs (≥ 100/cell) and the existence of a saturation level of SCE formation (or detection) in a given cell. These workers also showed that SCE elevations could be detected in bone marrow, as well as blood cells, from patients with Bloom's syndrome (207), and that there was little direct correlation between the distribution of SCEs and structural aberrations in chromosomes from patients with this disease.

Patterson et al. (171) reported that cells from patients with ataxia telangiectasia exhibited a reduced ability to excise DNA bases damaged by high energy radiation. More recent studies (38) have indicated that the X-ray survival of cells from ataxia telangiectasia patients is well below normal, while survival of cells from heterozygotes was intermediate between that of cells from normal and diseased individuals. However, cells from ataxia telangiectasia patients show normal baseline SCE levels (61, 77), as well as a normal SCE response after exposure to X-irradiation, MMC, EMS, and adriamycin (60).

Cells from patients in complementation groups A, B, C, and D of xeroderma pigmentosum, another hereditary disease with a predisposition for neoplasia (46), hyperreact to UV-irradiation (14, 46, 199), or alkylating agents (251), undergoing a much greater increase in SCEs than do identically treated normal cells. Xeroderma pigmentosum cells which exhibit SCE hyperinducibility also have a reduced ability to excise alkylation products (e.g., 6-0-methylguanine) (70). This is compatible with the idea that SCE results from DNA damage that has not been removed. However, the relative inducibility of SCE and chromosome breaks in xeroderma pigmentosum cells depends strongly on the type of DNA damage involved (245, 251), and, for damage other than that caused by UV light, on other factors, such as SV40 transformation.

It is interesting to note that, in xeroderma pigmentosum, a hyperinducibility in SCEs correlates with a hyperinducibility of mutations by similar agents (149). Conversely, in Fanconi's anemia, the hypoinducibility in SCEs, more marked with mitomycin C than with ethylmethane sulfonate, is accompanied by a decrease in the ability of both of these alkylating agents to induce mutations (54). Thus, even though SCE may reflect only a small fraction of the total damage caused to DNA, it is intriguing to speculate that the SCE-inducing component of this damage might ultimately prove to be important biologically.

SCE frequencies have been examined in a number of other human diseases, as well as in a variety of chromosomal abnormalities, generally with weakly positive or negative results. For example, SCE formation in Cockayne's syndrome appears to be normal (39), while SCEs in Down syndrome have been reported to be normal in one study (253), elevated in a second (142), and normal in the presence of only BrdU, but hyperinducible by X-rays and mitomycin C (20), in a third. Similarly, SCE frequencies in most individuals with either supernumerary or structurally abnormal chromosomes appear to be normal (5, 220), although isolated examples of abnormal karyotypes with slightly elevated SCEs have been observed (5). Some of these discrepancies may reflect an apparent, very slight age dependence of SCE frequencies (106), different culture conditions, e.g., sera types (104), or, perhaps more likely, variations between SCEs in different normal individuals or different samplings from the same individual (5, 44, 52, 126).

The Mechanism of SCE Formation

Various approaches have been used to investigate the mechanism of SCE formation. Kato (101) has examined SCE inducibility in unsynchronized Chinese hamster cells that were allowed to incorporate BrdU for one cycle and grow a second cycle in the presence or absence of BrdU. SCE induction at a time approximating the last few hours of the second S phase was effected by irradiation with near-UV light. Only a small additional increase in SCE was observed in those cells which had incorporated BrdU for the second S phase, prompting the suggestion that SCE induction might have multiple pathways, at least one of which was independent of the degree of BrdU substitution. However, if SCE induction in a particular chromosome region requires DNA synthesis following damage (138), then only the regions that had not replicated a second time at the time of irradiation would be susceptible to SCE induction. These would be unifilarily substituted with BrdU, independent of the growth protocol used, and no difference in SCE induction would be expected in the two types of cells, whatever the specific mechanisms involved.

Kato (101) also examined the effect of caffeine on SCE induction, and found it to inhibit induction in cells which had undergone one round of BrdU incorporation, but to stimulate SCE induction, during the second S phase, in cells which incorporated BrdU for two cycles. Interpretation of this result will depend on the chromosomal location of these additional SCEs. Kato has previously reported (96) that caffeine inhibited SCE induction by UV in Chinese hamster cells, prompting analogy with postreplication repair, while other workers have observed either a potentiation (241) or an inhibition (240) of SCE frequencies with caffeine (also see Table 3.3). Vogel and Bauknecht (240) stressed the importance of the toxicity of caffeine and its effect on selection of metaphases for scoring. Recently Ishii and Bender (93) have determined that SCE potentiation by caffeine requires that the caffeine be added with, or soon after, the SCE inducer. Caffeine may well exert multiple effects which might be very difficult to dissect.

SCE induction, like mutagenesis (187), may also be influenced by agents, e.g., cysteine (5), capable of trapping free radicals. However, these results, like those in which the enzymes superoxide dismutase and catalase protect cells from chromosome breakage (164, 165, 166), probably deal more with the chemistry of the inducing agent than with alterations in cellular response to the damage induced. The mechanism by which selenium, which under some conditions can induce SCEs (182), can in some cases suppress SCE induction (94) is not at present clear.

Evidence for Sister Chromatid Interchange at the DNA Level

Two types of experimental approaches have been used to search for DNA exchanges which correspond to SCEs. Both utilize cells which have incorporated BrdU for less than one cycle, and thus contain DNA substituted in only one strand. Following sister strand exchange, junctions of substituted and unsubstituted polynucleotide should result and appear as material of intermediate density in alkaline CsCl gradients (186). The Holliday model for DNA recombination (83) also predicts segments of heavy-heavy (and light-light) DNA at interchange sites in neutral CsCl gradients.

Rommelaere and Miller-Faures (186) reported the detection of Chinese hamster DNA with intermediate density in alkaline CsCl gradients. However, most of this material exhibited rapid renaturation following neutralization, a result expected for crosslinked DNA. If DNA from the Chinese hamster cells was centrifuged in neutral CsCl, approximately 0.1% of the material exhibited density greater than that of hybrid, heavy-light (HL) DNA and interpretable as containing segments of bifilarly substituted, heavy-heavy (HH) DNA. The amount of this DNA was increased 4-fold by UV-irradiation (100 ergs/mm^2) prior to BrdU incorporation, but the amount of DNA detected was more than 10 times that expected from the number of SCE in these cells.

Moore and Holliday (157) similarly detected 0.5% "heavy-heavy" DNA from rapidly growing CHO cells cultured not quite one cycle in medium containing BrdU. Mitomycin C, when administered in highly toxic amounts (1 µg/ml) five hours prior to harvest, appeared to increase both "heavy-heavy" DNA and SCE. Again, the amount of "heavy-heavy" DNA was much more than expected for the number of SCE observed.

Loveday (146) has repeated the Moore and Holliday experiments, using synchronized CHO cells which had incorporated BrdU for one cycle. These cells exhibited a small amount (0.4 ± 0.2%) of DNA banding with a density expected for HH DNA, but this was not increased by addition of sufficient mitomycin C (0.03 µg/ml, at the start of S) to more than triple the SCE frequency (Fig. 3.9). Significantly, the dense DNA persisted after a subsequent round of replication in the absence of BrdU (calling its bifilar substitution into question), and material with a similar density shift from the main band DNA was seen in cells which incorporated ^{3}H-dT (but not BrdU). While Loveday's data do not rule out the existence of the "heavy-heavy" DNA predicted by the Holliday model, they suggest that the biochemical evidence thus far claimed for this DNA is very weak, and that additional experiments are

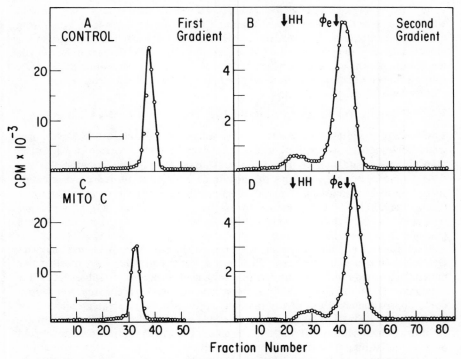

Figure 3.9 Density gradient analysis of CHO DNA after one cycle of BrdU incorporation. Synchronized CHO cells were released into medium containing ^{3}H-BrdU (5 mCi/ml, 2×10^{-5} M), and Colcemid (0.45 µg/ml) was added after 8 hours. DNA was isolated from metaphase cells 4 hours later, sheared and centrifuged to equilibrium in CsCl. Three drop fractions were collected and aliquots (10%) were counted: (A) control; (C) 0.3 µg/ml mitomycin C added at the time of release from G_1-S. The indicated fractions were recentrifuged with ^{14}C-labeled DNA and 2 drop fractions were collected onto filters. The second gradients are shown in (B) no mitomycin C; (D) + mitomycin C. The arrows mark the expected position of HH DNA (HH) and the actual position of DNA. Total radioactivity (CPM): (A) 1.1×10^6, with 6.6×10^3 in dense DNA (0.65%); (C) 6.7×10^5, with 4.9×10^3 in dense DNA (0.73%) (146).

necessary to clarify the chemical events associated with DNA interchange during SCE formation.

Summary

Methodology for sister chromatid exchange detection is now well-developed. BrdU-fluorochrome or Giemsa techniques have greatly simplified cytological SCE analysis. SCE formation can be studied in cultured cells, in intact animals, or in combined systems in which cells from treated animals are cultured in vitro or chemicals are activated by microsomal preparations before exposure to cultured cells. Alterations in SCE frequencies in any of these systems can serve as sensitive indices of the interaction of mutagen-carcinogens with chromosomes.

Most mutagen-carcinogens are potent inducers of SCEs, while a few others increase chromosome breakage. The combination of SCE and chromosome aberration as a test for clastogens thus has few "false negatives". Conversely, there is no convincing example yet of a compound which is highly effective at inducing SCEs that is not mutagenic and/or carcinogenic in at least some system.

In vitro SCE analysis can also be used clinically, for differentiation of human chromosome fragility diseases, and it may prove useful for monitoring chromosome damage in cells from patients exposed to clastogenic agents during chemotherapy. Information about the mechanism of SCE induction by chemical and physical agents is still rudimentary, and little is known about the molecular abnormalities underlying most human chromosome fragility diseases or about the causes for alterations in SCE formation in these diseases. Present empirical applications of SCE analysis should increase as more is understood about the basic mechanism of SCE formation.

Acknowledgments

Research in the authors' laboratory reviewed in this article was supported by grants (GM 21121 and HD 04807, CD-36F, and 1-353, from the National Institutes of Health, the American Cancer Society, and the National Foundation March of Dimes, respectively). S. Latt and K. Loveday are recipients of a Research Career Development Award (GM 00122, from the National Institute of General Medical Sciences) and a postdoctoral fellowship (PF-1223, from the American Cancer Society), respectively.

We thank Dr. Stephen Bloom for making available to us some of his unpublished data on SCE induction in chick embryos.

Literature Cited

1. Aaronson, M. W.; Nichols, W. W.; Miller, R. C.; Meadows, A. T. Sister chromatid exchange in childhood cancer. Lake Yamanaka SCE Conference; July 1978.
2. Abe, S.; Sasaki, M. Chromosome aberrations and sister chromatid exchanges in Chinese hamster cells exposed to various chemicals. J. Nat. Canc. Inst. 58:1635–41; 1977.
3. Abe, S; Sasaki, M Studies in chromosomal aberrations and sister chromatid exchanges induced by chemicals. Proc. Japan Acad. 53:46–49; 1977.
4. Abramovsky, I.; Vorsanger, G.; Hirschhorn; K. Sister chromatid exchange induced by X-ray of human lymphocytes and the effect of L-cysteine. Mutat. Res. 50:93–100; 1978.
5. Alhadeff, B.; Cohen, M. M. Frequency and distribution of sister chromatid exchanges in human peripheral lymphocytes. Israel J. Med. Sci. 12:1440–47; 1976.
6. Allen, J. W.; Latt, S. A. Analysis of sister chromatid exchange formation in vivo in mouse spermatogonia as a new test system for environmental mutagens. Nature 260:449–51; 1976.
7. Allen, J. W.; Latt, S. A. In vivo BrdU-33258 Hoechst analysis of DNA replication kinetics and sister chromatid exchange formations in mouse somatic and meiotic cells. Chromosoma 58:325–40; 1976.
8. Allen J. W.; Shuler, C. F.; Mendes, R. W.; Latt, S. A. A simplified technique for in vivo analysis of sister chromatid exchanges using 5-bromodeoxyuridine tablets. Cytogenet. Cell Genet. 18:231–7; 1977.
9. Allen, J. W.; Shuler, C. F.; Latt, S. A. BrdU tablet methodology for in vivo studies of DNA synthesis. Somat. Cell Genet. 4:393–405; 1978.

10. Alves, P.; Jonasson, J. New staining method for the detection of sister chromatid exchanges in BrdU-labelled chromosomes. J. Cell Sci. 32:185-95; 1978.
11. Ames, B. N.; McCann, J.; Yamasaki, E. Methods for detecting carcinogens and mutagens with the salmonella/mammalian-microsome mutagenicity test. Mutat. Res. 31:347-64; 1975.
12. Auerbach, A. D.; Wolman, S. R. Susceptibility of Fanconi's anemia fibroblasts to chromosome damage by carcinogens. Nature 261:494-96; 1976.
13. Auerbach, A. D.; Wolman, S. R. Carcinogen-induced chromosome breakage in Fanconi's anemia heterozygous cells. Nature 271:69-71; 1978.
14. Bartram, C. R.; Koske-Westphal, T.; Passarge, E. Chromatid exchange in ataxia telangiectasia, Bloom's syndrome, Werner's syndrome, and xeroderma pigmentosum. Ann. Hum. Genet. 40:79-86; 1976.
15. Bauknecht, Th.; Vogel, W.; Bayer, U.; Wild, D. Comparative in vivo mutagenicity testing by SCE and micronucleus induction in mouse bone marrow. Human Genetics 35: 299-307; 1977.
16. Bayer, U.. The in vivo induction of sister chromatid exchanges in the bone marrow of the Chinese hamster. II. N-nitrosodiethylamine (DEN) and N-isopropyl-α-(2-methylhydrazino)-p toluamide (Natulan), two carcinogenic compounds with specific mutagenicity problems. Mutat. Res. 56: 305-9; 1978.
17. Bayer, U.; Bauknecht, Th. The dose dependence of sister chromatid exchanges in the in vivo bone marrow test with Chinese hamsters induced by 3-hydrocarbons. Experientia 33:25; 1977.
18. Beek, B.; Obe, G. The human leukocyte test system. VI. The use of sister chromid exchanges as possible indicators for mutagenic activities. Humangenetik 29:127-34; 1975.
19. Benedict, W. F.; Banerjee, A.; Venkatesan, N. Cyclophosphamide induced oncogenic transformation, chromosomal breakage, and sister chromatid exchange following microsomal activation. Cancer Res. 38:2922-24; 1978.
20. Biederman, B.; Bowen, P. Sister chromatid exchanges in Down syndrome. Mamm. Chrom. Newsletter 18:12; 1977.
21. Bigger C. A. H.; Tomaszewski, J. W.; Dipple, A. Differences between products of binding of 7,12-dimethylbenz (α) anthracene to DNA in mouse skin and in a rat liver microsomal system. Biochem. Biophys. Res. Commun. 80:229-35; 1978.
22. Bloom, S. Chick embryos for detecting environmental mutagens. In Hollaender, A.; DeSerres, F., eds. Mutagens V. Vol. 5 Plenum Press; 1978: 203-32.
23. Bloom, S. E.; Hsu, T. C. Differential fluorescence of sister chromatids in chicken embryos exposed to 5-bromodeoxyuridine. Chromosoma 51: 261-67; 1975.
24. Boobis, A. R.; Reinhold, C.; Thorgiersson, S. S. Induction of aryl hydrocarbon (benzo (α) pyrene) hydroxylase and 2-acetylaminofluorene N-hydroxylase by polycyclic hydrocarbons in regenerating liver from inbred strains of mice. Biochem. Pharm. 26:1501-5; 1977.
25. Bostock, C. J.; Christie, S. Analysis of the frequency of sister chromatid exchange in different regions of chromosomes of the kangaroo rat. Chromosoma 56:275-87; 1976.
26. Brewen, J. G.; Peacock, W. J. Restricted rejoining of chromosomal subunits in aberration formation. A test for subunit in aberration formation. A test for subunit dissimilarity. Proc. Nat. Acad. Sci. USA 62:389-94; 1968.
27. Brewen, J. G.; Peacock, W. J. The effect of tritiated thymidine on sister chromatid exchange in a ring chromosome. Mutat. Res. 7:433-40; 1969.
28. Brown, R. L.; Crossen, P. E. Increased incidence of sister chromatid exchanges in Rauscher leukemia virus infected mouse embryo fibroblasts. Exper. Cell Res. 103:418-20; 1976.
29. Brown, R. F.; Wu, Y.. Induction of sister chromatid exchanges in Chinese hamster cells by chlorpropamide. Mutat. Res. 56:215-17; 1977.
29A. Bryant, E.M.; Huehn, H.; Martin, G.M. Normalization of sister chromatid exchange frequencies in Bloom's syndrome by euploid cell hybridization. Nature 279:795-96; 1979.
30. Burgdorf, W.; Kurvink, K.; Cerevenka, J. Elevated sister chromatid exchange rate in lymphocytes of subjects treated with arsenic. Human. Genet. 36:69-72; 1977.
31. Burkholder, G. D. Reciprocal Giemsa staining of late DNA replicating regions produced by low and high pH sodium phosphate. Exper. Cell Res. 111:489-92; 1978.
32. Carrano, A. V.; Johnston, G. R. The distribution of mitomycin C-induced sister chromatid exchanges in the euchromatin and heterochromatin of the Indian muntjac. Chromosoma 64:97-107; 1977.

33. Carrano, A. V.; Thompson, L. H.; Lindl, P. A.; Minkler, J. L. Sister chromatid exchange as an indicator of mutagenesis. Nature 271:551-53; 1978.
34. Carrano, A. V. Sister chromatid exchange: relation to gene mutation, repeated chemical exposures in vivo and standardization for human monitoring. Lake Yamanaka Chromosome Conference on SCE; July 1978.
35. Carrano, A. V.; Wolff, S. Distribution of sister chromatid exchanges in the euchromatin and heterochromatin of the Indian muntjac. Chromosoma 53:361-69; 1975.
36. Carter, D. M.; Wolff, K.; Schnedl, W. J. 8-methoxypsoralen and UVA promote sister chromatid exchanges. Invest. Dermatol. 67:548-51; 1976.
36A. Cassel, D. M.; Latt, S.A. Relationship between DNA adduct formation and sister chromatid exchange induction by ^{3}H-8-methoxypsoralen in Chinese hamster ovary cells. Exper. Cell Res. 128:15-22; 1980.
37. Chaganti, R. S. K.; Schonberg, S.; German, J. A manyfold increase in sister chromatid exchanges in Bloom's syndrome lymphocytes. Proc. Nat. Acad. Sci. USA 71:4508-12; 1974.
38. Chen, P. C.; Lavin, M. F.; Kidson, C.; Moss, D. Identification of ataxia telangiectasia heterozygotes, a cancer prone population. Nature 274:484-86; 1978.
39. Cheng, W. S.; Tarone, R. F.; Andrews, A. D.; Whang-Peng, J. S.; Robbins, J. H. Ultraviolet light-induced sister chromatid exchanges in xeroderma pigmentosum and in Cockayne's syndrome lymphocyte cell lines. Cancer Res. 38:1601-9; 1978.
40. Cole, R. S. Repair of DNA containing intrastrand crosslinks in *Escherichia coli:* sequential excision and recombination. Proc. Nat. Acad. Sci. USA 70:1064-68; 1973.
41. Comings, D. E. Isolabelling not compatible with single stranded model. Nature New Biol. 229:24-25; 1971.
42. Comings, D. E. The distribution of sister chromatids at mitosis in Chinese hamster cells. Chromosoma 29:428-33; 1970.
43. Craig-Holmes, A. P.; Shaw, M. W. Effects of six carcinogens on SCE frequency and cell kinetics in cultured human lymphocytes. Mutat. Res. 46:375-84; 1977.
44. Crossen, P. E.; Drets, M. E.; Arrighi, F. E.; Johnston, D. A. Analysis of the frequency and distribution of sister chromatid exchanges in cultured human lymphocytes. Human Genetics 35:345-52; 1977.
45. Daryzynkiewicz, Z.; Andreeff, M.; Traganos, F.; Sharpless, T.; Melamed, M. R. Discrimination of cycling and noncycling lymphocytes by BUdR-suppressed acridine orange fluorescence in a flow cytometric system. Exper. Cell Res. 115:31-36; 1978.
46. DeWeerd-Kastelein, E. A.; Keijzer, W.; Rainaldi, G.; Boostma, D. Induction of sister chromatid exchanges in xeroderma pigmentosum cell after exposure to ultraviolet light. Mutat. Res. 45:253-61; 1977.
47. DeRaat, W. K. The induction of sister chromatid exchanges by cyclophosphamide in the presence of differently induced microsomal fractions of rat liver. Chem. Biol. Int. 19:125-31; 1977.
47a. DeRaat, W. K. Induction of sister chromatid exchanges by styrene and its presumed metabolite styrene oxide in the presence of rat liver homogenate. Chem. Biol. Interactions 20:163-70; 1978.
48. Dutrillaux, B.; Laurent, C.; Couturier, J.; Lejeune, J. Coloration des chromosomes humains par l'acridine orange apres traitement par 5-bromodeoxyuridine. C.R. Acad. Sci. (D) Paris 276:3179-81; 1973.
49. Dutrillaux, B.; Fosse, A. M.; Prieur, M.; LeJeune, J. Analyses des echanges de chromatides dans les cellules somatiques humaines. Chromosoma 48:327-40; 1974.
50. Evans, L. A.; Kevin, M. J.; Jenkins, E. C. Human sister chromatid exchange caused by methylazoxymethanol acetate. Mutat. Res. 56:51-58; 1977.
51. Faed, M. J. W.; Mourelatos, D. Enhancement by caffeine of sister chromatid exchange frequency in lymphocytes from normal subjects after treatment by mutagens. Mutat. Res. 49:437-40; 1978.
52. Falek, A.; Madden, J. J.; Shafer, D. A. Interindividual differences in mutagenic sensitivity in human lymphocytes. Amer. J. Hum. Genet. 30:118A; 1978.
53. Farrel, S. A.; Worton, R. C. Chromosome loss is responsible for segregation at the HPRT locus in Chinese hamster cell hybrids. Somat. Cell Genet. 3:539-51; 1977.
54. Finkelberg, R.; Buchwald, M.; Siminovich, L. Decreased mutagenesis in cells from patients with Fanconi's anemia. Amer. J. Hum. Genet. 29:42a; 1977.
55. Finkelberg, R.; Thompson, M. W.; Siminovich, L. Survival after treatment with EMS,

X-rays, and mitomycin C of skin fibroblasts from patients with Fanconi's anemia. Amer. J. Hum. Genet. 26:30a; 1974.

56. Fujiwara, Y.; Tatsumi, M. Repair of mitomycin C damage to DNA in mammalian cells and its impairment in Fanconi's anemia cells. Biochem. Biophys. Res. Commun. 66:592-98; 1975.

57. Fujiwara, Y.; Tatsumi, M.; Sasaki, M. S. Cross-link repair in human cells and its possible defect in Fanconi's anemia cells. J. Mol. Biol. 113:635-49; 1977.

58. Funes-Cravioto, F.; Kolmodin-Hedman, B.; Lindsten, J.; Nordenskjolo, M.; Apata Gayon, G.; Lambert, B.; Norberg, G.; Olin, R.; Swensson, A. Chromosome aberrations and sister chromatid exchanges in workers in chemical laboratories and a retoprinting factory and in children of laboratory workers. Lancet ii:322-25; 1977.

59. Furukawa, M.; Sirianni, S. R.; Tan, J. C.; Huang, C. C. Sister chromatid exchanges and growth inhibition by the flame retardant Tris (2,3 dibromopropyl phosphate) in Chinese hamster cells. J. Nat. Canc. Inst. 60:1179-81; 1978.

60. Galloway, S. M. Ataxia telangiectasia: the effects of chemical mutagens and X-rays on SCE in blood lymphocytes. Mutat. Res. 45:343-49; 1977.

61. Galloway, S. M.; Evans, H. J. Sister chromatid exchange in human chromosomes from normal individuals and patients with ataxia telangiectasia. Cytogenet. Cell Genet. 15:17-29; 1975.

62. Galloway, S.; Wolff, S. The relationship between chemically induced sister chromatid exchanges and chromatid breakage. Mutat. Res.; 61:297-307; 1979.

63. Gatti, M.; Pimpinelli, S.; Santini, G.; Olivierei, G. Lack of spontaneous sister chromatid exchange (SCE) in somatic cells of *Drosophila melanogaster*. Genetics 91:255-74; 1979.

64. Gebhart, E.; Kappauf, H. Bleomycin and sister chromatid exchanges in human lymphocyte chromosomes. Mutat. Res. 58:121-24; 1978.

65. German, J. Genes which increase chromosomal instability in somatic cells and predispose to cancer. Prog. in Medical Genetics 8:61-101; 1972.

66. German, J.; Schonberg, S.; Loue, E.; Chaganti, R. S. K. Bloom's syndrome. IV. Sister chromatid exchanges in lymphocytes. Amer. J. Hum. Genet. 29:248-55; 1977.

67. Gianelli, F.; Benson, P. F.; Pawsey, S. A.; Polani, P. E. Ultraviolet light sensitivity and delayed DNA-chain maturation in Bloom's syndrome fibroblasts. Nature 265:466-69; 1977.

68. Gibas, Z.; Limon, J. Isolabeling of the long arm of the human Y chromosome demonstrated by the FPG technique. Chromosoma 69:113-20; 1978.

69. Gibson, D. A.; Prescott, D. M. Induction of sister chromatid exchanges in chromosomes of rat kangaroo cells by tritium incorporated into DNA. Exper. Cell Res. 74:397-402; 1972.

70. Goth-Goldstein, R. Repair of DNA damage by alkylating carcinogens is defective in xeroderma pigmentosum-derived fibroblasts. Nature 267:81-92; 1977.

71. Goto, K.; Akematsu, T.; Shimazu, H.; Sugiyama, T. Simple differential Giemsa staining of sister chromatids after treatment with photosensitive dyes and exposure to light and the mechanism of staining. Chromosoma 53:223-30; 1975.

72. Goto, K.; Maeda, S.; Kano, Y.; Sugiyama, T. Factors involved in differential Giemsa staining of sister chromatids. Chromosoma 66:351-59; 1978.

73. Gratzner, H. G.; Pollack, A.; Ingram, D. J.; Leif, R. C. Deoxyribonucleic acid replication in single cells and chromosomes by immunologic techniques. J. Histochem. Cytochem. 24:34-39; 1976.

74. Grisham, J. W. *In* Zimmerman, A. M.; Padilla, G. M.; Cameron, I. Z. eds. Drugs and cell cycle. New York: Academic Press; 1973: 95-136.

75. Hand, R.; German, J. Bloom's syndrome: DNA replication in cultured fibroblasts and lymphocytes. Human Genetics 38:297-306; 1977.

76. Hansteen, I. L.; Hilllestad, L.; Thiis-Evensen, E.; Heldas, S. S. Effects of vinyl chloride in man; a cytogenetic follow-up study. Mutat. Res. 51:271-78; 1978.

77. Hatcher, N. H.; Brinson, P. S.; Hook, E. B. Sister chromatid exchanges in ataxia telangiectasia. Mutat. Res. 35:333-36; 1976.

78. Hayashi, K.; Schmid, W. The rate of siser chromatid exchanges parallel to spontaneous chromosome breakage in Fanconi's anemia and to trenimon-induced aberrations in human lymphocytes and fibroblasts. Humangenetik 29:201-6; 1975.

79. Heddle, J. A.; Whissel, D.; Bodycote, J. D. Changes in chromosome structure induced by radiation: a test of the two chief hypotheses. Nature 221:159-60; 1969.
80. Henderson, P. Th.; Kersten, K. J. Metabolism of drugs during rat liver regeneration. Biochem. Pharmacol. 19:2343-51; 1970.
81. Herreros, B.; Giannelli, F. Spatial distribution of old and new chromatid subunits and frequency of chromatid exchanges in induced human lymphocyte endoreduplications. Nature 182:286-88; 1967.
82. Hollander, D. H.; Tockman, M. S.; Liang, Y. W.; Borgaonkar, D. S.; Frost, J. K.: Sister chromatid exchanges in the peripheral blood of cigarette smokers and in lung cancer patients; and the effect of chemotherapy. Hum. Genet. 44:167-71; 1978.
83. Holliday, R. A mechanism for gene conversion in fungi. Genet. Res. 5:282-304; 1964.
84. Hsu, T. C.; Pathak, S. Differential rates of sister chromatid exchanges between euchromatin and heterochromatin. Chromosoma 58:269-73; 1976.
85. Huang, C. C.; Furukawa, M. Sister chromatid exchanges in human lymphoid lines cultured in diffusion chambers in mice. Exper. Cell Res. 111:458-61; 1978.
86. Hunke, M. H.; Carpenter, N. J. Effects of diphenylhydantoin on the frequency of sister chromatid exchanges in human lymphocytes. Amer. J. Hum. Genet. 30:83A; 1978.
87. Hutchinson, F. The lesions produced by ultraviolet light in DNA containing 5-bromouracil. Quart. Rev. Biophys. 6:201-46; 1973.
88. Huttner, K. M.; Ruddle, F. H. Study of mitomycin C-induced chromosomal exchange. Chromosoma 56:1-13; 1975.
89. Igali, S.; Bridges, B. A.; Ashwood-Smith, M. J.; Scott, B. R. Mutagenesis in *E. coli*. IV. Photosensitization to near-UV by 8-methoxypsoralen. Mutat. Res. 9:20-30; 1970.
90. Ikushima, T. Role of sister chromatid exchanges in chromatid aberration formation. Nature 268:235-36; 1977.
91. Ikushima, T.; Wolff, S. Sister chromatid exchanges induced by light-flashes to 5-bromodeoxyuridine and 5-iododeoxyuridine-substituted Chinese hamster chromosomes. Exper. Cell Res. 87:15-19; 1974.
91a. Ishidate, M.; Odashima, S. Chromosome tests with 134 compounds on Chinese hamster cells in vitro—a screening for chemical carcinogens. Mutat. Res. 48:337-54; 1977.
92. Ishii, Y.; Bender, M. Factors influencing the frequency of mitomycin C-induced sister chromatid exchanges in 5-bromodeoxyuridine substituted human lymphocytes in culture. Mutat. Res. 51:411-18; 1978.
93. Ishii, Y.; Bender, M. Caffeine inhibition of prereplication repair of mitomycin C-induced DNA damage in human peripheral lymphocytes. Mutat. Res. 51:419-25; 1978.
94. Jacobs, M. M. Inhibitory effects of selenium on 1,2-diemthylhydrazine and methylazoxymethanol colon carcinogenesis. Cancer 40:2557-64; 1977.
95. Kaplan, J. C.; Zamansky, G. B.; Black, P. H.; Latt, S. A. Parallel induction of sister chromatid exchanges and infectious virus from SV-40 transformed cells by alkylating agents. Nature 271:662-63; 1978.
96. Kato, H. Induction of sister chromatid exchanges by UV light and its inhibition by caffeine. Exper. Cell Res. 82:382-90; 1973.
97. Kato, H Spontaneous sister chromatid exchanges detected by BudR-labelling method. Nature 251:70-72; 1974.
98. Kato, H. Induction of sister chromatid exchanges by chemical mutagens and its possible relevance to DNA repair. Exper. Cell Res. 85:239-47; 1974.
99. Kato, H. Possible role of DNA synthesis in function of sister chromatid exchanges. Nature 252:739-41; 1974.
100. Kato, H. Is isolabelling a false image? Exper. Cell Res. 89:416-20; 1974.
101. Kato, H. Mechanisms for sister chromatid exchanges and their relation to the production of chromosome aberrations. Chromosoma 59:179-91; 1977.
102. Kato, H. Spontaneous and induced sister chromatid exchanges as revealed by the BudR-labelling method. Int. Rev. Cytology 37:55-95; 1977.
103. Kato, H.; Sandberg, A. A. Effects of herpes simplex virus on sister chromatid exchange and chromosome abnormalities in human diploid fibroblasts. Exper. Cell Res. 109: 423-27; 1977.
104. Kato, H.; Sandberg, A. A. The effect of sera on sister chromatid exchanges in vitro. Exper. Cell Res. 109:445-48; 1977.

105. Kato, H.; Shimada, H. Sister chromatid exchanges induced by mitomycin C: a new method of detecting DNA damage at the chromosomal level. Mutat. Res. 28:459-64; 1975.

106. Kato, N.; Stich, N. F. Sister chromatid exchanges in aging and repair-deficient human fibroblasts. Nature 260: 447-48; 1976.

107. Kihlman, B. S. Sister chromatid exchanges in *Vicia faba*. II. Effects of thiotepa, caffeine, and 8-ethoxy caffeine in the frequency of SCEs. Chromosoma 51:11-18; 1975.

108. Kihlman, B. A.; Kronberg, D. Sister chromatid exchanges in *Vicia faba* I. Demonstration by a modified fluorescence plus Giemsa (FPG) technique. Chromosoma 51:1-10; 1975.

109. Kihlman, B. A.; Sturelid, S. Effects of caffeine on the frequencies of chromosomal aberrations and sister chromatid exchanges induced by chemical mutagens in root tips of *Vicia faba*. Hereditas 88:35-41; 1978.

110. Kim, M. A. Chromatid austausch und heterochromatin veränderungen menschlicher chromosomen nach BrdU-markierang. Humangentik 25:179-88; 1974.

111. Kinsella, A. R.; Radman, M. Tumor promoter induces sister chromatid exchanges: relevance to mechanisms of carcinogens-S. Proc. Nat. Acad. Sci. USA 75:6149-53; 1978.

112. Kligerman, A. D.; Bloom, S. E. Sister chromatid differentiation and exchanges in adult mudminnows (*Umbra limi*) after in vivo exposure to 5-bromodeoxyuridine. Chromosoma 56: 101-9; 1976.

113. Knuutila, S.; Helmined, E.; Vuopio, P.; de la Chapella, A. Sister chromatid exchanges in human bone marrow cells. I. Control subjects and patients with leukemia. Hereditas 88: 189-96; 1978.

114. Knuutila, S.; Maki-Paakkanen, J.; Kahkonen, M.; Hookanen, G. An increased frequency of chromosomal changes and SCEs in cultured blood lymphocytes of 12 subjects vaccinated against small pox. Human Genet. 41:89-96; 1978.

115. Kram, D.; Schneider, E. L. Reduced frequencies of mitomycin C-induced sister chromatid exchanges in AKR mice. Human Genetics 41:45-51; 1978.

116. Kurvink, K.; Bloomfield, C. D.; Cervenka, J. Sister chromatid exchange in patients with viral disease. Exper. Cell Res. 113:450-53; 1978.

117. Kurvink, K.; Bloomfield, C. D.; Keenen, K. M.; Levitt, S.; Cervenka, J. Sister chromatid exchange in lymphocytes from patients with malignant lymphoma. Hum. Genet. 44:137-44; 1978.

118. Lambert, B.; Hansson, K.; Lindsten, J.; Sten, M.; Werelius, B. Bromodeoxyuridine-induced sister chromatid exchanges in human lymphocytes. Hereditas 83:163-74; 1976.

119. Lambert, B.; Linblad, A.; Nordenskjold, M.; Werelius, B. Increased frequency of sister chromatid exchanges in cigarette smokers. Hereditas 88:147-49; 1978.

120. Lambert, B.; Morad, M; Bredberg, A.; Swanbeck, G.; Thyresson-Hok, M. Sister chromatid exchanges in patients treated with psoralen and UV light. Mutat. Res. 46:228-29; 1977.

121. Langenbach, R.; Freed, H. J.; Raveh, D.; Huberman, E. Cell specificity in metabolic activation of aflatoxin B and Benzo (α) pyrene to mutagens for mammalian cells. Nature 276:277-80; 1978.

122. Latt, S. A. Microfluorometric detection of deoxyribonucleic acid replication in human metaphase chromosomes. Proc. Nat. Acad. Sci. USA 70:3395-99; 1973.

123. Latt, S. A. Localization of sister chromatid exchanges in human chromosomes. Science 185:74-76; 1974.

124. Latt, S. A. Sister chromatid exchanges, indices of human chromosome damage and repair: detection by fluorescence and induction by mitomycin C. Proc. Nat. Acad. Sci. USA 71:3162-66; 1974.

125. Latt, S. A. Longitudinal and lateral differentiation of metaphase chromosomes based on the detection of DNA synthesis by fluorescence microscopy. *In* Pearson, P. L.; Lewis, K. R., eds. Chromosomes today. Vol. 5. New York: John Wiley and Sons; 1976: 367-94.

126. Latt, S. A.; Juergens, L. Determinants of sister exchange frequencies in human chromosomes. *In* Hood, E. B.; Porter, I., eds. Population cytogenetics. New York: Academic Press; 1976: 217-36.

127. Latt, S. A.; Loveday, K. S. Characterization of sister chromatid exchange induction by 8-methoxypsoralen plus near-UV light. Cytogenet. Cell Genet. 21:184-200; 1978.

128. Latt, S. A.; Stetten, G. Spectral studies on 33258 Hoechst and related bisbenzimidazole dye, useful for fluorescent detection of deoxyribonucleic acid synthesis. J. Histochem. Cytochem. 24:24-33; 1976.

129. Latt, S. A.; Wohlleb, J. C. Optical studies of the interaction of 33258 Hoechst with DNA, chromatin, and metaphase chromosomes. Chromosoma 52:297-316; 1975.

130. Latt, S. A.; George, Y. S.; Gray, J. W. Flow cytometric analysis of BrdU-substituted cells stained with 33258 Hoechst. J. Histochem. Cytochem. 25:297-334; 1977.

131. Latt, S. A.; Allen, J. W.; Rogers, W. E.; Juergens, L. A. In vitro and in vivo analysis of sister chromatid exchange formation. In Handbook of mutagenicity test procedures. Amsterdam: Elsevier/North Holland Biomed. Press; 1977: 275-91.

132. Latt, S. A.; Allen, J. W.; Shuler, C.; Loveday, K. S.; Monroe, S. H. The detection and induction of sister chromatid exchanges. In Sparkes, R. S.; Comings, D. E.; Fox, C. F., eds. Molecular human cytogenetics. VII ICN-UCLA Symposium on Molecular and Cellular Biology. New York: Plenum; 1977: 315-34.

133. Latt, S. A.; Allen, J. W.; Stetten, G. In vitro and in vivo analysis of chromosome structure replications, and repair using BrdU-33258 Hoechst techniques. In International Cell Biology, 1976-1977. New York: Rockefeller Univ. Press; 1977.

134. Latt, S. A.; Davidson, R. L.; Lin, M. S.; Gerald, P. S. Lateral asymmetry in the fluorescence of human Y chromosomes stained with 33258 Hoechst. Exper. Cell Res. 87:425-29 1974.

135. Latt, S. A.; Munroe, S. H.; Disteche, C.; Rogers, W. E.; Cassell, D. M. Uses of fluorescent dyes to study chromosome structure and replication. In DeLaChapelle, A.; Sorsa, M., eds. Chromosomes today. Vol. 6. Elsevier; 1977: 27-36.

136. Latt, S. A.; Schreck, R. R.; Loveday, K. S.; Shuler, C. F. In vitro and in vivo analysis of sister chromatid exchange. Pharmacol. Rev. 30:501-35; 1979.

137. Latt, S. A.; Stetten, G.; Juergens, L. A.; Buchanan, G. R.; Gerald, P. S. Induction by alkylating agents of sister chromatid exchanges and chromatid breaks in Fanconi's anemia. Proc. Nat. Acad. Sci. USA 72:4066; 1975.

138. Latt, S. A.; Stetten, G.; Juergens, L. A.; Willard, H. F.; Scher, C. D. Recent developments in the detection of deoxyribonucleic acid synthesis by 33258 Hoechst fluorescence. J. Histochem. Cytochem. 23:493-505; 1975.

139. Lavappa, K. S.; Yerganian, G. Spermatogonial and meiotic chromosomes of the Armenian hamster *Cricetulus migratius*. Exper. Cell Res. 61:159-72; 1970.

140. Lau, Y. F.; Hittleman, W. N.; Arrighi, F. E. Sister chromatid differential staining pattern in prematurely condensed chromosomes. Experientia 32:917-18; 1976.

141. Legator, M. S.; Malling, H. V. The host-mediated assay, a practical procedure for evaluating potential mutagenic agents in mammals. In Hollaender, A., ed. Chemical mutagens. Vol. 2. New York: Plenum; 1971: 569-89.

142. Lezana, E. A.; Bianchi, N. O.; Bianchi, M. S.; Zabala-Suarez, J. E. Sister chromatid exchanges in Down syndrome and normal human beings. Mutat. Res. 45:85-90; 1977.

143. Lehmann, A. R.; Bridges, B. A. DNA repair. Essays in Biochem. 13:71-119; 1977.

144. Lin, M. S.; Comings, D. E.; Alfi, O. S. Optical studies of the interaction of 4′,6-diamidino-2-phenylindole with DNA and metaphase chromosomes. Chromosoma 60:15-25; 1977.

145. Lin, M. S.; Latt, S. A.; Davidson, R. L. Microfluorometric detection of asymmetry in the centromeric region of mouse chromosomes. Exper. Cell Res. 86:392-94; 1974.

145a. Loveday, K. S.; Latt, S. A. The effect of a tumor promoter, 12-0-tetradecanoyl-phorbol-13-acetate (TPA) on sister chromatid exchange formation in cultured Chinese hamster cells. Mutat. Res. 67:343-48; 1979.

146. Loveday, K. S.; Latt, S. A. Search for DNA interchange corresponding to sister chromatid exchanges in Chinese hamster ovary cells. Nuc. Acid Res. 5:4087-104; 1978.

147. McCann, J.; Choi, E.; Yamaski, E.; Ames, B. N. Detection of carcinogens as mutagens in the salmonella/microsome test; assay of 300 chemicals. Proc. Nat. Acad. Sci. USA 72:5135-39; 1975.

148. McKenzie, N. H.; Hall, S. H. Conventional aberration and sister chromatid exchange analysis of human lymphocytes exposed to ozone in vivo. Amer. J. Hum. Genet. 30:73A; 1978.

149. Maher, V. M.; Ouelette, L. M.; Curren, R. D.; McCormick, J. J. Frequency of ultraviolet light-induced mutations is higher in xeroderma pigmentosum variant cells than in normal cells. Nature 261:593-95; 1976.

150. Marin, G.; Prescott, D. M. The frequency of sister chromatid exchanges following exposure to varying doses of ^{3}H-thymidine or X-rays. J. Cell Biol. 21:159-67; 1964.

151. Marquardt, H.; Bayer, U. The induction in vivo of sister chromatid exchanges in the bone marrow of the Chinese hamster. Mutat. Res. 56:169-76; 1977.

152. Martin, C. N.; McDermid, A. D.; Garner, R. C. Testing of known carcinogens and noncarcinogens for their ability to induce unscheduled DNA synthesis in HeLa cells. Cancer Res. 38:2621-27; 1978.

153. Matsushima, T.; Sawamura, M.; Umezawa, K.; Sugimura, T. Induction of SCE by quercetin and suppression of SCE by elastatinal, a microbial protease. Meeting on SCEs. Lake Yamanaka, Japan; July 1978.

154. Mazrimas, J. A.; Stetka, D. G. Direct evidence for the role of incorporated BudR in the induction of sister chromatid exchanges. Exper. Cell Res. 117:23-30; 1978.

155. Meyn, M. S.; Rossman, T.; Troll, W. A protease inhibitor blocks SOS functions in *Escherichia coli*: antipain prevents repressor inactivation, ultraviolet mutagenesis, and filamentous growth. Proc. Nat. Acad. Sci. USA 74:1152-56; 1977.

156. Miller, R. C.; Aaronson, M. M.; Nichols, W. W. Effects of treatment on differential staining of BrdU-labeled metaphase chromosomes; three way differentiation of M_3 chromosomes. Chromosoma 55:1-11; 1976.

157. Moore, P. D.; Holliday, R. Evidence for the formation of hybrid DNA during mitotic recombination in Chinese hamster cells. Cell 8:573-79; 1976.

158. Morgan, W. F.; Corssen, P. E. The frequency and distribution of sister chromatid exchanges in human chromosomes. Human Genetics 38:271-78; 1977.

159. Mourelatos, D.; Faed, J. J. W.; Johnson, B. E. Sister chromatid exchanges in human lymphocytes exposed to 8-methoxypsoralen and long-wave UV radiation prior to incorporation of bromodeoxyuridine. Experientia 33:1091-93; 1977.

160. Natarajan, A. T.; Tates, A. D.; Van Buul, P. P. W.; Meijers, M.; DeVogel, N. Cytogenetic effects of mutagens/carcinogens, after activation in a microsomal system in vitro. I. Induction of chromosome aberrations and sister chromatid exchanges by diethylnitrosamine (DEN) and dimethynitrosamine (DMN) in CHO cells in the presence of rat liver microsomes. Mutat. Res. 37:83-90; 1976.

161. Nevstad, N. P. Sister chromatid exchanges and chromosome aberrations induced in human lymphocytes by the cytostatic drug adriamycin, in vivo and in vitro. Mutat. Res. 57:253-58; 1978.

162. Nichols, W. W.; Bradt, C. I.; Toji, L. H.; Godley, M.; Segawa, M. Induction of sister chromatid exchanges by transformation with simian virus 40. Cancer Res. 38:960-64; 1978.

163. Nilsson, K.; Ponten, J. Classification and biological nature of established human hematopoetic cell lines. Int. J. Cancer 15:321-41; 1975.

164. Nordenson, I.; Beckman, G.; Beckman, L. The effect of superoxide dismutase and catalase on radiation-induced chromosome breaks. Hereditas 80:125-26; 1976.

165. Nordenson, I. Effect of superoxide dismutase and catalase on spontaneously occurring chromosome breaks in patients with Fanconi's anemia. Hereditas 86:147-50; 1977.

166. Nordenson, I. Chromosome breaks in Werner's syndrome and their prevention in vitro by radical-scavenging enzyme. Hereditas 87:151-54; 1977.

167. Obe, G.; Ristow, H. Acetaldehyde, but not ethanol induces sister-chromatid exchanges in Chinese hamster cells in vitro. Mutat. Res. 56:211-13; 1977.

168. Pal, K.; Tierney, B.; Grover, P. L.; Sims, P. Induction of sister chromatid exchanges in Chinese hamster ovary cells treated in vitro with non-K-region dihydrodiols of 7-methylbenz(α)anthracene and benzo(α)pyrene. Mutat. Res. 50:367-75; 1978.

169. Palitti, F.; Becchetti, A. Effect of caffeine on sister chromatid exchanges and chromosomal aberrations induced by mutagens in Chinese hamster cells. Mutat. Res. 45:157-59; 1977.

170. Pathak, S.; Ward, O. G.; Hsu, T. C. Rate of sister chromatid exchanges in mammalian cells differing in diploid numbers. Experientia 33:875-76; 1977.

171. Patterson, M. C.; Smith, B. P.; Lohman, P. H.; Anderson, A. K.; Fishman, L. Defective excision repair of X-ray damaged DNA in human (ataxia telangiectasia) fibroblasts. Nature 260:444-46; 1976.

172. Peacock, W. J. Replication, recombination, and chiasmata in *Gonices austiaclasics*. Genetics 65:593-617; 1970.

173. Pera, F.; Mattias, P. Labelling of DNA and differential sister chromatid staining after BrdU treatment in vivo. Chromosoma 57:13-18; 1976.

174. Perry, P. Use of sister chromatid exchange techniques for cytological detection of mutagen carcinogen exposure. Mutat. Res. 46:205 (Abstracts); 1977.
175. Perry, P. Chemical mutagens and sister chromatid exchange. In Hollaender, A.; deSerres, F. eds.Chemical mutagens. Vol. 6. New York: Plenum; 1979.
176. Perry, P.; Evans, H. J. Cytological detection of mutagen-carcinogen exposure by sister chromatid exchange. Nature 258:121-24; 1975.
177. Perry, P. E.; Searle, C. E. Induction of sister chromatid exchanges in Chinese hamster cells by the hair dye constituents 2-Nitro-p-phenylene Diamine and 4-Nitro-l-phenylenediamine. Mutat. Res. 56:207-10; 1977.
178. Perry, P.; Wolff, S. New Giemsa method for differential staining of sister chromatids. Nature 261:156-58; 1974.
179. Poon, P. K.; O'Brien, R. L.; Parker, J. W. Defective DNA repair in Fanconi's anemia. Nature 250:223-25; 1974.
180. Popescu, N. C.; Turnbull, D.; DiPaolo, J. A. Sister chromatid exchanges/chromosome aberration analysis with the use of several carcinogens and noncarcinogens. J. Nat. Canc. Inst. 59:289-93; 1977.
181. Raposa, T. Sister chromatid exchange studies for monitoring DNA damage and repair capacity after cytostatics in vitro and in lymphocytes of leukaemic patients under cytostatic therapy. Mutat. Res. 57:241-51; 1978.
182. Ray, J. H.; Altenburg, L. C. Cytogenetic effects of activated selenium. Amer. J. Hum. Genet. 30:92A; 1978.
183. Remsen, J. F.; Cerutti, P. A. Deficiency of gamma-ray excision repairs in skin fibroblasts from patients with Fanconi's anemia. Proc. Nat. Acad. Sci. USA 73:2419-23; 1976.
184. Renault, G.; Pot-Deprun, J.; Chouroulinkov, I. Induction d'echanges entre chromatides soeurs in vivo sur les cellules de moelle osseuse de souris AKR. C.R. Acad. Sci. D 286:887-90; 1978.
185. Ristow, H.; Obe, G. Acetaldehyde induces crosslinks in DNA and causes sister chromatid exchanges in human cells. Mutat. Res. 58:115-19; 1978.
186. Rommelaere, J.; Miller-Faures, A. Detection by density equilibrium centrifugation of recombinant-like DNA molecules in somatic mammalian cells. J. Mol. Biol. 98:195-218; 1975.
187. Rosin, M. P.; Stich, H. F. The inhibitory effect of cysteine on the mutagenic activities of several carcinogens. Mutat. Res. 54:73-81; 1978.
188. Rudiger, H. W.; Kohl, F.; Mangeles, W.; VonWichert, P.; Bartram, C. R.; Wohler, W.; Passarge, E. Benzpyrene induces sister chromatid exchanges in cultured human lymphocytes. Nature 262:290-92; 1976.
188A. Rudiger, H.W.; Haenisch, F.; Metzler, M.; Oesch, F.; Glatt, H.R. Metabolites of diethylstilbesterol induce sister chromatid exchange in cultured human fibroblasts. Nature 281:392-94; 1979.
189. Rupp, W. D.; Howard-Flanders, P. Discontinuities in the DNA synthesis in an excision defective strain of *Escherichia coli* following ultraviolet irradiation. J. Mol. Biol. 31:291-304; 1968.
190. Rupp, W. D.; Wilde, C. E., III; Reno, D. L.; Howard-Flanders, P. J. Exchanges between DNA strains in ultraviolet-irriated *Escherichia coli*. Mol. Biol. 61:25-44; 1971.
191. Sakanishi, S.; Takayama, S. Reverse differential staining of sister chromatid after substitution with BUdR and incubation in sodium phosphate solution. Exper. Cell Res. 115:448-50; 1978.
192. Sasaki, M. Is Fanconi's anemia defective in a process essential to the repair of DNA crosslinks? Nature 257:501-3; 1975.
193. Sasaki, M. S. Sister chromatid exchange and chromatid inter-change as possible manifestation of different DNA repair processes. Nature 269:623-25; 1977.
194. Sasaki, M. S.; Tonomura, A. A high susceptibility of Fanconi's anemia to chromosome breakage by DNA cross-linking agents. Cancer Res. 33:1829-35; 1973.
195. Scheres, J. M. J. C.; Hustinx, T. W. J.; Ruttem, F. J.; Merkx, G. F. M. "Reverse" differential staining of sister chromatids. Exper. Cell Res. 109:466-68; 1977.
196. Schneider, E. L.; Chaillet, J.; Tice, R. In vivo BrdU labelling of mammalian chromosomes. Exper. Cell Res. 100:396-99; 1976.
197. Schneider, E. L.; Monticove, R. E. Cellular aging and sister chromatid exchange. II. Effect of in vitro passage of human fetal lung fibroblasts on baseline and mutagen-induced sister chromatid exchange frequency level. Exper. Cell Res. 115:269-76; 1978.

198. Schneider, E. L.; Sternberg, H.; Tice, R. R. In vivo analysis of cellular replication. Proc. Nat. Acad. Sci. USA 74:2041-44; 1977.

199. Schonwald, A. D.; Bartram, C. R.; Rudiger, H. W. Benzpyrene-induced sister chromatid exchanges in lymphocytes of patients with lung cancer. Hum. Genet. 36:261-64; 1977.

199A. Schreck, R.R.; Latt, S.A. Comparison of benzo (a) pyrene metabolism and sister chromatid exchange induction in mice. Nature 288:407-8, 1980.

200. Schreck, R. R.; Paika, I. J.; Latt, S. A. In vivo induction of sister chromatid exchanges SCE in liver and marrow cells by drugs requiring metabolic activation. Mutat. Res. 64:315-28; 1979.

201. Schvartzman, J.; Cortes, F. Sister chromatid exchanges in *Allium cepa*. Chromosoma 62:119-31; 1977.

202. Schwartz, A. L.; Coles, F. S.; Fiedorek, F.; Matthews, D.; Paika, I.; Frantz, I. D.; Latt, S. A. Effect of phototherapy on sister chromatid exchange in premature infants. Lancet ii:157-58; 1978.

203. Schwarzacher, H. G.; Schnedl, W. Endoreduplication in human fibroblast cultures. Cytogenetics 4:1-18; 1965.

204. Shafer, D. A. Replicative bypass model of sister chromatid exchanges, implications for Bloom's syndrome and Fanconi's anemia. Human Genetics 39:177-90; 1977.

205. Shiriashi, Y. Chromosome aberrations induced by monomeric acrylamide in bone marrow and germ cells of mice. Mutat. Res. 57:313-24; 1978.

206. Shiriashi, Y. The sister chromatid exchange and the DNA repair replication in human chromosomes. Proc. Japan Acad. 54: Ser. B 179-82; 1978.

207. Shiriashi, Y.; Freeman, A. I.; Sandberg, A. A. Increased sister chromatid exchange in bone marrow and blood cells from Bloom's syndrome. Cytogenet. Cell Genet. 17:162-73; 1976.

208. Shiriashi, Y.; Sandberg, A. A. Effects of mitomycin C on normal and Bloom's syndrome cells. Mutat. Res. 49:239-48; 1978.

209. Shiriashi, Y.; Sandberg, A. A. The relationship between sister chromatid exchanges and chromosome aberrations in Bloom's syndrome. Cytogenet. Cell Genet. 18:13-23; 1977.

210. Shuler, C. F.; Latt, S. A. Sister chromatid exchange induction resulting from systemic, topical and systemic-topical presentations of carcinogens. Cancer Res. 39:2510-14; 1979.

211. Sirianni, S. R.; Huang, C. C. Sister chromatid exchange induced by promutagens/carcinogens in Chinese hamster cells cultured in diffusion chambers in mice. Proc. Soc. Exper. Biol. Med. 158:269-74; 1978.

212. Solomon, E.; Bobrow, M. Sister chromatid exchanges: A sensitive assay of agents damaging human chromosomes. Mutat. Res. 30:273-78; 1975.

213. Stetka, D.; Carrano, A. V. The interaction of Hoechst 33258 and BrdU-substituted DNA in the formation of sister chromatid exchanges. Chromosoma 63:21-31; 1977.

214. Stetka, D. G.; Minkler, J; Carrano, A. V. Induction of long-lived chromosome damage as manifested by sister chromatid exchange in lymphocytes of animals exposed to mitomycin-C. Mutat. Res. 51:383-96; 1978.

215. Stetka, D. G.; Wolff, S. Sister chromatid exchanges as an assay in genetic damage induced by mutagenic-carcinogens I. In vivo test for compounds requiring metabolic activation. Mutat. Res. 41:333-42; 1976.

216. Stetka, D. G.; Wolff, S. Sister chromatid exchanges as an assay for genetic damage induced by mutagenic-carcinogens. II. In vitro test for compounds requiring metabolic activation. Mutat. Res. 41:343-50; 1976.

217. Statten, G.; Latt, S. A.; Davidson, R. L. 33258 Hoechst enhancement of the photosensitivity of bromodeoxyuridine-substituted cells. Somat. Cell Genet. 2:285-90; 1976.

218. Stetten, G.; Davidson, R. L.; Latt, S. A. 33258 Hoechst enhances the selectivity of the bromodeoxyuridine-light method of isolating conditional lethal mutants. Exper. Cell Res. 108:447-52; 1977.

219. Stoll, C.; Borgaonkar, D.; Levy, J. M. Effect of vincristine on sister chromatid exchanges of normal human lymphocytes. Cancer Res. 36:2710-13; 1976.

220. Stoll, C.; Borgaonkar, D. S.; Bigel, P. Sister chromatid exchanges in balanced translocation carriers and in patients with unbalanced karyotypes. Human Genet. 37:27-32; 1977.

221. Swift, M. Fanconi's anemia in the genetics of neoplasia. Nature 230:370-73; 1971.
222. Swift, M. Malignant neoplasms in heterozygous carriers of genes for certain autosomal recessive syndrome. *In* Mulvihill, J. J.; Miller, R. W.; Fraumeni, Jr., J. F., eds. Genetics of human cancer. New York: Raven Press; 1977: 209-15.
223. Swift, M.; Sholman, L.; Perry, M.; Chase Malignant neoplasms in the families of patients with ataxia telangiectasis. Cancer Res. 36:209-16; 1976.
224. Takayama, S.; Sakanishi, S. Differential Giemsa staining of sister chromatids after extraction with acids. Chromosoma 64:109-15; 1977.
225. Takehisa, S.; Wolff, S. Induction of sister chromatid exchanges in Chinese hamster cells by carcinogenic mutagens requiring metabolic activation. Mutat. Res. 45:263-70; 1977.
226. Takehisa, S.; Wolff, S. The induction of sister chromatid exchanges in Chinese hamster ovary cells by prolonged exposure to 2-acetylaminofluorene and S-9 mix. Mutat. Res. 58:103-6; 1978.
227. Taylor, J. H. Sister chromatid exchanges in tritium-labelled chromosomes. Genetics 43:515-29; 1958.
228. Taylor, J. H. Distribution of tritium-labelled DNA among chromosomes during meiosis. I. Spermatogenesis in the grasshopper. J. Cell Biol. 25:57-67; 1965.
229. Taylor, J. H.; Woods, P. S.; Hughes, W. L. The organization and duplication of chromosomes as revealed by autoradiographic studies using tritium-labelled thymidine. Proc. Nat. Acad. Sci. USA 43:122-28; 1957.
230. Tease, C. Cytological detection of crossing-over in BudR-substituted meiotic chromosomes using the fluorescent plus Giemsa technique. Nature 272:823-24; 1978.
230A. Thompson, L.H.; Baker, R.M.; Carrano, A.V.; Brookman, K.W. Failure of the phorbol ester 12-0-tetradecanoyl-13-acetate to enhance sister chromatid exchange, mitotic segregation, or expression of mutations in Chinese hamster cells. Cancer Res. 40:3245-51; 1980.
231. Thust, R.; Warzok, R.; Grund, E.; Mendel, J. Use of human-liver microsomes from kidney-transplant donors for the induction of chromatid aberrations and sister chromatid exchanges by means of precarcinogens in Chinese hamster cells in vitro. Mutat. Res. 51:397-402; 1978.
232. Tice, R.; Chaillet, J.; Schneider, E. L. Evidence derived from sister chromatid exchanges of restricted rejoining of chromatid sub-units. Nature 256:642-44; 1975.
233. Tice, R.; Windler, G.; Rary, J. M. Effect of cocultivation on sister chromatid exchange frequencies in Bloom's syndrome and normal fibroblast cells. Nature 273:538-40; 1978.
234. Treiff, N. M.; Cantell-Fort, G.; Smart, V. B.; Kempen, R. R.; Kilian, D. J. Appraisal of fluorometric assay hydrocarbon (Benzo(α)pyrene) hydroxylase in cultured human lymphocytes. Brit. J. Cancer 38:335-38; 1978.
235. Trosko, J. E.; Chu, E. H. Y.; Carrier, W. C. The induction of thymidine dimers in ultraviolet-irradiated mammalian cells. Radiat. Res. 24:667-72; 1965.
236. Ueda, N.; Uenaka, H.; Akematsu, T.; Sugiyama, T. Parallel distribution of sister chromatid exchanges and chromosome aberrations. Nature 262:581-83; 1976.
236A. Umezawa, K.; Sawamura, M.; Matsushima, T.; and Sugimura, T. Inhibition of chemically induced sister chromatid exchanges by elastatinal. Chem. Biol. Interact. 24:107-10; 1979.
237. Utakoji, T.; Hosoda, K. High-concentration thymidine and sister chromatid exchanges in Chinese hamster cells in vitro. Lake Yamanaka SCE Conference; July 1978.
238. van Buul, P. P. W.; Natarajan, A. T.; Verdegaal-Immerzeel, A. M. Suppression of the frequencies of sister chromatid exchanges in Bloom's syndrome fibroblasts by cocultivation with Chinese hamster cells. Hum. Genet. 44:187-89; 1978.
239. Vogel, W.; Bauknecht, T. Differential chromatid staining by in vivo treatment as a mutagenicity test system. Nature 260:448-49; 1976.
240. Vogel, W.; Bauknecht, Th. Effects of caffeine on sister chromatid exchange (SCE) after exposure to UV light or triaziquone studies with a fluorescence plus Giemsa (FPG) technique. Human Genet. 40:193-98; 1978.
241. Waksvik, H.; Brogger, A.; Stene, J. Psoralen/UVA treatment and chromosomes. I. Aberrations and sister chromatid exchange in human lymphocytes in vitro and synergism with caffeine. Human Genet. 38:195-207; 1977.

242. Walen, K. H. Spatial relationships in the replication of chromosomal DNA. Genetics 51:915-29; 1965.
243. Walker, A. P.; Dumars, K. W. Commonly used pediatric drugs, sister chromatid exchanges, and the cell cycle. Amer. J. Hum. Genet. 110A 1978.
244. Witkin, E. M. Ultraviolet mutagenesis and inducible DNA repair in *Escherichia coli.* Bacteriol. Rev. 40:869-907; 1976.
245. Wolff, S. Sister chromatid exchanges. Ann. Rev. Genet. 11:183-201; 1977.
246. Wolff, S.; Perry, P. Differential Giemsa staining of sister chromatids and the study of sister chromatid exchanges without autoradiography. Chromosoma 48:341-53; 1974.
247. Wolff, S.; Perry, P. Insights of chromatid structure from sister chromatid exchange ratios and the lack of both isolabelling and heterolabelling as determined by the FPG technique. Exper. Cell Res. 93:23-30; 1975.
248. Wolff, S.; Rodin, B. Saccharin-induced sister chromatid exchanges in Chinese hamster and human cells. Science 200:543-45; 1978.
249. Wolff, S.; Bodycote, J.; Painter, R. B. Sister chromatid exchanges induced in Chinese hamster cells by UV-irradiation at different stages of the cell cycle: the necessity of cells to pass through S. Mutat. Res. 25:73-81; 1974.
250. Wolff, S.; Bodycote, J.; Thomas, G. H.; Cleaver, J. E. Sister chromatid exchanges in xeroderma pigmentosum cells that are defective in DNA excision repair on postreplication repair. Genetics 81:349-55; 1975.
251. Wolff, S.; Rodin, B.; Cleaver, J. E. Sister chromatid exchanges induced by mutagenic carcinogens in normal and xeroderma pigmentosum cells. Nature 265:345-47; 1977.
252. Yamamoto, M.; Miklos, G. L. G. Genetic studies on heterochromatin in *Drosophila melanogaster* and their implications for the functions of satellite DNA. Chromosoma 66:71-98; 1978.
253. Yu, C. W.; Borgaonkar, D. S. Normal rate of sister chromatid exchange in Down syndrome. Clin. Genet. 11:397-401; 1977.
254. Yunis, J. J.; Sanchez, O. High resolution of human chromosomes. Science 191:1268-70; 1976.
255. Yunis, J. J.; Sawyer, J. R.; Ball, D. W. The characterization of high-resolution G-band chromosomes of man. Chromosoma 67:293-307; 1978.
256. Zack, G. W.; Rogers, W. E.; Latt, S. A. Automatic measurement of sister chromatid exchange frequency. J. Histochem. Cytochem. 25:741-53; 1977.
257. Zakharov, A. F.; Egolina, N. A. Differential spiralization along mammalian mitotic chromosomes. I. BUdR-revealed differentiates in Chinese hamster chromosomes. Chromosoma 38:341-55; 1972.
258. Zimmerman, A. M.; Stich, H.; San, R. Nonmutagenic action of cannabinoids in vitro. Pharmacology 16:333-43; 1978.

4

Root Tips of Vicia faba as a Material for Studying the Induction of Chromosomal Aberrations and Sister Chromatid Exchanges

by B. A. KIHLMAN*

Abstract

Root tips of the broad (field) bean, *Vicia faba,* constitute an inexpensive material for cytogenetic studies which is easy to grow and handle and which is available all year round. The mitotic frequency in the root meristem is high and the chromosome number low (2n = 12). The large size of the chromosomes makes scoring of both chromosomal aberrations and sister chromatid exchanges easy and accurate. In the reconstructed karyotypes described in the appendix, all the sick chromosome pairs are easily identifiable. Methods for growing roots from seeds; treatment procedures; and the preparation, staining, and scoring of slides are described. The staining methods include both differential staining of sister chromatids for the analysis of sister chromatid exchanges and conventional Feulgen staining for the study of chromosomal aberrations. Finally, the advantages and disadvantages of *Vicia faba* root tips in comparison with other cytogenetic test materials are discussed.

Introduction

Before *in vitro* cultures of animal cells became readily available, most information on the effects of mutagenic agents on chromosome structure was obtained in plant materials. Thus, for instance, Sax (1940), Lea (1946), Catcheside (1948) and Giles (1954) used microspores of Tradescantia species in their classical studies on

*Department of General Genetics, University of Uppsala, S-750 07 Uppsala 7, Sweden.

radiation-induced chromosomal aberrations. The large chromosomes of these species in combination with a low chromosome number in the haploid microspores made them an excellent material for cytogenetic studies at a time when ionizing radiation was the main agent for producing chromosomal aberrations. Micropores, however, cannot easily be exposed to known concentrations of chemical substances. Therefore, when the common occurrence and general importance of the chemical mutagens were realized during the decade following their detection in the 1940s, a system had to be found that made it possible to study the effects of both chemical and physical mutagens.

The obvious choice of plant tissue for this purpose is the root meristem. Root tips are easy to grow and handle, they are available all year round, and in the meristem a large number of dividing cells may readily be obtained. Since the root-tip cells are directly exposed to the chemicals, the effect of known concentrations can be studied. Roots grown from bulbs of the common onion, *Allium cepa* ($2n = 16$), a classical material for chromosome studies, were introduced by Levan (1949) for standardized tests of the effects of chemicals on chromosomes and cell division. Roots of the broad bean, *Vicia faba* ($2n = 12$), which were used in radiation experiments as early as in 1913 (Read 1959), soon became an even more popular a material. It is hardly an exaggeration to say that for some years *Vicia* root tips were the most commonly used system for studies on the effects of mutagenic agents on chromosome structure. It is not so any more; today, cytogenetic testing of mutagenic activities is carried out mainly in in vitro cultures of mammalian cells. But even today there are workers, like the authors of the present chapter and its appendix, who feel that the *Vicia* system offers sufficient advantages for its retention as their main material for mutagenicity testing.

The Material

As shown in Figure 4.1, a root-tip cell of *Vicia faba* contains five pairs of chromosomes of approximately the same size and provided with subterminal centromeres (S-chromosomes) and one chromosome pair with median centromere (M-chromosome), the diploid (2n) chromosome number thus being 12. In the M-chromosome, a large satellite is separated from the rest of the chromosome by a secondary (nucleolar) constriction. The ratio of the length of the M-chromosome to the mean length of the S-chromosomes is approximately 2.3:1. If therefore, aberrations induced by mutagenic agents were randomly distributed between the chromosomes, the 10 S-chromosomes should contain 2.3 times as many aberrations as the 2 M-chromosomes. Since aberrations, particularly those induced by chemical substances, are rarely randomly distributed, the S/M ratio usually diverges considerably from 2.3:1. In fact, after treatment with many mutagens, certain chromosome regions are involved in aberration formation much more often than expected on the basis of their relative metaphase length. Furthermore, these "hot spots" for aberration formation are frequently mutagen-specific.

In the standard karyotype of *Vicia faba,* the close similarity in size and structure of the five pairs of S-chromosomes makes it very difficult to identify hot spots in these chromosomes. To overcome this difficulty, Michaelis and Rieger (1971) have developed reconstructed karyotypes in which all of the six chromosome pairs are

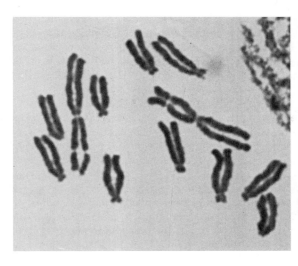

Figure 4.1 The chromosome complement of *Vicia faba* (2n = 12).

easily identifiable. In an appendix to this chapter, Schubert et al. have described these reconstructed karyotypes and their usefulness for the investigation of certain specific problems.

Sturelid (1971) has estimated the mean duration of the mitotic cycle in lateral roots of the field bean, *Vicia faba* minor (Weibulls Åkerböna "Primus") to be 19.7 hr. at 20°C. The stages of active division, M (mitosis) lasted 2 hr. Of the 17.7 hr. of interphase, 8.1 were spent in S (the period of DNA and chromosome replication), 4.1 in G_1 (the period between telophase and S) and 5.5 in G_2 (the period between S and prophase).

Previously, Evans and Scott (1964), using primary roots of eleven-day-old seedlings of *Vicia faba* (var. Suttons' Prolific Longpod), had found the following mean durations of the various parameters of the mitotic cycle at 20°C: M = 2.0 hr., G_1 = 4.9 hr., S = 7.5 hr., and G_2 = 4.9 hr., giving a total mitotic cycle time of 19.3 hr.

Method for Growing Roots from Seeds

The growth process in initiated by soaking of the seeds. In our laboratory, seeds are soaked in tap water at 20°C for a period of at least 6 but not more than 12 hr. Soaking may be preceded by disinfection which consists of a short immersion (3-10 min.) in a decanted 5 % calcium hypochlorite solution.

After soaking, the seeds are allowed to germinate at 20°C in moist Perlite (Deutsch. Perlite AG, Dortmund) or in moist coarse vermiculate. We prefer Perlite, which is an expanded, water-absorbing, silicious material derived from pumice. When the seedlings have grown 3-5 cm long primary roots, they are ready to be transferred to the water tank. In connection with the transfer, the seed coat and the shoot are removed. If lateral roots are to be used, the tip of the primary root must be removed as well, since the production of laterals is stimulated by the removal of the primary root meristem.

The water in the tank should be fresh, well-aerated, and of constant temperature (preferably 19–20°C) during the growth period. In our laboratory, these requirements are met by growing the roots in thermo-regulated, running tap water. When tap water cannot be used (for instance, when it is too heavily chorinated), roots may be grown in a salt solution such as Hoagland's (0.005M KNO_3; 0.005M $Ca(NO_3)_2$; 0.002M $MgSO_4$; 0.001M KH_2PO_4; 0.5 % iron tartrate, 1 ml per liter solution).

After 24 hr. in the water tank, primary roots are ready for use. Lateral roots can be exposed to mutagenic agents when they are 1–2 cm long. The roots of a sufficient number of seedlings have reached this length when they have been growing for four days in the water tank. When lateral roots are to be used for studies on the frequency of SCE, they should be exposed to the 5-bromodeoxyuridine solution after three days in the water tank.

With the exception mentioned in the preceding sentence, roots are grown for studies on sister chromatid exchanges (SCEs) in the same way as for studies on chromosomal aberrations. Depending on the effect to be studied, there are differences in the methods of treatment, as well as in the preparation and scoring of slides. In sections dealing with these subjects, some of the instructions and considerations are general, that is, applicable both to chromosomal aberrations and sister chromatid exchanges, whereas others are valid only for one of the two types of effect.

Treatment of Root Tips with Mutagenic Agents

General

When physical and chemical agents are tested for their capacity to affect the structural organization of chromosomes, it should be borne in mind that the effect is dependent not only on the type and the dose of the agent, but that it may also be influenced by factors such as the stage of the cell cycle at the time of exposure, the oxygen tension during treatment, and, for chemical substances, the pH and the temperature of the treatment solution. Therefore, it is important that the treatments are carried out under controlled conditions. Well-aerated, buffered solutions should be used and the temperature should be kept constant. Both treatments and recovery should take place in the dark.

Since aqueous solutions of chemical substances are often unstable, it is advisable to use freshly prepared solutions. For the same reason the solutions should not be exposed to strong light or high temperatures when being prepared.

Treatments with chemical substances may be carried out in vials or tubes containing at least 25 ml of solution. Primary roots of three seedlings or lateral roots of one seedling can be treated in such vials. The number of seedlings to be used for each treatment should be at least six in experiments with primary roots and at least two in experiments with lateral roots. When the effect on the frequency of SCE is studied, it is advisable to increase the number of seedlings per treatment, since the quality of the preparations may vary considerably among roots from different beans.

Chemical substances should be tested over a wide range of concentration but it is advisable to keep the period of treatment as short as possible. Short treatment periods make it easier to keep the experimental conditions constant and to determine the stage(s) of the mitotic cycle during which the effect is produced. A thorough test should also include a treatment period of 24 hr. or more, since there are substances that require an extended period of treatment to be effective. When, however, roots are exposed for a longer period of time, the cells obtained for analysis represent a very heterogeneous population with respect to the stage of the cell cycle at the time of the treatment.

When the effects of radiations are studied, the seedlings may be irradiated in open petri dishes containing enough water to keep the roots moist. In experiments with laterals, the roots in front of and behind the main root should be removed to ensure that all roots of the seedlings receive the same dose. If the irradiations are to be carried out at controlled oxygen tensions, special treatment vials (Kihlman 1959) are recommended.

The effect is usually scored in metaphase. To collect metaphases and facilitate analysis, the roots are treated with colchicine before fixation. Colchicine prevents the formation of a mitotic spindle and delays the separation of the sister chromatids into daughter chromosomes. As a result, there is an accumulation of cells at a stage (c-metaphase) corresponding to metaphase in untreated cells. A suitable treatment is to expose the roots for 2–3 hr. at 19–20°C to an aerated solution containing 0.02 – 0.05 % colchicine.

For fixation of roots, a simple fixative such as acetic-alcohol can usually be used with satisfactory results. The fixative consists of 3 parts absolute alcohol and 1 part of glacial acetic acid. The alcohol can be either ethanol or methanol, although the latter is preferable. The best results are obtained when the fixative is freshly prepared and cool (4–10°C). Although the fixation is completed within 15–20 min., the preparation of permanent slides is facilitated by leaving the roots in the fixative overnight at 4°C. If kept in a freezer, roots can be stored in the fixative for several days.

The choice of time interval between treatment and fixation will be discussed separately for chromosomal aberrations and sister chromatid exchanges.

Chromosomal Aberrations

It was previously mentioned that the effect of mutagenic agents may be influenced by factors such as oxygen tension, pH and temperature of the treatment solution and the stage of the cell cycle at the time of exposure. For chromosomal aberrations such influences are illustrated by Figures 4.2–4.5 and by Table 4.1.

Figure 4.2 shows the effect of oxygen tension on the frequencies of chromosomal aberrations produced by 8-ethoxycaffeine (EOC) and methylphenylnitrosamine (MPNA). EOC is completely inactive in the absence of oxygen and reaches its maximum effect at an oxygen concentration corresponding to that of air. Like EOC, MPNA is inactive in the absence of oxygen, but in contrast to the effect of EOC, that of MPNA increases when the oxygen concentration is increased above the concentration of air. A dependence of the aberration frequency on oxygen concentration has been demonstrated for many chemical and physical mutagens (for a review, see Kihlman 1961). For X-rays, the dose required to produce a given frequency of aberrations is three times higher under anaerobic than under aerobic conditions.

The frequency of chromosomal aberrations produced by most chemical mutagens is strongly dependent on the temperature during treatment. Usually, more aberrations are produced at high than at low temperature. However, methylated oxypurines, such as caffeine, 8-methoxycaffeine (MOC) and EOC, are exceptions to

86 Root Tips of *Vicia Faba* as Study Material

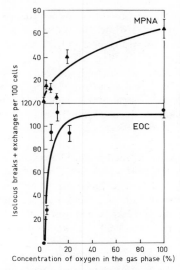

Figure 4.2 The influence of oxygen concentration on the frequencies of chromosomal aberrations obtained after treatments with methylphenylnitrosamine (MPNA) and 8-ethoxycaffeine (EOC). (After Kihlman 1966).

this rule. As shown in Figure 4.3, these agents are most active at temperatures between 10 and 17°C. Above 20°C the effect rapidly decreases, being nearly zero at 30°C.

Figure 4.4 shows how the frequency of chromosomal aberrations produced by maleic hydrazide varies with the pH of the treatment solution. The effect obtained at pH 4.7 is more than five times stronger than that obtained at pH 7.3. Such an influence of the pH of the treatment solution on the effect is expected for substances with acid properties.

Finally, Figure 4.5 and Table 4.1 illustrate how the types and frequencies of chromosomal aberrations are influenced by the time between treatment and fixation for three different types of mutagenic agents. As shown in Figure 4.5, a methylated oxypurine such as 1,3,7,9-tetramethyluric acid (TMU) is active only in cells that are in late interphase or prophase at the time of the treatment. The effect consists of "subchromatid"- and chromatid-type aberrations. In Figure 4.5, quite another type of effect is also illustrated. Tris(1-aziridinyl)-phosphine oxide or tepa, the substance producing this effect, is a trifunctional alkylating agent. Like other alkylating and arylalkylating agents, tepa is capable of inducing lesions in chromosomal DNA at all stages of the cell cycle. These lesions, however, require DNA and chromosome replication in order to be transformed into chromosomal aberrations; that is, the effect is S-dependent. Therefore, only cells exposed during or before S contain aberrations in the first mitosis after treatment. The aberrations produced by these agents are always of the chromatid type.

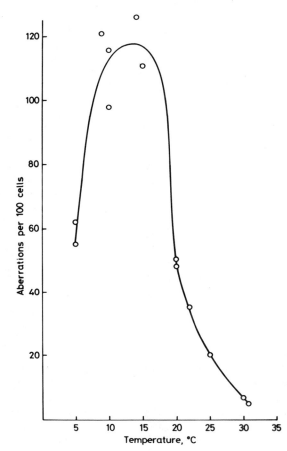

Figure 4.3 The influence of treatment temperature on the frequency of chromosomal aberrations produced by 8-methoxycaffeine (From Kihlman and Kronborg 1972).

Figure 4.4 The influence of the pH of the treatment solution on the frequency of chromosomal aberrations produced by the herbicide maleic hydrazide. (From Kihlman 1971, by permission of Plenum Publishing Corporation, New York.)

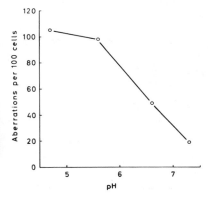

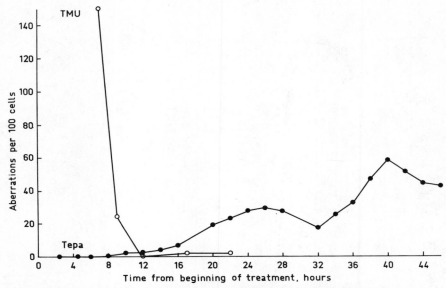

Figure 4.5 Frequencies of chromosomal aberrations obtained at various times after two hr. treatments with 2×10^{-2}M 1,3,7,9-tetramethyluric acid (TMU) and 7.5×10^{-4}M tepa.

In contrast to alkylating agents, ionizing radiation produces "subchromatid" aberrations in cells exposed during prophase, chromatid-type aberrations in G_2 and S cells, and chromosome-type aberrations in cells exposed during G_1. The effect does not require DNA and chromosome replication to be expressed as chromosomal aberrations, that is, the effect of ionizing radiation is S-independent and appears in the first mitosis after the exposure. Although most typically produced by ionizing radiation, a rather similar effect is obtained with long-wave UV on chromosomes with bromouracil-substituted DNA (Kihlman et al. 1978) and with certain antibiotics, such as bleomycin and streptonigrin. Table 4.1 shows the types and frequencies of aberrations obtained at 4 and 20 hr. after treatments with streptonigrin.

The effects illustrated in Figure 4.5 and Table 4.1 show that different agents may

Table 4.1 Types and Frequencies of Chromosomal Aberrations Obtained at 4 and 20 hr. after 1 hr. Treatments with Streptonigrin.

Concentration of streptonigrin μM	Time between treatment and fixation hr.	Abnormal metaphases %	Aberrations per 100 cells					Total aberrations per 100 cells
			Breaks			Exchanges		
			Chromatid	Isochromatid	"Subchromatid"	Chromatid	Chromosome	
7.5	4	38.7	8.0	5.3	18.0	6.7	0.0	38.0
10.0	20	52.0	3.0	27.0	0.0	46.0	4.0	80.0

produce aberrations at different stages of the cell cycle. The agents may also affect the duration of the cell cycle differently. Aberrations produced by agents with S-independent effects appear in dividing cells within an hour after exposure. With increasing time between exposure and fixation, both the types and the frequencies of aberrations produced by these agents change. On the other hand, when roots are exposed to agents with S-dependent effects, the appearance of aberrations in metaphase is delayed and the peak frequency may not be obtained until the second mitosis after exposure (about 40 hr. after the beginning of treatment). After treatment with these agents, however, there is little variation with time in the *types* of aberrations produced.

Since the time factor is so important for the types and frequencies of chromosomal aberrations produced by mutagenic agents, several fixation times should be used. It is advisable that the test should include at least one experiment in which roots are fixed at short intervals and over a period long enough to cover at least one, preferably two, whole mitotic cycles.

Sister Chromatid Exchanges

The method used in our laboratory for demonstrating SCE in *Vicia faba* is a modification of the fluorescence plus Giemsa (FPG) or 'harlequin' technique developed by Perry and Wolff (1974). The technique is based on the fact that the staining properties of chromosomes are altered when the thymine component of their DNA is replaced by 5-bromouracil (BrUra). A chromatid that has incorporated BrUra into both strands of its DNA (a 'BB'-chromatid) stains more weakly than a chromatid that has incorporated BrUra into only one strand of its DNA (a "TB"-chromatid), and a TB-chromatid, in turn, does not stain as strongly as a completely unsubstituted "TT"-chromatid.

Figure 4.6 demonstrates the principle of the technique. When cells are grown for one cell cycle in the presence of 5-bromodeoxyuridine (BrdUrd), the DNA in both sister chromatids of the metaphase chromosomes has one BrUra-substituted strand, and so are of the TB-constitution. Since the sister chromatids have the same constitution, they cannot be distinguished by differential staining. Consequently, the SCEs that may have occurred during the first cell cycle cannot be detected at this stage. Sister chromatids of different constitution, however, can be obtained at the next mitosis (second mitosis after labelling), either by allowing the roots to grow continuously in the presence of BrdUrd, or by growing them in the presence of the natural DNA component thymidine. The constitution of the sister chromatids will be TB and BB in the former case, TT and TB in the latter. They now stain differently and the SCEs formed both during the first and the second cell cycle after the beginning of labelling are made visible.

Since 1975, when the FPG-technique was modified for *Vicia faba* (Kihlman and Kronborg, 1975), it has been used in quite a large number of experiments. In these experiments, the mean number of SCEs per cell in untreated control roots varies between 19 and 26. The frequency of SCE in cells which have been growing for two cell cycles in the presence of BrdUrd (sister chromatids of the TB and BB constitution) does not seem to differ markedly from that in cells that have been exposed to BrdUrd only during the first cell cycle (sister chromatids of the TT and TB constitution).

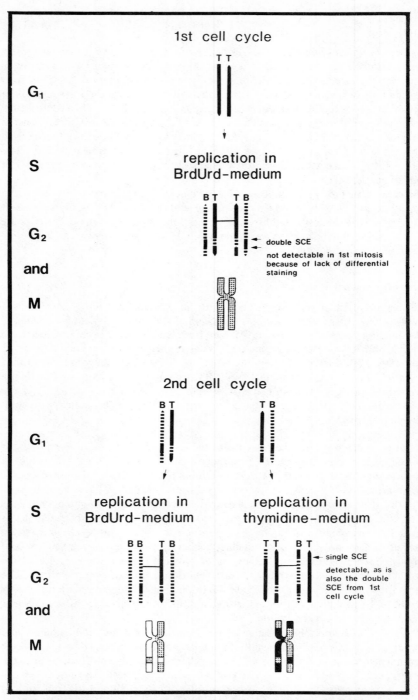

Figure 4.6 Diagram demonstrating the principle of the BrdUrd-labelling technique. (From Kihlman et al. 1978, by permission of Elsevier Publishing Company, Amsterdam.)

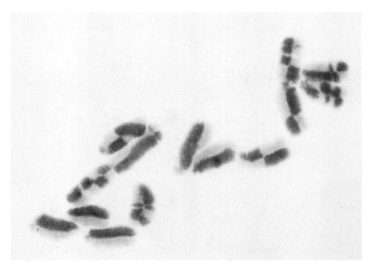

Figure 4.7 Metaphase cell of *Vicia faba* stained according to the FPG-technique. The sister chromatids are of the TT and TB constitution and the cell contains 27 SCEs. (From Kihlman and Sturelid 1978.)

A frequency of SCE between 19 and 26 per cell may seem high in comparison with the control values of 5–10 SCEs per cell which frequently are observed in materials such as cultured Chinese hamster and human cells. If, however, the frequency of SCE is related to the DNA content per cell, the outcome of the comparison is quite different. A root-tip cell of *Vicia faba* contains about 44 picograms of DNA (Baetcke et al. 1967), whereas a diploid Chinese hamster cell contains 8.3 and diploid human cell 7.3 picograms of DNA (Bachmann 1972). Consequently, the frequency of SCE per picogram of DNA will be 0.43 – 0.59 for *Vicia faba,* 0.60 – 1.20 for Chinese hamster cells and 0.68 – 1.37 for human cells.

Figure 4.7 shows a metaphase cell of *Vicia faba* stained according to the FPG-technique. The sister chromatids are of the TT and TB constitution. The cell contains 27 SCEs.

Over the last few years the effects of a number of chemical and physical mutagens on the frequency of SCE have been studied in *Vicia faba*. Figure 4.8 shows that a strong increase in the frequency of SCE was obtained with maleic hydrazide and with all the mono- , di- , and trifunctional alkylating agents tested. As producers of chromosomal aberrations, all these agents have S-dependent effects. No significant increase in the frequency of SCE was obtained with bleomycin, which as a producer of chromosomal aberrations has an S-independent effect. At the concentrations used in these experiments, bleomycin produced a considerably higher frequency of chromosomal aberrations than maleic hydrazide and the alkylating agents tested (Kihlman and Sturelid 1978). Similarly X-rays, which like bleomycin produce

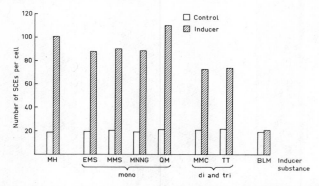

Figure 4.8 The effect of various mutagenic agents (inducers) on the frequencies of sister chromatid exchanges (SCEs) in root tips of *Vicia faba*. (After Kihlman and Sturelid 1978).

chromosomal aberrations independently of DNA and chromosome replication, had little effect on the frequency of SCE even at dosages producing relatively high frequencies of chromosomal aberrations (data not shown in Figure 4.8). These findings are in agreement with observations made by other authors in mammalian cells (e.g., Perry and Evans 1975).

The capacity to produce chromosomal aberrations independently of DNA and chromosome replication, however, is not necessarily always correlated with a low effect on the frequency of SCE. The exposure of cells with BrUra-substituted chromosomes to long-wave UV results in a strong increase in the frequency of SCE, but this treatment also produces chromosomal aberrations independently of DNA and chromosome replication (Kihlman et al. 1978). This suggests that the chromosome-damaging effect of long-wave UV on BrUra-substituted chromosomes contains an S-dependent component which is correlated with the capacity to increase the frequency of SCE. In any case, it can be concluded that the ability to increase the frequency of SCE is a sensitive indicator of the capacity of an agent to produce S-dependent chromosome damage, but not necessarily of its capacity to produce S-independent chromosome damage.

Figure 4.9 shows a root-tip cell of *Vicia faba* containing about 80 SCEs as a result of treatment with the alkylating agent tris(1-aziridinyl)-phosphine sulphide (thiotepa). It should be noted in the figure that some of the SCEs are so close together that the distance between them does not spread across the whole width of the chromatid when the chromosome is contracted at metaphase.

The fact that the production of chromosomal aberrations may be influenced by a number of different factors has already been discussed. Whether the induction of SCE is influenced by pH, temperature, and oxygen tension is not known. Since the lesions responsible for SCEs and chromosomal aberrations are likely to be produced by the same active species (which need not be identical with the agents applied) and since it is the penetration, formation, and/or reactivity of these species that is influenced by pH, temperature and oxygen tension, there is no reason for expecting

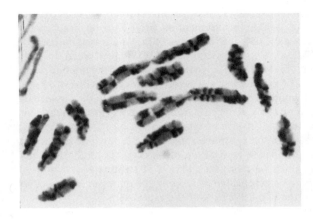

Figure 4.9 Root-tip of Vicia faba containing about 80 SCEs as a result of treatment with the trifunctional alkylating agent thiotepa. (From Kihlman 1975a, by permission of Springer Verlag, Berlin.)

the two effects to be influenced in a different way. Nevertheless, it may be worthwhile to investigate this problem.

Regarding the dependence of the induction of SCE on cell cycle stage, information is already available. Evidently, the formation of SCE is S-dependent, although the lesions giving rise to SCEs may be induced at any stage of the cell cycle (Kato 1974; Wolff et al. 1974; Kihlman 1975a).

The experimental procedure we have found suitable for studying the induction of SCE in lateral root tips of *Vicia faba* is the following (Kihlman and Kronborg 1975; Kihlman et al. 1978).

Lateral roots of *Vicia faba* minor (Weibulls Åkerböna "Primus") are exposed for 16 hours (one round of DNA replication) to an aqueous (tap water) solution containing 100 μM BrdUrd, 0.1 μM 5-fluorodeoxyuridine (FdUrd) and 5 μM uridine (Urd). This solution is referred to as the BrdUrd-solution. During the second round of DNA replication, the roots are grown either in the presence of thymidine or in the BrdUrd-solution. In the former case, the sister chromatids will have the TT and TB constitution, in the latter they will have the TB and BB constitution. Since in *Vicia* a better differentiation is obtained between TT and TB chromatids than between TB and BB chromatids, we usually grow the roots during the second cell cycle in a solution containing thymidine. The thymidine solution contains 100 μM thymidine and 5 μM Urd dissolved in tap water.

Since the effects of mutagens on the frequency of SCE can be analyzed only in fully differentiated cells, roots should be fixed when the frequency of such cells is high. In our experiments, the duration of the (second cell cycle) period in thymidine or BrdUrd varies between 19 and 23 hr., depending on the inhibitory effect of the mutagen on cell cycle progression. The last three hours before fixation, the roots are exposed to 0.05 % colchicine.

The reason for the addition of FdUrd to the BrdUrd-solution is that a sufficiently good differentiation between sister chromatids cannot be obtained by growing the roots in the presence of BrdUrd alone. FdUrd, the monophosphate of which is an inhibitor of the in vivo synthesis of thymidylic acid from deoxyuridylic acid, stimulates the incorporation of the thymidine analogue BrdUrd into chromosomal DNA. (At FdUrd concentrations below 0.1 μM, this stimulation proved to be insufficient, whereas at concentrations above 0.1 μM, the frequency of SCE increased markedly). In the presence of 0.1 μM FdUrd, a good differentiation is obtained when the concentration of BrdUrd exceeds 50 μM. The relative irreversibility of the inhibitory action of FdUrd on the activity of thymidylate synthetase makes it necessary to grow the roots in the presence of thymidine during the second cell cycle. In the absence of thymidine, the mitotic rate is strongly reduced. Uridine is added to the treatment solutions to counteract any possible adverse effects of FdUrd on RNA synthesis.

The routine in our laboratory is to expose the roots to the (known or suspected) mutagen after the first round of DNA replication in the presence of the BrdUrd-solution. For chemical mutagens, we then use a period of treatment of 1 – 2 hr. However, in some cases it may be practicable to prolong the mutagen treatment so that it covers one or both of the two cell cycles of the experiment.

During the whole treatment period, care is taken not to expose the roots to light. For this reason, treatments are carried out in glass tubes (containing 40 ml of solution), covered on the outside with black tape. The main root of the seedlings is passed through a small hole in a black rubber disk resting on top of the tubes. The treatments take place in the dark at 20°C.

Preparation of Slides and Staining

Chromosomal Aberrations

For the analysis of chromosomal aberrations, slides are conveniently prepared as Feulgen squashes. There are different ways of doing this; our variant of the technique, which has been published previously (Kihlman 1971, 1975b), is as follows:

The roots, which have been kept overnight in the fixative at 4°C, are transferred to room-tempered distilled water and left there for a couple of minutes. They are then hydrolyzed in 1 M hydrochloric acid (HCl) at 60°C for 6–7 min. After hydrolysis, the roots are transferred to leuco-basic fuchsin, i.e., the Feulgen stain. Ready-made leuco-basic fuchsin is commercially available but can easily be prepared from basic fuchsin and potassium pyrosulphite (potassium metabisulphite), $K_2S_2O_5$, according to the description given by Darlington and La Cour (1969). When staining is complete, which at room temperature takes 45–60 min., the roots are transferred onto a clean slide. The number of roots to be used for each preparation depends on the type of root. The meristem of one primary root or three lateral roots usually gives a suitable number of cells. On the slide, the dark-stained tips, which contain the meristem, are separated from the rest of the roots. They are then crushed in a drop of 45 % acetic acid with the flat end of an aluminium rod and squashed under a coverslip. The pressure is applied under several thicknesses of blotting paper, allowing no sideways movement of the coverslip.

Preparations are made permanent according to the dry-ice procedure, published by Conger and Fairchild (1953). The slide is placed on a block of dry ice and when the material is thoroughly frozen (which takes only a couple of minutes), the coverslip is popped off by slipping a razor blade or scalpel under one of its corners.

The material sticks to the slide which is quickly passed to absolute ethyl alcohol (two changes). From the second change of alcohol, the preparation can be directly mounted in Euparal or Diaphane, using a clean coverslip. If Canada balsam is used as mounting medium, the slide is passed from alcohol to xylene (two changes) before mounting.

Sister Chromatid Exchanges

For the analysis of SCEs, slides are prepared essentially according to the procedure of Kihlman and Kronborg (1975), but with the modifications subsequently described by Kihlman et al. (1978).

After fixation overnight in alcohol-acetic acid at 4°C, the roots are rinsed in a 0.01 M citric acid ($C_6H_8O_7.H_2O$)-sodium citrate ($Na_3C_6H_5O_7.2\ H_2O$) buffer, pH 4.7, and incubated for two hours at 27°C with 0.5 % pectinase (Sigma, from *Aspergillus niger*) dissolved in the same buffer.

The pectinase treatment is terminated by transferring the roots to distilled water, and squash preparations are made in 45 % acetic acid on clean slides, coated with a mixture of 10:1 gelatine and chrome alum (Riedel-De Haën A G Seelze, Hannover). The procedure for making the preparations is the same as that previously described for Feulgen squashes. The coverslips are removed on dry ice and the preparations passed via absolute (two changes), 85, 70, 50, and 30% ethyl alcohol to distilled water.

To reduce the cytoplasmic staining and improve the contrast between sister chromatids, the preparations are treated with RNase in the following way: 200 µl of an RNase solution consisting of 1 mg of Ribonuclease-A from bovine pancreas (Sigma) dissolved in 10 ml 0.5 × SSC (0.075 M NaCl + 0.0075M $Na_3C_6H_5O_7 \cdot 2\ H_2O$) is placed onto the slides. The preparations are covered with a 24 × 40 mm coverslip and incubated in a moist chamber for one hour at 27°C.

After the RNase treatment, the preparations are rinsed in 0.5 × SSC and stained for 20 min. with a solution of 33258 Hoechst prepared in the following way: 1 mg of the fluorochrome compound is dissolved in 1 ml ethyl alcohol and 0.1 ml of this solution is added to 200 ml 0.5 × SSC. The stained preparations are rinsed and mounted in 0.5 × SSC, the coverslips being sealed with rubber cement. At this stage it may be convenient to check the differentiation by fluorescence microscopy.

In the next step, we expose the preparations for 15 min. to long-wave UV at a distance of 6 cm from a radiation source consisting of two Philips 40-W TL09 Black light fluorescent lamps. (The TL09 lamps produce light primarily in the 320 – 380 nm range). The slides are then stored for one day under ordinary light and temperature conditions in the laboratory and for an additional three days at 4°C in the dark. After storage, the slides are incubated for 30 min. at 65°C (TB – BB chromosomes) or for 60 min. at 55°C (TT – BB chromosomes) in 0.5 × SSC.

The slides are then thoroughly rinsed in 0.017M phosphate buffer, pH 6.8, and stained for six min. in a solution containing 3 % Giemsa (Gurr's improved R66) in the same buffer. After staining, the preparations are rinsed, first in a phosphate buffer, then in distilled water, and air-dried. The dry preparations are dipped in xylene and mounted in Canada balsam.

Scoring of Slides

General

Scoring of both chromosomal aberrations and sister chromatid exchanges is usually carried out in metaphase cells which have been collected by exposing the root tips to 0.05 % colchicine for the last three hours before fixation. It is recommended that slides be masked and coded before analysis.

Chromosomal Aberrations

For scoring of chromosomal aberrations in bean root tips, the use of Feulgen-stained preparations is recommended. In metaphase cells collected by colchicine treatment, it is possible to distinguish not only between "sub-chromatid"-, chromatid-, and chromosome-type aberrations, but also between the various types of breaks and exchanges that are produced by mutagenic agents. The main types of these aberrations have been described and illustrated in an earlier publication (Kihlman 1971). There is, therefore, no need to discuss them here.

An approximate idea about the amount of chromosome damage produced may be obtained by scoring cells in anaphase. The effect can then be expressed as percentage of abnormal anaphases or as fragments and bridges per (100) anaphase(s). As a rule, it is not possible to distinguish between the various types of aberrations or to make a detailed analysis of the effect. Nevertheless, the more rapid and simple anaphase scoring can be useful in preliminary experiments when the main purpose is to find out whether or not an agent is capable of producing chromosome damage. Furthermore, there is one type of aberration that can be scored more easily at anaphase than at metaphase and that is the so-called sub-chromatid exchange (Fig. 4.10A) which during anaphase gives rise to a characteristic side-arm bridge or "pseudochiasma" (Fig. 4.10B).

The number of cells to be analyzed per treatment and fixation depends to some extent on the magnitude of the effect: the lower the frequency of aberrations, the more cells should be analyzed. However, even when the effect is relatively high, that is, over 30 aberrations per 100 cells, it is advisable to analyze at least 100 and preferably 200–300 cells.

Sister Chromatid Exchanges

In *Vicia,* SCEs are scored in metaphase cells collected by colchicine treatment. In comparison with the analysis of chromosomal aberrations, the scoring of SCEs is easy as long as their frequency is not too high. When a cell contains more than four times the control level of SCEs (that is, more than 80 SCEs) only a rough estimate of their frequency can be given. After treatments with mutagenic agents, the switches of staining intensity may occur very close to each other within the sister chromatids. As a result, portions of different staining intensity may be found that range in size from a minute dot to a narrow line running across the entire width of the chromatid (Kihlman 1975; Schwartzman et al. 1978). Obviously, they are the result of double SCEs occurring in close proximity to each other and should be scored as such.

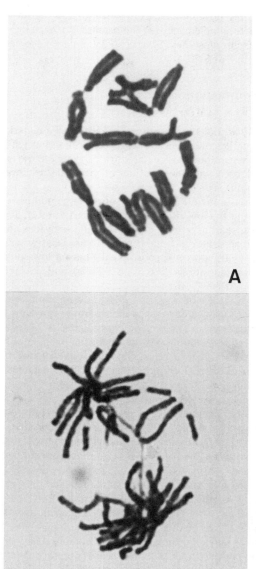

Figure 4.10 The appearance of a "sub-chromatid" exchange at metaphase (A) and anaphase (B). (A from Kihlman and Kronborg 1975; B from Kihlman 1977, by permission of Elsevier Publishing Company, Amsterdam.)

SCEs should be scored only in fully differentiated chromosomes and, as far as possible, in intact cells with a complete complement of chromosomes. For each treatment, a minimum of 25 cells or 300 chromosomes (50 M- and 250 S-chromosomes) should be analyzed for their frequency of SCE.

Advantages and Disadvantages of *Vicia faba* Root Tips as a Test System

To list the advantages first, *Vicia faba* root tips constitute an inexpensive material which is easy to grow and handle (sterile conditions are not necessary) and which is available all year round. No expensive equipment is required for carrying out the experiments. Root tips can be exposed to known concentrations of chemical substances under controlled environmental conditions.

For cytogenetic studies, the material is favourable because of the large number of dividing cells in the root meristem and the low chromosome number ($2n = 12$). The large size of the chromosomes makes scoring of both chromosomal aberrations and sister chromatid exchanges easy and accurate. In comparison with artificial systems such as in vitro cultures of animal and plant cells, root-tip cells have a low spontaneous frequency of chromosomal aberrations (less than 0.5 %) and a stable chromosome number.

Perhaps the most obvious disadvantage of root-tip cells in comparison with cultured mammalian cells is that the former, in contrast to the latter, are surrounded by walls containing cellulose and pectic substances as main constituents. In order to be able to make squash preparations and to obtain a good separation of the cells in the root meristem, the pectic substances in the middle lamella of the cell wall must be dissolved by hydrolysis with acids or enzymes. Both types of hydrolysis are liable to change the morphology of the chromosomes. By acid hydrolysis, basic proteins of the chromosomes are extracted and the DNA component is altered and partly degraded. The commercially available pectinases are not 100 % pure but contaminated with other enzymes which may affect both the protein and nucleic acid components of the chromosomes. The cell walls also contribute to make it difficult or impossible to use root tips for studies on the effects of radiations with low penetration capacity (e.g., UV).

Another disadvantage is that in comparison with a culture of mammalian cells, the cells in the root meristem represent a very heterogenous population, having mitotic cycles of different duration and, perhaps, different sensitivities to mutagenic agents.

As a method for detecting agents that may be of a genetic risk to man, the root-tip technique has the disadvantage of not allowing for mammalian metabolism. On the other hand, when root-tip cells and cultured mammalian cells are compared as test systems, there appears to be a rather good correspondence between the results obtained with the two materials. As a rule, an agent that produces chromosomal aberrations and/or sister chromatid exchanges in one of the two materials does the same in the other. There are, however, exceptions such as maleic hydrazide, which is completely inactive in cultured mammalian cells (e.g., Barnes et al. 1957; Perry and Evans 1975), although it is an extremely efficient producer of chromosomal aberrations and sister chromatid exchanges in root tips of *Vicia faba* (McLeish 1953; Kihlman and Sturelid 1978). Very likely, the different effects in plant and mammalian cells reflect differences in the way maleic hydrazide is being metabolized in the two systems. Unfortunately, little is known about the metabolism of maleic

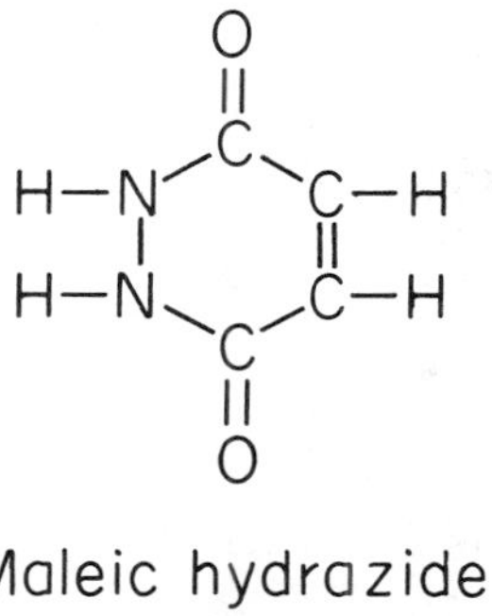

Maleic hydrazide

Figure 4.11 Maleic hydrazide or 1,2-dihydro-3,6-pyridazinedione, a herbicide that produces chromosomal aberrations and sister chromatid exchanges in *Vicia* root tips, but not in cultured mammalian cells.

hydrazide and nothing about the active species formed from the herbicide in root-tip cells.

Obviously, there are quantitative differences in the responses of the two types of material to mutagenic agents. Thus, Chinese hamster cells appear to be more sensitive than root-tip cells to most alkylating agents, although they are more resistant than root-tip cells to X-rays (Sturelid 1971; Sturelid and Kihlman 1975). The materials further differ regarding the relative frequencies of the various types of chromosomal aberrations produced by mutagenic agents. Generally, the aberrations produced in cell cultures of the Chinese hamster (perhaps the most popular material for in vitro testing of mutagenic agents) are characterized by a greater incompleteness: i.e., breaks predominate and exchanges are few, whereas it is the other way round in *Vicia* root tips. These differences at the chromosomal level may reflect differences in the nature and/or efficiency of mechanisms for repairing DNA lesions.

Finally, as for the outlook of the *Vicia* root-tip system in studies on genetic toxicology, it is the opinion of the author of this chapter that in spite of the disadvantages mentioned, the system has enough advantages for making its use in conjunction with, and as a complement to, other test systems not only justified but recommendable.

Literature Cited

Bachmann, K. 1972. Genome size in mammals. Chromosoma 37:85–93.
Baetcke, K. P.; Sparrow, A. H.; Nauman, C. H.; Schwemmer, S. S. 1967. The relationship of DNA content to nuclear and chromosome volumes and to radiosensitivity (LD_{50}). Proc. Nat. Acad. Sci. USA 58:533–40.
Barnes, J. M.; Magee, P. N.; Boyland, E.; Haddow, A.; Passey, R. D.; Bullough, W. S.; Cruickshank, C. N. D.; Salaman, M. H.; Williams, R. T. 1957. The nontoxicity of maleic hydrazide for mammalian tissues. Nature 180:62–64.
Catcheside, D. G. 1948. Genetic effects of radiations. Adv. Genet. 2:271–358.
Conger, A. D.; Fairchild L. M. 1953. A quick-freeze method for making smear and squash slides permanent. Stain Technol. 28:281–83.
Darlington, C. D.; La Cour, L. F. 1969. The Handling of chromosomes, 5th ed. London: George Allen and Unwin Ltd.

Evans, H. J.; Scott, D. 1964. Influence of DNA synthesis on the production of chromatid aberrations by X-rays and maleic hydrazide in *Vicia faba*. Genetics 49:17-38.
Giles, N. H. 1954. Radiation-induced chromosome aberrations in *Tradescantia*. *In* Hollander, A., ed., Radiation biology. I (pt. 2) 713-61, New York: McGraw-Hill Book Company, Inc.
Kato, H. 1974. Possible role of DNA synthesis in formation of sister chromatid exchanges. Nature 252:739-41.
Kihlman, B. A. 1959. The effect of respiratory inhibitors and chelating agents on the frequencies of chromosomal aberrations produced by X-rays in *Vicia*. J. Biophys. Biochem. Cytol. 5:481-90.
Kihlman, B. A. 1961. Biochemical aspects of chromosome breakage. Adv. Genet. 10:1-59.
Kihlman, B. A. 1966. Actions of chemicals on dividing cells, Englewood Cliffs, New Jersey: Prentice-Hall, Inc.
Kihlman, B. A. 1971. Root tips for studying the effects of chemicals on chromosomes. *In* Hollaender, A., ed. Chemical mutagens: principles and methods for their detection. Vol. 2 489-514, New York: Plenum Press.
Kihlman, B. A. 1975a. Sister chromatid exchanges in *Vicia faba*. II. Effects of thiotepa, caffeine and 8-ethoxycaffeine on the frequency of SCEs. Chromosoma 51:11-18.
Kihlman, B. A. 1975b. Root tips of *Vicia faba* for the study of the induction of chromosomal aberrations. Mutat. Res. 31:401-12.
Kihlman, B. A. 1977. Caffeine and chromosomes. Amsterdam: Elsevier Scientific Publishing Company.
Kihlman, B. A.; Kronborg, D. 1972. Caffeine, caffeine derivatives, and chromosomal aberrations. V. The influence of temperature and concentration on the induced aberration frequency in *Vicia faba*. Hereditas 71:101-18.
Kihlman, B. A.; Kronborg, D. 1975. Sister chromatid exchanges in *Vicia faba*. I. Demonstration by a modified fluorescent plus Giemsa (FPG) technique. Chromosoma 51:1-10.
Kihlman, B. A.; Natarajan, A. T.; Andersson, H. C. 1978. Use of the 5-bromodeoxyuridine-labelling technique for exploring mechanisms involved in the formation of chromosomal aberrations. I. G_2 experiments with root tips of *Vicia faba*. Mutat. Res. 52:181-98.
Kihlman, B. A.; Sturelid, S. 1978. Effects of caffeine on the frequencies of chromosomal aberrations and sister chromatid exchanges induced by chemical mutagens in root tips of *Vicia faba*. Hereditas 88:35-41.
Lea, D. E. 1946. Actions of radiations on living cells. Cambridge: The University Press.
Levan, A. 1949. The influence on chromosomes and mitosis of chemicals, as studied by the Allium test. Proc. Eighth Internat. Congr. Genet., Stockholm. Hereditas (Suppl. Vol.): 325-37.
McLeish, J. 1953. The action of maleic hydrazide in *Vicia*. Symposium on Chromosome Breakage. Heredity 6 (Suppl.): 125-47.
Michaelis, A.; Rieger, R. 1971. New karyotypes of *Vicia faba* L. Chromosoma 35:1-8.
Perry, P.; Evans, H. J. 1975. Cytological detection of mutagen-carcinogen exposure by sister chromatid exchange. Nature 258:121-25.
Perry, P.; Wolff, S. 1974. New Giemsa method for the differential staining of sister chromatids. Nature 251:156-58.
Read, J. 1959. Radiation biology of *Vicia faba* in relation to the general problem. Oxford: Blackwell Scientific Publications.
Sax, K. 1941. Types and frequencies of chromosomal aberrations induced by X-rays. Cold Spring Harbor Symp. Quant. Biol. 9:93-101.
Schvartzman, J. B.; Cortés, F.; López-Sàez, J. F. 1978. Sister subchromatid exchanged segments and chromosome structure. Exp. Cell Res. 114:443-46.
Sturelid, S. 1971. Chromosome-breaking capacity of tepa and analogues in *Vicia faba* and Chinese hamster cells. Hereditas 68:255-76.
Sturelid, S.; Kihlman, B. A. 1975. Enhancement by methylated oxypurines of the frequency of induced chromosomal aberrations. II. Influence of temperature on potentiating activity. Hereditas 80:233-46.
Wolff, S.; Bodycote, J.; Painter, R. B. 1974. Sister chromatid exchanges induced in Chinese hamster cells by UV-irradiation of different stages of the cell cycle: the necessity for cells to pass through S. Mutat. Res. 25:73-81.

APPENDIX:
Reconstructed Karyotypes as a Tool for Investigating Differential Chromosomal Mutagen Sensitivity

I. SCHUBERT, R. RIEGER and A. MICHAELIS

Michaelis and Rieger (1971) have provided and used structurally reconstructed karyotypes of *V. faba* that allow investigation in more detail than in the standard karyotype of some aspects of the chromosome breaking activity of physical and chemical agents. Karyotype reconstruction consisted of selecting seedlings with spontaneous or mutagen-induced reciprocal translocations or inversions. By selfing of plants that contained heterozygous translocations or inversions. By selfing of plants that contained heterozygous translocations or inversions the corresponding aberrations became homozygous. Systematic crossings of plants with single translocations or inversions resulted in multiply reconstructed karyotypes (Fig. 4.12). In this special material all of the six chromosome pairs are easily identifiable; in the standard karyotype, it is rather difficult to interdistinguish the five acrocentric chromosome pairs because of their close similarity in size and structure.

Using these reconstructed karyotypes, two new routes of investigation became possible:

1. To look for mutagen-specifities of inter- and intrachromosomal distribution of induced chromatid aberrations, i.e., for mutagen-specific differential "sensitivity" of certain chromosome regions as measured by their frequency of involvement in induced chromatid aberrations. The following results were obtained: a) a number of chromosome regions proved to be aberration 'hot spots', i.e., were significantly more frequently involved in aberrations than expected on the basis of their relative metaphase length. These segments contain heterochromatin as evident from the presence of Giemsa marker bands (Fig. 4.13) and late replicating DNA (Döbel et al. 1973, 1978); b) the actual expression of the hot spot character of these potential hot spot segments was found to be, at least in part, mutagen-specific (Rieger et al. 1975, 1977); c) in addition, it was observed that mutagens inducing chromatid aberrations independently of the DNA synthesis phase of the cell cycle, i.e., showing nondelayed effects (Kihlman 1966) as, e.g., different kinds of ionizing irradiations or bleomycin, resulted in much less specific inter- and intrachromosomal distribution patterns of chromatid aberrations than mutagens showing delayed effects (Schubert and Rieger 1977); d) in some cases, regions less frequently involved in induced chromatid aberrations than expected on the basis of their relative metaphase length have been recognized (Schubert and Rieger 1976).

*Zentralinstitut für Genetik und Kulturpflanzenforschung der Akademie der Wissenschaften der DDR - 4325 Gatersleben

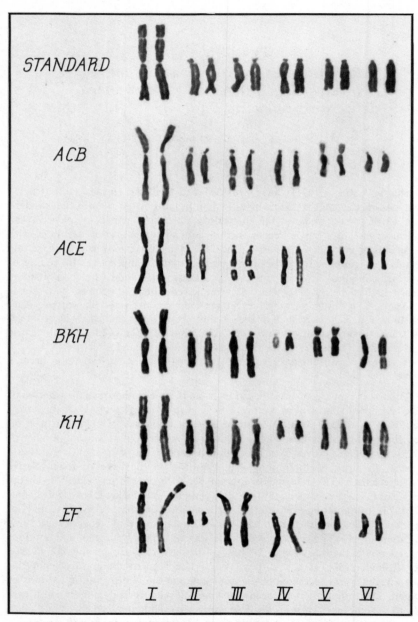

Figure 4.12 Karyograms of the standard and five differently reconstructed karyotypes of *Vicia faba* (Michaelis and Rieger 1971). (A = translocation between chromosomes I and III; B = pericentric inversion in chromosome V; C = translocation between chromosomes I and VI; E = translocation between chromosomes IV and V; F = translocation between chromosomes II and III; H = translocation between chromosomes III and IV; K = translocation between chromosomes I and VI.)

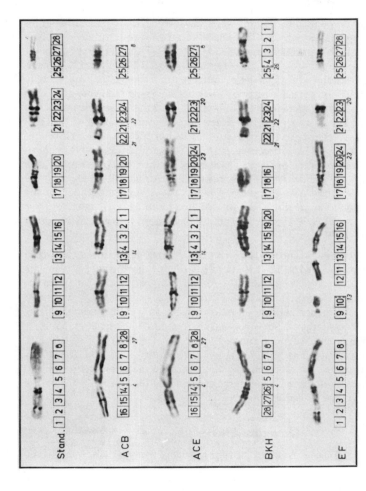

Figure 4.13 Chromosome samples of the standard and four reconstructed karyotypes of *Vicia faba* (compare Fig. 4.12) showing Giemsa marker bands (Döbel et al. 1973), which are indicative of the presence of heterochromatin. For investigation of inter- and intrachromosomal aberration distribution, each karyotype has been subdivided into 28 segments. The sequence of segments and the marks with exposed numbers demonstrate the position of translocation and inversion points (segment 2 represents the secondary constriction, the other segments marked by broken lines represent centromeres and include – with respect to the acrocentric chromosomes – the short arms of the chromosomes). On the basis of their relative length, it is possible to calculate whether the involvement in aberrations of the individual segments is random or significantly above or below the confidence limits for length proportional aberration distribution.

2. By comparison of inter- and intrachromosomal distribution patterns of induced chromatid aberrations in differently reconstructed karyotypes after treatment with the same mutagen, the influence of segment transposition on the mutagen-sensitivity of the whole system as well as on single chromosome regions can be tested. The main results obtained via this experimental route may be summarized as follows: a) Some of the karyotype reconstructions were found to be accompanied by an increase of the mutagen sensitivity of the whole karyotype while other karyotype reconstructions resulted in decreased mutagen sensitivity of the karyotype in question, when compared to the standard karyotype (Rieger and Michaelis 1972; Heindorff 1976). This phenomenon might be of interest for testing mutagenic activity quantitatively. b) The hot spot character of some chromosome segments proved to be independent of their position in the karyotype. c) In other cases, however, the transposition of the hot spot segment into a new chromosome position was accompanied by a significant change of the expression of its hot spot character (Rieger et al. 1977; Schubert et al. 1979; Kaina et al. 1979). d) The modified involvement in chromatid aberrations of some hot spot segments after their transposition to new chromosomal positions was neither directly responsible for parallel changes in the mutagen-sensitivity of the whole karyotype nor was the new neighborhood of the segments in question alone the direct and immediate reason for the change of their mutagen-sensitivity as measured by the amount of aberration clustering (Schubert et al. 1979; Kaina et al. 1979).

As opposed to the intrachromosomal distribution of mutagen-induced chromatid aberrations, mutagen-induced SCEs (with the exception of the nucleolus organizing secondary constriction, which appeared as SCE hot spot) were found to be distributed according to length. This indicates that SCE distribution is independent of the presence of heterochromatin in chromosome regions, of the mutagen being used for their induction, and of karyotype reconstruction (Schubert et al. 1978).

The same scheme for karyotype reconstruction is now followed for the development of a similar system of karyotypes in barley (Nicoloff et al. 1975) and may perhaps serve to compare the mutagen-specificity of gene mutation frequency and aberration frequency in definite chromosome regions and their dependence on the position in the karyotype of certain chromosome regions.

Literature Cited

Döbel, P.; Rieger, R.; Michaelis, A. 1973. The Giemsa banding patterns of the standard and four reconstructed karyotypes of *Vicia faba*. Chromosoma (Berl.) 43:409-22.

Döbel, P.; Schubert, I.; Rieger, R. 1978. Distribution of heterochromatin in a reconstructed karyotype of *Vicia faba* as identified by banding and DNA-late replication patterns. Chromosoma (Berl.) 69:193-209.

Heindorff, K. 1976. Untersuchungen zur Karyotypstruktur und Cytostasan-Sensibilität rekonstruierter Karyotypen von *Vicia faba* L. Diplomarbeit, Martin-Luther-Universität, Halle-Wittenberg.

Kaina, B.; Rieger, R.; Michaelis, A.; Schubert, I.: 1979. Effects of chromosome repatterning in *Vicia faba* L. IV. Chromosome constitution and its bearing on the frequency and distribution of chromatid aberrations in *Vicia faba*. Biol. Zbl. 98:272-83.

Kihlman, B. A. 1966. Actions of chemicals on dividing cells. New Jersey: Prentice-Hall, Inc.

Michaelis, A.; Rieger, R. 1971. New karyotypes of *Vicia faba* L. Chromosoma 35:1-8.

Nicoloff, H.; Rieger, R.; Künzel, G.; Michaelis, A. 1975. Non-random intrachromosomal distribution of chromatid aberrations induced by alkylating agents in barley. Mutat. Res. 33:149-52.

Rieger, R.; Michaelis, A. 1972. Effects of chromosome repatterning in *Vicia faba* L. I. Aberration distribution, aberration spectrum, and karyotype sensitivity after treatment with ethanol of differently reconstructed chromosome complements. Biol. Zbl. 91:151-69.
Rieger, R.; Michaelis, A.; Schubert I.; Döbel, P.; Jank, W. 1975. Non-random intrachromosomal distribution of chromatid aberrations induced by X-rays, alkylating agents, and ethanol in *Vicia faba*. Mutat. Res. 27:69-79.
Rieger, R.; Michaelis, A.; Schubert, I.; Kaina, B. 1977. Effects of chromosome repatterning in *Vicia faba* L. II. Aberration clustering after treatment with chemical mutagens and X-rays as affected by segment transposition. Biol. Zbl. 96:161-82.
Schubert, I.; Rieger, R. 1976. Nonrandom intrachromosomal distribution of radiation-induced chromatid aberrations in *Vicia faba*. Mutat. Res. 35:79-90.
Schubert, I.; Rieger, R. 1977. On the expressivity of aberration hot spots after treatment with mutagens showing delayed or nondelayed effects. Mutat. Res. 44:327-36.
Schubert, I.; Sturelid, S.; Döbel, P.; Rieger, R. 1978. Intrachromosomal distribution patterns of mutagen-induced SCEs and chromatid aberrations in reconstructed karyotypes of *Vicia faba*. Mutat. Res. 59:27-38.
Schubert, I.; Rieger, R.; Michaelis, A. 1979. Effects of chromosome repatterning in *Vicia faba* L. III. On the influence of segment transposition on differential "mutagen sensitivity" of *Vicia faba* chromosomes. Biol. Zbl. 98:13-20.

5

Insect Cells for Testing Clastogenic Agents

by MARY ESTHER GAULDEN AND JAN C. LIANG*

Abstract

Insect cells as test systems for assaying clastogenic agents are evaluated in terms of the "ideal cell" and are found to have a number of advantageous features. Those insects with chromosomes considered suitable for detection of aberrations are outlined on the basis of whether their formal genetics has or has not been studied. The various types and sources of insect cells, including tissue culture cell lines, are given. Details are presented for a squash technique that permits rapid preparation of semipermanent slides for chromosome analysis. The present limitations of chromosome banding in insect cells are described. Some insect cells which have been used to test for clastogenic chemicals are listed. It is concluded that the full potential of insects, the most plentiful of organisms, for mutagenic studies has yet to be realized.

Introduction

Almost one hundred years have elapsed since Waldeyer (1888) introduced the term "chromosome" to describe the prominent structures of the cell nucleus. Soon thereafter an insect was the source of material with which the field of cytogenetics was inaugurated by the cytologist, W. S. Sutton (1902). It was he who, building on the observations and theoretical considerations of Montgomery (1901), Wilson (1902), and Boveri (1902), used the cells of the large lubber grasshopper, *Brachystola magna*, to first present a convincing argument that chromosomes "constitute the physical basis of the Mendelian law of heredity."

*Radiation Biology Section, Department of Radiology, University of Texas Health Science Center at Dallas, 5323 Harry Hines Blvd., Dallas TX 75235.

In rapid succession, a number of fundamental genetic and cytogenetic principles were established with insect chromosomes, for example, chromosome determination of sex (McClung 1902), breakage and reunion of chromosomes during meiotic prophase (Janssens 1909), intimate synapsis of homologues at zygotene (Wilson 1912), independent assortment of chromosomes at meiosis (Carothers 1913), and four-strandedness of meiotic bivalents (Bridges 1916). Meriting special attention is the fruit fly, the study of which gave great impetus to genetics and cytogenetics, beginning with T. H. Morgan's work in 1910. Of particular relevance to the present volume was the report of the mutagenic action of X-rays in *Drosophila* by Muller in 1927. This had been preceded in 1919 by Mohr's investigations on the effects of radiation on cells of an insect, the long-horned grasshopper, *Decticus verrucivorus*.

Insect cell studies have continued to yield new and important information about the structure and function of chromosomes in spite of the trend in recent years for research to be focused on mammalian cells. Several references containing much information on insect genes and chromosomes should be noted at the outset: *Advances in Genetics, Annual Review of Entomology, Animal Cytology and Evolution* (White 1973), *Handbook of Genetics,* Vol. 3 (King 1975). Because many insects over a broad range of species have been examined cytogenetically, no attempt is made in this review to be all-inclusive. Rather, we will highlight some of the advantages insect cells offer for the study of and testing for clastogenic agents, with emphasis on directing the reader to representative literature. Discussion will be limited to the typical mitotic and meiotic chromosomes and will not, therefore, include the giant polytene chromosomes of insects whose analysis is much more time-consuming. Analysis of mutagen effects on chromosomes is currently done with intact cells, but mass-isolated chromosomes will probably be used in the future when cytofluorometric methods are perfected so that quantitative fractionation of normal and aberrant chromosomes can be achieved. Such technical advances will speed the testing of agents for clastogenicity.

The Ideal Cell

What is considered to be an "ideal" cell for study depends to some extent on the motivation of the investigator. Mouse and human chromosomes compared with the chromosomes of many other organisms are small, numerous and not always morphologically distinguishable without banding, but for obvious reasons they have been avidly investigated and to good avail.

The "best" clastogen test system yields reproducible data on chromosome aberration frequencies over a wide range of mutagen doses with economy of both time and money. To gain perspective, let us enumerate some characteristics of the ideal cell for clastogen testing, so that we can see how insect cells measure up to it.

The ideal cell would have long chromosomes with localized rather than diffuse centromeres; the chromosomes would be few in number and would be morphologically distinguishable, even when uniformly stained. Long chromosomes have been reported to show more breakage and reunion than do short ones in some cell types (Ross and Cochran 1975; Evans 1976) but not in others (Savage et al. 1973). Constrictions, gaps, and breaks seem to be more easily detected, or to occur more frequently, in long than in short chromosomes. Fragmentation is more easily detected microscopically in a chromosome with a single centromere. Few and morphologically distinguishable chromosomes permit the rapid identification of aberrant ones, thereby reducing the time that has to be spent on scoring procedures.

A short cell cycle reduces the time required to obtain data, not only on mutagens whose effects are demonstrable within one cell cycle, but especially on those mutagens requiring several cell cycles for full expression of effects.

The ideal cell would show no spontaneous chromosome aberrations, thus reducing or obviating the need for scoring untreated cells.

Large numbers of rapidly dividing cells make it relatively easy to obtain a dose response of chromosome aberration frequency over a wide range of doses of a given agent.

The source of the cells should be plentiful and readily accessible year-round. If the source is the organism, it should be easy to maintain in the laboratory, and maintenance costs should be low. If the source is cells in tissue culture, the cell lines should be "immortal" with an invariable karyotype, preferably haploid or diploid.

The chromosomes would be easily banded by the various techniques that identify different regions of individual chromosomes. This would permit localization of "hot spots," i.e., regions of chromosomes showing nonrandom, increased frequencies of break points.

The chromosomes of the ideal cell would show no heteromorphisms (polymorphisms), which are heritable morphological variations consistent in an individual but variable among the individuals of a population; their biological significance is not always obvious (White 1973). As the term is currently used, heteromorphic regions are particular regions of chromosomes that contain different types of highly redundant DNA in variable amounts (Verma and Dosik 1980). Hetermorphism may increase the overall length of a chromosome, and thereby slow down scoring of cells from several individuals, because a heteromorphic chromosome must be distinguished from an aberrant one and unequivocal identification requires banding methods.

The ideal cell would be large and would have a "hollow spindle," that is, the points of chromosome attachment would occur only at the edge of the spindle. These characteristics make for easy spreading of metaphase chromosomes on squashing, so that hypotonic medium treatment is not necessary. Treatment with a spindle inhibitor can also be dispensed with unless large numbers of blocked metaphases are required.

Ideally, the formal genetics of the organism should have been studied so that a large number of specific genes localized at defined positions on given chromosomes would be known. This would permit some estimation of the genetic consequences of chromosome breakage, especially if there were hot spots of breaks. As mutagen and carcinogen testing progresses in the future to more sophisticated stages, the genetic background of the organisms used will probably play a greater role than at present in the extrapolation from and application of cytogenetic data on experimental cells to man.

Organisms

Of the known organisms in the world, insects are the largest group: there are estimated to be 2.5 to 3 million species of insects, about one-third of which have been described (Lokki and Saura 1980). Of the latter, the chromosomes of a relatively small proportion have been reported on and an even smaller number have been investigated genetically. Attention will be focused here on some of those species which have chromosomes deemed suitable for detection of aberrations. Such insects may be arbitrarily divided into those which have been examined genetically and those which have not. Laboratory breeding methods have of course been worked out for those insects used for mutation studies; details of the methods and sources of stocks can be found in the references cited below. Smith (1966, 1967) gives methods for raising some insects whose genetics has not been determined. We also include brief mention of a few insects for which mutations have been described but whose chromosomes are not suitable for aberration analysis; such organisms offer excellent material for studying somatic cell mutation, e.g., an eye pigment mutation can be detected in a single ommatidium of *Tribolium* (Sokoloff 1975).

Genetically Investigated Insects

The known genetic characteristics of invertebrates are with a few exceptions limited to the insects. Understandably, they are mainly species of economic import; in some cases, attempts to control insect population size genetically (Whitten and Foster 1975) and in others, resistance to insecticides (Ross and Cochran 1975) have spurred genetic research.

Drosophila. By far the most thoroughly investigated genetic system in insects (yea, in animals) is that of the fruit fly. The genetics of this organism has been studied unabated for almost three quarters of a century. The large number of known gene mutations, more than 3,000, together with more than 1,500 delineated chromosome changes (Lindsley and Grell 1968) make *Drosophila* an unparalleled organism for eukaryotic genetic and cytogenetic studies. Most of the chromosome aberrations have, however, been analyzed with the giant polytene chromosomes that are commonly found in the salivary glands and that are "naturally" banded. The neuroblast cells of the larval brain have been the preferred source for ordinary metaphase chromosomes, but as Kaufmann (1934) has noted, the relatively laborious preparation of slides for neuroblast observations has been a deterrent to their use. More recently, established cell lines in tissue culture have made somatic cells of *Drosophila* more attractive for cytogenetic studies. *D. melanogaster* is the species most studied; it has four distinctive pairs of chromosomes (Fig. 5.1A). Clayton and Wheeler (1975) have described mitotic metaphase chromosome configurations of approximately 500 species and subspecies or about 40% of the known species of *Drosophila*. Laboratory maintenance of the fruit fly is easy and culture methods have been thoroughly described by Ashburner and Thompson (1978).

Blatella germanica L. The German cockroach is the most important pest of the 4,000 known species of this insect, having a world-wide distribution. Even though it might not be considered by some as the most pleasant of laboratory animals, this species offers acceptable cytogenetic material, as do other species (Cohen and Roth, 1970, have described the chromosomes of 106 species of cockroaches). The chromosomes are metacentric and submetacentric, and their number is $2n = 23$ in the males (XO) and $2n = 24$ in females (XX) (Cochran and Ross 1967; Cohen and Roth 1970). The longest autosome and the X chromosome are ca. 3 μm in length and the shortest ca. 1 μm; the X chromosome is 2.5 μm long. One possible difficulty with cockroach

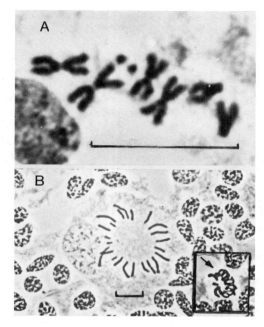

Figure 5.1 (bars represent 10 μm) A: The four distinctive pairs of chromosomes in a squash of a larval neuroblast of *Drosophila melanogaster* (from Lewis and Riles 1960). B: Metaphases of a neuroblast and a small embryonic cell (arrow at lower right corner) from a squash of an untreated embryo of the grasshopper, *Chortophaga viridifasciata* (male: 23,XO).

chromosomes is that the centromere position may be difficult to ascertain in cells not treated with colchicine. The lengths of the meiotic chromosomes correspond well to those of the mitotic chromosomes; five of the meiotic bivalents are morphologically distinguishable (Cochran and Ross 1969). For *B. germanica*, 60 mutant stocks with 12 linkage groups and over 20 translocation stocks have been isolated and maintained in the laboratory (Ross and Cochran 1975).

Anopheles and other mosquitoes. As noted in an extensive review of 103 species of mosquitoes by Kitzmiller (1976), six is the diploid number for the vast majority of mosquitoes, with two pairs of autosomes and a smaller pair of sex chromosomes (female XX; male XY). The autosomes in the brain cells of larvae are usually 6 to 7 μm long. In *Anopheles*, malarial vector, a few morphological, biochemical, and insecticide resistance genes have been reported, and some assigned to linkage groups; considerable work has been devoted to chromosome studies of this genus. Several homozygous translocation stocks have been produced (Baker et al. 1980; Coluzzi and Kitzmiller 1975). More gene mutants are known in *Aedes* (vector of yellow fever, dengue, and hemorrhagic fever) than in *Anopheles*, including a gene for susceptibility to parasites; a number of translocations has been synthesized (Rai and Hartberg 1975). For *Culex*, vector of filariasis, Barr (1975) lists 47 mutants in *C. pipiens*, with many assigned to linkage groups.

Lucilia cuprina. The Australian sheep blowfly has $n = 6$, $2n = 12$ with five autosome pairs and one sex chromosome pair (female XX; male XY) (Ullerich 1963). The autosomes range in length from approximately 4 to 5.5 μm; the X is approximately 6 μm long and the Y 2.5 μm. Forty-eight visible mutations are listed in six linkage groups for this organism by Whitten et al. 1975.

Musca domestica. The house fly has $n = 6$ (Perje 1948), $2n = 12$ (Boyes and Naylor 1962). There are five pairs of distinguishable meta- and submetacentric chromosomes and a pair of sex chromosomes (female XX; male XY). In the brain cells, the autosomes are relatively long, ranging from 9 to 14 μm; the X is about 14 μm and the Y 2.5 μm (Wagoner 1967). Close to 200 mutants have been described by Milani (1967, 1975).

Several species of insects which have been subjects for considerable genetic analysis do not have particularly favorable chromosomes for analysis because of large number and/or small size:

Tribolium castaneum and *T. confusum,* flour beetles. The chromosome number of *T. castaneum* is $2n = 20$ (female XX, male XY) and of *T. confusum,* $2n = 18$ (XX or XY). The chromosomes of both species are small, the longest autosome being only about 2.5 μm long (Smith 1952a). One interesting feature of the pachytene chromosomes of *T. confusum* is the large block of heterochromatin associated with the centromere (Smith 1952b). Over 100 mutants have been described in the two species (Sokoloff 1975).

Habrobracon and *Mormoniella*, parasitic wasps on cereal-infesting moths (Cassidy 1975). In *Habrobracon* $2n = 20$ and in *Mormoniella* $2n = 10$ (Johnson and Ray 1972). In both genera the female is diploid, arising from fertilized eggs, and the male is usually haploid, developing from unfertilized eggs (Torvik-Greb 1935). The chromosomes of both genera are small, the longest in *Habrobracon* being about 2.5 μm and that of *Mormoniella* 3.5 μm. A number of mutants have been reported for both genera (see also Cassidy 1975).

Bombyx mori, silkworm, has $n = 28$ and $2n = 56$ with small and approximatley oval chromosomes, which may be holokinetic (diffuse centromere). Lepidoptera in general have large numbers of chromosomes which at mitotic anaphase behave as though they are holokinetic (Maeki 1980); whether this situation obtains at meiosis is not firmly established (White 1973). More than 100 visible and biochemical mutants in 22 linkage groups are known in *B. mori* (Tazima et al. 1975).

Ephestia kühniella, Mediterranean meal moth, a flour mill pest, has $2n = 60$, with small and either round or oval chromosomes, probably with diffuse centromeres. In oocytes the individual pachytene chromosomes are distinguishable (Traut and Mosbacher 1973). About 40 mutants have been described in *Ephestia*, but the numerous chromosomes have made linkage studies difficult (Caspari and Gottlieb 1975).

Apis mellifera, the honey bee, has $2n = 32$. Queens and workers are diploid and drones are usually haploid. The chromosomes are tiny (Woyke and Skowronek 1974). More than 3 dozen mutant genes have been reported for *Apis* (Rothenbuhler 1975; Goncalves and Stort 1978).

Insects Not Genetically Investigated

The species of insects whose chromosomes are suitable for aberration analysis but whose formal genetics has not been studied are too many to enumerate. Many of them have diploid chromosomes numbers of 12 or less. In addition to those listed in

Tables 5.1 and 5.2 are several which can be singled out as possible experimental material and which may not be well known.

Even though grasshoppers have long been a favorite subject for cytologic investigators, their mitotic chromosomes are usually not morphologically distinguishable. One notable exception is *Dichroplus silveiraguidoi* Lieb, a neotropical species which was collected and first reported by Saez (1957). It has four pairs of readily identifiable chromosomes. In prometaphase of spermatogonia and ovariole wall cells, the shortest autosome is about 10 μm long and the longest is about 15 μm; in neuroblasts they are probably larger. The habitat of this species is limited to a small area of northern Uruguay which Saez and Perez-Mosquera (1977) characterize as having "scarce xenophillic plants over a rocky ground." They reported difficulty in maintaining adults in the laboratory; with improved breeding methods now available, it might be possible to establish a breeding colony.

Some ant species have few and long chromosomes, e.g., *Myrmecia pilosula*, with $n = 5$. In prometaphases of the prepupae ganglia of *M. sp. cf. fulvipes*, the two longest chromosomes are 15 μm or more in length; these and the other chromosomes show what appears to be spontaneous banding (Imai et al. 1977).

One of the longest nonpolytene chromosomes described in animals is the X in several alticid beetles (Coleoptera). In *Alagoasa extrema*, the X chromosome at metaphase I of spermatocytes is 45 μm in length; the autosomes (10 pairs) are 1.5-2.5 μm long. In a related species, *Walterianella venusta*, the X is 33 μm, the Y is 17 μm, and the autosomes (22 pairs) are about 1 μm long. The sex chromosomes in these beetles do not synapse (Virkki 1964).

Tissues

In Situ

The tissues commonly used for analysis of insect chromosomes are testes, embryos, and larval brains or ganglia. Used to a lesser extent are ovaries and other tissues obtained from larvae or nymphs, such as hepatic caeca, hypodermis, intestinal muscles, fat body, wing buds, etc. (Smith 1943), and a few tissues from adults, e.g., gastric caeca and ovarian follicle cells. Gonadal tissue from pupae or nymphs usually provides better material than that from adults for chromosome studies. Both mitotic (gonial cells) and meiotic chromosomes can be found in the younger life forms, and this probably accounts for the fact that the chromosomes of many insects have been studied primarily in the gonads.

The cells of the early stages of embryogenesis, such as those in the blastoderm, are in some insects preferable for obtaining somatic chromosomes. Because this stage sometimes occurs prior to oviposition—e.g., in *Cimex*, the bedbug (Ueshima 1967)—dissection of females is required. Large number of cells can be much more readily obtained with those insects in which all embryogenesis takes place after oviposition. Podlasek and Hunt (1978) obtained good chromosome preparations from the cells of young embryos (12-18 hr. after ovipostion) of *Oncopeltus fasciatus* Dallas, the greater milkweed bug; the cells can be easily separated from the yolk by centrifugation (10 min. at 500g) after manual removal of the chorion. These investigators report that the isolated cells continue to divide in Waddington's insect Ringer's solution. Although the small size (0.85-2.2 μm) and holocentricity of *On-*

Table 5.1 Some Insects Used to Investigate Banding of Mitotic and Meiotic Chromosomes

Insect	Banding method[a]							Reference
	C	Q	H	G	R	N	Other	
Orthoptera:								
Schistocerca gregaria	+			+				Fox et al. 1973
Schistocerca gregaria	+							Brown & Wilmore 1974
Schistocerca gregaria	+					+		Hagele 1979
Myrmeleotettix maculatus	+							Gallagher et al. 1973
Dissosteira carolina	+							Bregman 1973
Bryodema tuberculata	+	+	+					Klasterska et al. 1974
Gryllus argentinus	+							Drets & Stoll 1974
Dichroplus silveiraguidoi	+							Cardoso et al. 1974
Tettigidea lateralis	+							Fontana & Vickery 1975
Caledia captiva	+							Shaw et al. 1976
Chortoicetes terminifera	+			+				Webb 1976
Chortoicetes terminifera				+				Webb & Komarowski 1976
Chortoicetes terminifera	+			+				Webb & Neuhaus 1979
Warramaba virgo	+			+				Webb et al. 1978
Phaulacridium vittatum	+			+				Webb & Westerman 1978
Spathosternum prasiniferum	+		+					Das et al. 1979
Acheta domesticus	+	+	+	+				Ali 1979
Acrididae 23 species	+							King & John 1980
Hemiptera:								
Triatoma infestans				+				Maudlin 1974
Heteroptera:								
Carbula humerigera	+							Muramoto 1980
Dolycoris baccurum	+							
Diptera:								
Aedes species	+							Motara & Rai 1977
Anopheles species		+		+				Tiepolo et al. 1975
Anopheles species							+[b]	Mezzanotte et al. 1979a
Anopheles species			+					Bonaccorsi et al. 1980
Culiseta longiareolata							+[b]	Mezzanotte et al. 1979b
Chrysomya species	+							Ullerich 1976
Sacrophaga bullata	+							Samols & Swift 1979a
Sacrophaga bullata		+						Samols & Swift 1979b
Lucilia cuprina	+	+	+					Bedo 1980
Musca domestica				+				Lester et al. 1979
Drosophila melanogaster		+						Vosa 1970
Drosophila melanogaster	+	+						Dolfini 1974a, b
Drosophila melanogaster	+	+			+			Lemeunier et al. 1978
Drosophila melanogaster	+	+						Halfer et al. 1980
Drosophila species		+	+					Holmquist 1975
Drosophila species		+	+					Gatti et al. 1976
Drosophila nasutoides	+	+	+					Wheeler & Altenburg 1977
Samoaia leonensis		+						Ellison & Barr 1972

[a] The banding method indicated does not imply that bands were observed in all chromosomes or that a band fine-structure was obtained comparable to that in higher vertebrate chromosomes subjected to the same treatment (see text).
[b] Coriphosphine-O

Table 5.2 Some Chemical Compounds Reported to Induce Chromosome Aberrations in Insect Cells (exclusive of inherited aberrations)[a]

Mutagen	Exposure Dose	Exposure Time	Organism	Reference
Apholate[b]	10 ppm	1-4 d	Aedes aegypti[j]	Rai 1964
Tepa[c]	5-40 ppm	3 d	Culex pipiens fatigans[j]	Tadano & Kitzmiller 1969
Apholate Metepa[d] Hempa[e]	5-1000 ppm	3-4 d	Culex pipiens fatigans[j]	Grover et al. 1973
MMS vapor[f]	No data	25-35 min.	Drosophila melanogaster[j]	Gatti et al. 1975
Caffeine	$5 \times 10^{-4} - 2 \times 10^{-2}$ M	2 hr.	Drosophila melanogaster[j]	De Marco and Cozzi 1980[m]
NaOH, pH 9 HCl, pH 3,4	No data	4-72 hr.	Oxya velox[k] Spathosternum prasiniferum[k]	Manna & Mukherjee 1966
$CaCl_2$	1%	4-24 hr.	Spathosternum prasiniferum[k]	Parida et al. 1972
Versene[g]	10^{-3} M	31 hr.	Spathosternum prasiniferum[k]	Saha & Chakrabarty 1973
Barbital	0.5, 0.7%	24 hr.	Spathosternum prasiniferum[k]	Das & Mukherjee 1977
Methanol	0.3, 0.5%	30 hr.	Oxya velox[k]	Saha & Khudabaksh 1974
Actinomycin D	60 μg	30 hr.	Schistocerca gregaria[k]	Jain & Singh 1967
Penicillin	5 mg	40 hr.	Phleoba antennata[k]	Parida 1972
Methyl mercury hydroxide	$8 \times 10^{-6} - 10^{-3}$ mg	2-72 hr.	Stethophyma grossum[k]	Klasterska & Ramel 1978
4NQO[h]	1-20 μM	1 hr.	Chortophaga viridifasciata[j]	Liang & Gaulden unpubl.[m]
MNNG[i]	1-20 μM	1 hr.	Chortophaga viridifasciata[j]	Liang & Gaulden unpubl.[m]
Adriamycin	0.1-2 μM	1 hr.	Chortophaga viridifasciata[j]	Liang & Gaulden unpubl.[m]
Bleomycin	0.1-2 μM	1 hr.	Chortophaga viridifasciata[j]	Liang & Gaulden unpubl.[m]
Cyclophosphamide	200-4000 μM	1 hr.	Chortophaga viridifasciata[j]	Liang & Gaulden unpubl.[m]
LSD	20 μg/ml	5 hr.	Anteraea eucalypti[l]	Quinn 1970[m]
Mitomycin C	0.01%	24 hr.	Poecilocerus pictus[k]	Mohanta & Parida 1975

[a] The cells were exposed in vivo except in three studies for which they were exposed in vitro. For the in vivo exposures, the doses given are the amounts per animal. Squash preparations were used by all investigators.
[b] Apholate: 2,2,4,4,6,6-hexahydro-2,2,4,4,6,6-nexakis (1-aziridinyl)-1,3,5,2,4,6-triazatriphosphorine
[c] Tepa: tris (1-aziridinyl) phosphine oxide
[d] Metepa: tris (2-methyl-1-aziridinyl) phosphine oxide
[e] Hempa: hexamethyl phosphoric triamide
[f] MMS: methyl methane sulfonate
[g] Versene: ethylene diamine tetracetic acid
[h] 4NQO: 4-nitroquinoline-1-oxide
[i] MMNG: N-methyl-N'-nitro-N-nitrosoguanidine
[j] Neuroblasts
[k] Spermatocytes
[l] Cultured ovarian cells
[m] in vitro exposure

copeltus chromosomes make them unsuitable for aberration analysis, the method applied to other insect embryos may be useful for clastogen studies.

Chromosome size, i.e., length and width, has been shown in several organisms, not surprisingly, to be under genic control and to show Mendelian segregation (Darlington 1937); in *Matthiola*, chromosome length is controlled by a single gene pair, the long-chromosome condition being recessive (Lesley and Frost 1927). Variations in chromosome size also occur within a single organism and appear to be controlled to a considerable extent by cell size. Such variations have been reported for a number of insect cells. A striking example is found in the *Triatominae,* whose chromosomes as revealed in the gonial cells are small, holokinetic, and are not, for the most part, morphologically distinguishable (Ueshima 1966). The medical importance of one species, *Triatoma infestans*, the vector of *Trypanosoma cruzi* responsible for Chagas' disease, has undoubtedly been responsible for persistent investigation of its genetics and cytogenetics. Maudlin (1974) has recently found that the chromosomes of neuroblasts in *T. infestans* embryos are as long as 10 μm, appear to have localized centromeres, and have distinct G-banding patterns which permit identification of the individual 11 pairs of chromosomes.

Brown (1965) demonstrated a significant difference in chromosome size in two cell types of some thelytokous coccids: the chromosomes of embryonic cells are half as wide but up to seven times as long as those of egg follicle cells. The chromosomes of imaginal disc cells in *Drosophila* larvae are smaller than those in the ganglia cells (Lewis and Riles 1960). According to Boyes and Wilkes (1953), the chromosomes of Tachinid fly pupal cells are consistently so small that analysis of them is "difficult or impossible," whereas the chromosomes of early third instar larvae range from 2 to 7 μm in length (within a cell). A differential chromosomal variation among cells of the same tissue has been found by Boyes and Naylor (1962), who compared the total autosomal length with that of the two heterochromatic sex chromosomes in larval cells of the house fly (not treated with colchicine): the X chromosome is longest in short complements and shortest in long complements.

We call attention to the possible differences in chromosome size to emphasize the need to search for the cell type, even within one organism, that meets the needs of the investigator. In general, the largest and most easily spread mitotic chromosomes are found in the neuroblasts of insect embryos and larvae. Because there is little or no tough connective tissue in young embryos, and because the chromosomes are well spaced on a hollow spindle of a large cell like the neuroblast, the chromosomes can be spread without overlapping when the cell is squashed. This is well illustrated in Figure 5.1B in which the chromosomes of a neuroblast can be compared with those of a smaller cell in a squash preparation.

Tissue Culture Cells

Quantitative analysis of the effects of chemical clastogens is best achieved under conditions in which the cells are essentially surrounded by the agent. For this reason, tissue culture cells are ideal material.

Goldschmidt in 1915 was the first to try culturing insect cells; he observed maturation but not mitosis in the testicular cells of the moth *Samia cecropia* L., suspended in hemolymph. His results, as well as those of Frew (1928), were obtained with what is now known as organ cultures. Success in culturing insect cells and the development of cell lines came much later. This is dramatized by the fact that the first review on the subject did not appear until 1959 (Day and Grace), in which it was pointed out that not a single line of insect cells had yet been established. Matters were a little better by 1971 (Brooks and Kurtti) when a total of 14 lines from four orders (Lepidoptera, Diptera, Hemiptera, and Blattaniae) were known but none of them was available from the American Type Culture Collection (ATCC). As these authors have noted, the lack of knowledge of the composition of insect tissue fluids inhibited investigators in developing media that would support long-term cell growth and survival. Much of the motivation for obtaining insect cell lines derives from mission-oriented work on those insects with economic and medical import, both good (e.g., silkworms) or bad (e.g., mosquitoes), so the emphasis has not been on organisms with necessarily favorable chromosomes for cytogenetic studies. But, as we shall see, some cell lines with good chromosomes have been developed.

The most recent reviews (Schneider and Blumenthal 1978; Marks 1980) reveal considerable progress in the development of culture media and insect cells lines, which is reflected in the publication of four compilations on the subject in the 1970s (cf. Marks 1980). The ATCC now lists in its catalogue three cell lines, one derived from ovarian tissue of the moth *Antheraea eucalypti* and two from larval cells of the mosquitoes *Aedes aegypti* and *A. albopictus*. The ATCC recently initiated a service designed to help investigators locate sources of cell lines in private laboratories: Cell Sources Information Bank, ATCC, 12301 Parklawn Drive, Rockville, Maryland 20852 (Shannon 1977).

We wish to call attention to some points that may be of special interest for the use of insect tissue culture cells for research on clastogenic agents. In working with cell lines, it is important to use cloned cells to minimize selection pressures. As noted by Schneider and Blumenthal (1978), the definition of an "established line" for invertebrate tissue culture has not yet been set as has been done for vertebrate cells. They propose that a standard for invertebrate cell lines comparable to that used for human fibroblast-like lines would be cells that "demonstrate the potential to be subcultured indefinitely in vitro" (Fedoroff 1967), that is, a minimum of 40 subcultures at 5- to 7-day intervals.

Cell lines of four insects with chromosomes suitable for clastogen studies have been established: mosquito, *Drosophila*, cockroach, and *Triatoma*. Mosquito cell lines were among the first to be established (Grace 1966; Singh 1967), and some of them continue to be used. The generation time (27°C) for lines of two species has been determined: *Aedes albopictus*, 8 hr. and *A. aegypti*, 12 hr. (Baim and Mukherjee 1978). The three pairs of morphologically distinguishable chromosomes in cells of these lines vary in length from 12 to 24 μm (Bianchi et al. 1972). There is somatic pairing in mosquito, as there is in *Drosophila*, which aids in analyzing chromosome aberrations, especially exchanges.

The first success in culturing *Drosophila melanogaster* cells was reported by Horikawa and Fox in 1964, but it was not until 1969 that Echalier and Ohanessian and Kakpakov et al. reported long-term culture of these cells. Subsequently, other cell lines were established; 31 were listed in 1978 (Schneider and Blumenthal) and additional ones continue to be reported (e.g., Halfer 1980). That *Drosophila* tissue culture cells are no longer an oddity is evidenced by their use not only for cytogenetic studies and chromosome banding but also for somatic hybridization (e.g., Wyss 1979) and biochemical studies (Schneider and Blumenthal 1978). The Kc line of

Echalier and Ohanessian is one of the more widely used; the average durations of the phases of its cell cycle (26° C) have been determined: $G_1 = 1.8$ hr., $S = 10.0$ hr. and $G_2 = 7.2$ hr., with a total generation time of 18.8 hr. (Dolfini et al. 1970).

Cell lines have been established from two genera of cockroaches, *Blatella germanica* and *Periplaneta americana* (cf. Kurtii 1976), but progress has not been as rapid in their use as with *Drosophila* and mosquito cell lines. The chromosomes of the cultured cells are good for cytogenetic studies (Philippe and Landureau 1975).

Insect tissue culture cells have not been used to test clastogens, perhaps because they have one distinct disadvantage that must be kept in mind, namely, the spontaneous occurrence of aneuploidy, polyploidy, and chromosome aberrations in them (Bianchi et al. 1972; Dolfini 1971, 1974b; Philippe and Landureau 1975; Halfer 1980). Perhaps with more transfers, it will be possible to isolate highly stable, diploid clones. A line notable for its stability is a haploid *Drosophila* one reported by Debec (1978) to be 95% haploid after 30 transfers (one year). Haploidy is usually a nonviable condition in animal somatic cells, but it is normal for some adult forms of insects, so other haploid insect cell lines may be established in the future.

For isolation of chromosomes en masse, tissue culture cells are ideal, and Hanson (1978) has used *Drosophila* cell lines (and pregastrulal embryos) for this purpose.

Pudney and Lanar (1977) have obtained an established cell line from embryos of *Triatoma infestans*. It remains to be seen, though, whether the size of the chromosomes in the tissue cultures is as optimal for chromosome studies as those of neuroblasts (Maudlin 1974).

Organ Culture

Although short-term insect organ cultures have long been used, the development of methods for their long-term culture has not kept pace with that for tissue culture of insect cells (Martin and Schneider 1978). Nevertheless, for testing of chemical mutagens, organ cultures provide an in vitro technique that is preferable to the in vivo ones (injection or feeding) for exposures of short duration. When the tissue mass is small, the medium containing the mutagen is accessible to the cells, so that one can be more confident that the concentration of mutagen reaching the cell is close to that of the medium than is the case with the in vivo exposure.

We have found exposure of whole isolated grasshopper embryos (membranes and yolk removed: Carlson and Gaulden 1964) to be quite satisfactory for testing the effects of chemical mutagens on the chromosomes of neuroblasts that lie at the surface of the embryo. The small ganglia of *Drosophila* larvae can also be exposed in vitro (De Marco and Cozzi 1980).

Cross and Shellenbarger (1979) have recently used improved culture media for detailed studies of meiosis and spermiogenesis in single cysts from *Drosophila melanogaster* testes over a 72 hr. period. This technique might be useful for the study of clastogen effects on meiotic chromosomes, because the combined duration of meiotic divisions I and II is estimated by these authors to be only several hours at room temperature. It is not clear from the data whether this includes all of prophase I.

Srdic and Jacobs-Lorena (1978) have reported that fully mature *Drosophila* eggs develop when a single egg chamber from an ovary is transplanted into adults, including males. This method provides an opportunity to expose cells at various stages of oogenesis to chemical mutagens in vitro and follow their subsequent response under in vivo conditions. It would permit comparison of the effects of in vivo exposure with those of in vitro exposure, but it would be laborious.

Methods

We will discuss those points and methods which may be helpful to workers not familiar with cytogenetic procedures, special attention being given to insect cells and to chromosome banding techniques which have recently been applied to these cells.

Techniques designed specifically for studying uniformly stained insect chromosomes have been described by a number of investigators, including Smith (1943), Lewis and Riles (1960), French et al. (1962), Carlson and Gaulden (1964), Rai (1967), and Crozier (1968). The reader is also referred to two general compilations for other methods: Sharma and Sharma (1972) and Darlington and La Cour (1975).

For rapid routine analysis of chromosomes, the technique of choice is the squash or the smear. As pointed out by Sharma and Sharma in 1972, the terms "squash" and "smear" are often used interchangeably, but they actually involve different methods of tissue preparation even though both result ideally in a single layer of fixed-stained cells in which chromosomes can be viewed without overlying material. The literal meanings of the terms convey the correct impressions. For a squash preparation, fixed-stained cells are squashed or flattened by force between the slide and the cover glass. Preparation of a smear entails the spreading of cells onto the slide, either as fresh or fixed material pressed out of a tissue or as fixed single cells in suspension. The selection of one or the other technique depends on the type of tissue used. The squash is ideal for a tissue, such as an embryo, whose cells are loosely held together in situ. The smear is best used for tissues, such as the testes, which contain connective tissue but from which can be squeezed the cells of interest.

The squash is the most widely used technique, and was originally developed by Stevens (1908) for temporary preparations of fly chromosomes: 45% aceto-carmine (formulated by Schneider in 1880) served as the simultaneous fixer-stain and vaseline as the cover glass sealer. Improvements were later made, beginning with Belling (1926) and McClintock (1929), who used cells prefixed in acetic-alcohol (to improve chromosome staining) and who devised ways to make the slides permanent. The technique we currently use for analyzing the effects of chemical mutagens on the chromosomes of grasshopper embryos (*Chortophaga viridifasciata* de Geer) can serve as a guide for use with other cells; it exemplifies a speedy procedure for making semipermanent slides and is a combination of the thick and thin squash methods previously described by Carlson and Gaulden (1964). We use it for analyzing late anaphase neuroblasts for acentric chromosome fragments (Gaulden and Read 1978; Liang and Gaulden 1980).

1. Separate 14-day old embryos from their yolk and membranes with small knives (surgical needles honed to desired shape). Treat with mutagen. Fix in 3 parts freshly made absolute methyl alcohol: 1 part glacial acetic acid for 1 minute. Methyl, unlike ethyl, alcohol does not harden embryos; therefore they can be left in fixative overnight if necessary.

2. Stain 5 minutes in aceto-carmine (prepared in open container so that the final solution is only about 30% acetic acid; stronger concentrations dissociate the embryo).

3. Place one embryo in the center of a slide with a minimum of stain; if the embryo is curled, it should be straightened with a needle. Add 2 drops of a modification of Zirkle's (1937) mounting medium (3 parts aceto-carmine, 3 parts distilled water, 2 parts white Karo syrup).

4. Lower cover glass (22 mm^2) over embryo so as to exclude air bubbles.

5. Place under a "squasher" (Fig. 5.2) and apply enough pressure to spread and flatten the cells to a 1-cell thickness. Practice quickly establishes the amount of pressure that will flatten but not break the cells. Blot excess mounting medium from the edges of the cover glass, taking care to insure that the cover glass does not move sideways, as this will break and roll cells.

6. Set aside at room temperature for 10 minutes to allow water and acetic acid to evaporate at the edge of the cover glass so that the syrup forms a seal. The slides can now be examined with a high dry objective. (We routinely use a 25 X objective with 12.5 X oculars for scanning a 40 X or 100 X objective for scoring aberrations in neuroblasts.) Let stand overnight before using with an oil immersion objective.

The most time-consuming part of this method is the manual separation of each embryo from its membranes. Even so, dissection of 10 embryos (used to test one concentration of an agent) can be completed in 20 minutes. This plus fixing, staining, and squashing equals a total time of 1 hr. to prepare 10 slides, exclusive of embryo exposure to the test agent.

The semi-permanent squash method is not only a fast way to prepare chromosomes but prevents loss of cells in processing. If the slides are kept in an air-conditioned room, i.e., with no extreme changes of temperature, they will remain in good shape for several years. At high humidity, the Karo residue will absorb water and become soft. It is easy to make the slides permanent by brushing the cover glass edge with a mounting medium such as balsam, Permount, or clear finger nail polish.

If large numbers of metaphases are required, the embryos are placed in 0.05% Colcemid or 0.001% colchicine for 2 hr. at 38°C and then in 0.8% sodium citrate for 10 minutes at room temperature prior to fixation in 3:1 methyl alcohol:acetic acid for 1–1.5 hr. Embryos are stained in aceto-carmine for 2 minutes and then squashed as described above. This method separates the sister chromatids of metaphase chromosomes so that chromatid aberrations can be easily detected.

C-mitotic agents may be used to block metaphases in nymph or adult insect gonads by injection into the hemocoel (John and Lewis 1957).

Time of exposure to a C-mitotic agent should be kept as short as commensurate with obtaining a reasonable number of C-metaphases, because the chromosomes in blocked cells continue to contract. Longer exposure times will increase the number of cells with shortened chromosomes, which may reduce the ease with which chromosome aberrations can be identified. Ford and Woollam (1963) found in colchicine-treated mouse cells that the length of the longest chromosome varied among cells from 3 to 11 μm and that of the shortest form 0.75 to 3 μm. Although part of the variation may be related to differences in cell size, our experience suggests that much of it is caused by varying times in colchicine (Gaulden and Carlson 1951). In this connection, it is advisable when testing an unknown mutagen to examine first some cells which have not been exposed to Colcemid or colchicine to determine whether the test agent by itself has any effect on spindle function.

If it is necessary to remove the cover glass for further treatment, such as for

Insect Cells for Testing Clastogenic Agents   121

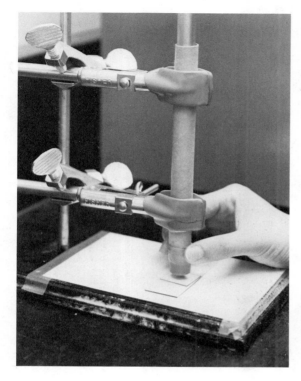

Figure 5.2 A squasher used to make a flattened monolayer preparation of the cells in a grasshopper embryo.

chromosome banding or autoradiography, a modification of the squash method is used as follows (Thornton and Gaulden 1971). The fixed embryo is squashed in aceto-carmine with filter paper over the cover glass to blot the stain extruded at the edges. The slide is immediately pressed onto a flat piece of dry ice and held in position while freezing for 30–60 sec. (Conger and Fairchild 1953). The cover glass is popped off by sliding a sharp scapel under one corner. The slide is removed from the dry ice and placed immediately without thawing in 95% alcohol, then in absolute alcohol followed by xylene (5 minutes in each). A cover glass is applied with Permount. To insure that the cells stick to the slide, siliconized cover glasses may be used. It is also advisable to use slides subbed (Boyd 1955) as follows. Prepare a solution of 0.5 l cold distilled water, 0.25 g potassium sulfate, 2.5 g gelatin (add last slowly while stirring to prevent clumping). Heat to 60° C with continuous stirring. Cool to room temperature and remove any surface bubbles with a pipette; dip slides and set on end on paper towel to dry. Subbed slides can be stored in dust-free boxes for future use.

Phase contrast optics enhance the microscopic image of chromosomes in aceto-carmine squashes or smears. Because small cytoplasmic inclusions may be mistaken for minute fragments, cells containing them should also be viewed with bright-field illumination so that such elements can be unequivocally identified as chromatin or not (Fig. 5.3A).

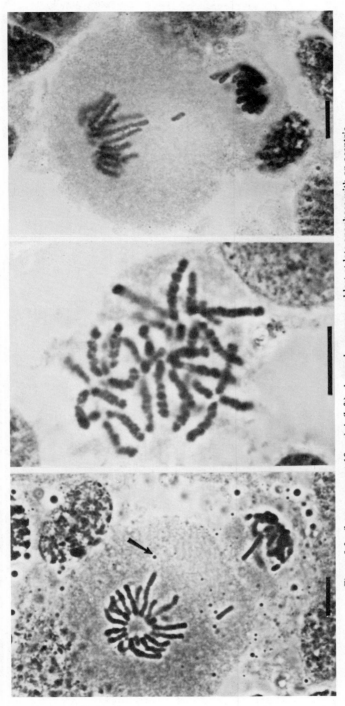

Figure 5.3 (bars represent 10 μm) A (left): A grasshopper neuroblast at late anaphase with an acentric fragment (induced at an earlier stage by X-rays). Stained with acetocarmine and photographed with phase microscopy. The arrow indicates a nonchromatin element of unknown origin which is not visible with bright-field illumination. The neuroblast normally divides unequally. B (middle): Spontaneous banding of grasshopper chromosomes in a prometaphase neuroblast which was only treated with 0.05% Colcemid (2 hr. 37°C) and hypotonic medium (0.8% sodium citrate 10 min., 25°C). From a conventional aceto-carmine squash. C (right): A grasshopper neuroblast at very early telophase with an acentric fragment induced by 1μM of 4NQO (1hr. in vitro exposure + 3hr. in medium without mutagen, 38°C); acetocarmine squash.

Banding

Chromosome banding offers an investigator an invaluable means for localizing chromosome breaks when data on uniformly stained chromosomes suggest that a mutagen is acting primarily at specific regions.

The longitudinal differentiation of chromosomes into distinctive bands (chromomeres?) was first described for the polytene chromosomes of *Drosophila melanogaster* by Painter in 1933. The use of bands for identifying chromosomes and for localizing genes and break points in this organism began with the publication by Bridges in 1935 of detailed band maps for each chromosome. It was not until almost 35 years later that banding of mitotic chromosomes was achieved (Caspersson et al. 1968); with the publication of a Q-banded human karyotype in 1970 by Caspersson et al. began a new and productive era in human cytogenetics. The resolution of the banding patterns and the number of bands (greater than 5,000 in four chromosomes: Lefevre 1976) observed in the polytene chromosomes of *Drosophila* have yet to be obtained in mitotic cells of this or any other organism. However, the method used for high-resolution G-banding of human prometaphase chromosomes by Francke and Oliver (1978) provides the largest number of bands so far observed in non-polytene chromosomes (721 euchromatic plus 62 variable heterochromatic bands in the haploid complement) and holds promise for investigators of other cells.

Unlike the polytene chromosomes of Diptera, which band readily in ordinary aceto-carmine squashes and whose details are minimally improved by electron microscopy (Sorsa and Saura 1980), mitotic and meiotic chromosome preparations must be pretreated and stained in specific ways to yield good bands. The various banding techniques developed for human chromosomes have been applied to the chromosomes of several orders of insects, some of which are presented in Table 5.1.

The most commonly used techniques for banding insect chromosomes involve minor modifications of those devised by the following: Sumner et al. (1971), Arrighi and Hsu (1971), and Sumner (1972), for C-bands; Caspersson et al. (1968) for Q-bands; Seabright (1971) and Webb's (1976) modification for G-bands; Latt (1973) for H-bands. For making smears of a suspension of cells, the air-drying methods of Evans et al. (1964) and Crozier (1968) seem to be preferred.

The results in most instances are cytologically disappointing and have mainly been useful for distinguishing and interrelating genera, species, and races of a given type of insect. They can be summarized as follows. The most frequently reported and the most prominent bands in mitotic cells are those obtained with C-banding methods. The usual finding is illustrated by *Drosophila* in which C-bands occur at the centromeres and in very large blocks in one or more chromosomes, especially the Y and the large heterochromatic chromosomes of some species (Wheeler and Altenburg 1977; Lemeunier et al. 1978; Halfer et al. 1980). One or more whole chromosomes maybe uniformly stained. Q- and H-band patterns are reported to be different (Holmquist 1975). Some investigators, however, report that Q- and H-bands more or less coincide with the segments or whole chromosomes that are stained by C-banding (Vosa 1970; Dolfini 1974 a,b; Gatti et al. 1976). All of these results are interpreted as showing the importance of heterochromatin in the evolution of *Drosophila*; the same is true for the Acridoid grasshoppers (King and John 1980).

In some organisms, such as grasshoppers, the C-bands may occur not only at the centromeres but also on one or both sides of the centromeres and at the telemeres. In addition there may be one to three well-defined interstitial C-bands in some chromosomes of the complement (e.g., Shaw et al. 1976). Lester et al. (1979) have reported no success with C-banding of *Musca* chromosomes, but they obtained bands on some chromosomes with the trypsin-Giemsa technique.

Only a few investigators have reported G-banding of insect chromosomes, i.e., the bands typically obtained with trypsin pretreatment of mammalian chromosomes. This procedure produced good reproducible bands in the B chromosomes of the locust *Chortoicetes* (Webb and Neuhaus 1979), and in some of the chromosomes of the housefly *Musca* (Lester et al. 1979), the house cricket *Acheta* (Ali 1979), and *Triatoma* (Maudlin 1974). In the grasshopper *Warramaba*, Webb et al. (1978) showed that the G-bands coincide with the few small nonstaining gaps observed after C-banding; in *Schistocerca* G- and C-banding methods stain the same bands (Fox et al. 1973).

Some of the most striking banding patterns (in both number and resolution) in insects are those observed in Australian ants by Imai et al. (1977), especially in cells with extended chromosomes. They are referred to by the authors as "C-bands" even though they were obtained with *no* pretreatment: squashes stained with Giemsa produced dark staining pericentromeric bands in some chromosomes and slightly less-dark to light staining bands in all chromosome arms. As there is nothing unusual about the squashing technique (Imai and Kubota 1972), these bands might best be labeled "spontaneous." An occasional spontaneously banded cell may be obtained in other organisms as, for example, in the neuroblasts of *Chortophaga* (Fig. 5.3B).

Banding of meiotic chromosomes in a few insects has been attempted, but the results are not significantly different from those obtained with mitotic cells (e.g., Fox et al. 1973; Klasterska et al. 1974; Fontana and Vickery 1975; Das et al. 1979).

The technique of choice for localizing break points in chromosomes is the one that resolves the most bands. In mammalian cells, pretreatment with trypsin followed by Giemsa staining, i.e., G-banding, is usually preferred. Thus it is surprising that many of the investigators who report banding of insect chromosomes do not mention attempting to get G-bands; perhaps they were not sure that a lack of success represented the real situation. Is the predominant large-block band pattern revealed thus far in insects an inherent property of these chromosomes? Insect chromosomes are not unique in being relatively refractory to G-banding: plant and amphibian chromosomes share this characteristic. In fact, the first report on Q-banding by Caspersson et al. (1968) revealed that fluorescent bands were limited to one or several per chromosome in two plants (*Vicia* and *Trillium*), whereas a number of bands was observed in Chinese hamster chromosomes. Can the difficulties encountered in banding insect chromosomes be related to the degree of coiling or condensation, to large amounts of DNA per genome, to the relatively large amount of repetitive DNA or heterochromatin evidently present in many insects, or to DNA base pair composition? Greilhuber (1977) and King and John (1980) have reviewed some of the evidence which bears on these points, and no clear-cut answers appear to be immediately forthcoming. The latter authors conclude that "C-banding does not appear to define anything within the genome in a unique fashion." This could also be said to apply to G-banding. The situation appears to be a little better with respect to Q- and H-banding in that they seem to correlate to some extent with the base composition of DNA (Holmquist 1975; Gatti et al. 1976).

It is obvious that the data available at present do not permit unequivocal delineation of the nature of the chemistry and organization of chromatin and the changes occurring in them that underlie the G- and C-banding procedures. At least some portions of the gap in our knowledge between the nucleosome and the condensed metaphase chromosome will probably have to be filled in before the relation of banding to chromosome structure and the response of chromosomes to mutagens can be fully understood. Such deficiencies, which apply to all cell types, should not however deter us from continuing to try to obtain better banding in insect chromosomes. It is possible that some technical breakthrough will come to our rescue. This approach is not farfetched, because the detail observed in the first published human banded karyotype (Caspersson et al. 1968) is minimal compared to that attainable with more recent techniques (Francke and Oliver 1978). A review of the literature gives one the impression that banding of insect chromosomes is in its infancy compared to that of human chromosomes.

If insect chromosomes can spontaneously show a large number of bands (Fig. 5.3B and Imai et al. 1977), it is tempting to speculate that the cause of the paucity of pretreatment-induced G-bands may be technical in nature. What aspects of the banding techniques might be altered in attempts at increasing the number of observable bands and improving their resolution in insect cells? Greilhuber (1977) notes two relevant points: (1) the number of G-bands visible in human chromosomes at metaphase is much lower than that at prophase, and (2) if at metaphase, human chromosome 1 were as contracted as is the *Vicia* M chromosome, the former would be reduced to 1/10 its normal size. He reasons, therefore, that the lack of G-bands in *Vicia,* and other non-G-banding chromosomes, may be caused by a high degree of condensation that obliterates the banding. If correct, the use of prophase insect cells should yield G-banding; obtaining well-spread chromosomes will present a problem in those species with a number of long chromosomes, but the idea is well worth pursuing.

That C-bands are the most frequently reported type in insects might be due to a technical response rather than to an inherent characteristic of the chromosomes. As Webb (1976) has pointed out, C-banding requires much more severe procedures than does G-banding. Thus the milder G-banding procedures may not be as effective. Webb also calls attention to the fact that C-banding techniques sometimes produce a mixture of G- and C-type bands in grasshopper cells. How cells on the same slide differ so as to give different results remains to be determined.

Many investigators of insect chromosomes do not comment on the aging of slides before pretreatment for banding. Webb (1976) reports that aging is detrimental, but details of the aging method are not given. In view of the fact that Francke and Oliver (1978) got optimal high-resolution banding of human chromosomes with controlled, specific aging conditions (95°C for 10–20 min.), attention to this detail might prove productive for insect chromosomes.

Dissection of *Drosophila* salivary glands in saline rather than in fixative has been reported by Beerman (1962) to alter band resolution of the polytene chromosomes. It is possible that the solutions to which insect mitotic cells are subjected before fixation, including the hypotonic ones used to swell cells, may alter the response of the chromosomes to the banding techniques.

For permanent mounting of banded chromosomes stained with Giemsa, Rønne et al. (1977) recommend using the epoxy resin medium formulated by Spurr (1969). It is a low viscosity medium developed for embedding specimens for electron microscope sectioning and does not contain an organic solvent characteristic of the usual microscopic mounting media. It does not, therefore, result in loss of band resolution with time, i.e., a blurring of band edges, which is evidently caused by diffusion of the stain. If slides are stored in the dark, fading of bands induced by light is prevented.

Until the question can be answered as to whether the current state of the art of banding insect chromosomes represents the ultimate "truth" or is subject to improvement by technical means, we urge investigators to report fully the details of their banding methods. This may help speed resolution of the issue. We also suggest that for clarity the terms "G-banding" and "C-banding" be limited to those bands obtained by the pretreatment methods used to define these types of bands in human chromosomes.

Differential Chromatid Staining

The differential staining of chromatids after 5-bromodeoxyuridine (BrdU) incorporation, widely used to detect mutagen-induced sister chromatid exchanges (SCE) in a variety of organisms, has not yet been used to any extent with insect cells. One question not resolved with data on other cell types is whether euchromatin and heterochromatin show consistent differences in SCE frequency. Dolfini (1978) has examined the frequency and localization of spontaneous SCE with two established cell lines of *Drosophila melanogaster*, an organism in which the two types of chromatin have been well delineated both cytogeneticaly and genetically. She found that the number of SCE varied not only with respect to the type of chromatin but also with sex: no SCE were observed in the heterochromatic regions of chromosomes, and in euchromatin the frequency was much higher in XY than in XX cells. Further, the number of SCE in the X chromosomes was significantly greater than in the autosomes. She also found that the frequency of SCE/cell increased with increasing doses of BrdU, as has been observed in other cells. These observations form an interesting data base against which the effects of mutagens on SCE frequency can be tested in this and other insects, especially in view of the relatively large proportions of heterochromatin in the genomes of some of them.

Differential staining of sister chromatids by BrdU offers an excellent method with which to examine the chiasmatype theory of the relation of chiasmata to crossing-over in meiotic bivalents, but the results with mammals have not met expectations (Allen and Latt 1976). Recently the locust has been used to good advantage to address this question: the large monochiasmate bivalents were examined after implantation of BrdU tablets in the abdomens of males (Tease 1978; Tease and Jones 1978, 1979; Jones and Tease 1979). Excellent evidence is presented supporting the position that a chiasma results directly from recombination between non-sister chromatids. These investigators also examined SCE in bivalents and conclude that the special meiotic circumstances that promote exchange between non-sister chromatids do not cause an increased frequency of SCE.

Use of Insect Cells for Testing Chemicals

Chemosterilants were among the chemicals first tested on insect cells. Before 1958, the only method of inducing sterility in male insects was by ionizing radiation. This procedure, however, caused serious damage to the somatic tissue of many males and hampered their viability and sexual competitiveness. It was hoped that chemicals could sterilize without causing the same unwanted side effects. Numerous chemosterilants, mostly alkylating agents, were identified within a few years, among which apholate, tepa, metepa, and hempa (see Table 5.2) received the most attention. Labrecque and Fye (1978) reviewed the effects of these chemosterilants on insect cells and pointed out that they were not only mutagenic but also clastogenic. Because many types of chromosome aberrations, e.g., dicentrics, anaphase bridges, acentric fragments, rings, breaks, and stickiness, were observed in both mitotic and meiotic cells, the use of these compounds as field insect chemosterilants was regarded as unacceptable.

Spermatocytes of various species of grasshoppers have been the prevalent type of cell with which the clastogenic properties of chemicals have been examined. In those studies cited in Table 5.2, the in vivo route of exposure was abdominal injection in all but one (Parida 1972), in which the chemical was added to food given starved adults. A variety of chemicals has been shown to induce chromosome aberrations in spermatocytes, including alkaline sodium hydroxide and acidic hydrochloride solution; metal compounds, such as calcium chloride and methyl mercury hydroxide; the chelating agent, versene; the sedative, barbital; methanol; the antimetabolite, actinomycin D; the antibiotics, penicillin and mitomycin C; and others.

The large neuroblasts of insect embryos have not been used much for testing clastogens. Neuroblasts in larval ganglia of *Drosophila melanogaster* were used by Gatti et al. (1975) to study the in vivo clastogenic action of MMS. Only chromatid type aberrations were induced by MMS, and the frequency observed in females was approximately 3 times that in males. Data obtained with X-rays permitted a good comparison of the effects of radiation with those of the chemical. MMS was found to have a greater specificity than X-rays for inducing breaks in the pericentromeric heterochromatin of the X chromosomes and the autosomes, whereas it was far less efficient than X-rays in breaking the heterochromatic Y chromosome. The types of chromatid exchanges induced by these two agents were also found to be different: symmetrical exchanges were most frequent after MMS treatment and asymmetrical ones were most frequent after X-ray treatment.

The neuroblast of the grasshopper embryo is being used in our laboratory and is proving to be a sensitive, rapid, and inexpensive in vitro system for detecting clastogens. It has a very short cell cycle: 4 hr. at 38°C. Direct-acting chemicals, such as 4NQO, MNNG, Adriamycin, and bleomycin, induce acentric chromosome fragments that can be easily scored at late anaphase and early telophase in these cells (Fig. 5.3C). For example, the fragment frequency induced by 1 hr. exposure (38°C) of embryos to 1μM 4NQO is equivalent to that induced by 4 R of X-rays (Gaulden and Read 1978). Preliminary data on indirect-acting chemicals, such as cyclophosphamide, reveal that acentric fragments are induced in grasshopper neuroblasts only when rat liver microsomes (S9 mix) or rat hepatocytes are present for metabolizing the compound to active intermediates.

Summary and Conclusions

Although no single type of cell from an insect, or any other organism for that matter, meets all the criteria of an "ideal cell" for clastogen studies, the cells and chromosomes of many insects have several advantageous characteristics. A number of them have chromosomes that are large, few in number, and morphologically distinguishable and that have very few if any heteromorphisms. The techniques for preparing the chromosomes for analysis are simple and fast. All types of chromosome aberrations, including the enigmatic gaps and breaks, can be readily identified. Large numbers of cells with short cell cycles are found in embryos. Several insects with defined genetics have chromosomes suitable for clastogen studies. At least one species, the grasshopper *Chortophaga*, has little or no spontaneous chromosome breakage (Gaulden and Read 1978), and this may be true for others. For exposure of tissue to mutagens either in vivo or via organ culture, an abundance of material can be obtained from insects maintained inexpensively in the laboratory. Lines of tissue culture cells from a few insects, notably *Drosophila* and mosquito, are available for in vitro studies, and they promise to be eventually as useful as those of mammalian origin have proved to be. The number and resolution of G-bands in insect chromosomes is less than desirable; further work is needed to determine whether this is a technical or inherent problem.

The insects, the most abundant of organisms, have so far been underutilized for studies of the clastogenic effects of chemical agents. One group of compounds that immediately comes to mind and that has had tremendous impact on many facets of society is pesticides. Data obtained in the last few years confirm some of the fears that were expressed earlier about the hazards of these and other chemicals in the environment (e.g., Epstein and Legator 1971). We believe that insect cells have considerable potential for the screening of a variety of chemical agents to determine if they affect chromosomes, and that the data so obtained will be helpful in evaluating hazards to man.

Acknowledgments

This chapter is dedicated to Dr. J. Gordon Carlson, who introduced us directly (MEG) and indirectly (JCL) to insect chromosomes.

We thank Mss. Catherine Smith, Dianna Hallford, and Judy Alexander and Mr. Jerry Cheek for assistance in preparing the manuscript and Ms. Elizabeth S. Von Halle of the Environmental Mutagen Information Center (Oak Ridge National Laboratory) for help with the literature search.

Literature Cited

Ali, S. 1979. Chromosomes of the house cricket *Acheta domesticus* (Gryllidae: Orthoptera). Ind. J. Exp. Biol. 17:1038-40.
Allen, J. W.; Latt, S. A. 1976. In vivo BrdU-33258 Hoechst analysis of DNA replication kinetics and sister chromatid exchange formation in mouse somatic and meiotic chromosomes. Chromosoma 58:325-40.
Arrighi, F. E.; Hsu, T. C. 1971. Localization of heterochromatin in human chromosomes. Cytogenetics 10:81-86.
Ashburner, M.; Thompson, J. N., Jr. 1978. The laboratory culture of *Drosophila*. In Ashburner, M; Wright, T. R. F., eds. The Genetics and Biology of Drosophila. Vol. 2A. New York:Academic Press; 1-109.
Baim, A. S.; Mukherjee, A. B. 1978. The cell cycle of an established cell line of the mosquito *Aedes aegypti*. Can. J. Genet. Cytol. 20:373-76.
Baker, R. H.; Sakai, R. K.; Perveen, A.; Raana, K. 1980. Isolation of a homozygous translocation in *Anopheles culicifacies*. J. Heredity 71:25-28.
Barr, A. R. 1975 *Culex*. In King, R. C. ed. Handbook of Genetics. Vol. 3. New York: Plenum Press; 347-75.

Bedo, D. G. 1980. C, Q and H-banding in the analysis of Y chromosome rearrangements in *Lucilia cuprina* (Wiedeman) (Diptera: Calliphoridae). Chromosoma 77:299-308.
Beermann, W. 1962. Riesenchromosomen. Protoplasmatologia VI/C. Wien: Springer; 1-161.
Belling, J. 1926. The iron-acetocarmine method of fixing and staining chromosomes. Biol. Bull. 50:160-62.
Bianchi, N. O.; Sweet, B. H. Ayres, J. P. 1972. Chromosome characterization of three cell lines derived from *Aedes albopictus* (Skuse) and *Aedes aegypti* (L.). Proc. Soc. Exp. Biol. Med. 140: 130-34.
Bonaccorsi, S.; Santini, G. Gatti, M.; Pimpinelli, S.; Coluzzi, M. 1980. Intraspecific polymorphism of sex chromosome heterochromatin in two species of the *Anopheles gambiae* complex. Chromosoma 76:57-64.
Boveri, T. 1902. Über mehrpolige Mitosen als Mittel zur Analyse des Zellkerns. Verhandl. d. phys.-med. Gesellsch. zu Wurzb. 35(n.F.):67-90.
Boyd, G. A. 1955. Autoradiography in Biology and Medicine. New York: Academic Press; 209.
Boyes, J. W.; Wilkes, A. 1953. Somatic chromosomes of higher diptera. Can. J. Zool. 31:125-65.
Boyes, J. W.; Naylor, A. F. 1962. Somatic chromosomes of higher diptera. VI. Allosome-autosome length relations in *Musca domestica* L. Can. J. Zool. 40:777-84.
Bregman, A. A. 1973. Lateral subunits in C-bands of unreplicated chromosomes of *Dissosteira carolina* (L.) (Orthoptera: Acrididae). Can. J. Genet. Cytol. 15:757-61.
Bridges, C. B. 1916. Non-disjunction as proof of the chromosome theory of heredity. Genetics 1:1-52, 107-63.
Brooks, M. A.; Kurtti, T. J. 1971. Insect cell and tissue culture. Ann. Rev. Entomol. 16:27-52.
Brown, S. W. 1965. Chromosomal survey of the armored and palm scale insects (Coccoidea: Diaspididae and Phoenicococcidae). Hilgardia 36:189-294.
Brown, A. K.; Wilmore, P. J. 1974. Location of repetitious DNA in the chromosomes of the desert locust (*Schistocerca gregaria*). Chromosoma 47:379-83.
Cardoso, H.; Saez, F. A.; Brum-Zorrilla, N. 1974. Location, structure, and behavior of C-heterochromatin during meiosis in *Dichroplus silveiraguidoi* (*Acrididae-Orthoptera*). Chromosoma 48:51-64.
Carlson, J. G.; Gaulden, M. E. 1964. Grasshopper neuroblast techniques. Methods in Cell Physiol. 1:229-76.
Carothers, E. E. 1913. The Mendelian ratio in relation to certain Orthopteran chromosomes. J. Morph. 24:487-511.
Caspari, E. W.; Gottlieb, F. J. 1975. The Mediterranean meal moth, *Ephestia kühniella*. In King, R. C., ed. Handbook of Genetics. Vol. 3. New York: Plenum Press; 125-47.
Caspersson, T.; Farber, S.; Foley, G. E.; Kudynowski, J.; Modest, E. J.; Simonsson, E.; Wagh, U.; Zech, L. 1968. Chemical differentiation along metaphase chromosomes. Exp. Cell Res. 49:219-22.
Caspersson, T.; Zech, L.; Johansson, C.; Modest, E. J. 1970. Identification of human chromosomes by DNA-binding fluorescent agents. Chromosoma 30:215-27.
Cassidy, J. D. 1975. The parasitoid wasps, *Habrobracon* and *Mormoniella*. In King, R. C., ed. Handbook of Genetics. Vol. 3. New York: Plenum Press; 173-203.
Clayton, F. E.; Wheeler, M. R. 1975. A catalog of *Drosophila* metaphase chromosome configurations. In King, R. C., ed. Handbook of Genetics. Vol. 3. New York: Plenum Press; 471-512.
Cochran, D. G.; Ross, M. H. 1967. Preliminary studies of the chromosomes of twelve cockroach species (Blattaria: Blattidae, Blattellidae, Blaberidae). Ann. Entomol. Soc. Am. 60:1265-72.
Cochran, D. G.; Ross, M. H. 1969. Chromosome identification in the German cockroach. J. Hered. 60:87-92.
Cohen, S.; Roth, L. M. 1970. Chromosome numbers of the Blattaria. Ann. Entolmol. Soc. Am. 63:1520-47.
Coluzzi, M; Kitzmiller, J. B. 1975. Anopheline mosquitoes. In King, R. C., ed. Handbook of Genetics. Vol. 3. New York: Plenum Press; 285-309.
Conger, A; Fairchild, L. 1953. A quick-freeze method for making smear slides permanent. Stain Technol. 28:281-83.
Cross, D. P.; Shellenbarger, D. L. 1979. The dynamics of *Drosophila melanogaster* spermatogenesis in in vitro cultures. J. Embryol. Exp. Morph. 53:345-51.

Crozier, R. H. 1968. An acetic acid dissociation, air-drying technique for insect chromosomes, with aceto-lactic orcein staining. Stain Technol. 43:171-73.
Darlington, C. D. 1937. Recent Advances in Cytology. 2nd ed. Philadelphia: Blakiston Co.
Darlington, C. D.; La Cour, L. F. 1975. The Handling of Chromosomes. 6th ed. New York: John Wiley & Sons.
Das, R. K.; Mukherjee, S. 1977. Barbital induced meiotic chromosome aberrations in grasshopper. Geobios. 4:42-43.
Das, B. C.; Raman R.; Sharma, T. 1979. Chromosome condensation and Hoechst 33258 fluorescence in meiotic chromosomes of the grasshopper *Spathosternum prasiniferum* (Walker). Chromosoma 70:251-58.
Day, M. F.; Grace, T. D. C. 1959. Culture of insect tissues. Ann. Rev. Entomol. 4:17-38.
Debec, A. 1978. Haploid cell cultures of *Drosophila melanogaster*. Nature 274:255-56.
DeMarco, A.; Cozzi, R. 1980. Chromosomal aberrations induced by caffeine in somatic ganglia of *Drosophila melanogaster*. Mutation Res. 69:55-69.
Dolfini, S. 1971. Karyotype polymorphism in a cell population of *Drosophila melanogaster* cultured in vitro. Chromosoma 33:196-208.
Dolfini, S. F. 1974a. The distribution of repetitive DNA in the chromosomes of cultured cells of *Drosophila melanogaster*. Chromosoma 44:383-91.
Dolfini, S. F. 1974b. Spontaneous chromosome rearrangements in an established cell line of *Drosophila melanogaster*. Chromosoma 47:253-61.
Dolfini, S. F. 1978. Sister chromatid exchanges in *Drosophila melanogaster* cell lines in vitro. Chromosoma 69:339-47.
Dolfini, S.; Courgeon, A. M.; Tiepolo, L. 1970. The cell cycle of an established line of *Drosophila melanogaster* cells in vitro. Experientia 26:1020-21.
Drets, M. E.; Stoll, M. 1974. C-banding and non-homologous associations in *Gryllus argentinus*. Chromosoma 48:367-90.
Echalier, G.; Ohanessian, A. 1969. Isolement, en cultures in vitro, de lignees cellulaires de *Drosophila melanogaster*. C.R. Acad. Sci. (Paris) 268:1771-73.
Ellison, J. R.; Barr, H. J. 1972. Quinacrine fluorescence of specific chromosome regions. Chromosoma 36:375-90.
Epstein, S. S.; Legator, M. S. 1971. The Mutagenicity of Pesticides. Cambridge, Mass.: MIT Press.
Evans, H. J. 1976. Cytological methods for detecting chemical mutagens. Chemical Mutagens 4:1-29.
Evans, E. P.; Breckon, G.; Ford, C. E. 1964. An air-drying method for meiotic preparations from mammalian testes. Cytogenet. 3:289-94.
Fedoroff, S. 1967. Proposed usage of animal tissue culture terms. Exp. Cell Res. 46:642-48.
Fontana, P. G.; Vickery, V. R. 1975. The B-chromosome system of *Tettigidae lateralis* (Say). II. New karyomorphs, patterns of pycnosity and Giemsa-banding. Chromosoma 50:371-91.
Ford, E. H. R.; Woollam, D. H. M. 1963. A study of the mitotic chromosomes of mice of the Strong A line. Exp. Cell Res. 32:320-26.
Fox, D. P.; Carter, K. C.; Hewitt, G. M. 1973. Giemsa banding and chiasma distribution in the desert locust. Heredity 31:272-82.
Francke, U.; Oliver, N. 1978. Quantitative analysis of high-resolution trypsin-Giemsa bands on human prometaphase chromosomes. Hum. Genet. 45:137-65.
French, W. L.; Baker, R. H.; Kitzmiller, J. B. 1962. Preparation of mosquito chromosomes. Mosquito News 22:377-83.
Frew, J. G. H. 1928. A technique for the cultivation of insect tissues. J. Exp. Biol. 6:1-11.
Gallagher, A.; Hewitt, G.; Gibson, I. 1973. Differential Giemsa staining of heterochromatic B-chromosomes in *Myrmeleotettix maculatus* (Thunb.) (*Orthoptera: Acrididae*). Chromosoma 40:167-72.
Gatti, M.; Pimpinelli, S.; De Marco, A.; Tanzarella, C. 1975. Chemical induction of chromosome aberrations in somatic cells of *Drosophila melanogaster*. Mutat. Res. 33:201-12.
Gatti, M.; Pimpinelli, S.; Santini, G. 1976. Characterization of *Drosophila* heterochromatin. I. Staining and decondensation with Hoechst 33258 and quinacrine. Chromosoma 57:351-75.

Gaulden, M. E.; Carlson, J. G. 1951. Cytological effects of colchicine on the grasshopper neuroblast in vitro with special reference to the origin of the spindle. Exp. Cell Res. 2:416-33.
Gaulden, M. E.; Read, C. B. 1978. Linear dose-response of acentric chromosome fragments down to 1 R of X-rays in grasshopper neuroblasts, a potential mutagen test system. Mutat. Res. 49:55-60.
Goldschmidt, R. 1915. Some experiments on spermatogenesis in vitro. Proc. Natl. Acad. Sci. USA 1:220-22.
Goncalves, L. S.; Stort, A. C. 1978. Honey bee improvement through behavioral genetics. Ann. Rev. Entomol. 23:197-213.
Grace, T. D. C. 1966. Establishment of a line of mosquito (*Aedes aegypti* L.) cells grown in vitro. Nature (London) 211:366-67.
Greilhuber, J. 1977. Why plant chromosomes do not show G-bands. Theor. Appl. Genet. 50:121-24.
Grover, K. K.; Pillai, M. K. K.; Dass, C. M. S. 1973. Cytogenetic basis of chemically induced sterility in *Culex pipiens fatigans* (Wiedemann). I. Chemosterilant-induced damage in the somatic chromosomes. Cytologia 38:21-28.
Hagele, K. 1979. Characterization of heterochromatin in *Schistocerca gregaria* by C- and N-banding methods. Chromosoma 70:239-50.
Halfer, C.; Privitera, E.; Barigozzi, C. 1980. A study of spontaneous chromosome variations in seven cell lines derived from *Drosophila melanogaster* stocks marked by translocations. Chromosoma 76:201-18.
Hanson, C. V. 1978. Mass isolation of metaphase chromosomes from *Drosophila melanogaster*. In Ashburner, M.; Wright, T. R. F., eds. The Genetics and Biology of Drosophila. Vol. 2A. New York: Academic Press; 140-45.
Holmquist, G. 1975. Hoechst 33285 fluorescent staining of *Drosophila* chromosomes. Chromosoma 49:333-56.
Horikawa, M.; Fox, A. S. 1964. Culture of embryonic cells of *Drosophila melanogaster* in vitro. Science 145:1437-39.
Imai, H. T.; Crozier, R. H.; Taylor, R. W. 1977. Karyotype evolution in Australian ants. Chromosoma 59:341-93.
Imai, H. T.; Kubota, M. 1972. Karyological studies of Japanese ants (*Hymenoptera, Formicidae*). III. Karyotypes of nine species in *Ponerinae, Formicinae,* and *Myrimicinae*. Chromosoma 37:193-200.
Jain, H. K.; Singh, U. 1967. Actinomycin D induced chromosome breakage and suppression of meiosis in the locust *Schistocerca gregaria*. Chromosoma 21:463-71.
Janssens, F. A. 1909. La theorie de la chiasmatypie. Nouvelle interprétation des cinèses de maturation. Cellule 25:389-411.
John, B.; Lewis, K. R. 1957. Studies on *Periplaneta americana*. I. Experimental analysis of male meiosis. Heredity 11:1-22.
Johnson, C. D.; Ray, D. T. 1972. Chromosome identification in *Mormoniella*. J. Hered. 63:217-18.
Jones, G. H.; Tease, C. 1979. Analysis of exchanges in differentially stained meiotic chromosomes of *Locusta migratoria* after BrdU-substitution and FPG staining. III. A test for chromatid interference. Chromosoma 73:85-91.
Kakpakov, V. T.; Gvozdev, V. A.; Platova, T. P.; Polukarova, L. G. 1969. Establishment in vitro of embryonic cell lines of *Drosophila melanogaster*. Soviet Genetics 5:1647-55.
Kaufmann, B. P. 1934. Somatic mitoses of *Drosophila melanogaster*. J. Morphol. 56:125-53.
King, R. C., ed. 1975. Handbook of Genetics. Vol. 3. Invertebrates of Genetic Interest. New York: Plenum Press.
King, M.; John, B. 1980. Regularities and restrictions governing C-band variation in Acridoid grasshoppers. Chromosoma 76:123-150.
Kitzmiller, J. B. 1976. Genetics, cytology, and evolution of mosquitos. Adv. Genet. 18:315-433.
Klasterska, I.; Natarajan, A. T.; Ramel, C. 1974. Heterochromatin distribution and chiasma localization in the grasshopper *Bryodema tuberculata* (Fabr.) (*Acrididae*). Chromosoma 44:393-404.

Klasterska, I.; Ramel, C. 1978. The effect of methyl mercury hydroxide on meiotic chromosomes of the grasshopper *Stethophyma grossum*. Hereditas 88:255-62.
Kurtti, T. J. 1976. Phenotypic variations of cell lines from cockroach embryos. In Maramorosh, K., ed. Invertebrate Tissue Culture. New York: Academic Press; 39-56.
Labrecque, G. C.; Fye, R. L. 1978. Cytogenetic and other effects of the chemosterilants tepa, hetepa, apholate, and hempa in insects (a review). Mutat. Res. 47:99-113.
Latt, S. A. 1973. Microfluorometric detection of DNA replication in human metaphase chromosomes. Proc. Natl. Acad. Sci. (USA) 70:3395-99.
Lefevre, Jr., G. 1976. A photographic representation and interpretation of the polytene chromosomes of *Drosophila melanogaster* salivary glands. In Ashburner, M.; Novitski, E., eds. The Genetics and Biology of Drosophila. Vol. 1A. New York: Academic Press; 31-66.
Lemeunier, F.; Dutrillaux, B.; Ashburner, M. 1978. Relationships within the *Melanogaster* subgroup species of the genus *Drosophila (Sophophora)*. III. The mitotic chromosomes and quinacrine fluorescent patterns of the polytene chromosomes. Chromosoma 69:349-61.
Lesley, M. M.; Frost, H. B. 1927. Mendelian inheritance of chromosome shape in Matthiola. Genetics 12:449-60.
Lester, D. S.; Crozier, R. H.; Shipp, E. 1979. G-banding patterns of the housefly, *Musca domestica*, autosomes and sex chromosomes. Experientia 35:174-75.
Lewis, E. B.; Riles, L. S. 1960. A new method of preparing larval ganglion chromosomes. Drosophila Information Service 34:118-19.
Liang, J. C.; Gaulden, M. E. 1980. Neuroblasts of the grasshopper embryo as a new mutagen test system. Environ. Mutagenesis 2:275.
Lindsley, D. L.; Grell, E. H. 1968. Genetic Variations of Drosophila Melanogaster. Carnegie Inst. Wash. Publ. No. 627.
Lokki, J.; Saura, A. 1980. Polyploidy in insect evolution. In Lewis, W. H., ed. Polyploidy. Biological Relevance. New York: Plenum Press; 277-312.
Maeki, K. 1980. The kinetochore of the Lepidoptera. I. Chromosomal features and behavior in mitotic and meiotic-I cells. Proc. Japan Acad. 56(B):152-56.
Manna, G. K.; Mukherjee, P. K. 1966. Spermatocyte chromosome aberrations in two species of grasshoppers at two different ionic activities. Nucleus (Calcutta) 9:119-31.
Marks, E. P. 1980. Insect tissue culture: an overview, 1971-1978. Ann. Rev. Entomol. 25:73-101.
Martin, P.; Schneider, I. 1978. *Drosophila* organ culture. In Ashburner, M; Wright T.R.F., eds. The Genetics and Biology of Drosophila. Vol. 2A. New York: Academic Press; 219-59.
Maudlin, I. 1974. Giemsa banding of metaphase chromosomes in triatomine bugs. Nature 252:392-93.
McClintock, B. 1929. A method for making aceto-carmine smears permanent. Stain Technol. 4:53-56.
McClung, C. E. 1902. The accessory chromosome — sex determinant? Biol. Bull. 3:43-84.
Mezzanotte, R.; Ferrucci, L.; Marchi, A. 1979a. Y chromosome in the sibling species *Anopheles atroparvus* (van Thiel, 1927) and *A. labranchiae* (Falleroni, 1926) (Diptera: Culicidae): differential behavior of the short arm after acid-alkaline treatment and Coriphosphine-O staining. Experientia 35:312-13.
Mezzanotte, R.; Ferrucci, L. Marchi, A. 1979b. Light-induced banding (LIB) in metaphase chromosomes of Culiseta longiareolata (Diptera: Culicidae). Genetica 51:149-52.
Milani, R. 1967. The genetics of *Musca domestica* and of other Muscoid flies. In Wright, J. W.; Pal, R., eds. Genetics of Insect Vectors of Disease. New York: Elsevier Publ. Co.; 315-69.
Milani, R. 1975. The house fly, *Musca domestica*. In King, R. C., ed. Handbood of Genetics. Vol. 3. New York: Plenum Press; 377-99.
Mohanta, B. K.; Parida, B. B. 1976. Spermatocyte chromosome aberrations induced by antibiotics in grasshoppers. II. Mitomycin C. Sci. Cult. 42:158-59.
Mohr, O. L. 1919. Mikroskopische Untersuchungen zu Experimenten über den Einfluss der Radiumstrahlen und der Kältewirking auf die Chromatinreifung und das Heterochromosom bei *Decticus verruccivorus* (♂). Arch. mikroskop. Anat. Entwicklungsmech. (Abt.II) 92:300-68.

Montgomery, T. H. 1901. A study of the chromosomes of the germ cells of metazoa. Trans. Am. Phil. Soc. 20:154-236.
Morgan, T. H. 1910. Sex-limited inheritance in *Drosophila*. Science 32:120-22.
Motara, M. A.; Rai, K. S. 1977. Chromosomal differentiation in two species of *Aedes* and their hybrids revealed by Giemsa C-banding. Chromosoma 64:125-32.
Muller, H. J. 1927. Artificial transmutation of the gene. Science 66:84-87.
Muramoto, N. 1980. A study of the C-banded chromosomes in some species of Heteropteran insects. Proc. Japan Acad. 56(B):125-30.
Painter, T. S. 1933. A new method for the study of chromosome rearrangements and the plotting of chromosome maps. Science 78:585-86.
Parida, B. B. 1972. Spermatocyte chromosome aberrations induced by antibiotics in grasshoppers. I. Penicillin. Sci. Cult. 38:523-25.
Parida, B. B.; Das, C. C.; Baig, S. 1972. Calcium chloride-induced meiotic chromosome aberrations in grasshopper *Spathosternum prasiniferum*. Curr. Sci. 41:457-58.
Perje, A. M. 1948. Studies on the spermatogenesis in *Musca domestica*. Hereditas 34:209-32.
Philippe, C.; Landureau, J. C. 1975. Culture de cellules embryonnaires et d-hemocytes de blatte d'origine parthenogenetique. Exp. Cell Res. 96:287-96.
Podlasek, S.; Hunt, L. M. 1978. A simple technique for preparing mitotic chromosome spreads from Hemipteran embryos. J. Heredity 69:419-20.
Pudney, M; Lanar, D. 1977. Establishment and characterization of a cell line (BTC-32) from the triatomine bug, *Triatoma infestans* (Klug) (Hemitera: Reduviidae). Ann. Trop. Med. Parasitol. 71:109-18.
Quinn, C. 1970. LSD: Chromosomal breaks in *Antheraea eucalypti* (Scott) (Lepidoptera: Saturniidae). Entomol. News 81:241-42.
Rai, K. S. 1964. Cytogenetic effects of chemosterilants in mosquitoes. I. Apholate-induced aberrations in the somatic chromosomes of *Aedes aegypti* L. Cytologia 29:346-53.
Rai, K. S. 1967. Techniques for the study of cytogenetics and genetics of vectors. In Wright, J. W.; Pal, R., eds. Genetics of Insect Vectors of Disease. New York: Elsevier Publ. Co.; 673-701.
Rai, K. S.; Hartberg, W. K. 1975. *Aedes*. In King, R. C., ed. Handbook of Genetics. Vol. 3. New York: Plenum Press; 311-45.
Rønne, M.; Boye, H. A.; Sandermann, J. 1977. A mounting medium for banded chromosomes. Hereditas 86:155-58.
Ross, M. H.; Cochran, D. G. 1975. The German cockroach, *Blatella germanica*. In King, R. C., ed. Handbook of Genetics. Vol. 3. New York: Plenum Press; 35-62.
Rothenbuhler, W. C. 1975. The honey bee, *Apis mellifera*. In King, R. C., ed. Handbook of Genetics. Vol. 3. New York: Plenum Press; 165-72.
Saez, F. A. 1957. An extreme karyotype in an Orthopteran insect. Am. Naturalist 91:259-64.
Saez, F. A.; Perez-Mosquera, G. 1977. Structure, behavior, and evolution of the chromosomes of Dichroplus silveiraguidoi (Orthoptera: Acrididae). Genetica 47:105-13.
Saha, A. K.; Chakrabarty, A. 1973. Cytological effect of versene on the production of dicentric bridges in spermatocytic chromosomes of grasshopper *Spathosternum prasiniferum*. Indian J. Exp. Biol. 11:351-52.
Saha, A. K.; Khudabaksh, A. R. 1974. Chromosome aberrations induced by methanol in germinal cells of grasshopper, *Oxya velos fabricius*. Indian J. Exp. Biol. 12:72-75.
Samols, D.; Swift, H. 1979a. Genomic organization in the flesh fly *Sacrophaga bullata*. Chromosoma 75:129-43.
Samols, D.; Swift, H. 1979b. Characterization of extrachromosomal DNA in the flesh fly *Sarcophaga bullata*. Chromosoma 75:145-59.
Savage, J. R. K.; Watson, G. E.; Bigger, T. R. L. 1973. The participation of human chromosome arms in radiation-induced chromatid exchange. In Wahrman, J.; Lewis, K. R., eds. Chromosomes Today. Vol. 4. New York: John Wiley & Sons; 267-76.
Schneider, A. 1880. Ueber Befruchtung/der thierischen Eier/, Carus. Zool. Anzeiger 3: 252-56, 426-27.
Schneider, I.; Blumenthal, A. B. 1978. *Drosophila* cell and tissue culture. In Ashburner, M.; Wright, T. R. F., eds. The Genetics and Biology of Drosophila. Vol. 2A. New York: Academic Press; 265-315.
Seabright, M. 1971. A rapid banding technique for human chromosomes. The Lancet 2:971-72.

Shannon, J. E. 1977. The Cell Source Information Bank (CSIB) referral service. In Vitro 13:335.
Sharma, A. K.; Sharma, A. 1972. Chromosome Techniques. 2nd ed. London: Butterworths.
Shaw, D. D.; Webb, G. C.; Wilkinson, P. 1976. Population cytogenetics of the genus *Caledia* (*Orthoptera: Acridinae*). II. Variation in the pattern of C-banding. Chromosoma 56:169-90.
Singh, K. R. P. 1967. Cell cultures derived from larvae of *Aedes albopictus* (Skuse) and *Aedes aegypti* (L.). Curr. Sci. 36:506-8.
Smith, C. N. 1966. Insect Colonization and Mass Production. New York: Academic Press.
Smith, C. N. 1967. Mass-breeding procedures. In Wright, J. W.; Pal, R. eds. Genetics of Insect Vectors of Disease. New York: Elsevier Publ. Co.; 653-72.
Smith, S. G. 1943. Techniques for the study of insect chromosomes. Can. Entomol. 75:21-34.
Smith, S. G. 1952a. The cytology of some tenebrionoid beetles (Coleoptera). J. Morphol. 91:325-63.
Smith, S. G. 1952b. The evolution of heterochromatin in the genus *Tribolium (Tenebrionidae: Coleoptera)*. Chromosoma 4:585-610.
Sokoloff, A. 1975. The flour beetles, *Tribolium castaneum* and *Tribolium confusum*. In King, R. C., ed. Handbook of Genetics. Vol. 3. New York: Plenum Press; 149-63.
Sorsa, V; Saura, A. O. 1980. Electron microscopic analysis of the banding pattern in the salivary gland chromosomes of *Drosophila melanogaster*. Division 1 and 2 of X. Hereditas 92:73-83.
Spurr, A. R. 1969. A low-viscosity epoxy resin embedding medium for electron microscopy. J. Ultrastruc. Res. 26:31-43.
Srdic, Z.; Jacobs-Lorena, M. 1978. *Drosophila* egg chambers develop to mature eggs when cultured in vivo. Science 202:641-43.
Stevens, N. M. 1908. A study of the germ cells of certain Diptera, with reference to the heterochromosomes and the phenomena of synapsis. J. Exp. Zool. 5:359-74.
Sumner, A. T. 1972. A simple technique for demonstrating centromeric heterochromatin. Exp. Cell Res. 75:304-6.
Sumner, A. T.; Evans, H. J.; Buckland, R. A. 1971. New techniques for distinguishing between human chromosomes. Nature New Biology 232:31-32.
Sutton, W. S. 1902. On the morphology of the chromosome group in *Brachystola magna*. Biol. Bull. 4:24-39.
Tadano, T.; Kitzmiller, J. B. 1969. Chromosomal aberrations induced by the chemosterilant tepa in *Culex pipiens fatigans* (Wiedemann). Pak. J. Zool. 1:93-96.
Tazima, Y.; Doira, H.; Akai, H. 1975. The domesticated silkmoth, *Bombyx mori*. In King, R. C., ed. Handbook of Genetics. Vol. 3. New York: Plenum Press; 63-124.
Tease, C. 1978. Cytological detection of crossing-over in BUdR substituted meiotic chromosomes using the fluorescent plus Giemsa technique. Nature 272:823-24.
Tease, C.; Jones, G. H. 1978. Analysis of exchanges in differentially stained meiotic chromosomes of *Locusta migratoria* after BrdU-substitution and FPG staining. I. Crossover exchanges in monochiasmate bivalents. Chromosoma 69:163-78.
Tease, C.; Jones, G. H. 1979. Analysis of exchanges in differentially stained meiotic chromosomes of *Locusta migratoria* after BrdU-substitution and FPG staining. II. Sister chromatid exchanges. Chromosoma 73:75-84.
Thornton, J.; Gaulden, M. E. 1971. Relation of X-ray-induced thymidine uptake to chromosome reversion at prophase in grasshopper neuroblasts. Int. J. Radiat. Biol. 19:65-78.
Tiepolo, L.; Fraccaro, M.; Landani, U.; Diaz, G. 1975. Homologous bands on the long arms of the X and Y chromosomes of *Anopheles atroparvus*. Chromosoma 49:371-74.
Torvik-Greb, M. 1935. The chromosomes of Habrobracon. Biol. Bull. 68:25-34.
Traut, W.; Rathjens, B. 1973. Das W-Chromosom von *Ephestia kuehniella (Lepidoptera)* und die Ableitung des Geschlechtschromatins. Chromosoma 41:437-46.
Ueshima, N. 1966. Cytotaxonomy of the *Triatominae* (Reduviidae: Hemiptera). Chromosoma 18:97-122.
Ueshima, N. 1967. Supernumerary chromosomes in the human bed bug, *Cimex lectularius* Linn. (*Cimicidae: Hemiptera*). Chromosoma 20:311-31.
Ullerich, F. H. 1963. Geschlechtschromosomen und Geschlechtsbestimmung bei einigen Calliphorinen (Calliphoridae, Diptera). Chromosoma 14:45-110.

Verma, R. A.; Dosik, H. 1980. Human chromosomal heteromorphisms: nature and clinical significance. Internat. Rev. Cytol. 62:361-83.
Virkki, N. 1964. On the cytology of some neotropical chrysomelids (Coleoptera). Ann. Acad. Sci. Fenn. Series A, IV:1-24.
Vosa, C. G. 1970. The discriminating fluorescense patterns of the chromosomes of *Drosophila melanogaster*. Chromosoma 31:446-51.
Wagoner, D. E. 1967. Linkage group-karyotype correlation in the house fly determined by cytological analysis of X-ray induced translocations. Genet. 57:729-39.
Waldeyer, W. 1888. Über Karyokinese und ihre Beziehungen zu den Befruchtungsvorgängen. Arch. Mikroscop. Anat. 32:1-122.
Webb, G. C. 1976. Chromosome organization in the Australian plague locust, *Chortoicetes terminifera*. I. Banding relationships of the normal and supernumerary chromosomes. Chromosoma 55:229-46.
Webb, G. C.; Komarowski, L. 1976. Haplo-diploid locust embryos arising by accidental thelytoky in *Chortoicetes terminifera* investigated by G-banding. Chromosoma 55:247-51.
Webb, G. C. Neuhaus, P. 1979. Chromosome organization in the Australian plague locust, *Chortoicetes terminifera*. II. Banding variants of the B-chromosome. Chromosoma 70:205-38.
Webb, G. C.; Westerman, M. 1978. G- and C-banding in the Australian grasshopper, *Phaulacridium vittatum*. Heredity 41:131-36.
Webb, G. C.; White, M. J. D.; Contreras, N.; Cheney, J. 1978. Cytogenetics of the parthenogenetic grasshopper, *Warramaba* (formerly *Moraba*) *virgo* and its bisexual relatives. IV. Chromosome banding studies. Chromosoma 67:309-39.
Wheeler, L. L.; Altenburg, L. C. 1977. Hoechst 33258 banding of *Drosophila nasutoides* metaphase chromosomes. Chromosoma 62:351-60.
White, M. J. D. 1973. Animal Cytology and Evolution. 3rd ed. Cambridge: Cambridge University Press.
Whitten, M. J.; Foster, G. G. 1975. Genetical methods of pest control. Ann. Rev. Entomol. 20:461-76.
Whitten, M. J.; Foster, G. G.; Arnold, J. T.; Konowalow, C. 1975. The Australian sheep blowfly, *Lucilia cuprina*. In King, R. C., ed. Handbook of Genetics. Vol. 3. New York: Plenum Press; 401-18.
Wilson, E. B. 1902. Mendel's principles of heredity and the maturation of the germ-cells. Science 16:991-93.
Wilson, E. B. 1912. Studies on chromosomes. VIII. Observations on the maturation-phenomena in certain Hemiptera and other forms, with considerations on synapsis and reduction. J. Exp. Zool. 13:345-49.
Woyke, J; Skowronek, W. 1974. Spermatogenesis in diploid drones of the honeybee. J. Apicult. Res. 13:183-90.
Wyss, C. 1979. TAM selection of *Drosophila* somatic cell hybrids. Somatic Cell Genet. 5:29-37.
Zirkle, C. 1937. Aceto-carmine mounting media. Science 85:528.

6
Detection of Sister Chromatid Exchanges In Vivo Using Avian Embryos

by STEPHEN E. BLOOM*

Abstract

An in vivo system for the detection of sister chromatid exchange (SCE) in the prenatal period has been perfected and applied to the problem of screening environmental chemicals for potential mutagenicity. This screening system uses the 3-4 day chick embryo in ovo as test material. The 6-7 day embryo is also included to gain information on mutagenicity as a function of developmental stage. SCE and breakage are studied in cells obtained from the limb buds, allantoic sac, and blood. The background rate of SCE is 1.3/metaphase and breakage rate is about zero. The chick embryo cytogenetic test (CECT) was evaluated for its sensitivity and accuracy for detecting mutagens as well as for economy and application as a screen for chemicals of various kinds including water-insoluble agents. Some 50 chemicals were chosen for study in the CECT including 21 mutagens and 29 nonmutagens. This selection included direct-acting mutagens as well as compounds that required metabolic activation for action. Different modes of action on DNA were represented as well as a broad spectrum of mutagenic potencies. All 29 nonmutagens were negative in the CECT; all mutagens but three caused increases in the SCE rate. One of these exceptions, bleomycin, caused substantial breakage, but the two others, gentian violet and Fyrol FR-2, failed to increase the rate of SCE or breakage. Ranking of chemical mutagens detected in the CECT indicates a high positive correlation between SCE induction capacity and mutagenic potency. All indirect-acting mutagens gave dose-dependent increases in SCE, with the 6-day response greater than that observed for 4-day embryos. These results indicate the presence of a drug metabolizing enzyme system for the early chick embryo. The use of the CECT for research and genetic toxicology testing in agriculture, industry, and medicine is discussed.

The avian embryo is one of the most useful sources of living material for investigating problems in agriculture, biology, and medicine (69). Investigators in the areas of developmental biology (70, 79), genetics (1), toxicology (64, 65), teratology

*Department of Poultry Science and Division of Biological Sciences, Cornell University, Ithaca, NY 14853.

(33, 67, 80), tumor biology (72), virology (41, 69, 71), immunology (27, 60, 81, 89), and nutrition (38) have gained much useful information from experimental studies with the chick embryo system (CES). Comparative studies have included embryos of other avian species such as Japanese quail, turkey, pheasant, duck, herring gull, and ringdove. Embryos of *Gallus domesticus* have been attractive for study because they are higher vertebrates, any developmental stage can be studied, and genetically defined material is available. Injection of chemical solutions, exposure to radiation, tissue transplantation, cell transfer by parabiosis, and yolk replacement are some experimental manipulations that have been performed on chick embryos (16, 60, 69, 80, 83, 89). An added bonus is that early development in the chick shows striking morphological similarity to that in mammals, including humans (7).

Despite the experimental advantages of this compact in ovo system and frequent use by biologists, little knowledge of the chromosomes and their molecular composition was provided prior to 1960. However, a number of interested and courageous cytogeneticists (and geneticists who switched) entered this fertile field. As a result, accurate karyotypes of chicken and numerous other avian species have been presented (8, 9, 73, 74). The nature of avian DNA and the locations of heterochromatin have been studied with banding and molecular hybridization techniques (14, 85, 86, 87, 93). The nucleolar organizer region has been localized on a pair of microchromosomes (21).

The cellular DNA content is given at 2.4 pg for chicken (26). About half of this DNA is packed in chromosomes 1–5, the macrochromosomes. The Z-sex chromosome is the fifth largest in the karyotype; the female is the heterogametic sex containing a small W-sex chromosome easily identified by its late replication and C-banding (14). This ability to resolve avian chromosomes by morphology and molecular composition has allowed the use of avian material for such studies as cytogenetic evaluation of abortuses, experimental induction of chromosomal rearrangements, taxonomic relationships, gene mapping, gene transfer, elucidation of the mechanism of parthenogenesis, intersexuality in chickens, and cytogenetic damage caused by radiation and chemical mutagens (2, 10, 11, 12, 13, 15, 16, 17, 28, 48, 58, 61, 68, 87, 96). Experimental cytogenetics with avian material is now a reality.

In my laboratory, an extensive cytogenetic screening program was undertaken to determine the full range of types and frequencies of spontaneous chromosome aberrations in chicks. Abnormalities such as haploidy and triploidy were correlated with phenotype alterations such as retardation of growth. For the 12 strains of chickens tested, an aberration rate of 2.4% (heteroploidy, aneuploidy and rearrangements) was found, and estimates of prezygotic, fertilization and postfertilization errors were determined (13, 15). From this screening of over 6,500 embryos, and other work on Marek's tumors (11) and autoimmune thyroiditis (12, 58), it was observed that the rate of structural chromosomal aberrations such as breaks and gaps was close to zero. This observation on low rate of breakage led us to investigate the question of whether chromosome damage could in fact be detected in chick chromosomes.

Our studies with X-rays (a potent physical clastogen) showed clearly that chromosome breaks and gaps could be detected with certainty in the chick macrochromosomes (25). With an increase in the total exposure, a proportional increase in the frequency of one-hit aberrations (gaps and breaks) was observed. Next, the clastogenicity of direct-acting chemical mutagens was examined. Ethyl methanesulfonate (EMS) and methyl methanesulfonate (MMS) produced toxic responses in the 4-day chick embryo test but failed to induce much breakage. It was hard to imagine that EMS and MMS were not producing damage of some sort to chicken DNA.

Other cytogenetic assays were, therefore, sought that would aid in detecting mutagenicity (or repair thereof) at the subchromosomal level. To this end, the 5-bromodeoxyuridine (BrdU) method for revealing sister chromatid exchange (SCE) was adapted to the 4-day CES. Latt showed that sister chromatids of each human metaphase chromosome could be differentiated if cells were cultured in the presence of BrdU for two cell cycles, and then 33258 Hoechst-stained chromosome preparations were observed by fluorescence microscopy (59). This technique allowed very precise observation of SCEs. Bloom and Hsu obtained sister chromatid differentiation (SCD) in chick embryo chromosomes exposed to BrdU in vivo for 26 hours (18). Since Latt and Kato (49, 59) reported significant increases in SCE upon exposure of mammalian cells to low amounts of mitomycin C and EMS, respectively, we wondered if the SCE rate would be elevated in chick embryo chromosomes at levels of mutagens that failed to show chromosome breakage. A dosage of 25 µg EMS per embryo was sufficient to cause elevation in the rate of SCE. As dose increased, a linear increase in the SCE response was observed (19). This exciting result provided the stimulus for further studies on the relationship between mutagenicity and SCE using the chick embryo (16).

This chapter reports on the effects of 21 known chemical mutagens and clastogens on chick embryo chromosomes including both structural damage and SCE. A series of nonmutagens was also evaluated to determine whether an exact positive correlation exists between production of point mutations and mutagenic action (SCE and breakage) in the chick embryo cytogenetic test (CECT). The relationship between known mutagenic potency and ability to induce SCE in the chick was investigated. Lastly, the question of whether indirect-acting mutagens-carcinogens would produce increased rates of SCE was examined.

Chick Embryo Cytogenetic Test

The Chick embryo cytogenetic test (CECT) consists of two developmental stages and measures three different events, i.e., SCE, chromosome breakage, and cytotoxicity. In one test, cytogenetic damage is assessed at an early stage, between days 3 and 4 of incubation. At this stage, the limb buds and the allantoic sac are excellent for study because of their high mitotic activity and ease of sampling. No definite liver is present at day 4, so metabolic alteration of chemical compounds must occur (and we show that it does) throughout the embryonic system. In another test, cytogenetic damage is determined in the period six to seven days of incubation (D.I.), a time when the embryo is more highly differentiated and a liver is present. The most favorable tissue for cytological examination at this time is the red blood cell (RBC) which divides in circulation.

For the experiments described here, the Cornell K-strain (24) was used as the source of fertile eggs. Uniform eggs are produced that average 59 g in weight. Fertility and hatchability are both over 90%, and this genetic strain is highly resistant to Marek's disease.

Incubation and Embryology

Eggs were set in a Robbbins incubator with temperature at 37.5°C and humidity at 85%. Eggs were mechanically rotated hourly in this forced-air incubation system. Prior to incubation, fertile eggs were stored at 15°C for not more than seven days. The general procedure was to initiate incubation of a batch of eggs (usually 90) at a given time for the experiment. Incubation time was counted from the moment of setting in the incubator to the time of removal. For the early CECT, the laboratory routine consisted of setting eggs on Fridays at 2 p.m., and experimental manipulations such as BrdU and mutagen injections were performed on Mondays (2 p.m.) after 3 D.I. At 3 D.I., normal fertile eggs were selected by candling with bright light. Infertiles, deads, and eggs with abnormally placed air cells were discarded. Chromosome preparations were made using the 4 D.I. embryos recovered at 12 noon on Tuesdays. For the late CECT, eggs were set on Tuesdays at 2 p.m., injections performed on Mondays (2 p.m.) at 6 D.I., and 7-day embryos were recovered for study on Tuesdays at 2 p.m.

Details of the course of development are well-known for the chick. The hourly and daily developmental changes have been documented with descriptions, photographs, and drawings (23, 32, 47, 79).

Administration and Timing of Chemical Reagents

Chemical compounds can be applied to the chick embryo at the two selected developmental periods of 3-4 or 6-7 D.I. Most of the test results presented are from the 3-4 CECT. At this stage, normal-growing embryos were easily identified by candling, and this material was excellent for cytogenetic analysis. Also, dramatic changes in embryo size and form occur between days 3 and 4, facilitating the detection of toxic effects. After much of the results were in, we discovered that testing at 6-7 D.I. was possible. Additional work was performed at this stage to test the theory that with a more developed organ system, a greater response to indirect-acting mutagens would be obtained. This idea is not unreasonable since biochemical studies on chick embryos of this age show the presence of drug metabolizing enzyme systems (78).

Application of Chemicals

Previously we reported two methods of application of chemical solutions to chick embryos, the air-cell method and the window method (16, 18). In our evaluation of chemicals, we chose the former method to avoid unnecessary trauma to the CES. In the air-cell method, applied to the early and late CES, a 0.5 cm^2 portion of shell overlying the air cell is removed with forceps, and the test solution is dropped onto the inner shell membrane (ISM) from an Eppendorf micropipette. The hole is covered with ½" wide Johnson & Johnson waterproof adhesive tape. Amounts in the range of 10-100 μl are used. Since the embryo is positioned just below the ISM (vertical position), solutions are rapidly absorbed by the embryo proper, the surrounding vascular network, or both. Although the exact proportion of the original dose reaching the embryo is not known (some is lost in the ISM and albumen), calculations should be made on a dose/embryo weight basis rather than on a dose/egg weight basis. It is pertinent to note here that exact dosimetry for most systems including mammals is uncertain due to numerous variables such as absorption, distribution, and binding to proteins and lipids.

Dosing the Embryo

The early embryo can tolerate small amounts of various solvents such as water, Hanks' balanced salt solution (HBSS), 15-30% ethanol, and acetone. For most work, 50-100 µl injections were given, but with acetone, 20 µl was the maximum amount. These volumes did not interfere with growth over the experimental period, and successive injections could be given without inducing undue trauma. Purified and/or sterile reagents were used to avoid contamination. Water was purified through a Milli-Q system, and solutions were filtered through 0.25 µm Millipore filters before use.

In the CECT, a series of chemical treatments is applied sequentially and at precise times (Tables 6.1 and 6.2). For chromosome-breakage studies, the test chemical is given to the 74-hour embryo, and two hours prior to sacrifice (hour 94), Colcemid is applied to arrest cells at metaphase (Table 6.1). Such breakage studies are performed in the 6-7-day CECT as shown in Table 6.2. For SCE studies, BrdU is injected before the mutagen to produce SCD.

Test Chemicals

Some 50 chemical compounds were evaluated in the CECT (Table 6.3). Among these, 21 are known mutagens, 17 are nonmutagens, and 12 are miscellaneous but probable nonmutagens (4, 31, 44, 46, 53, 63, 92). Of the mutagens, 13 are also carcinogenic. All compounds were evaluated in the 3-4-day CECT and four were also tested at the 6-7-day stage including the flame retardants tris-BP (22) and Fyrol FR2 (36), 2-acetylaminofluorene (2-AAF), and aflatoxin B_1 (AT-B_1). All solutions were made fresh before use and compounds were dissolved in the appropriate solvents at

Table 6.1 Experimental Protocol for Treating 3-4 Day Chick Embryos with Test Chemicals and BrdU to Reveal Sister Chromatid Exchanges

Embryo age (hrs. of incubation)	Experimental manipulation	Chemical amounts (concentrations)
0	Start egg incubation	—
70	Select normal embryos by candling eggs	—
72	Inject BrdU[a]	100 µl (150 µg)
74	Inject test chemical	10-50 µl
92	Inject Colcemid[b]	20 µl (0.2 µg)
94	Remove embryos from eggs	—

[a] Omit BrdU to study only chromosome breakage.
[b] Omit Colcemid and BrdU to measure toxic and teratogenic effects of test chemical itself. Incubation is continued until hatching or until embryo dies.

Table 6.2 Experimental Protocol for Treating 6–7 Day Chick Embryos with Test Chemicals and BrdU to Reveal Sister Chromatid Exchanges

Embryo age (hrs. of incubation)	Experimental manipulation	Chemical amounts (concentrations)
0	Start egg incubation	—
142	Select normal embryos by candling eggs	—
144	Inject BrdU[a]	100 μl (1 mg)
146	Inject test chemical	10–50 μl
166	Inject Colcemid[b]	100 μl (2 μg)
168	Remove embryos from eggs	—

[a] Omit BrdU to study only chromosome breakage.
[b] Omit Colcemid and BrdU to measure toxic and teratogenic effects of test chemical itself. Incubation is continued until hatching or until embryo dies.

various concentrations (Table 6.4 and 6.5). The highest doses approach the cytotoxic level. Low doses were included to detect a no-response level. A wide variety of types of chemicals was selected representing different modes of action on DNA including alkylation, intercalation, and base substitution. Both direct-acting compounds and those requiring metabolic activation for action were included. Among the non-mutagens, solvents, tissue culture components, antibiotics, buffers, and stains were tested.

Use of Bromodeoxyuridine to Reveal Sister Chromatid Exchanges

The interaction of chemical mutagens with the eukaryote genome can be monitored through quantitative analysis of SCE (51, 94). If mitotic cells are exposed to BrdU for two cell cycles, the chromatids show differential staining (SCD) with fluorochromes (29, 50, 59) or special Giemsa methods (55, 57, 77, 95, 98) (Fig. 6.1). With these procedures, SCEs are clearly demarcated. Chemical mutagens cause significant increases in the rate of SCE (19, 42, 52, 54, 59, 76, 84). Advantages of the BrdU-SCD technique include: (1) ease in identifying and scoring SCEs; (2) small number of cells needed to detect a response to mutagens; (3) simultaneous measurement of SCE rate and cell cycle time; and (4) sensitivity of the BrdU technique to low levels of chemicals.

BrdU was applied to the chick embryo by the air cell method according to the schedules in Tables 6.1 and 6.2. The optimal injection of 100 μl was given to deliver BrdU concentration of 150 μg and 1 mg for the 3-4 D.I. and 6-7 D.I. systems, respectively.

To obtain clear and complete SCD in numerous metaphase cells, four experimental parameters must be controlled. First, enough BrdU must be administered to obtain sufficient incorporation into the replicating chromosomes. Too little BrdU results in weak SCD and sometimes incomplete labelling of chromatids. At 25 μg, SCD is visible by fluorescence, but the weak contrast makes scoring SCEs difficult. Very dramatic SCD is obtained at 150 μg BrdU for both the fluorescence and Giemsa methods. Higher levels do not improve contrast between chromatids, and above 150

Table 6.3 Classification of Chemicals Evaluated in the Chick Embryo Cytogenetic Test

Classes	Mutagens[a]	Carcinogens	Nonmutagens[a]	Miscellaneous
Alkylating agents	ethyl methanesulfonate (EMS)	EMS		
	methyl methanesulfonate (MMS)	MMS		
	cyclophosphamide (CP)	CP		
	trimethylphosphate (TMP)			
Antibiotics	bleomycin (BM)			
	mitomycin C (MC)	MC		
			neomycin (NM)	
			penicillin (PC)	
			streptomycin (SM)	
			tetracycline (TC)	
				erythromycin (EM)
Base analogue	5-bromodeoxyuridine (BrdU)			
Buffers			Sorenson's	
			McIlvaine's	
Culture components			McCoy's 5a medium	
			fetal calf serum (FCS)	
			heparin	
Dyes	acridine orange (AO)	AO		
	ethidium bromide (EB)			
	gentian violet (GV)			
	neutral red (NR)			
	quinacrine mustard (QM)			
				basic fuchsin
				eosin yellowish
				fast green
				haematoxylin
				33258 Hoechst
				methyl blue
				orcein, natural
				phenol red
				quinacrine HCl
Flame retardants	tris-(2,3-dibromo-propyl)phosphate (tris-BP)	tris-BP		
	tris-(1,3-dichloro-2-propyl)phosphate (Fyrol FR2)			
Fungal toxins	aflatoxin B_1 (AT-B_1)	AT-B_1		
	sterigmatocystin (SMC)	SMC		
Solvents and salt solutions			pure water	
			Hanks' balanced salt solution	
			acetone	
			dimethyl sulfoxide (DMSO)	
			ethanol	
			1-butanol	
			ethylene glycol	
				2 × sodium chloride sodium citrate (SSC)

continued

Table 6.3 Continued)

Classes	Mutagens[a]	Carcinogens	Nonmutagens[a]	Miscellaneous
Aromatic amines	2-aminoanthracene (2-AA)	2-AA		
	2-acetylaminofluorene (2-AAF)	2-AAF		
	o-tolidine (TD)	TD		
Polycyclic aromatics	3,4-benzo(a)pyrene (BaP)	BaP		
	9,10-dimethyl-1,2-benzanthracene (DBA)	DBA		
			anthracene	
Miscellaneous oils				lens immersion oil (34% PCB)

[a] Mutagens are those compounds that give a positive result in one or more test systems usually including the Ames *Salmonella* test, *Drosphila*, in vivo mammalian tests. Nonmutagens are negative in these tests.

Table 6.4 Dosages of Chemical Mutagens Evaluated in the Chick Embryo Cytogenetic Test

Chemical	Carrier	Range of doses (μg per embryo)[a]	Source of chemical
Mitomycin C	water	0.01-0.1	Sigma
Aflatoxin B_1	acetone	0.05-0.2	Aldrich
Cyclophosphamide	water	0.05-1	ICN-K & K
2-Acetylaminofluorene	acetone	0.25-250	Aldrich
Neutral red	water	0.1-250	Bacto
2-Aminoanthracene	ethanol/TW80	1-250	Sigma
9,10-dimethyl-1,2-benzanthracene	acetone	4-80	Sigma
3,4-Benzo(α)pyrene	acetone	1.8-143	Sigma
Ethidium bromide	30% ethanol	1-250	Sigma
Gentian violet	water	0.5-100	W. H. Curtin
Quinacrine mustard	water	1-50	GIBCO
Sterigmatocystin	30% ethanol	1.3-62.5	Sigma
Bleomycin	water	3.3-50	Bristol
o-tolidine	30% ethanol	10-500	Sigma
EMS	30% ethanol	12.5-300	ICN-K & K
MMS	15% ethanol	12.5-150	ICN-K & K
Acridine orange	water	10-200	ICN-K & K
5-Bromodeoxyuridine	water	25-500	Sigma
Tris-BP	acetone	50-5,000	Great Lakes
TMP	water	750-6000	Aldrich
Fyrol FR2	acetone	8-5000	Stauffer

[a] In initial toxicity tests, a wide range of doses was tested. For cytogenetic tests, the LD_{80-90} was used as the high dose, and 3 to 4 lower doses were selected. For each chemical a low or no-response dose was sought.

μg, the SCE rate increases beyond the desired level of one or two SCEs per metaphase. Second the typical SCD pattern (Fig. 6.1) will occur only if cells are exposed to BrdU for two cell cycles and hence two complete S-periods. Shorter exposures produce replication patterns, and if a third S-period is included, a typical three-cycle pattern is revealed. Third, if the cell population under study is mostly synchronous, numerous metaphases with SCD can be recovered. This facilitates scoring of SCEs. Interestingly enough, we find a high proportion (about 70%) of metaphases with SCD in limb bud and allantoic cells from chick embryos exposed to BrdU for 22 hours. Finally, the SCD pattern is influenced by the staining methodology. The Hoechst fluorescence technique is more sensitive to BrdU substitution into chromosomes than the Giemsa dye methods.

Cell Cycle Estimates

The BrdU technique for differentiating chromatids offers a simple method for measuring the cell-cycle times in a particular cell population. The SCD pattern appears after two cell cycles. In the chick embryo, SCD is revealed in 70–80% of metaphases by 20 hours post-BrdU injection. At 24 hours post-BrdU, the three-cycle SCD pattern appears in about 6% of metaphases. Thus, most cells have a 10–12 hour mitotic cycle, but some are more rapid, at 8 hours. The effects of test chemicals on the mitotic cycle can be measured.

Cytogenetic Testing with Other Avian Species

Theoretically, it should be possible to apply the BrdU-SCD procedure, developed for the chick embryo, to other avian species. Modifications of the chick procedure would be needed to adjust for differences in egg characteristics and also developmental differences such as growth rate.

Embryos of the herring gull *Larus argentatus* were selected for study because of the poor hatching rate of populations on various islands of Lake Ontario (34). Numerous chemical pollutants were detected in such gull eggs including PCBs, DDE, hexachlorobenzene, mirex, photomirex, and dieldrin (34, 35). We wondered if any of these pollutants were mutagenic to the gull embryos and if this could be domonstrated by examining the rate of SCE. After some trial and error work, the following technical protocol for studying SCE was adopted:

1. At hour 0, set gull eggs in incubator at 37.5°C.
2. At hour 84, inject eggs with 50 μl (500 μg) of BrdU.
3. At hour 106, inject eggs with 40 μl of Colcemid (10 μg/ml).
4. At hour 108, crack-out embryos, and treat with hypotonic and fix as given for the chick embryo technique. The slide preparatory technique is the same as for the chick.

The karyotype of *L. argentatus* consists of 10 pairs of macrochromosomes (includes the ZW pair) and 23 pairs of microchromosomes. The rate of SCE for a group of 13 embryos from the Maritime Provinces (clean area) was 2.1 SCE/cell, or $0.12 \pm .04$ SCE/macrochromosome. Preliminary results on a small sampling of eggs from polluted areas showed no SCE increase. Further studies are in progress.

Table 6.5 Dosages of Nonmutagens and Other Compounds in the Chick Embryo Cytogenetic Test

Chemical	Carrier	Range of doses per embryo[a]	Source of chemical
Anthracene	acetone	4–80 μg	Sigma
Water	–	10–100 μl	Milli-Q pure
Hanks' BSS	–	10–100 μl	GIBCO
15% and 30% ethanol	water	10–50 μl	Commercial Solvents
0.5%, 1% DMSO	water	10–50 μl	Sigma
Acetone	–	1–10 μl	Mallinckrodt
15% n-butyl alcohol	water	1–10 μl	Mallinckrodt
50% ethylene glycol	water	1–10 μl	Matheson, Coleman & Bell
Sorenson's buffer, pH 6.8	water	10–50 μl	J. T. Baker
McIlvaine's buffer, pH 4.4	water	10–20 μl	J. T. Baker
McCoy's 5a medium	–	1–10 μl	GIBCO
Fetal calf serum	–	1–10 μl	GIBCO
Heparin	water	165–500 μg	Sigma
Tetracycline	water	10–1000 μg	Sigma
Neomycin	water	125–500 μg	GIBCO
Penicillin G	water	165–500 units	GIBCO
Streptomycin	water	165–500 μg	GIBCO
Miscellaneous[b]			
Basic fuchsin	20% ethanol	20–500 μg	Chroma-Gesellschaft
Eosin yellowish	water	10–1000 μg	J. T. Baker
Fast green FCF	water	10–1000 μg	Harleco
Hoechst 33258	water	10–1000 μg	Hoechst
Haematoxylin	water	10–1000 μg	National Aniline
Methyl blue	water	10–1000 μg	Harleco
Orcein, natural	20% ethanol	10–1000 μg	Gurr
Phenol red	water	10–1000 μg	GIBCO
Quinacrine HCl	water	10–1000 μg	ICN-K & K
2 X SSC	water	10–20 μl	Mallinckrodt
Erythromycin	water	200–1000 μg	Abbot
Lens immersion oil (34% PCB)	acetone	5–70 μg	R. P. Cargille

[a] A series of 3–5 levels of each compound was used in the ranges shown.
[b] These compounds are probable nonmutagens.

Chromosome Preparations from Avian Embryos

Squash Preparations

Excellent chromosome spreads from various embryonic tissues can be obtained by a simple squashing procedure (20). We have used this method to study the types and frequencies of spontaneous and X-ray-induced chromosome aberrations in early chick embryos (15, 25). One of the major advantages of tissue squashing is that maximal recovery of metaphase cells is obtained. We now use, however, the solid-tissue (ST) technique because it produces chromosome preparations compatible with various banding techniques including SCD by fluorescence and Giemsa. In the studies reported here, we used the ST technique exclusively.

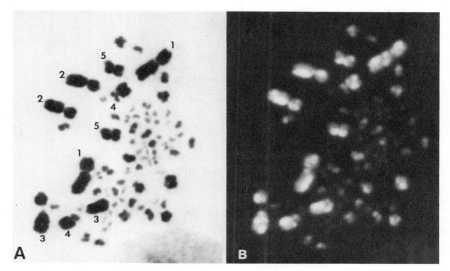

Figure 6.1 A metaphase cell from a 7-day chick embryo exposed to 0.1 μg of the mutagen-carcinogen aflatoxin B_1 over two cell cycles. Embryo also received BrdU to reveal sister chromatid differentiation: (A) Cell stained with 4% Giemsa. No chromosome aberrations are visualized in this case; (B) Cell stained with 33258 Hoechst and examined under fluorescent microscope. Numerous sister chromatid exchanges are visualized by this procedure. Approximately 20 SCEs can be counted among the largest ten chromosomes.

Solid-Tissue Preparations

The ST technique, developed for meiotic and embryonic tissues of mammals (30, 66, 88), is ideal for obtaining high-quality spreads from chick embryos. The procedure is simple and rapid, requires few pieces of equipment, and is ideal for routine screening experiments. Metaphase cells are easily located in the rings that are formed on air-drying. The ST preparations can be stained by all the well-known banding techniques. The ST protocol is as follows:

1. Expose the 4-day or 7-day embryo to Colcemid (see Tables 6.1 and 6.2 for concentrations) for two hours at the normal incubation temperature.
2. Remove embryo from egg and place in 2–3 ml 0.9% sodium citrate (hypotonic) for 30 min. at room temperature.
3. Fix embryo in ethanol:acetic acid (3:1) for one hour and then in fresh fix overnight. Storage in fixative is possible for up to two weeks if refrigerated.
4. Remove tissue (allantoic sac, limb bud, etc.) and place in a well-slide containing several drops of 50% acetic acid. Mince tissue gently for no more than 1 min. to facilitate the formation of a cell suspension.
5. Using a microhematocrit capillary tube (Sherwood BLU-TIP HR1 8889-301506 (75 × 1.2 mm)) equipped with a rubber bulb, withdraw one capillary tube full of the suspension and expel it onto a clean slide heated to 47° on a slide warmer.

6. Quickly draw the suspension back into the capillary tube, leaving a ring of cells approximately 1 cm in diameter on the slide.

7. Repeat steps 5 and 6, producing two or three rings per slide. This will yield a sufficient number of C-metaphases for analysis of an embryo.

The ST technique may be modified to meet particular needs:

1. Raising the temperature increases the number of cells deposited near the circumference of the ring and can improve chromosome spreading. Temperatures much above 50°C can distort chromosome morphology and cause scattering of the chromosomes.

2. Increased time in acetic acid can improve spreading of chromosomes. Excessive treatment times produce broken cells and scattering of chromosomes.

3. Higher concentrations of acetic acid (60%) can be used to increase the dispersion of the tissue into a cell suspension.

4. Increasing the size of the rings usually leads to better spreading of chromosomes.

Staining Techniques

For studies of gross chromosome damage, ST preparations are stained in 4% Giemsa prepared in 0.01 M phosphate buffer, pH 6.8–7.0. After being rinsed in distilled water and air-dried, slides are placed in Xylene for 5 min. Eukitt or other suitable medium is used for mounting slides.

In some cases, it may be desirable to locate chromosome breaks with respect to heterochromatin. We recommend the G- and C-banding techniques summarized in a recent note by Hsu et al. (45). Chicken chromosomes have a pattern of G-bands of the type seen in mammals (87, 93). In exceptional preparations, very light C-bands appear at the centromeres of the macrochromosomes in addition to the usual microchromosomal staining.

Fluorescence and Giemsa Staining to Achieve Sister Chromatid Differentiation

SCD can be visualized in chick embryos by fluorescence of 33258 Hoechst or by a special Giemsa technique. While the fluorescence method is sensitive to low BrdU levels, the image fades rapidly (30–60 sec.) on exposure to UV light. It is possible to reduce the fading (or quenching) by lowering the pH of the mounting fluid from 7 to 5.5–6.0. For photography, the pH is lowered further to 4.5–5.0.

Where high SCE rates are produced, more time is needed for scoring and photographing aberrations. We have adopted a protocol that allows both fluorescence and Giemsa methods to be performed on the same cytological preparation. The slide is checked briefly under fluorescence to determine whether adequate SCD is present, and if it is, the slide is then made permanent by the Giemsa method. All embryos receive 150 μg BrdU (air-cell), so that SCD will appear using both staining techniques.

The staining protocol for SCD in chick embryo chromosomes is as follows:

1. Stain in 0.5 μg/ml 33258 Hoechst in deionized water (Coplin jar) for about 10 min. Rinse slide in water and air-dry for several min.

2. Mount preparation in McIlvaine's buffer, pH 7.0, to check degree of SCD; pH 6.0 to score SCEs; pH 4.5 to photograph. The same slide can be repeatedly remounted at different pHs by rinsing with water and air-drying after each buffer treatment.

3. To produce permanent preparations, mount the slide with several drops of 225 µg/ml Hoechst in McIlvaine's buffer, pH 7. The coverslip is floated over the buffer (37).

4. Expose the mounted slide to fluorescent light (17" Sylvania white lamp, 15 W) for 30 min. Slides are placed immediately above two lamps and a third lamp is placed directly above the slides. Slides are checked after 15 min. and additional buffer is added if they begin to dry out.

5. Wash slides in distilled water and air-dry.

6. Rinse in 95% ethanol for 15 min. and air-dry.

7. Stain in 4% Giemsa (0.01 M phosphate buffer, pH 7) for about 10 min. Air-dry.

8. Mount with Eukitt or other synthetic media.

Fluorescence of Hoechst-stained preparations is observed using a Leitz Ortholux II microscope (with Ploemopak 2.2) equipped with a 100 W mercury vapor lamp, UG1 filter, and TK 400 dichroic mirror for excitation and a K430 barrier filter. Photographs were taken through a 100× achromatic objective using Leitz Orthomat camera and Ilford Pan F extra fine grain film. Photographs of Giemsa-stained metaphases were taken with the same Orthomat system using a Leitz tungsten illuminator, a green substage filter, and Kodak High Contrast Copy film.

Analysis of Chromosome Aberrations

The diploid complement of *Gallus domesticus* contains some 78 chromosomes, ranging in size from about 8 µm to barely visible by light microscopy (73, 74, 90, 97). The first five pairs (macrochromosomes) are large and distinctive enough to permit identification and detection of aberrations without the aid of a photokaryotype (11, 16, 18, 25). Pairs 1-5 contain about 50% of the nuclear DNA per cell. In photographs, pairs 6-10 can also be identified. The W sex chromosome, present only in females, is similar in size and shape to chromosomes in pair 9. When the C-banding technique is used, however, the W chromosome is differentially stained.

While it is theoretically possible to study damage to about 70% of the DNA (pairs 1-10), in practice, chromosome breaks, gaps, and rearrangements can be scored with assurance only in pairs 1-5. This is true of SCE scoring as well, but in photomicrographs, we can see exchanges in chromosomes as small as pair 8.

Rates of chromosome breakage can be expressed as aberrations per metaphase or, more precisely, as aberrations per chromosome (25). When the latter statistic is used, the number of Z sex chromosomes must be recorded. The male is ZZ, and has ten scorable chromosomes (targets); the female (ZW) has nine targets, the W chromosome being too small to score.

Haploid and polyploid complements are readily detected in chick embryos (11, 13, 15, 17). Identification of euploidy (including haploidy) is facilitated by nucleolar counts and measurements of nuclear size (10). Aneuploid detection is limited to pairs 1-10.

Scoring of SCEs is done by counting the number of exchange points on each chromosome (usually pairs 1-5) in the metaphase (Fig. 6.1). Here again, data may be expressed as SCEs per metaphase or per chromosome. For routine screening studies, we calculate SCEs per metaphase, and incorporate equal numbers of male (ZZ) and female (Z) embryos in the control and treated classes. Even this precaution may be discarded, since large numbers of SCEs are seen in each cell from treated embryos, and only 7% of the SCEs in a cell occur on the Z sex chromosome.

One problem is scoring SCEs at the centromere. Many such exchanges turn out to be subcentromeric on careful examination. In other cases, it is clear that each chromatid is continuous on one side through the centromere. In still other cases, it is not clear whether there is a centromeric SCE or a twisting of the chromatids. When there is such doubt, we do not score an SCE.

Rate of Sister Chromatid Exchange

The spontaneous rate of SCE has been studied in embryos from five Cornell experimental chicken strains (C, K, S, N, and P) and one commercial source (Rich-Glo, El Campo, Texas). The mean background rate of SCE measured in various experiments during the last two years is 1.2 exchanges per metaphase (range: 0.8–3.0 SCEs). No strain differences in SCE are apparent.

Several characteristics typify chromosome preparations from control embryos prepared for SCE studies. There is a high mitotic index (20–25%), about 70% of the metaphases show the typical two-round SCD pattern, many metaphases show no SCEs, and SCEs, when they occur, are distributed one to each chromosome. In the presence of mutagens, practically all SCE metaphases show a typcial "checkerboard" pattern (Fig. 6.1B). It is usually possible to differentiate controls from treated embryos by examination of just two or three metaphases.

Induction of Sister Chromatid Exchanges by Chemical Mutagens

Using the SCE and breakage cytogenetic assay, chemical mutagens were discriminated from nonmutagens with a high degree of accuracy. However, SCE analysis was far more reliable an indicator of mutagenicity than chromosome breakage. SCEs were easy to score, and the presence of mutagens could usually be determined from examination of just a few cells. Such cells were characterized by the presence of multiple exchanges within each macrochromosome (Fig. 6.1B). Of the 21 mutagens, only five were classified as definite clastogens, with five possible clastogens, five nonclastogens, and five not determined (Table 6.6). With SCE analysis, 86% (18) of the mutagens were detected. With both SCE and breakage, 90% (19) of the mutagens were detected. All 17 nonmutagens and 12 miscellaneous (probable nonmutagens) failed to induce SCE or breakage in our study.

Of the three mutagens that failed to induce SCE, bleomycin caused chromosome breakage, Fyrol FR2 is a weak mutagen in the Ames *Salmonella* test (36), and gentian violet is a clastogen in vitro (6) but is converted to a nonmutagenic but cytotoxic form after metabolism (5). Whether Fyrol FR2 is a mutagen in higher animals and man is not known. Our test suggests that it probably is not. Gentian violet must be considered a potent toxicant and direct-acting mutagen.

Direct-acting chemical mutagens were effective SCE inducers. Neutral red was the

Table 6.6 Induction of Sister Chromatid Exchanges by Chemical Mutagens in the Chick Embryo Cytogenetic Test

Chemical	Lowest dosage of chemical giving significant SCE increase[a] (μg)	Rate of SCE/cell at highest dose[b] (mean ± S.D.)	Classification of chemical for:	
			SCE	breakage[c]
Aflatoxin B_1[d]	0.05	12.9 ± 2.5	+	N.D.
Mitomycin C	0.05	5.2 ± 1.7	+	+
Cyclophosphamide	0.5	13.6 ± 1.2	+	+
2-Aminoanthracene	1.0	13.0 ± 2.4	+	±
Sterigmatocystin	1.3	10.7 ± 3.2	+	+
3,4-Benzo(α)pyrene	1.8	7.0 ± 2.2	+	N.D.
2-Acetylaminofluorene	2.5	9.7 ± 1.6	+	±
9,10-Dimethyl-1,2-benzanthracene	4.0	16.5 ± 2.5	+	N.D.
Neutral red	5	18 ± 1.7	+	+
o-Tolidine	10	6.3 ± 1.0	+	±
MMS	12.5	9.5 ± 0.3	+	±
Quinacrine mustard	25	8.1 ± 4.8	+	±
Tris-BP	25	8.5 ± 1.4	+	N.D.
Ethidium bromide	50	4.3 ± 0.1	+	N.D.
EMS	100	8.6 ± 1.8	+	–
Acridine orange	100	7.9 ± 0.5	+	+
5-Bromodeoxyuridine	500	4.6 ± 0.1	+	–
TMP	1500	7.5 ± 1.4	+	–
Bleomycin	N.S.	2.2 ± 0.8	–	+
Fyrol FR2[d]	N.S.	2.2 ± 0.5	–	–
Gentian violet	N.S.	2.4 ± 0.1	–	–

[a] A 3-fold or higher increase in the rate of SCE above background (1.2 SCE/cell) is considered a significant increase. Student's t-test was performed on all data, and SCE values shown are all significantly different from control as at $P < 0.01$. For most chemicals a rate of 4 SCE/cell or higher was used to indicate the dosage causing a significant increase. N.S. = no significant increase in SCE at any dose.
[b] This dose refers to the high doses given for each chemical in Table 6.4.
[c] + signifies a definite positive; ±, a marginal effect; –, no increase above control; and N.D. is not determined. Compounds are listed in order of potency for inducing SCE.
[d] Data shown for these two chemicals are from the 6–7 day CECT.

most potent of these followed by MMS, QM, EB, EMS, AO, BrdU, and TMP. This order of mutagens represents a dosage range of 5 μg (NR) to 1500 μg (TMP).

The indirect-acting chemical mutagens or promutagens were the most potent SCE inducers with AT-B_1, MMC, CP, and 2-AA active at dosages as low as 0.05 to 1.0 μg per embryo (Table 6.6). Active in the range of 1.3 to 4.0 μg was SMC, BaP, 2-AAF, and DBA. The o-tolidine and tris-BP caused increased SCE beginning at 10.0 and 25 μg, respectively.

If SCE is related to damage to the DNA, then a dose-response relationship, linear at least for the direct-acting mutagens, is expected. Such was the case for these compounds (16), save for EB and BrdU, which did not increase the SCE rate above 4.6 per cell. These two compounds, therefore, cause significant but borderline increases in exchange frequency. In general, as dose increased so did the SCE level for the class of pro-mutagens (Fig. 6.2). However, a clear linear response was not always apparent. Additional studies are in progress to produce additional data on this point.

An extremely interesting finding was an age-related SCE response to the mutagen

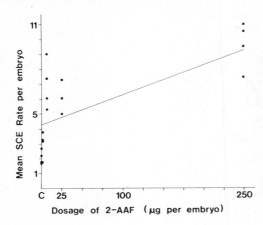

Figure 6.2 The effect of 2-acetylaminofluorene on the rate of sister chromatid exchange in the 3-4-day chick embryo test. The increased yield of SCE with 2-AAF indicates metabolic activation of this promutagen. The rate of SCE was elevated at the low dose of 0.25 µg of 2-AAF, and at 2.5 µg a rate of 7 SCE/cell was observed. The r value for this dose-response line is 0.73. Each dot on the graph represents the mean rate of SCE per cell per embryo, with 25 metaphases scored for each.

and carcinogen AT-B_1. At the 3-4 day stage, only a slight increase in the SCE rate was found, but at the 6-7 day stage, AT-B_1 was the most potent inducer of SCE! Such an age-related effect was not apparent for the flame retardant tris-BP (Fig. 6.3).

The relative potencies for inducing SCE is shown in Table 6.6. AT-B_1 is the number one inducer and TMP is the weakest. The eight top SCE inducers are also potent animal carcinogens.

The maximum number of SCEs per metaphase is perhaps an indicator of the tolerance of the macrochromosomes (pairs 1-5) for DNA damage and repair. AT-B_1, tris-BP, and NR induced the highest level of about 35 scorable SCEs/cell, or about 4 SCE/macrochromosome. This number may be higher since small SCEs would escape resolution by fluorescence microscopy.

Discussion and Implications of Results

The original goal in initiating these studies on chemical induction of SCEs was to determine if chromosomes of the chicken were more or less sensitive than other species to the mutagenic action of chemicals. From testing some 50 chemical compounds, it appears that the avian genome is not substantially more resistant to insult than other genomes. For example, we compared the chick embryo test results with those of the Ames *Salmonella* test (63). Chemicals causing point mutation also induce SCE. Ten of the Ames-tested mutagens were ranked according to the number of revertants/plate/1 µg and then compared with the ranking in the chick-SCE test determined from both dose-response data and the number of SCEs/1 µg of mutagen. In both of these test systems, aflatoxin B_1 ranked as the most potent mutagen. Cyclophosphamide, which ranked second for SCE, ranked ninth for point mutation. The five top-ranked mutagens, excluding CP, were so for both tests, and these are AT-B_1, 2-AA, SMC, BaP, and 2-AAF. These highly genotoxic compounds include the mold toxins and the polycyclic aromatic hydrocarbons, also known for their potency for inducing cancer. MMS, EB, EMS, and AO were in the lower ranking in both the Ames system and the SCE test. However, AO ranked higher as a mutagen than as an inducer of SCE.

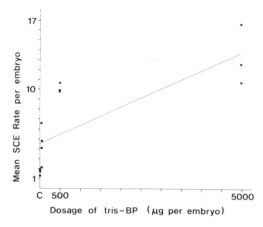

Figure 6.3 The effect of the flame retardant tris-BP on the rate of sister chromatid exchange in the 6–7-day chick embryo test. The SCE rate was elevated to 4.4/cell at the low dose of 50 µg. The r value for this dose-response line of 0.77. Each dot on the graph represents the mean rate of SCE/cell/embryo.

In general terms, the results show that the induction of SCE in the chick embryo in vivo test is related to the ability of a chemical to induce point mutations in DNA. This is true for both direct-acting mutagens and those requiring metabolic activation for mutagenesis and carcinogenesis. When SCE and breakage were both included in our embryo assay, 90% of mutagens were detected. None of the nonmutagens turned up positive for SCE or breakage.

Our results are informative with respect to those mutagens that turned up negative in the chick test. Gentian violet, a potent clastogen in vitro (6) was not mutagenic in the CES where a metabolic system is operative. Later studies with mammalian cell culture and Ames *Salmonella* systems, incorporating a liver S9 microsomal fraction, confirmed the reduction in mutagenicity upon metabolism (5). For GV, the chick test served to predict what would happen for the in vivo situation. However, this case illustrates the tremendous importance of tier testing to include both in vivo and in vitro systems. In the case of bleomycin, no SCEs were induced, but substantial chromosome breakage occurred. We, therefore, recommend that both SCE and breakage assays be included in all cytogenetic testing of chemicals.

Finally, Fyrol FR2 was not detected as a mutagen in the CECT. In the Ames test this compound is a weak mutagen (36). Is it possible that this flame retardant is not a mutagen in vertebrate cells? Only further tier testing can decide the issue.

Since the study of induction of SCE in the embryo incorporated a broad spectrum of chemical types, we could ask whether SCE is related to a particular mode of interaction with chromosomal DNA. Known alkylating agents, cross-linkers, and those chemicals that intercalate in DNA all produced increases in SCE. It appears then that SCE is reflective of multiple forms of damage to chromosomes. In terms of the utility of SCE as a screening device for chemical mutagens, this finding is fortunate. In addition, since increasing amounts of BrdU cause increased rate of SCE, base substitution is another mode of action leading to exchange. At the level of the whole chromosome, SCE appears to be a reflection of repair. Perhaps many forms of damage to DNA lead to SCE. This does not mean that the chromosome is necessarily repaired without mistakes. Small changes at the level of the nucleotide bases would not be visualized at the cytological level. Further studies are needed to elucidate the mechanism(s) leading to SCE and the molecular changes that occur as a result.

The testing of the 21 mutagens in the CECT forms an important data base for future work with this system. The response of one of the most potent mutagens-carcinogens, AT-B_1, has been determined as well as that of one of the weakest, TMP. The former is active as 0.05 μg/embryo and the latter at 1500 μg/embryo (i.e., 30,000-fold difference in potency). The other compounds form a spectrum of doses at many points in between. Unknown compounds can now be tested and compared with existing data.

Our results have important implications for studies on toxicity of environmental pollutants on avian embryos. Embryos as early as 3 D.I. appear to possess a metabolic machinery capable of converting chemicals to toxic and mutagenic forms. This is shown clearly for a number of compounds including 2-AAF, CP, 2-AA, and other hydrocarbons. Also, GV was very toxic in the chick test and as little as 20 μg proved embryotoxic (5). Also of interest is the change in metabolism that might be expected as development proceeds. Preliminary results suggest that aflatoxin B_1 becomes more mutagenic in the older embryo, i.e., by 6 D.I. Published work on microsomal enzyme systems of chicks shows the presence of drug-metabolizing enzymes in embryos of various ages, but no one has yet documented this in detail in the very early embryo.

The avian embryo protected by shell, membranes, and albumen may be quite susceptible to agents that become toxic upon metabolism. Such environmental pollutants may occur in eggs because of transfer from the hen during egg maturation (40) or perhaps due to direct penetration of chemicals that are applied to the shell surface (43). Various pesticides and other pollutants have been detected in eggs (34, 35) and oil applied in small amounts to the shell cause embryo deaths and teratogenicity (43).

Application of the Chick Embryo Cytogenetic Test

The chick embryo cytogenetic test may eventually have important uses in agriculture, industry, and medicine. In the agricultural sphere, this test would be useful for assessing the genetic toxicity of poultry health products such as antibiotics used in feed, medications for the treatment of diseases, and pesticides used to control insects in the poultry house and on the birds. If a chemical is found to be genotoxic in the CECT, this would warn of the possible induction of mutations in breeder flocks. In addition, a positive result in the chick test would signal a warning of potential human hazard.

In the industrial setting, the avian embryo test might be attractive as a rapid prescreen of chemicals for potential mutagenicity, embryo toxicity, and also teratogenicity. The test is relatively fast and definitely not as expensive as in vivo mammalian tests. The CECT should be amenable to automation similar to that achieved with the manufacture of vaccines. In that case, large numbers of eggs are handled by machine including the injection of solutions.

In the area of medicine, the CECT might be a very useful screen for the in vivo effects of drugs such as chemotherapeutic agents, anesthetics, sedatives, and other materials included with primary medications.

In the broad scheme of things, the CECT fits in the category of tier two testing.

The Ames test, at tier one, establishes mutagenic potential of chemicals with and without activation. At the second tier, the CECT would extend the information to effects in a eukaryote (and a vertebrate). Additional information on mutagenic activity of the test compound as effected by metabolism could be obtained quite rapidly using the 6-7 D.I. test. From both the Ames test and the CECT, the mutagenic potency would be reasonably well established. These data would be invaluable in performing tier three testing, to include perhaps in vivo mammalian SCE and breakage tests (3, 62, 75, 82, 91). Testing in a fish, the mudminnow, could be included where appropriate (56). The CECT also complements in vitro cytogenetic tests which are very valuable in determining any direct effects of the test compound.

Finally, the avian egg may be a very useful monitor of the environment in two respects. First, embryos from bird populations such as herring gulls and ringdoves may be examined cytogenetically to determine the presence of mutagenic materials accumulated in the wild. Second, environmental contaminants or materials containing such pollutants can be applied to avian embryos for assessment of genetic effects.

Acknowledgments

This chapter is dedicated to Alexis L. Romanoff.

The author wishes to thank Dr. Donald Lisk, Pesticide Residue Lab, Department of Food Science, Cornell University, for the gift of tris-BP and Fyrol FR2 and Dr. T. C. Hsu for supplying the dyes gentian violet, neutral red, methyl blue, eosin yellowish, haematoxylin, basic fuchsin, fast green, and natural orcein. The following persons contributed to the laboratory investigations: Dr. A. D. Kligerman, Dr. C. Goodpasture, Ms. B. Haller, Ms. E. G. Polakoff, Mr. P. Shalit, and Ms. C. F. Daniels. The author thanks Ms. Daniels for providing the drawings for Figures 6.2 and 6.3, and Ms. D. C. Wittner and Ms. J. Byrnes for suggestions concerning the manuscript.

The research presented in this chapter was supported in part by the U.S. Department of Agriculture under CSRS Project 412.

Literature Cited

1. Abbott, U. K.; Yee, G. W. Avian gentics. *In* King, R. C. ed. Handbook of genetics Vol. 4. New York: Plenum Press; 1976: 151-200.

2. Akiyama, Y.; Kato, S. Two cell lines from lymphomas of Marek's disease. Biken J. 17: 105-16; 1974.

3. Allen, J. W.; Latt, S. A. Analysis of sister chromatid exchange formation in vivo in mouse spermatogonia as a new test system for environmental mutagens. Nature (London) 260: 449-51; 1976.

4. Ames, B. N.; Durston, W. E.; Yamsaki, E.; Lee, F. D. Carcinogens are mutagens: A simple test system combining liver homogenates for activation and bacteria for detection. Proc. Nat. Acad. Sci. USA 70: 2281-85; 1973.

5. Au, W.; Butler, M. A.; Bloom, S. E.; Matney, T. S. Further study of the genetic toxicity of gentian violet. Mutat. Res. 66:103-112; 1979.

6. Au, W.; Pathak, S.; Collie, C. J.; Hsu, T. C. Cytogenetic toxicity of gentian violet and crystal violet on mammalian cells in vitro. Mutat. Res. 58:269-76; 1978.

7. Balinsky, B. I. *An introduction to embryology.* Philadelphia: W.B. Saunders Co.; 1960.

8. Benirschke, K.; Hsu, T. C. *Chromosome atlas: fish, amphibians, reptiles, and birds.* New York: Springer-Verlag; 1971.

9. Bloom, S. E. A current list of chromosome numbers and karyotype variations in the avian subclass *Carinatae*. J. Hered. 60: 217–20; 1969.
10. Bloom, S. E. Haploid chicken embryos: Evidence for diploid and triploid cell populations. J. Hered. 61: 147–49; 1970.
11. Bloom, S. E. Marek's disease: Chromosome studies of resistant and susceptible strains. Avian Dis. 14: 478–90; 1970.
12. Bloom, S. E. Cytogenetic studies of hereditary autoimmune thyroiditis in chickens. J. Hered. 62: 186–88; 1971.
13. Bloom, S. E. Chromosome abnormalities in chicken (*Gallus domesticus*) embryos: Types, frequencies, and phenotypic effects. Chromosoma 37: 309–26; 1972.
14. Bloom, S. E. Current knowledge about the avian W chromosome. BioScience 24:340–44; 1974.
15. Bloom, S. E. The origins and phenotypic effects of chromosome abnormalities in avian embryos. *Proceedings of the 15th World's Poultry Congress,* New Orleans. 316–20. Washington, D.C.: McGregor and Warner; 1974.
16. Bloom, S. E. Chick embryos for detecting environmental mutagens. *In* Hollaender, A.; de Serres, F. J. eds. Chemical mutagens: principles and methods for their detection. Vol. 5; 1978: 203–32.
17. Bloom, S. E.; Buss, E. G. Triploid-diploid mosaic chicken embryo. Science 153:759–60; 1966.
18. Bloom, S. E.; Hsu, T. C. Differential fluorescence of sister chromatids in chicken embryos exposed to 5-bromodeoxyuridine. Chromosoma 51: 261–67; 1975.
19. Bloom, S. E.; Hsu, T. C. Detection of sister chromatid exchanges in chick embryos exposed to bromodeoxyuridine. Mutat. Res. 38: 401–2; 1976.
20. Bloom, S. E.; Povar, G.; Peakall, D. B. Chromosome preparations from the avian allantoic sac. Stain Technol. 47: 123–27; 1972.
21. Bloom, S. E.; Shalit, P.; Bacon, L. D. Chromosomal localization of nucleolus organizers in the chicken. Genetics 88: 13; 1978.
22. Blum, A.; Ames, B. N. Flame-retardant additives as possible cancer hazards. Science 195: 17–23; 1977.
23. Calvert, J. A. Avian embryo development. Ministry of Agriculture, Fisheries and Food, Bull. No. 23. London: Her Majesty's Stationery Office; 1972.
24. Cole, R. K. Leukosis control through genetics. Proceedings of the Poultry Health Conference, University of New Hampshire: 59–72; 1967.
25. Cormier, J. M.; Bloom, S. E. An in vivo study of the effects of X-radiation on the chromosomes of chick allantoic membranes. Mutat. Res. 20: 77–85; 1973.
26. DeRobertis, E. D. P.; Nowinski, W. W.; Saez, F. A. General cytology. 3rd ed. Philadelphia: W.B. Saunders Co.; 1960.
27. Dietert, R. R.; Sanders, B. G. Expression of an onco-developmental antigen among avian species. J. Exp. Zoo. 206(1): 17–24; 1978.
28. Donner, L; Chyle, P.; Sainerová, Malformation syndrome in *Gallus domesticus* associated with triploidy. J. Hered. 60: 113–15; 1969.
29. Dutrillaux, B.; Fosse, A. M.; Prieur, M.; Lejeune, J. Analyse des échanges de chromatides dans les cellules somatique humaines: Traitement au BUdR (5-bromodéoxyuridine) et fluorescence bicolore par l'acridine orange. Chromosoma 48: 327–40; 1974.
30. Evans, E. P.; Burtenshaw, M. D.; Ford, C. E. Chromosomes of mouse embryos and new-born young: Preparations from membranes and tail tips. Stain Technol. 47: 229–34; 1972.
31. Fishbein, L; Flamm, W. G.; Falk, H. L. Chemical mutagens. New York: Academic Press; 1970.
32. Freeman, B. M.; Vince, M. A. Development of the avian embryo. London: Chapman and Hall; 1974.
33. Gebhart, D. O. E. The use of the chick embryo in applied teratology. *In* Woolam, D. H. M. ed. Advances in teratology. Vol. 5. New York: Academic Press; 1972:97–111.
34. Gilbertson, M.; Fox, G. A. Pollutant-associated embryonic mortality of great lakes herring gulls. Environ. Pollut. 12: 211–16; 1977.
35. Gilman, A. P.; Peakall, D. B.; Hallett, D. J.; Fox, G. A.; Norstrom, R. J. The herring

gull as a monitor of the great lakes contamination. *In* Animal models and wildlife as monitors. Intern. Symp. Pathobiology of Environ. Pollut. Storrs, CT; 1977.

36. Gold, M. D.; Blum, A.; Ames, B. N. Another flame retardant, tris (1,3-dichloro-2-propyl)-phosphate, and its expected metabolites are mutagens. Science 200: 785-87; 1978.

37. Goto, K; Akematsu, T.; Shimazu, H; Sugiyama, T. Simple differential Giemsa staining of sister chromatids after treatment with photosensitive dyes and exposure to light and the mechanism of staining. Chromosoma 53: 223-30; 1975.

38. Grau, C. R. Avian embryo nutrition. Fed. Proc. 27(1): 185-92; January-February 1968.

39. Grau, C. R.; Klein, N. W.; Lau, T. L. Total replacement of the yolk of chick embryos. J. Embryol. Exp. Morph. 5: 210-14; 1957.

40. Grau, C. R.; Roudybush, T.; Dobbs, J.; Wathen, J. Altered yolk structure and reduced hatchability of eggs from birds fed single doses of petroleum oils. Science 195: 779-81; 1977.

41. Guntaka, R. V.; Mahy, B. W. J.; Bishop, J. M.; Varmus, H. E. Ethidium bromide inhibits appearance of closed circular viral DNA and integration of virus-specific DNA in duck cells infected by avian sarcoma virus. Nature (London) 253: 507-11; 1975.

42. Hayashi, K.; Schmid, W. The rate of sister chromatid exchanges parallel to spontaneous chromosome breakage in Fanconi's anemia and to Trenimon-induced aberrations in human lymphocytes and fibroblasts. Humangenetik 29: 201-6; 1975.

43. Hoffman, D. J. Embryotoxic effects of crude oil in Mallard ducks and chicks. Toxicol. Appl. Pharmacol.; in press, 1979.

44. Hollaender, A., editor Chemical mutagens: principles and methods for their detection. Vols 1-5. New York: Plenum Press; 1971-78.

45. Hsu, T. C.; Arrighi, F. E.; Pathak, S.; Shirley, L.; Stock, A. D. Technical notes. Mammalian Chromosomes Newsletter 15: 88-96; 1974.

46. Hsu, T. C.; Collie, C. J.; Lusby, A. F.; Johnston, D. A. Cytogenetic assays of chemical clastogens using mammalian cells in culture. Mutat. Res 45: 233-47; 1977.

47. Huettner, A. F. Comparative embryology of the vertebrates. New York: MacMillan; 1954.

48. Kao, F. Identification of chick chromosomes in cell hybrids formed between chick erythrocytes and adenine-requiring mutants of Chinese hamster cells. Proc. Nat. Acad. Sci. USA 70: 2893-98; 1973.

49. Kato, H. Is isolabeling a false image? Exper. Cell Res. 89: 416-20; 1974.

50. Kato, H. Spontaneous sister chromatid exchanges detected by a BUdR-labelling method. Nature (London) 251:70-72; 1974.

51. Kato, H. Spontaneous and induced sister chromatid exchanges as revealed by the BUdR-labelling method. Int. Rev. Cytol. 49: 55-97; 1977.

52. Kato, H.; Shimada, H. Sister chromatid exchanges induced by mitomycin C: A new method of detecting DNA damage at chromosomal level. Mutat. Res. 28: 459-64; 1975.

53. Kihlman, B. A. *Actions of chemicals on dividing cells.* Englewood Cliffs, NJ: Prentice-Hall; 1966.

54. Kihlman, B. A. Sister chromatid exchanges in *Vicia faba.* II. Effects of thiotepa, caffeine, and 8-ethoxycaffeine on the frequency of SCEs. Chromosoma 51: 11-18; 1975.

55. Kihlman, B. A.; Kronborg, D. Sister chromatid exchanges in *Vicia faba.* I. Demonstration by a modified fluorescent plus Giemsa (FPG) technique. Chromosoma 51: 1-10; 1975.

56. Kligerman, A. D.; Bloom, S. E. Sister chromatid differentiation and exchanges in adult mudminnows (*Umbra limi*) after in vivo exposure to 5-bromodeoxyuridine. Chromosoma 56: 101-9; 1976.

57. Korenberg, J. R.; Freedlender, E. R. Giemsa technique for the detection of sister chromatid exchanges. Chromosoma 48: 355-60; 1974.

58. Korf, B. R.; Bloom, S. E. Cytogenetic and immunologic studies in chickens with autoimmune thyroiditis. J. Hered. 65: 219-22; 1974.

59. Latt, S. A. Sister chromatid exchanges, indices of human chromosome damage and repair: Detection by fluorescence and induction by mitomycin C. Proc. Nat. Acad. Sci. USA 71: 3162-66; 1974.

60. Lazzarini, A. A., Jr. Immunological effects of multiple experimental embryonal parabiosis in birds. New York Acad. Sci. 87: 133-39; 1960.

61. Leung, W. C.; Chen, T. R.; Dubbs, D. R.; Kit, S. Identification of chick thymidine

kinase determinant in somatic cell hybrids of chick erythrocytes and thymidine kinase-deficient mouse cells. Exper. Cell Res. 95: 320–26; 1975.

62. Marquardt, H.; Bayer, U. The induction in vivo of sister chromatid exchanges in the bone marrow of the Chinese hamster. I. The sensitivity of the system (methyl methanesulphonate). Mutat. Res. 56: 169–76; 1977

63. McCann, J.; Choi, E.; Yamasaki, E.; Ames, B. N. Detection of carcinogens as mutagens in the Salmonella/microsome test: assay of 300 chemicals. Proc. Nat. Acad. Sci. USA 72: 5135–39; 1975.

64. McLaughlin, J.; Marliac, J. P.; Verrett, M. J.; Mutchler, M. K.; Fitzhugh, O. G. The injection of chemicals into the yolk sac of fertile eggs prior to incubation as a toxicity test. Toxicol. Appl. Pharmacol. 5: 760–71; 1963.

65. McLaughlin, J., Jr.; Marliac, J. P.; Verrett, M. J.; Mutchler, M. K.; Fitzhugh, O. G. Toxicity of fourteen chemicals as measured by the chick embryo method. Am. Ind. Hyg. Assoc. J. 25: 282–84; 1964.

66. Meredith, R. A simple method for preparing meiotic chromosomes from mammalian testis. Chromosoma 26: 254–58; 1969.

67. Miller, R. W. The discovery of human teratogens, carcinogens, and mutagens: Lessons for the future. In Hollaender, A.; de Serres, F.J. eds. Chemical mutagens: principles and methods for their detection. Vol. 5. New York: Plenum Press; 1978: 101–26.

68. Miller, R. C.; Fechheimer, N. S.; Jaap, R. G. Chromosome abnormalities in 16- to 18-hour chick embryos. Cytogenetics 10: 121–36; 1971.

69. Miner, R. W. The chick embryo in biological research. Ann. N.Y. Acad. Sci. 55: 37–344; 1952.

70. Moore, M. A. S.; Owen, J. J. T. Stem-cell migration in developing myeloid and lymphoid systems. Lancet 2: 658–59; 1967.

71. Neiman, P. E.; Purchase, H. G.; Okazaki, W. Chicken leukosis virus genome sequence in DNA from normal chick cells and virus-infected bursal lymphomas. Cell 4: 311–19; 1975.

72. Noguchi, P. D.; Johnson, J. B.; O'Donnell, R.; Petricciani, J. C. Chick embryonic skin as a rapid organ culture assay for cellular neoplasia. Science 199: 980–83; 1978.

73. Ohno, S. Sex chromosomes and microchromosomes of Gallus domesticus. Chromosoma 11: 484–98; 1961.

74. Owen, J. J. T. Karyotype studies on Gallus domesticus. Chromosoma 16: 601–8; 1965.

75. Pera, F.; Mattias, P. Labelling of DNA and differential sister chromatid staining after BrdU treatment in vivo, Chromosoma 57: 13–18; 1976.

76. Perry, P.; Evans, H. J. Cytological detection of mutagen-carcinogen exposure by sister chromatid exchange. Nature (London) 258: 121–25; 1975.

77. Perry, P.; Wolff, S. New Giemsa method for the differential staining of sister chromatids. Nature (London) 251: 156–58; 1974.

78. Powis, G.; Drummond, A. H.; McIntyre, D. E.; Jondorf, W. R. Development of liver microsomal oxidations in the chick. Xenobiotica 6: 69–81; 1976.

79. Romanoff, A. L. The avian embryo. New York: Macmillan; 1960.

80. Romanoff, A. L. Pathogenesis of the avian embryo. New York: Wiley-Interscience; 1972.

81. Sanders, B. G. Developmental disappearance of a fowl red blood cell antigen. J. Exp. Zool. 167: 165–78; 1968.

82. Schneider, E. L.; Chaillet, J. R.; Tice, R. R. In vivo BUdR labelling of mammalian chromosomes. Exper. Cell Res. 100: 396–99; 1976.

83. Searle, B. M.; Bloom, S. E. Parabiosis between binucleated red blood cell mutant (bn) and normal avian embryos. Poultry Science; In press 1979.

84. Solomon, E.; Bobrow, M. Sister chromatid exchanges: A sensitive assay of agents damaging human chromosomes. Mutat. Res. 30: 272–78; 1975.

85. Stefos, K.; Arrighi, F. E. The heterochromatic nature of the W chromosome in birds. Exper. Cell Res. 68: 228–31; 1971.

86. Stefos, K.; Arrighi, F. E. Repetitive DNA of Gallus domesticus and its cytological location. Exper. Cell Res. 83: 9–14; 1974.

87. Stock, A. D.; Arrighi, F. E.; Stefos, K. Chromosome homology in birds: Banding patterns of the chromosomes of the domestic chicken, ring-necked dove, and domestic pigeon. Cytogenet. Cell Genet. 13: 410–18; 1974.

88. Stock, A. D.; Burnham, D. B.; Hsu, T. C. Giemsa banding of meiotic chromosomes with description of a procedure for cytological preparations from solid tissues. Cytogenetics 11: 534-39; 1972.
89. Sundick, R. S.; Bloom, S. E.; Kite, J. H., Jr. Parabiosis between obese (OS) and normal strain chicken embryos. Clin Exp. Immunol. 14: 437-42; 1973.
90. Van Brink, J. M.; Ubbels, G. A. La question des hétérochromosomes chez les sauropsides: Oiseaux. Experientia 12: 162-64; 1956.
91. Vogel, W.; Bauknecht, T. Differential chromatid staining by in vivo treatment as a mutagenicity test system. Nature (London) 260: 448-49; 1976.
92. Vogel, F.; Röhrborn, G., editors. Chemical mutagenesis in mammals and man. New York: Springer-Verlag; 1970.
93. Wang, N.; Shoffner, R. N. Trypsin G- and C-banding for interchange analysis and sex identification in the chicken. Chromosoma 47: 61-69; 1974.
94. Wolff, S. Sister chromatid exchange. Ann. Rev. Genet. 11: 183-201; 1977.
95. Wolff, S.; Perry, P. Differential Giemsa staining of sister chromatids and the study of sister chromatid exchanges without autoradiography. Chromosoma 48: 341-53; 1974.
96. Wooster, W. E.; Fechheimer, N. S.; Jaap, R. G. Structural rearrangements of chromosomes in the domestic chicken: Experimental production by X-irradiation of spermatozoa. Can. J. Genet. Cytol. 19: 437-46; 1977.
97. Yamashina, M. Y. Karyotype studies in birds. I. Comparative morphology of chromosomes in seventeen races of domestic fowl. Cytologia 13: 270-96; 1944.
98. Zakharov, A. F.; Egolina, N. A. Differential spiralization along mammalian mitotic chromosomes.I. BUdR-revealed differentiation in Chinese hamster chromosomes. Chromosoma 38: 341-65; 1972.

7

The Use of Cytogenetics to Study Genotoxic Agents in Fishes

by A. D. KLIGERMAN*

Abstract

Although fishes make up the largest and most diverse group of vertebrates, they have been little utilized in cytogenetic investigations of genotoxic phenomena. Early workers were hampered by the lack of appropriate cytogenetic methodologies and poor choice of species. By applying modern methods of metaphase chromosome preparation for analyzing chromosome breakage and sister chromatid exchange and selecting species with small numbers of relatively large chromosomes, fishes can be used in in vivo investigations designed to detect genotoxic agents.

Fishes of the class Osteichthyes make up the largest and one of the oldest classes of vertebrates. The class consists of more than 20,000 species with origins dating back to the Ordovician period over 400 million years ago (Lagler et al. 1977). While fishes have been important laboratory animals for behavioral, biochemical, and physiological studies, their value in mutagenic and hence carcinogenic research has only recently been explored. In order to obtain an appreciation for the role fishes can play in such studies, a general understanding of the relationship between mutagenicity and carcinogenicity is necesary.

During the last decade evidence has accumulated implicating environmental agents such as chemicals, ionizing radiations, and viruses as causative factors in the majority of cancers (Higginson 1969; Epstein 1974; Miller 1978). Though the exact mechanisms involved in the initiation of neoplastic events are unknown, recent studies have lent strong support to the somatic mutation theory of carcinogenesis,

---

*Box 3156 Duke University Medical Center, Department of Pathology, Durham, NC 27710.
Present address: Chemical Industry Institute of Toxicology
        Box 12137 Research Triangle Park, NC 27709

originally proposed by Boveri (1914). Simplified, the modernized version of this theory states that one or more mutational events in somatic cells are the primary factors leading to cellular dedifferentiation and tumor formation.

While a few early investigators such as Demerec (1947) and Strong (1949) were able to show that some chemical carcinogens were indeed mutagenic, a good correlation between the two phenomena could not be demonstrated (Burdette 1955). This was in part due to the fact that most chemical carcinogens (procarcinogens) must be metabolically activated to their ultimate forms before displaying mutagenic and carcinogenic activity. As reviewed by Miller (1978), a good correlation could be found if direct-acting carcinogens were tested for mutagenicity in systems not possessing metabolic activation, or if procarcinogens were evaluated in systems having the capacity for metabolic activation. Studies by Ames and his group (Ames et al. 1973; McCann and Ames 1976) have shown that approximately 90% of the carcinogens tested in their *Salmonella* histidine revertant system are indeed mutagenic. This correlation lends support to the theory that both phenomena are in many cases initiated through electrophilic attack on the genetic material (Miller 1978).

Other evidence for the somatic mutation theory of carcinogenesis comes from studies of human genetic disorders such as Bloom's syndrome (BS), Fanconi's anemia (FA), ataxia telangectasia (AT), and xeroderma pigmentosum (XP). Cells from such patients can show deficiencies in DNA repair (XP, AT, FA), increased rates of chromosome breakage and instability (BS, AT, FA), and a highly elevated rate of sister chromatid exchange (BS) (Chaganti et al. 1974). Patients with these disorders are prone to elevated rates of cancer (Pitot 1978).

Direct evidence for the involvement of DNA damage in the initiation of neoplasia was shown by Hart and Setlow (1975) using a gynogenetic strain of fish, *Poecilia formosa*. Cells from this species were irradiated with UV light to produce pyrimidine dimers in the DNA. Some of the cells were then subjected to photoreactivation light, which breaks down pyrimidine dimers. Isogenic recipient fish injected with cells exposed just to UV light developed granulomas and thyroid carcinomas, while fish that received cells that were UV-irradiated and subjected to photoreactivation light showed a ten-fold reduction in levels of these lesions. Thus, DNA damage in the form of pyrimidine dimers was implicated as a direct cause of neoplasia.

Paralleling the recent advancements in understanding the mechanisms of carcinogenesis has been the development of numerous short-term test systems designed to detect the presence of environmental carcinogens through their mutagenic action. Most of these systems have been designed to analyze the effects of chemical agents on the genomes of prokaryotes, fungi, or mammalian cells. Only recently has the potential for using lower vertebrates, such as fishes, for the study of the mutagenic and carcinogenic action of chemicals been realized.

The aim of this paper is to explore some of the cytogenetic methods that can be used to study the effects of mutagenic agents on the genomes of fishes. In particular, past work will be reviewed, the advantages and disadvantages of using fishes for cytogenetic studies discussed, and methodologies described and suggested. Finally, recommendations for future laboratory and field work will be presented.

Historical Review

According to Svärdson (1945), Schwarz in 1887 was probably the first investigator to report an attempt to count accurately fish chromosomes. Early on, cytogeneticists realized that karyotyping common fish species was difficult, and many of the early papers contain grossly inaccurate reportings of diploid chromosome numbers. This is due mainly to the fact that most fish species have large numbers of small chromosomes which are difficult to spread and count accurately. Thus, the vast majority of cytogenetic studies with fishes did not progress beyond reporting species' chromosome numbers and construction and comparison of karyotypes.

However, some dedicated investigators did persevere and attempted to use fishes to study the cytogenetic effects of various agents. Most of these studies involved the use of ionizing radiations. As early as 1913, Oppermann (1913) found that irradiation of trout sperm prior to fertilization caused chromatin damage that resulted in elimination of the male pronucleus during the subsequent cleavage stages of the fertilized egg. A quarter of a century later, Solberg (1938) discovered that changes in the X-ray sensitivity of *Fundulus* embryos paralleled changes in the mitotic indices: the fewer cells in active mitosis at the time of irradiation, the lower the degree of injury displayed by the developing embryo.

With the rapid development of atomic energy after World War II, there was renewed interest in the mutagenic and clastogenic effects of ionizing radiations. Most of the studies involved irradiation of fish embryos or gametes and were undertaken in the Soviet Union, where there has been a great emphasis on experimental manipulation of piscine reproduction, especially radiation-induced gynogenesis and androgenesis.

Belyaeva and Pokrovskaia (1958, 1959) performed a series of experiments to observe what effects doses of X-radiation would have on the developing embryos of the loach, *Misgurnus fossilis*. They found that increased doses of radiation at the blastula stage led to a dose-dependent suppression of mitosis and increases in the percentage of anaphase cells displaying chromosome bridges and fragments. Anaphase and telophase (cleavage) were found to be the most sensitive stages for the production of radiation damage to the embryo, and a correlation between the extent of damage to the developing embryo and the degree of chromosome damage was noted. Similiar results were obtained by Vakhrameeva and Neifakh (1959) with the loach, and by Prokof'yeva-Bel'govskaya (1961) and Migalovskaya (1973a) using the Atlantic salmon, *Salmo salar*. A study by Pankova (1965) showed that irradiation of fertilized loach ova at the four blastomere stage led to an increase in the percentage of abnormal cell divisions (those with bridges and fragments) as the cells progressed from blastula to gastrula stages.

Investigations into the phenomena of radiation-induced androgenesis and gynogenesis by Romashov et al. (1960, 1963), Bakulina et al. (1962), and Romashov and Belyaeva (1964, 1966) with several fish species showed that the peak in mortality of embryos produced from irradiated sperm and nonirradiated eggs coincided with the peak number of chromosome aberrations seen at anaphase and telophase. Furthermore, doses of radiation to the sperm as low as 100 R caused increases in anaphase and telophase bridges which correlated well with gastrula viability. Chromosome damage could be found in tissues of ectodermal, mesodermal, and endodermal origin, and the damage persisted up through the 30-day larval stage. Analysis of tail fin epithelium showed that the more severely deformed larva had the higher percentages of abnormal cells.

In addition to studying the effects of X-irradiating fish embryos and gametes, Soviet scientists also have investigated the effects of chronic exposure of developing fish eggs to waterborne radionuclides. Studies by Tsytsugina (1972, 1973) showed that eggs of *Scorpaena porcus* that developed in water with radioactivity as low as 1.0×10^{-9} curies/liter ^{90}Sr-^{90}Y had significantly higher numbers of cells with chromosomal bridges and fragments than did control embryos. Phenotypically abnormal larva that developed in water containing low levels of ^{90}Y had higher incidences of chromosome aberrations than did normal appearing larva, but changes in water radioactivity levels over several orders of magnitude *did not* increase appreciably the rate of chromosomal damage. Migalovskaya (1973b) investigated the response of developing eggs of *Salmo salar* to water containing ^{90}Sr-^{90}Y. He found that chronic exposure to 2×10^{-5} curies/liter caused statistically significant increases in chromosome bridges, fragments, and multipolar mitoses in developing blastulas. The radionuclides did not have an appreciable effect on mitotic indices.

In apparent contradiction to these results is a study by Pechkurenkov (1970) that failed to show the low dose effects reported by Tsytsugina (1972; 1973) and Migalovskaya (1973b). He found that loach eggs that developed in water containing from 3.1×10^{-10} curies/liter to 1.5×10^{-4} curies/liter ^{90}Sr-^{90}Y had similar mitotic indices and numbers of chromosome aberrations (bridges and fragments) when compared to controls. However, significant increases in chromosome aberrations were found when overripe roe was incubated in water containing 1.4×10^{-4} curies/liter. Pechkurankov believes that this is due to faulty chromosome repair caused by low oxygen tensions in the body of the female fish.

Work on the effects of chemical mutagens on fish chromosomes has been very limited. The majority of studies have been carried out by Tsoi and his coworkers (Tsoi 1970; Tsoi 1974; Tsoi et al. 1975) during their investigations of chemically induced gynogenesis in *Salmo irideus* and *Coregonus peled*. It was found that dimethyl sulfate and nitrosomethyl urea induced chromatid and chromosome bridges in the embryos that developed from the fertilization of normal eggs and treated sperm. In similar studies with the carp, *Cyprinus carpio*, fertilization of ova with sperm treated with dimethyl sulfate, nitrosoethyl urea, or 1,4-bis-diazoacetyl butane produced chromosome abnormalities in the developing gastrulas.

In one of the few field studies undertaken, Longwell (1976) used the basic cytogenetic methodology employed by the Soviet scientists to analyze mackerel (*Scomber scombrus*) eggs collected from the surface film of the New York Bight. An average of one-third of all division figures scored from eggs collected in this highly polluted area were abnormal, showing misoriented spindles, chromosome stickiness, breakage, and loss.

Unfortunately, the aforementioned studies have been plagued by unrealistically high control levels of chromosome damage and the inability to obtain sound quantitative estimates of induced genetic damage because of the cytogenetic methods used and species studied. Only recently have the modern cytogenetic techniques of metaphase chromosome analysis and sister chromatid exchange been employed to investigate radiation- and chemically-induced damage to the genome of fishes. These techniques will be discussed in detail in later sections.

Advantages and Disadvantages of Using Fishes for Cytogenetic Studies of Mutagenesis

Small fishes have many advantages for use in laboratory and field investigations of mutagenesis and carcinogenesis (Table 7.1). In many cases, species can be used that are inexpensive to obtain and care for and that require only limited amounts of space

Table 7.1 Advantages of Using Small Fishes in Studies of Mutagens and Carcinogens

1. Inexpensive to maintain in a laboratory
2. Require limited space and equipment
3. Many species are fecund and mature rapidly
4. Possess an efficient microsomal system
5. Are very sensitive to the carcinogenic action of chemicals
6. Make excellent indicator organisms for field studies

and equipment. Thus, relatively large numbers of animals can be used in laboratories where space and funding are at a premium. In addition, fishes possess efficient metabolic activation systems (Stich and Acton 1976; Scarpelli 1977; Stott and Sinnhuber 1978) and are extremely sensitive to the carcinogenic effects of chemicals (Matsushima and Sugimura 1976). Fishes are also ideal for use as biological indicator organisms (Statham et al. 1976) since many are higher order predators that can concentrate xenobiotics in their body fluids and tissues. Some long-lived species are valuable for use in epidemiological studies of tumor incidences (Oishi et al. 1976; Stich et al. 1976).

However, there are some disadvantages of using fishes in cytogenetic studies of mutagenesis which have limited the value of the information obtained to date. The main difficulty is that most species of fishes have large numbers ($2n \geq 48$) of small chromosomes ($<5\mu$) making the resolution of chromosome damage quite arduous. As reviewed in the previous pages, most scientists relied upon scoring damage at anaphase and telophase using aceto-orcein or aceto-carmine squashes of fixed material. While this method of analysis may be suited for quick field studies (Longwell 1976) where the choice of species is limited, it is inefficient and gives misleading results in radiation studies and is not recommended for studies of mutagenesis (Evans 1976). The problems with this technique are compounded by the fact that most fish tissues have low mitotic indices, and since there is no reliable way to accumulate cells at anaphase and telophase, analysis is restricted to embryonic or larval tissues where cells are dividing rapidly.

Thus, the choice of fishes with poor karyotypes (e.g. *Misgurnus fossilis*, $2n = 100$ [Raicu and Taisescu 1972]; *Salmo salar*, $2n = 53-60$ [Roberts 1970]) coupled with the use of inadequate methodologies has limited the quantitative value of the data obtained in the majority of studies. These problems can be overcome through the use of species with small numbers of large chromosomes and the adoption of modern cytogenetic methods of metaphase analysis.

Cytogenetic Methodology

Most cytogenetic studies done today involve the examination of metaphase chromosomes. The study of chromosome damage at metaphase allows for the observation of a greater number of division figures and presents a more precise and detailed picture of the effects of clastogenic agents than does anaphase or telophase analysis.

In general, obtaining well-spread metaphase figures from fishes involves six steps (see Appendix for specific details):

1. *Choice of tissues for analysis.* In order to obtain adequate numbers of metaphases, mitotically active tissues must be chosen. Developing fish eggs or larvae provide an excellent source of rapidly dividing cells. Mitotically active tissues found in mature fishes include epithelia of the intestines, gills, and scale, hematopoietic tissues such as the kidneys and spleen, and gonadal tissues. For in vitro work, certain tissues can be induced to divide in culture. Ovarian and embryonic tissue seem to be most readily adaptable to cell culture, but fin, air bladder, testes, liver, kidneys, and spleen have been successfully propagated (Denton 1973). Some researchers have also reported success in culturing the white blood cell of fishes (Denton 1973; Kang and Park 1975; Thorgaard 1976).

2. *Metaphase accumulation.* To obtain cells at metaphase, it is necessary to stop cells from progressing into anaphase. Various spindle poisons, including colchicine and its synthetic analog, colcemid, are effective in blocking spindle formation, thus causing the accumulation of metaphase figures. Colcemid appears to be less toxic than colchicine, but it is also more expensive.

3. *Hypotonic treatment.* In order to insure adequate spreading of the chromosomes in metaphase figures, cells must be subjected to an appropriate hypotonic treatment for 20–45 minutes before they are fixed. Various hypotonic treatments have been used including distilled water, diluted tissue culture media, and dilute salt solutions (e.g., sodium citrate, potassium citrate, potassium chloride). Trial and error experimentation can determine the optimal hypotonic solution and duration of treatment which are best suited for the particular species and tissues under investigation.

4. *Tissue fixation.* To preserve chromosome morphology and insure adequate contrast upon staining, nonhardening, freshly prepared fixatives should be used. Currently, the only recommended fixatives for routine cytogenetic analyses are 3:1 ethanol- or 3:1 methanol-acetic acid. Tissues can be stored in fixative for months at 4°C without noticeably diminishing the quality of chromosome preparations obtainable.

5. *Slide preparation.* A variety of squashing, air-drying, hot-plate drying, and ignition methods have been devised for getting metaphases from the tissue onto a clean, glass microscope slide. Usually the type of tissue or tissue culture dictates the appropriate method for preparation. In other instances, cells from one species respond better to one method than cells from another species. Thus, obtaining good preparations can be a matter of trial and error. However, it is usually advisable to avoid harsh methods of slide preparation such as the squash and ignition methods since these tend to disrupt chromosome morphology. Denton's *Fish Chromosome Methodology* (1973) and a review paper by Blaxhall (1975) describe some of the older procedures while newer methods are given in papers by Gold (1974) and Kligerman and Bloom (1977).

6. *Staining and slide examination.* Unstained slides can be studied by using phase contrast microscopy, but generally it is easier to analyze stained preparations. Giemsa and aceto-orcein are the most commonly used stains for vertebrate chromosomes. For details of these staining methodologies see Kligerman and Bloom (1977) and

Chen (1970). Stained preparations should be permanently mounted using suitable mounting media and #1 coverslips. Examination and scoring of aberration should be done using a light microscope equipped with an oil immersion objective at a total magnification of at least 900×. The addition of photomicroscopy apparatus can greatly improve analysis.

Model Systems Approach to Fish Cytogenetics

By a model systems approach, it is meant that a few species of fishes that have the appropriate attributes are used to gather basic experimental cytogenetic information (e.g., sensitivity, dose-response, repair) that can be extrapolated to other fishes as a group, much as *Drosophilia melanogaster, Mus musculus,* and *Vicia faba* are being used as models for other groups of organisms. Once appropriate model systems are designed, hypotheses can be constructed and tested and reliable statements can be made about the cytogenetic effects of environmental agents on the genomes of fishes.

One of the most crucial considerations to be made when initiating in vivo cytogenetic studies, aside from the choice of an appropriate methodology, is the selection of an experimental animal. Unfortunately, in most previous studies, this choice was made on the basis of economic importance, availability, and past use of the species in other experimental situations. While these criteria can be important, they should not be the major reasons for selecting a particular species for cytogenetic work.

Table 7.2 lists some of the major factors that should be considered when choosing an organism for cytogenetic studies. Of primary importance is the karyotype of the animal. As mentioned earlier, most fishes possess large numbers of small chromosomes like those shown in Figure 7.1, making them less than desirable for clastogenic investigations. Ideally, the karyotype of the organisms should be stable and consist of small numbers of large chromosomes possessing well-defined centromeric regions. Large metacentric and submetacentric chromosomes are well-suited for breakage studies.

Table 7.3 lists some fish species that from the point of view of karyotype might be

Table 7.2 Criteria for Choosing an in vivo Cytogenetics Model Organism (modified from Kligerman 1980)

Species:
1. Should have a satisfactory karyotype
2. Should possess tissues that yield adequate numbers of well-spread metaphases
3. Should be able to withstand experimental conditions
4. Should be easy to obtain and maintain in the laboratory
5. Should be relatively small in size
6.[a] May be suitable for field studies
7.[a] May be adaptable for in vitro studies
8.[a] May be bred in the laboratory, ideally producing multiple broods/year

[a] Of secondary importance

Figure 7.1 Chromosomes of: (a) brook trout, *Salvelinus fontinalis;* (b) blue-gill, *Lepomis macrochirus;* (c) fathead minnow, *Pimephales promelas;* (d) central mudminnow, *Umbra limi* (Bar = 10 μ).

useful model aquatic organisms for cytogenetic studies. However, karyotype is not the only criterion necessary for deciding whether a particular species will suffice.

For in vivo studies, the animals should yield adequate numbers of well-spread metaphases from a diversity of tissues. This enables one to investigate mutagen transport, blood-tissue barriers, differential tissue sensitivities, and variations in the metabolic capabilities of tissues. The species chosen should be readily adaptable to laboratory conditions (e.g., hardy, easy to feed, small in size) and relatively inexpensive to obtain and maintain. Of secondary importance are such factors as the suitability of the animals for field studies, the possible use of its tissues in in vitro work, and the ease in which the organism can be raised in the laboratory.

If studies are going to be conducted solely in vitro, most of the above-mentioned criteria can be relaxed. The major factors to consider are the organism's karyotype and the adaptability of its tissues to cell culture. Once the cells are in culture, the karyotype must remain stable and show a low spontaneous rate of breakage and/or sister chromatid exchange.

An In Vivo Aquatic Model System

In vivo investigations offer the most realistic appraisal of damage to the genome of fishes caused by exposure to environmental mutagenic agents. By using intact

Table 7.3 Fish Species with Karyotypes Suitable for Clastogenic Studies (from Kligerman 1980)

Species	Diploid Chromosome Number	Reference
Ameca splendens	2n=26	Woodhead (1976)
Aphyosemion celiae	2n=20	Scheel (1972)
Aphyosemion christyi	2n=18	Scheel (1972)
Aphyosemion franzwerneri	2n=22	Scheel (1972)
Apteronotus albifrons	2n=24	Howell (1972)
Characodon lateralis	2n=24	Uyeno and Miller (1974)
Galaxias maculatus	2n=22	Merrilees (1975)
Sphaericthys osphoromoides	2n=16	Calton and Denton (1974)
Umbra limi	2n=22	Beamish et al. (1971)
Umbra pygmaea	2n=22	Beamish et al. (1971)

organisms, differential tissue sensitivity and metabolic capabilities as well as blood-tissue barriers can be revealed. Furthermore, the in vivo processes of chemical concentration, dilution, and excretion can be observed directly. In addition, in vivo systems can be used to measure acute as well as chronic effects, and correlations between mutagenicity, carcinogenicity, and toxicity can be attempted. These types of investigations would be difficult if not impossible to perform in vitro.

As already mentioned, only limited numbers of cytogenetic studies have attempted to assess the damage induced by chemical and physical agents on the genomes of fishes. In an effort to overcome some of the problems inherent in these previous cytogenetic investigations, Kligerman et al. (1975), Kligerman and Bloom (1976), and Kligerman (1979) have developed a model system for the study of cytogenetic damage in fishes. This system makes use of the central mudminnow, Umbra limi (Fig. 7.2); a small oviparous teleost found in mud-bottomed streams and ponds throughout the north central and eastern United States and southern Canada. This species is particularly noteworthy because of its karyotype, which consists of 22 large metacentric and submetacentric chromosomes (Fig. 7.3). Umbra limi meets the five primary criteria and the first two secondary criteria listed in Table 7.2. In the initial studies, Kligerman et al. (1975) found that the intestines, gills, and kidneys produced adequate numbers of well-spread metaphases. When mudminnows were subjected to 325 R of X-radiation 3.5-7.5 hours before sacrifice, chromatid breaks and gaps could be found in approximately 30% of the metaphases examined (Fig. 7.4). This could be compared to the spontaneous aberration rate of less than 0.1% in the controls. Though this experiment was hampered by the use of the squash methodology, it did demonstrate that fishes could be used in in vivo studies to visualize accurately radiation-induced chromosome damage. Encouraged by these findings, Prein et al. (1978) undertook a study of Rhine River water using the eastern mudminnow, Umbra pygmaea. They found that fish kept in Rhine River water for 3, 7, or 11 days showed significantly higher rates of chromosome breakage in gill cell preparations than did controls exposed to untreated ground water. However, Sugatt (1978) was not able to verify this in field studies with U. pygmaea in a similar water system downstream from the area Prein et al. investigated. In laboratory experiments, Sugatt (1978) did find that mudminnows exposed to Trenimon showed a dose-dependent increase in chromosome damage.

Figure 7.2 The central mudminnow, *Umbra limi*. (Actual size 7.6 cm). (From Kligerman et al. 1975.)

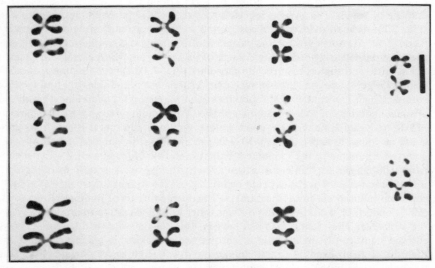

Figure 7.3 Karyotype of *Umbra limi*. (From Kligerman et al. 1975.) (Bar = 5 μ)

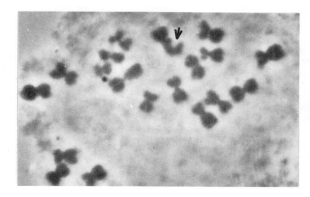

Figure 7.4 Chromatid break (see arrow) in a metaphase of a mudminnow exposed to 325 R of X-radiation. (From Kligerman et al. 1975).

However, chromosome breakage is not a very sensitive indicator of the presence of mutagenic agents. Many chemicals can cause significant genetic damage at concentrations that produce no visible structural aberrations. Thus, it was decided to incorporate the 5-bromodeoxyuridine (BrdU) method of studying sister chromatid exchange in vitro (Zakharov and Egolina 1972; Latt 1973; Kato 1974; Perry and Wolff 1974; Dutrillaux et al. 1974) into the in vivo mudminnow system.

In order to visualize SCEs, cells should go through two replications in the presence of BrdU. This causes the DNA in one chromatid of a metaphase chromosome to become bifilarly substituted with BrdU and the other chromatid to be singly substituted. Then, using the dye 33258 Hoechst and fluorescent microscopy (Latt 1973) or appropriate Giemsa staining procedures (Perry and Wolff 1974; Korenberg and Freedlender 1974; Goto et al. 1975), the two chromatids will stain contrasting colors enabling one to detect SCEs (Fig. 7.5a,b,c).

To visualize SCEs in the mudminnow, 500 μg/g BrdU were injected intraperitoneally (i.p.), and 33258 Hoechst-stained metaphases were examined for sister chromatid differentiation (SCD_2) characteristic of two rounds of replication in the presence of BrdU. Initial experiments (Kligerman and Bloom 1976) indicated that the cells of each tissue were highly asynchronous, but peak SCD_2 occurred after four and one-half to five and one-half days exposure to BrdU at 18°C. Varying the injected concentration of BrdU from 250 to 1000 μg/g did not have an appreciable effect on either SCD_2 or the rate of SCE.

For further experimentation (Kligerman 1979), the protocol shown in Figure 7.6 was adopted. Mudminnows were injected with 500 μg/g BrdU (Day 0), exposed to the chemical to be tested 24 hours later by either injection or addition to the water (Day 1), and sacrificed for SCE analyses of the gills, intestines, and kidneys four days later (Day 5). Baseline rates of SCE for these tissues were found to be quite low, with the gills having the lowest rate of SCE (2.0-3.3 SCEs/metaphase), the intestines having the highest rate (3.7-4.5 SCEs/metaphase), and the metaphases of the kidneys having an intermediate rate of exchange (2.6-3.4 SCEs/metaphase).

After it was shown that SCEs could be observed in metaphases of the mudminnows, experiments were undertaken to see if known mutagenic and carcinogenic agents would cause dose-dependent increases in the SCE rates.

Methyl methanesulfonate (MMS) is a direct-acting mutagenic carcinogen that is

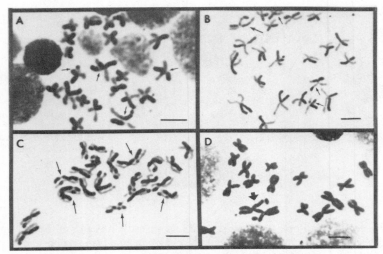

Figure 7.5 SCEs (see arrows) from MMS-exposed mudminnows. A) Renal metaphase from fish injected with 7 μg/g MMS. B) Intestinal metaphase from fish injected with 7 μg/g MMS. C) Metaphase from the intestines of a fish receiving 66 μg/g MMS having over 30 SCEs. D) Chromosome aberration (see arrow) showing a translocation and chromatid breaks from a fish given 103 μg/g MMS. (Arrows denote some of the chromosomes with multiple SCEs.) (Bar = 5 μ).

known to cause SCE in vitro (Perry and Evans 1975; Wolff et al. 1977) and in ovo (Bloom 1978). In order to determine if MMS could induce SCEs in fishes in vivo, mudminnows were injected i.p. with MMS on Day 1, and metaphases were analyzed for SCEs on Day 5. As shown in Figure 7.7, MMS induced a dose-dependent increase in SCEs in all tissues examined. Furthermore, significant increases in SCEs were seen at doses that did not produce chromosome breakage, and increases in chromosome breakage were not seen until high doses of MMS were administered (Fig. 7.5).

However, since most carcinogens are indirect-acting (i.e., must be metabolically activated), it was important to see if the mudminnow possesses the ability to convert a procarcinogen/promutagen into its active form. Cyclophosphamide (CP) was chosen since it is a promutagen/procarcinogen that will not induce SCE unless it is metabolized (Perry and Evans 1975; Allen and Latt 1976; Bloom 1978). Following the protocol shown in Figure 7.6, various amounts of CP were administered i.p. to mudminnows, and SCE analysis was performed. Dose dependent increases in SCEs were found in all tissues examined, with doses as small as 6 μg/g causing a doubling over baseline SCE rates (Fig. 7.8). The fact that CP could induce SCEs in the mudminnow was not surprising, since studies by Stich and Acton (1976), Scarpelli (1977), and Stott and Sinnhuber (1978) showed that livers from various fish species have the mixed-function oxidase enzymes necessary for metabolic activation.

In order for the mudminnow system to be of use in field studies or laboratory investigations mimicking normal exposure routes, it had to be demonstrated that mutagenic chemical agents present in the water could be taken up by the fish and

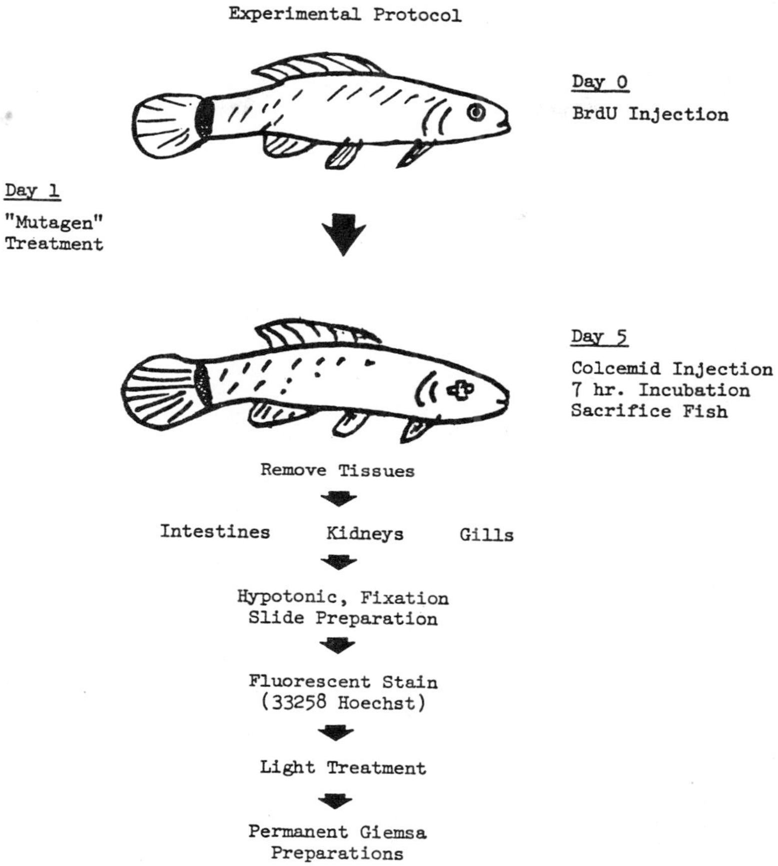

Experimental Protocol

Day 0
BrdU Injection

Day 1
"Mutagen"
Treatment

Day 5
Colcemid Injection
7 hr. Incubation
Sacrifice Fish

Remove Tissues

Intestines   Kidneys   Gills

Hypotonic, Fixation
Slide Preparation

Fluorescent Stain
(33258 Hoechst)

Light Treatment

Permanent Giemsa
Preparations

Figure 7.6 Experimental protocol used to study the effects of mutagens on the SCE rates of the tissues of the mudminnow. (From Kligerman 1979.)

lead to elevated SCE rates. Neutral red dye (NR) (3-amino-7-dimethyl-amino-2-methyl-phenazine HCl) was chosen for study since it was shown to be mutagenic in the Ames test (Longnecker et al. 1977) and could induce SCEs in the chick embryo system (Bloom 1978). In addition, NR is highly water soluble, displays low toxicity, and is taken up by living cells making it readily observable in the tissues of the organism.

One day after BrdU injection, mudminnows were exposed for four days to water concentration of NR ranging from 50 ppb to 12.5 ppm. Statistically significant increases in SCE rates were observed at water concentrations of 50 ppb, 100 ppb and 1 ppm for the gills, intestines, and kidneys, respectively (Fig. 7.9). Further analysis showed that the kidneys could concentrate NR over 500-fold and the gills over 100-fold the initial water concentrations.

Thus, an in vivo cytogenetic model system making use of a fish species is feasible.

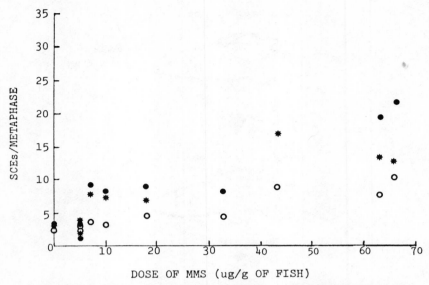

Figure 7.7 The effects of dose of MMS on the SCE rate of the gills (o), kidneys (*), and intestines (•). (From Kligerman 1979.)

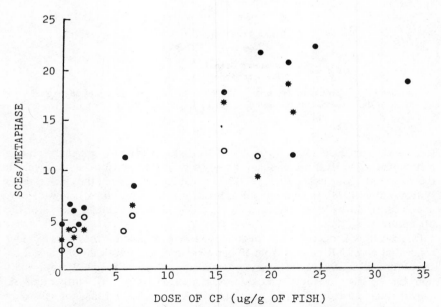

Figure 7.8 The effects of dose of CP on the SCE rate of the gills (o), kidneys (*), and intestines (•). (From Kligerman 1979.)

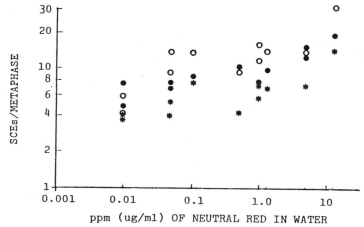

Figure 7.9 The effects of concentration of NR in the water on the SCE rate of the gills (o), kidneys (*), and intestines (•). (From Kligerman 1979.)

Results with the mudminnow system have shown that *Umbra* species can be used to study chromosome breakage both in the laboratory and field. In addition, through the use of SCE methodology, the action of both direct and indirect acting mutagenic carcinogens could be detected at concentrations that cause no significant chromosome breakage.

Nonetheless, these results are not sufficient to complete the validation of the system. Additional studies must be conducted to determine the sensitivity and response of the system to a wide diversity of genetically active substances as well as a large variety of substances that are known to be neither mutagenic nor carcinogenic. However, these initial results look promising.

In Vitro Cytogenetic Methods for Studying Mutagenesis in Fishes

In vitro systems cannot give as useful and as relevant a picture of the potential hazards posed by genetically active substances as can in vivo studies. However, cell culture methods offer many advantages. Once cell lines are established, they offer the investigator a continuous source of dividing cells displaying high mitotic indices. Furthermore, experiments can be controlled more easily and precisely than with in vivo systems, and problems encountered with feeding, and maintaining live animals are eliminated. The major drawback with tissue culture systems is trying to relate the data obtained in vitro with results that occur in vivo. With external penetrating ionizing radiations, this might not be too serious a problem since metabolism, transport, excretion, concentration, and dilution factors would not need to be taken into consideration. When chemical mutagenesis studies are attempted in vitro, the aforementioned processes become very important and difficult to mimic in vitro.

Woodhead (1976) has developed an effective system for studying the effects of radiation-induced chromosome damage in the cultured cells of *Ameca splendens*. Using cells derived from embryonic and ovarian tissues, he found that ^{60}CO induced a wide spectrum of chromosome damage and was able to construct acute dose-response curves from the data. Barker and Rackham (1979) have now incorporated the SCE methodology into the *Ameca splendens* system and have shown that ethyl methanesulfonate, methyl methanesulfonate, N-methyl-N-nitro-N-nitroso-guanidine, and mitomycin C induce dose-dependent increases in SCEs in vitro.

However, the use of in vitro methods to study chemically-induced cytogenetic damage will be limited unless a system is developed in which metabolic activation is coupled to the fish cells in culture. If this is accomplished, such a system might be useful as a quick preliminary screen to test water samples for mutagenic potential through chromosome breakage and SCE analysis. Positive results in vitro would warrant further testing in vivo.

Possible ways to circumvent the problem of activation would be to combine in vivo exposure with in vitro chromosome analysis. Blood from tissues exposed in vivo could be cultured in vitro or bile collected from fishes exposed to xenobiotics could be added to tissue culture.

Conclusions

The study of the mutagenic potential of water systems contaminated by man's activities is important not only for the quality of life found in the aquatic environment, but also for man himself since the aquatic environment is a source for human exposure to xenobiotics. Fishes provide an excellent source of material for the study of the mutagenic and carcinogenic potential of water samples since they are aquatic vertebrate organisms that can metabolize, concentrate, and store waterborne pollutants.

The direct analysis of neoplastic induction in fishes is important, but it is time consuming and expensive. With recent studies showing a strong relationship between an agent's mutagenic and carcinogenic capabilities, the use of mutagenic test systems to screen for carcinogenic agents has been gaining increased acceptance. A system that could detect the induction of point mutations at a number of loci in the fishes genome would be theoretically ideal. However, such a system would be time consuming and costly due to the large sample size needed to detect statistically significant increases in mutation frequencies. Thus, at present, cytogenetic methodologies offer the best opportunities for determining the effects genetically active substances have on the genomes of fishes.

Most of the clastogenic studies with fishes have been hampered by the choice of inappropriate experimental animals and poor cytogenetic methodologies. In addition, while chromosome breakage analysis can be used to study genomic damage caused by ionizing radiations, it is a fairly insensitive measure of the mutagenic capabilities of chemicals.

At present the best method for the analysis of genomic damage in fishes is the use of in vivo model systems for the detection of SCE. SCE analysis is a sensitive indicator of the presence of many mutagenic carcinogens. It is relatively easy to score, and requires much smaller sample sizes to obtain statistically significant results than does chromosome breakage or specific loci (point mutation) analyses.

The use of fish species to test for toxic, mutagenic, and carcinogenic potentials of water samples as well as to screen for genetically active compounds before they enter ecosystems offers great promise. As shown in Figure 7.10, fishes can be exposed to substances through a variety of routes, and both long-term and short-term endpoints

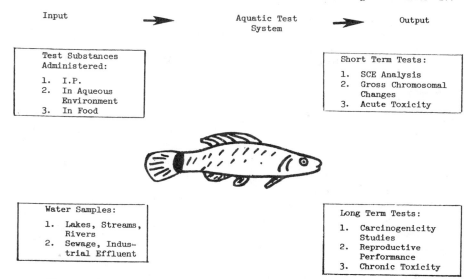

Figure 7.10 Model aquatic system for studying the effects of environmental pollutants. (From Kligerman 1979.)

monitored. However, one problem that continues to hinder research efforts, especially in industry and toxicology laboratories, is the heavy reliance upon the fathead minnow, blue gill, and trout as test organisms. While these are good species for use in toxicity and possible carcinogenecity studies, their karyotypes render them poor choices for cytogenetic work (Fig. 7.1a,b,c). The use of such species will likely end in frustration and abandonment of cytogenetic screening methods. Of utmost importance are choosing species with good karyotypes and using appropriate cytogenetic techniques. A new flexibility in outlook is needed to permit the modification or abandonment of outmoded methods and explore the use of new protocols.

To date, the number of investigators that have used cytogenetic methods in fishes to study mutagenicity is meager. However, recent studies that have shown the utility and feasibility of performing clastogenic and SCE analyses with fishes should create renewed interest in this area and lead to the development of improved methods to monitor water quality.

APPENDIX
In Vivo Piscine Cytogenetic Methodology (modified from Kligerman and Bloom 1977)

1. Select healthy, well-fed fishes for study.
2. Approximately 7 hr. prior to sacrifice, inject fishes with 25 µg/g colcemid or colchicine and return to well-aerated aquaria. For fishes too small to inject, allow them to swim for 6–8 hours in 0.005 to 0.01% well-aerated colchicine solution.

3. Sacrifice fish by decapitation and cut the gills, kidneys, intestines, and any other mitotically active tissues into small pieces (approximately 0.5 cm^3) and place each tissue in about 10 times its volume of 0.4% KCl hypotonic solution for 30 min.

4. Remove tissues from the hypotonic solution and place in 3:1 ethanol- or methanol-acetic acid fixative for 30 min. Remove fixative and replace with fresh fixative. Fix tissues for at least 2 hr. and preferably overnight.

5. After the tissues are adequately fixed, remove one or two pieces of tissue from the fixative, touch them briefly to absorbent paper to remove excess fixative, and immediately place them in the depression of a concave slide, and add two and three drops of 50% acetic acid.

6. Mince the tissue gently with forceps to form a cell suspension and return unsuspended pieces of tissue to fixative.

7. Using a microhematocrit capillary tube or a micropipette, withdraw approximately 30 µl of suspension and expel it onto a clean glass slide heated to 40-60°C on a slide warmer.

8. Approximately 5 sec. later, withdraw the suspension from the heated slide back into the capillary tube or pipette leaving behind a ring of dried cellular material approximately 1 to 2 cm in diameter.

9. Repeat steps 7 and 8 producing two or three rings/slide.

10. Stain slides in 4% Giemsa in 0.01 M phosphate buffer (pH 7) for 10 min.

11. Rinse slides in distilled water, air-dry, and place in xylene for at least 10 min.

12. Mount with suitable mounting media using #1 glass coverslips.

Literature Cited

1. Allen, J. W.; Latt, S. A. In vivo BrdU-33258 Hoechst analysis of DNA replication kinetics and sister chromatid exchange formation in mouse somatic and meiotic cells. Chromosoma 58:325-40; 1976.

2. Ames, B. N.; Durston, W. E.; Yamasaki, E.; Lee, F. D. Carcinogens are mutagens: a simple test system combining liver homogenates for activation and bacteria for detection. Proc. Nat. Acad. Sci. USA 70:2281-85; 1973.

3. Bakulina, E. D.; Pokrovskaia, G. L.; Romashov, D. D. On radiosensitivity of loach (*Misgurnus fossilis* L.) spermatozoa. Radiobiology 2: 135-47; 1962. AEC-tr-5428 (Transl. from Russian of Radiobiologiya 2:92; 1962).

4. Barker, C. J.; Rackham, B. D. The induction of sister chromatid exchanges in cultured fish cells (*Ameca splendens*) by carcinogenic mutagens. Mutat. Res. 68:381-87; 1979.

5. Beamish, R. J.; Merrilees, M. J.; Crossman, E. J. Karyotypes and DNA values for members of the suborder *Esocoidei (Osteichthyes:Salmoniformes)*. Chromosoma 34:436-47; 1971.

6. Belyaeva, V. N.; Pokrovskaia, G. L. Arrest of mitosis by X-rays at early developmental stages in loach spawn. Dokl. Akad. Nauk SSSR 119:149-55; 1958. (Transl. from Russian, p. 361).

7. Belyaeva, V. N.; Pokrovskaia, G. L. Changes in the radiation sensitivity of loach spawn during the first embryonic mitoses. Dokl. Akad. Nauk SSSR 125:192-95; 1959. (Transl. from Russian, p. 632).

8. Blaxhall, P. C. Fish chromosome techniques—a review of selected literature. J. Fish. Biol. 7:315-20; 1975.

9. Bloom, S. E. Chick embryos for detecting environmental mutagens. *In* Hollaender, A.; de Serres, F. J., eds. Chemical mutagens: principles and methods for their detection. Vol. 5. New York: Plenum Press; 1978: 203-32.

10. Boveri, T. Zur Frage der Entstehung maligner Tumoren. Fisher, Jena; 1914.

11. Burdette, W. L. The significance of mutation in relation to the origin of tumors: a review. Cancer Res. 15:201-26; 1955.

12. Calton, M. S.; Denton, T. E. Chromosomes of the chocolate gourami: a cytogenetic anomaly. Science 185:618-19; 1974.

13. Chaganti, R. S. K.; Schonberg, S.; German, J. A manyfold increase in sister chromatid exchanges in Bloom's syndrome lymphocytes. Proc. Nat. Acad. Sci. USA 71:4508-12; 1974.

14. Chen, T. R. Fish chromosome preparation: air-dryed displays of cultured ovarian cells in two killifishes (*Fundulus*). J. Fish. Res. Bd. Can. 27:158-61; 1970.

15. Demerec, M. Production of mutations in *Drosophila* by treatment with some carcinogens. Science 105:634; 1947.

16. Denton, T. E. Fish chromosome methodology. Springfield, IL: Charles C. Thomas, Publisher; 1973.

17. Dutrillaux, B.; Fosse, A. M.; Prieur, M.; Lejeune, J. Analyse des échanges de chromatides dans les cellules somatiques humaines. Traitement au BudR (5-bromodéoxyuridine) et fluorescence biocolore par l'acridine orange. Chromosoma 48:327-40; 1974.

18. Epstein, S. S. Environment determinants of human cancer. Cancer Res. 34:2425-35; 1974.

19. Evans, H. J. Cytological methods for detecting chemical mutagens. In Hollaender, A., ed. Chemical mutagens: principles and methods for their detection. Vol. 4. New York: Plenum Press; 1976: 1-29.

20. Gold, J. R. A fast and easy method for chromosome karyotyping in adult teleosts. Prog. Fish Cult. 36:169-71; 1974.

21. Goto, K.; Akematsu, T.; Shimazu, H.; Sugiyama, T. Simple differential Giemsa staining of sister chromatids after treatment with photosensitive dyes and exposure to light and the mechanism of staining. Chromosoma 53:223-30; 1975.

22. Hart, R. W.; Setlow, R. B. Direct evidence that pyrimidine dimers in DNA result in neoplastic transformation. In Hanawalt, P. C.; Setlow, R. B. eds., Molecular mechanisms for repair of DNA, Part B. New York: Plenum Press; 1975: 719-24.

23. Higginson, J. Present trends in cancer epidemiology. Can. Cancer Conf. 8:40-75; 1969.

24. Howell, W. M. Somatic chromosomes of the black ghost knifefish, *Apteronotus albifrons* (*Pisces: Apteronotidae*). Copeia No. 1:191-93; 1972.

25. Kang, Y. S.; Park, E. H. Leukocyte culture of the eel without autologous serum. Japan. J. Genetics 50:159-61; 1975.

26. Kato, H. Spontaneous sister chromatid exchanges detected by a BudR-labelling method. Nature 251:70-72; 1974.

27. Kligerman, A. D. Induction of sister chromatid exchanges in the central mudminnow following in vivo exposure to mutagenic agents. Mutat. Res. 64:205-17; 1979.

27a. Kligerman, A. D. The use of aquatic organisms to detect mutagens that cause cytogenetic damage. In Egami, N., ed. Radiation effects on aquatic organisms. Tokyo: Japan Scientific Societies Press; 1980: 241-52.

28. Kligerman, A. D.; Bloom, S. E. Sister chromatid differentiation and exchanges in adult mudminnows (*Umbra limi*) after in vivo exposure to 5-bromodeoxyuridine. Chromosoma (Berl.) 56:101-9; 1976.

29. Kligerman, A. D.; Bloom, S. E. Rapid chromosome preparations from solid tissues of fishes. J. Fish. Res. Bd. Can. 34:266-69; 1977.

30. Kligerman, A. D.; Bloom, S. E.; Howell, W. M. *Umbra limi*: a model for the study of chromosome aberrations in fishes. Mutat. Res. 31:225-33; 1975.

31. Korenberg, J. R.; Freedlender, E. F. Giemsa technique for the detection of sister chromatid exchanges. Chromosoma 48:355-60; 1974.

32. Lagler, K. F.; Bardach, J. E.; Miller, R. R.; Passino, D. R. Ichthyology. New York: John Wiley and Sons; 1977.

33. Latt, S. A. Microfluorometric detection of deoxyribonucleic acid replication in human metaphase chromosomes. Proc. Nat. Acad. Sci. USA 70:3395-99; 1973.

34. Longnecker, D. S.; Curphey, T. J.; Daniel, D. S. Mutagenicity of neutral red. Mutat. Res. 48:109-12; 1977.

35. Longwell, A. C. Chromosome mutagenesis in developing mackerel eggs sampled from the New York Bight. NOAA Technical Memorandum ERL MESA-7; 1976.

36. Matsushima, T.; Sugimura, T. Experimental carcinogenesis in small aquarium fishes. Prog. Exp. Tumor Res. 20:367-79; 1976.

37. McCann, J.; Ames, B. N. Detection of carcinogens as mutagens in the Salmonella/microsome test: assay of 300 chemicals: discussion. Proc. Nat. Acad. Sci. USA 73:950-54; 1976.

38. Merrilees, M. J. Karyotype of *Galaxias maculatus* from New Zealand. Copeia No. 1:176-78; 1975.

39. Migalovskaya, V. N. Effect of X-irradiation on the gametes and embryonal cells of the Atlantic salmon. *In* Sorokin, B. P., ed. Effect of ionizing radiation on the organism. AEC-tr-7418.Washington, D. C.: USAEC; 1973a: 100-112.

40. Migalovskaya, V. N. Chronic effect of strontium-90 + yttrium-90 on the frequency of chromosomal aberrations in the embryonic cells of the Atlantic salmon. *In* Sorokin, B.P., ed. Effect of ionizing radiation on the organism. AEC-tr-7418. Washington, D.C.: USAEC; 1973b: 89-99.

41. Miller, E. C. Some current perspectives on chemical carcinogenesis in human and experimental animals: Presidential Address. Can. Res. 38:1479-96; 1978.

42. Oishi, K.; Yamazaki, F.; Harada, T. Epidermal papillomas of flatfish in the coastal waters of Hokkaido, Japan. J. Fish. Res. Bd. Canada 33:2011-17; 1976.

43. Oppermann, K. Die Entwicklung von Forelleneiern nach Befruchtung mit Radioumbestrahlten Samenfaden. Archiv Für Mikroskopische Anatomie 83:307-23; 1913.

44. Pankova, N. Damage to the chromosomes in a series of cellular generations in irradiated loach embryos. Radiobiology 5:127-33; 1965. AEC-tr-6599 (Transl. from Russian of Radiobiologiya 5:248; 1965).

45. Pechkurenkov, V. L. Appearance of chromosome aberrations in larvae of the loach (*Misgurnus fossilis* L.), developing in solutions of strontium-90 and yttrium-90 of various activities. Soviet Genetics 6:1323-32; 1970.

46. Perry, P.; Evans, H. J. Cytological detection of mutagen-carcinogen exposure by sister chromatid exchange. Nature 258:121-25; 1975.

47. Perry, P.; Wolff, S. New Giemsa method for the differential staining of sister chromatids. Nature 251:156-58; 1974.

48. Pitot, H. C. Fundamentals of oncology. New York: Marcel Dekker, Inc.; 1978.

49. Prein, A. E.; Thie, G. M.; Alink, G. M.; Koeman, J. H.; Poels, C. L. M. Cytogenetic changes in fish exposed to water of the river Rhine. The Science of the Total Environment 9:287-91; 1978.

50. Prokof'yeva-Bel'govskaya, A. A. Radiation injury to chromosomes at early stages of development of the salmon. Tsitologiya 3:437-45; 1961. JPRS 12195 (Transl. from Russian).

51. Raicu, P.; Taisescu, E. *Misgurnus fossils*, a tetraploid-fish species. J. of Heredity 63:92-94; 1972.

52. Roberts, F. L. Atlantic salmon (*Salmo salar*) chromosomes and speciation. Trans. Amer. Fish. Soc. 99:105-11; 1970.

53. Romashov, D. D.; Belyaeva, V. N. Cytology of radiation gynogenesis and androgenesis in the loach (*Misgurnus fossilis* L.). Dokl. Akad. Nauk SSSR 157:503-6; 1964. (Transl. from Russian, p. 964).

54. Romashov, D. D.; Belyaeva, V. N. On the conservation of radiation damage to the chromosomes in fish embryogenesis. Soviet Genetics 2:1-10; 1966. (Transl. from Russian of Genetika 2:4; 1966.

55. Romashov, D. D.; Golovinskaia, K. A.; Belyaeva, V. N.; Bakulina, E. D.; Pokrovskaia, G. L.; Cherfas, N. B. Diploid radiation gynogenesis in fish. Biophysics 5:524-32; 1960.

56. Romashov, D. D.; Nikolyukin, N. I.; Belyaeva, V. N.; Timofeeva, N. A. Possibilities of producing diploid radiation-induced gynogenesis in sturgeons. Radiobiology 3:145-54; 1963. AEC-tr-5434. (Transl. from Russian of Radiobiologiya 3:104; 1963).

57. Scarpelli, D. G. General discussion. Ann. N.Y. Acad. Sci. 298:463-81; 1977.

58. Scheel, J. J. Rivuline karyotypes and their evolution (Rivulinae, Cyprinodontidae, Pisces). Zeitschrift für Zoologische Systematik und Evolutionsforschung 110:180-209; 1972.

59. Solberg, A. N. The susceptibility of *Fundulus heteroclitus* embryos to X-radiation. Journal of Exp. Zoo. 78:441-65; 1938.

60. Statham, C. N.; Melancon, M. J. Jr.; Lech, J. J. Bioconcentration of xenobiotics in trout bile: a proposed monitoring aid for some waterborne chemicals. Science 193:680-81; 1976.

61. Stich, H. F.; Acton, A. B. The possible use of fish tumors in monitoring for carcinogens in the marine environment. Prog. Exp. Tumor Res. 20:44-54; 1976.

62. Stich, H. F.; Acton, A. B.; Forrester, C. R. Fish tumors and sublethal effects of pollutants. J. Fish Res. Bd. Can. 33:1993-2001; 1976.

63. Stott, W. T.; Sinnhuber, R. O. Trout hepatic enzyme activation of aflatoxin B_1 in a mutagen assay system and the inhibitory effect of PCBs. Bulletin of Environ. Contam. and Toxic. 19:35-41; 1978.

64. Strong, L. C. The induction of mutations by a carcinogen. Proc. 8th Int. Congr. Genetics. Hereditas Suppl. 1949:486-99.

65. Sugatt, R. H. Chromosome aberrations in the eastern mudminnow (*Umbra pygmaea*) exposed in vivo to Trenimon or river water. Unpublished report MD-N+E 78/3 Central Laboratory TNO; 1978.

66. Svärdson, G. Chromosome studies on Salmonidae. Reports from the Swedish State Institute of Freshwater Fishery Research: Drottningholm #23, Stockholm 1-151; 1945.

67. Thorgaard, G. H. Robertsonian polymorphism and constitutive heterochromatin distribution in chromosomes of the rainbow trout (*Salmo gairdneri*). Cytogenet. Cell Genet. 17:174-84; 1976.

68. Tsoi, R. M. Effect of nitrosomethyl urea and dimethyl sulfate on sperm of rainbow trout (*Salmo irideus* Gibb.) and peled (*Coregonus peled* Gmel.). Dokl. Adad. Nauk SSSR 189:849-51; 1970. (Transl. from Russian, p. 411).

69. Tsoi, R. M. Chemical gynogenesis in *Salmo irideus* and *Coregonus peled*. Soviet Genetics 8:275-77; 1974.

70. Tsoi, R. M.; Men'shova, A. I.; Golodov, Y. F. Specificity of the influence of chemical mutagens on spermatozoids of *Cyprinus carpio* L. Soviet Genetics 10:190-93; 1975.

71. Tsytsugina, V. G. Effect of incorporated radionuclides on the chromosome apparatus of ocean fish. *In* Polikarpov, G. G., ed. Marine radioecology. AEC-tr-7299. Washington, D.C.:USAEC; 1972:157-65.

72. Tsytsugina, V. G.: Effect of incorporated radionuclides on the chromosomal apparatus of marine fish. *In* Sorokin, B. P., ed. Effect of ionizing radiation on the organism. AEC-tr-7418. Washington, D.C.: USAEC; 1973:157-65.

73. Uyeno, T.; Miller, R. R. Second discovery of multiple sex chromosomes among fishes. *In* Genetic studies of fish: II. New York: MSS Information Corp.; 1974:98-102.

74. Vakhrameeva, N. A.; Neifakh, A. A. Comparison of changes in radiation and thermal sensitivity during cleavage of the loach egg (*Misgurnus fossilis*). Dokl. Akad. Nauk SSSR 128:779-82; 1959. (Transl. from Russian, p. 429).

75. Wolff, S.; Rodin, B.; Cleaver, J. E. Sister chromatid exchanges induced by mutagenic carcinogens in normal and xeroderma pigmentosum cells. Nature 265:347-49; 1977.

76. Woodhead, D. S. Influence of acute irradition on induction of chromosome aberrations in cultured cells of the fish *Ameca spendens*. *In* Biological and environmental effects of low-level radiation. Vol. 1. Vienna: IAEA; 1976:67-76.

77. Zakharov, A. F.; Egolina, N. A. Differential spiralization along mammalian mitotic chromosomes. I. BUdR-revealed differentiation in Chinese hamster chromosomes. Chromosoma 38:341-65; 1972.

8

Human Peripheral Blood Lymphocyte Cultures: An In Vitro Assay for the Cytogenetic Effects of Environmental Mutagens

by K. E. BUCKTON and H. J. EVANS*

Abstract

Human peripheral blood provides a readily available source of cells that can be made to undergo mitotic division in culture and thereby provide metaphase cells for chromosome aberration and SCE scoring. This in vitro test system has been used to study the response of human cells to chromosome-damaging agents; the methods of culturing blood lymphocytes and of assaying for chromosome damage are described and discussed. Over 100 chemicals have been tested using this system and the results from these studies are summarised.

The ideal in vitro system for assaying for environmental agents that would damage human chromosomes following in vivo exposure should fulfil three requirements: (1) be a simple in vitro culture system that could provide large numbers of dividing cells and high quality chromosome preparations; (2) have the same sensitivity of response as the human genome; (3) contain metabolising systems (including activation, detoxification, and repair processes) identical to those that would be important, or relevant, to the tissues of interest in man—whether these be gonads or epithelial cells in skin, gut, etc., or other proliferating cells such as in the bone marrow. The very existence of differences in response between tissues indicates that no one system could fulfill all of these requirements, but perhaps the nearest approach would be the

*MRC Clinical and Population Cytogenetics Unit, Western General Hospital, Crewe Road, Edinburgh EH4 2XU.

human peripheral blood lymphocyte culture system. Such a system utilizes cells of human origin; they are readily available and are easily cultured to yield large numbers of well-spread mitotic figures; and the cells are able, to some extent, to activate certain mutagens that require to be metabolised before being transformed into a mutagenically active state, e.g., the polycyclic hydrocarbons such as benzo-(a)pyrene.

The use of human peripheral blood lymphocyte chromosomes as indicators of genetic damage in the somatic chromosomes of man stems from the successful application of the techniques of human cytogenetics in studies on chromosome damage in people, and in cells, exposed to ionizing radiations (19). In individuals exposed to ionizing radiations, blood samples taken after exposure provide populations of cells that are normally in a resting phase, but which can be made to divide in short-term culture and reveal any chromosome damage sustained as a consequence of exposure. In the same way, blood samples taken from the body have been exposed to radiation in vitro and the responses obtained from whole body irradiation in vitro and in vivo compared and shown to be qualitatively and quantitatively identical (7). For a given quality of radiation, there is a strict relationship between radiation dose and frequency of induced chromosome aberrations, so that in individuals accidentally exposed to radiation it is possible to extrapolate from aberration frequency in peripheral blood cells to obtain a realistic quantitative estimate of absorbed dose (60). This technique is now routinely used in cases of accidental exposure and provides a reliable monitoring system.

Chromosome damage can also be detected by cytogenetic studies on blood lymphocyte chromosomes of individuals exposed to chemical mutagens, e.g., patients exposed to cytotoxic drugs such as the alkylating agent cyclophosphamide (101) or workers occupationally exposed to mutagens such as those in the plastics industries, e.g., styrene oxide (65). The in vitro-in vivo quantitative equivalence of effect observed with penetrating ionizing radiations, where primary damage is independent of metabolism and strictly related to absorbed dose, cannot be expected with chemical mutagens, where the agent may be taken up and localised to a particular organ and where, for a given concentration, the mutagenic potency of an agent or its derivatives may differ widely between tissues. An ever present problem with in vitro mutagenicity test systems, therefore, is the relevance and applicability of in vitro results to in vivo expectations. This problem is not a disadvantage peculiar only to the in vitro blood lymphocyte system, but is common to all in vitro systems and presumably to a lesser extent (in some cases) to all nonhuman in vivo experimental systems. Notwithstanding the problems in quantitative extrapolation from test tube (or animal) to man, a human cell in vitro system is perhaps the nearest approach to a qualitatively informative and workable test system. In the present article, we consider the types of damage that are induced in human lymphocyte chromosomes exposed to chemical mutagens in vitro and the methods used to reveal such damage.

Blood Culture Technique

A basic culture method for obtaining chromosome preparations from human peripheral blood lymphocytes requires less than 1 ml of whole blood per culture. A standarized technique as used in our laboratory is as follows.

A sample of peripheral blood is taken and injected into a tube containing lithium heparin as an anticoagulant. A small amount (0.8 ml) of whole blood is then transferred to a bottle with a 25 ml capacity, containing 9 ml of liquid which consists of 84% of culture medium (Ham's F10), 15% of bovine serum, and 1% of phytohaemagglutinin (Burroughs Wellcome), also the antibiotics streptomycin 10 IU/ml and penicillin 100 IU/ml. The bottle containing culture medium and cells is then incubated at 37°C, preferably in a water bath, which maintains a more stable temperature than an incubator.

The sequence of events that occur during incubation is as follows. The small lymphocytes rapidly start synthesising RNA and enlarge; after about 18 hours the first cells start synthesising DNA (S-period). The S-period lasts approximately 20 hours, but, as the cells are nonsynchronous in their development, some cells will be found in S from about 18 hours onwards. After the S-period the cells enter the G_2 period which can last from 2 hours to 2 days. The first cells will start dividing at about 36 hours in culture, but this is very variable as between individual blood donors, i.e., there are "early growers" and "late growers" among the donors themselves. The time at which the first mitoses occur is also very dependent on the culture conditions. It has been found that if one requires the majority of cells in the first division in culture, then the cultures should be harvested at or before two days following their initiation; if second or later divisions are required, then the cultures should be harvested after longer intervals in culture. Two or three hours before harvesting the mitotic cells, colchicine (final concentration 5×10^{-7}M) is added to the culture in order to arrest the cells in mitosis.

Harvesting for chromosome preparations is carried out by removing the culture medium from the cells and exposing them to 0.075M KCl for 8 min. at room temperature, then removing the KCl and fixing the cells in 3 parts methyl alcohol to 1 part glacial acetic acid. This fixative should be changed several times to make sure that the cells are well-fixed and all the haemolised red cells are removed. To spread the cells on the slide, the cells in a suspension of fixative are dropped onto clean, dry slides and allowed to air-dry. The cells can then be stained immediately with either aceto orcein or Giemsa, or allowed to age for a few days before being subjected to the various treatments that may be necessary to obtain banded chromosomes.

Cytological Principles and Types of Damage Assayed

Our knowledge and approach in studying the types and consequences of alterations in chromosome structure induced following exposure to chemical mutagens is firmly based upon the wealth of information acquired from studies on the effects of ionizing radiations on chromosomes (19). Comparative studies show that the chromosome aberration, that is, the end-product scored, is similar in structure whether it is induced by a physical or chemical agent and is basically a result of a breakage and/or exchange of chromosome subunits. The number and distribution of such breakage/exchange events will depend upon various factors, viz., the nature of the mutagen; its mode of action in causing breakage/exchange; the development phase of the cell at the time of exposure; and whether breakage/exchange occurs before, during, or after the chromosome replication that is completed prior to observation. In addition, the efficiency of detecting induced aberrations will depend upon the selected time of cell sampling for cytological processing.

The aberrations are best observed in contracted metaphase chromosomes at the first (M_1) mitosis following their induction, but a proportion will result in chromatin bridges and fragments that may be seen, in the absence of colchicine, at the ensuing anaphase. Since a proportion of fragments are excluded from daughter telophase nuclei, these may in turn give rise to micronuclei that may be observed in interphase daughter cells. Some of the aberrations may be cell lethal, others may have little effect on viability and may be transmitted to daughter cells and be observed at the metaphase of the second (M_2) or subsequent mitosis following their induction. Cytological damage may, therefore, be detected in cells or their descendants sometime after exposure, but it is nevertheless clear that the best estimate of total aberration frequency is obtained by analyzing aberrations at the first metaphase following their induction. However, this may not always be possible in practice. For instance, when cells are exposed to ionizing radiations, the lesions that develop into chromosome aberrations do so within minutes or, at most, hours following exposure, so that virtually all the induced aberrations can be seen at the first mitosis (M_1) following irradiation. On the other hand, the lesions induced by many chemical mutagens that have a so-called delayed action (49) do not manifest themselves as aberrations until the damaged chromosomes proceed through a replication phase (21). With such agents, lesions induced in G_2 cells do not result in aberrations at M_1, but do so at the M_2 division. Moreover, not all lesions develop or interact to cause aberrations at the replication phase immediately following treatment, so that new aberrations may be generated in subsequent cell cycles. It may be difficult, therefore, to obtain accurate and comprehensive data on the levels of damage induced by chemical mutagens which have delayed effects, and it may be necessary to obtain estimates by analyzing cells at both M_1, M_2, and possibly subsequent divisions. Alternatively, and depending upon the cell systems used, valid information may be gained by counting the numbers of micronuclei produced over a given time period after exposure.

Chromosomal Aberrations in Metaphase Cells

The aberrations observed in metaphase cells are basically of two sorts distinguished on the basis of the unit of breakage/exchange. Those in which the structural changes involve the whole chromosome, i.e., both sister chromatids at identical loci, or chromosome-type aberrations, and those in which the unit of breakage/exchange is the half chromosome or chromatid that is, chromatid-type aberrations. One particular type of chromatid-type change that we shall consider separately is the sister chromatid exchange that does not result in an altered chromosome morphology, but which can be detected using special staining techniques. Sister chromatid exchanges are discussed in detail by Latt elsewhere in this volume and will therefore not be considered extensively in this chapter.

Chromosome-type aberrations. Chromosome-type aberrations are characteristic of damage sustained in G_1 cells which is translated into breakage/exchange events *prior* to chromosome replication. The aberrations induced include:

   1. Asymmetrical chromosome aberrations (Fig. 8.1)

Polycentrics. The main indication of chromosome damage is the "dicentric" which results from an exchange between two chromosomes giving a structure with two cen-

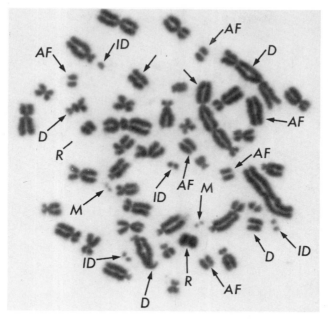

Figure 8.1 Human lymphocyte with all types of chromosome damage: D—Dicentric; T—Tricentric; R—Ring; AF—Acentric fragment; ID—Interstitial deletion; M—Minute.

tromeres and an acentric fragment. Much less frequently, a structure with three (tricentric) or more centromeres is produced.

Centric Rings. A centric ring is formed from a single chromosome which has been broken in both arms; the proximal segments join to form the ring and the distal segments join to form an acentric fragment. The number of rings induced is small compared to the number of dicentrics.

Acentric elements. These consist of three different categories: terminal deletions, interstitial deletions, and minutes. The distinction between these three aberration types is one of interpretation and they may be easily confused. A terminal deletion (acentric fragment) will result in a structure longer than the width of a chromatid and without a centromere with parallel-lying paired chromatids. An interstitial deletion has the appearance of an acentric ring, but the structure can be so small that the hole in the middle of the structure is not apparent. A minute is a very small interstitial deletion, also with paired chromatid structures, but smaller than the width of a chromatid.

2. Symmetrical chromosome exchanges (Fig. 8.2)

These are very difficult to detect on a conventionally stained chromosome preparation and are usually ignored in most assay systems because of the unreliability of scoring. A symmetrical chromosome exchange will result in either a reciprocal translocation or an inversion and, if seen as an aberration, will be apparent as one or more abnormal monocentrics. In order to improve the detection of symmetrical ex-

188 Human Peripheral Blood Lymphocyte Cultures

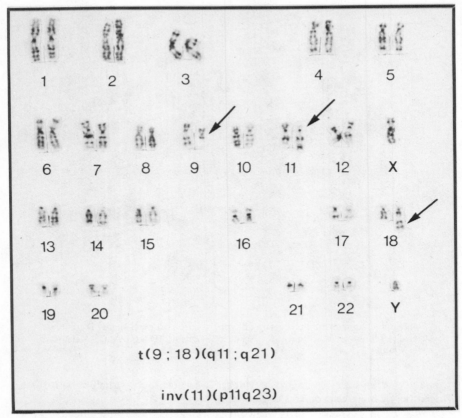

Figure 8.2 Karyotype of a human lymphocyte with a reciprocal translocation: t(9;18)(q11;q21); and a pericentric inversion: inv(11)(p11q23).

changes, a banded chromosome preparation can be used. It has been found that following exposure to irradiation the proportion of symmetrical:asymmetrical aberrations seen is 1 : 1 (6), but the time taken to analyze the cells in order to detect the symmetrical aberrations probably outweighs the benefits accrued therefrom. Shown in Figure 8.2 is a reciprocal translocation and pericentric inversion seen in a G-banded cell. A number of chromosome-banding techniques have been published and are available in the literature. Those commonly in use in this laboratory are a modification of the ASG technique (120) for G-banding, and the Sehested technique (103) for R-banding.

Chromatid-type aberrations (Fig. 8.3). Chromatid-type aberrations are characteristic of damage in single chromatids (sustained at any stage in the cell cycle) which are translated into breakage/exchange events *during* or *after* the replication of chromosomes in the S phase of the cycle. The independence of the chromatids with regard to breakage, and their close pairing within chromosomes, results in a

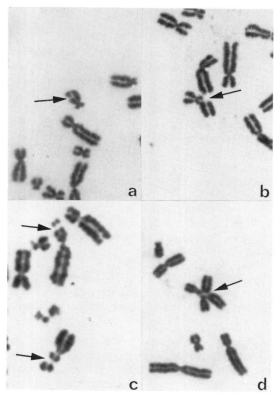

Figure 8.3 Parts of conventionally stained human chromosomes showing: (a) chromatid gap; (b) chromatid break; (c) isochromatid gap; (d) chromatid interchange.

somewhat wider variety of structural changes as compared with chromosome-type aberrations. In general, however, the aberrations are basically similar and include:

1. Achromatic lesion or gap (Fig. 8.3a)

A chromatid gap is an unstained region in the chromatid. On very careful examination fine threads can sometimes be seen running across the nonstaining region. Chromatid gaps occur in approximately 4% of blood lymphocyte cells from any normal individual not knowingly exposed to mutagens, and in most instances are a technical artifact; but exposure of an individual or cells to some chemicals or a virus infection can greatly increase their frequency. However, because of the technical artifact aspect of these chromatid aberrations, they are not reliable as an indicator of mutational damage unless seen in greatly increased numbers.

2. A chromatid break (Fig. 8.3b)

A chromatid break is similar to a chromatid gap, but the terminal part of the chromatid has been so displaced as to indicate that it is no longer attached in any way to the proximal part of the chromosome. Chromatid breaks are very infrequent, but may also be somewhat unreliable as indicators of mutational damage except if seen in greatly increased numbers.

3. Isochromatid gaps (Fig. 8.3c)

An isochromatid gap is similar to a chromatid gap, but affects both chromatids at identical loci. Again these chromatid aberrations can be technical artifacts and are unreliable unless they occur at greatly increased frequencies than would be found in unexposed cells.

4. Chromatid interchange (Fig. 8.3d)

Simple chromatid interchange involves an exchange between single chromatids in each of two chromosomes and results in a configuration having four arms which may be referred to as a "quadriradial". The exchange may be symmetrical, giving two new monocentric chromatids, or asymmetrical, giving a dicentric chromatid and an acentric fragment. Exchanges may involve homologous chromosomes at homologous loci—particularly involving C-band regions. Homologous equal exchange is a relatively rare event in normal individuals but is common in cells from patients with Bloom's syndrome (29). Certain chemical mutagens, such as mitomycin C, have been found to produce high frequencies of such homologous exchanges in human chromosomes (10).

5. Other chromatid aberrations

The chromatid aberrations described under 1-4 are those most commonly found in cells from a peripheral blood culture, but aberrations similar to those described for chromosome type changes can be found, such as acentric and centric ring chromatids, inversions, interstitial deletions, and duplications.

A point of importance in relation to chromatid-type aberrations follows from the fact that when a structurally altered chromatid proceeds through a replication phase, the chromatid with its alteration is duplicated. A chromatid-type aberration evident at the M_1 mitosis may therefore reappear as a duplicated chromosome-type at the ensuing M_2 division. Such aberrations have been termed "derived chromosome-type aberrations" (18). An awareness of the possible presence and origin of derived chromosome aberrations is particularly important in chemical mutagenesis, since most chemical agents produce chromatid-type aberrations and few directly induce chromosome-type aberrations in G_1 cells.

Sister Chromatid Exchanges in Metaphase Cells

Sister chromatid exchanges, or SCEs, involve a symmetrical exchange (in contrast to the asymmetrical exchange of iso-chromatid aberrations), at one locus, between sister chromatids, and thus do not result in any alteration in chromosome morphology. They were originally demonstrated by Taylor (122) to occur in plant somatic cell chromosomes observed at the M_2 mitosis following a pulse exposure to ^{3}H-thymidine during the S phase of the M_1 cycle. Autoradiographs of M_2 cells showed that isotopic labelling occurred on only one of the two sister chromatids at any given site along the lengths of the chromosomes, so that SCEs were evident as switches of radioactive label from one chromatid to its sister. It was later shown that most of the SCEs observed in autoradiographic experiments were induced by the endogenous radiation from the tritium; moreover, the incidence of these SCE events was increased if cells were exposed to UV light and certain chemical mutagens (20).

The development of staining techniques that result in sister chromatids staining differentially with the fluorescent dye Hoechst 33258 (55) or Giemsa (82,45) has allowed an easy visualisation of SCEs. Differential staining is usually obtained following exposure of the cells to 5-bromodeoxyuridine for two rounds of replication so that M_2 chromosomes consist of one chromatid unifilarly substituted with BrdUrd and the other bifilarly substituted. If cells are exposed to a wide spectrum of chemical mutagens, then the incidence of SCEs may be vastly increased above background levels (83,20,81).

For studying SCE induction, some modifications to the lymphocyte culture procedure is necessary. In our laboratory, we use tissue culture medium RPMI 1640, because it lacks thymidine. The culture technique is standard except that BrdUrd is added at the beginning of culture to give a final concentration of 25 μM. Immediately following the addition of 0.8 ml of whole blood to the liquid in the culture bottle, the bottle is wrapped in tinfoil to keep the cells in the dark, preventing photolysis of the BrdUrd containing DNA. In order to allow the cells to go through two rounds of replication in the presence of BrdUrd, they are not harvested until they have been at 37°C for three days. The harvesting and slide preparation are as previously described.

To stain the sister chromatids differentially, a modified FPG technique of Perry and Wolff (82) is used. The slides are stained in 0.5 μg ml^{-1} Hoechst 33258 for 15 min. rinsed, dried, and mounted in distilled water. The coverslips are sealed to prevent evaporation and the whole slide exposed to "black" light (366 nm) from an X-ray viewing box for 30 min. after which time the coverslips are removed and the slides incubated in 2 × SSC at 60°C for 30 min. The slides are then rinsed and stained in 4% Giemsa in pH 6.8 Sorensens buffer for 2–3 min., rinsed, dried, soaked in xylene, and mounted in DPX. The slides are then permanently stained and can be scored at leisure. Figure 8.4 shows a human metaphase treated by the FPG technique. This technique also permits a distinction to be made between those cells that are: in first division in culture (M_1) — these cells do not show any staining differentiation between the chromatids as both are unifilarly substituted with BrdUrd; in second division in culture (M_2) — these cells have chromosomes with one chromatid dark (unifilarly substituted) and one pale (bifilarly substituted); in third division in culture (M_3) — these cells have on average only half of the chromosomes differentially stained. The number of cells in first, second and third division can often be an important parameter in many experiments, even when SCEs are not being scored, as it gives information as to how many S periods the cells have been through postexposure and prior to observation. As with the incorporation of H^3 thymidine, the uptake of BrdUrd into the chromosomes causes a number of sister chromatid exchanges, variously described as the control, background, or "spontaneous" frequency of SCEs. The control frequency is dependent on the concentration of BrdUrd in the culture.

Bridges and Fragments in Anaphase Cells

A proportion of dicentric chromatid and chromosome exchanges will give rise to bridges at anaphase and these, together with any acentric fragments that have not been included in the two anaphase groups, are readily observed forms of damage. However, since only a proportion of aberrations result in anaphase abnormalities

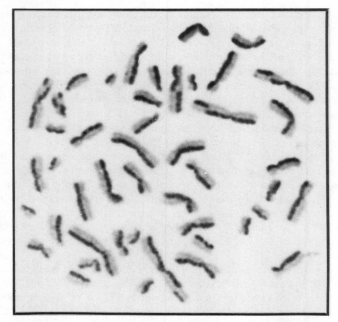

Figure 8.4 A cell from a human peripheral blood lymphocyte culture that has gone through two rounds of replication in the presence of BrdUdr. Note that each chromosome has one dark and one pale chromatid with several chromosomes having one or more exchanges between the dark and pale chromatids.

and since the anaphase stage is of short duration, the scoring of anaphase abnormalities is not a particularly efficient means of measuring induced damage. If anaphase abnormalities are to be scored, then colchicine must be omitted from the culture medium.

Micronuclei in Interphase Cells

Acentric fragments that are excluded from anaphase groups, and the resulting daughter nuclei, may round up in the cytoplasm and form separate micronuclei, each micronucleus containing one or more fragments. Micronuclei eventually disappear due to the action of cytoplasmic nucleases and proteases, but they persist for a considerable period of time. They may therefore be useful cytological indicators of genetic damage and indeed have three useful advantages: (1) in contrast to damage that can only be detected in visible chromosomes, i.e., in mitotic cells, micronuclei can be observed throughout the whole cell cycle so that a major proportion of the cell population may be analysed and little time is spent in scanning slides for cells in appropriate stages; (2) they are very simple to score, so that large numbers of cells can be analysed; (3) they provide a semipermanent record of damage, so that, as cells carrying chromosome damage proceed through the mitotic cycle, the number of cells with micronuclei increases and their frequency at any one time posttreatment is effectively an integration of the total damage expressed up to that time.

A disadvantage follows from what was said concerning anaphase bridges and fragments, since micronuclei are only formed from fragments (or whole chromosomes) that fail to participate in the normal anaphase movement: symmetrical aberrations that do not yield dicentric or fragment chromosomes are therefore not detected. Moreover, small fragments that are excluded from daughter nuclei may be difficult to see and may be obscured by the main nucleus in the cell.

For mutagen testing utilizing micronuclei as the end-point, the culture technique for blood lymphocytes is the standard method, but of course with colchicine omitted from the medium. Cultures are also incubated for a minimum of 72 to 96 hr. to allow a significant proportion of cells to proceed through mitosis to produce micronuclei, and are preferably sampled over two or three fixation times (36).

Methods, Problems, and Results with the Blood Lymphocyte Assay System

The methods used to expose blood lymphocytes to possible mutagens undergoing test are in themselves simple, but nevertheless have associated with them a number of problems. The cells must be kept in a saline solution or culture medium so that the test substance is presented as a solution—or possibly as a gas bubbled through the culture. If exposure occurs while in tissue culture medium, then the presence or absence of serum is important, since the agent may interact with serum proteins and so effectively reduce the concentration of mutagen. A second problem concerns the time of addition of the test agent and the duration for which it is present in the culture. Many mutagens are by their nature highly reactive and produce their damage in a short time following their introduction into the culture; thus in practice, although no active attempt may be made to remove the substance by washing the cells in saline, the exposure may nevertheless be an acute exposure. For example, sulphur mustard has a half-life of only a few minutes in culture medium and ethyl methanesulfonate a half-life of 2.2 hours in medium containing serum (43). Other substances may be active over longer periods of time, continually producing DNA lesions if present in the culture over long periods. Still other agents may even become more active as time goes on, due to the induction of enzyme systems in the cell which activate a substance that is relatively inactive in a nonmetabolized state. For example, it is believed that benzo(a)pyrene becomes more active after being present for 24 hr. or more in lymphocyte cultures because the presence of the substance induces the production of aryl hydrocarbon hydroxylase, which metabolizes benzo(a)pyrene to give reactive epoxides (12). Some potential mutagens may be completely inactive in producing chromosome damage in lymphocytes due to the absence of enzymes that may be present in other body cells, e.g., liver, and hence they are not transformed into their active metabolites in culture. An activating system containing liver-derived mixed-function oxidases (S9-mix) (2) may be added to the culture, or the chemical agent could be incubated in the presence of metabolizing liver cells in culture medium and the medium then introduced into the blood culture. However, very little work has been carried out on "activating" lymphocyte cultures (but see (62)), although S9-mix has been used to demonstrate enhanced activity of Aflatoxin B_1 in inducing SCE in cultured human lymphocytes (129). The substance is also active, at a lower efficiency, in the absence of S9.

In practice, most workers introduce a chemical mutagen into culture at the time of initiation of culture, or at the time of adding BrdUdr—if this substance is added 24

Table 8.1 Chemical Agents Studied for Their Effects[a] in Inducing Chromosome Aberrations and/or Sister Chromatid Exchanges in Human Lymphocytes Cultured in vitro.

Compound	Chromosome aberrations	SCEs	References
Acetylsalicylic acid	+		47
	−		64
Acetaldehyde	+	+	40, 96
Acetylaminofluorene (N-hydroxy-2-)	+	+	80, 92
Acridine orange		−	116
Adriamycin	+	+	25, 63, 71, 72, 127
Aflatoxin B$_1$		+	16, 89, 129
Alkeran		+	91
Allopurinol	−		115
9-Amino acridine derivatives		+	31
Aminopterin	+		113
Arsenic	+		85
(Potassium arsenate)	+		77
Azide (Na)	−		99
Azothioprene	+	+	33, 52
Barbiturates			
Phenobarbital	−		111
Secobarbital	−		111
Bensanthracene	+		46
Benzene	+	−	30, 67
Benzo(a)pyrene	+	+	12, 98
Bleomycin	+	−	17, 28, 35, 75, 88, 121
5-Bromodeoxyuridine	+	+	many references
Bis(2-ethyl hexyl)phthalate	+		110
Busulfan (Myleran)	+	+	23, 38, 91, 94
Butazolidin	+		133
Butylated hydroxytoluene			102
Cadmium	+		15
Caffeine	±	+	78, 131
		−	130
Corafur (nifurprazinum)	+		128
Chlorambucil	+	+	91, 93, 107, 114, 125
Chloramphenicol		−	78
Chromium chloride	+		44
Chromium trioxide	+		44
Clozapine	+		50
Cyclamate	+		117, 118
Cyclophosphamide		−	91
(with activation)	+	+	1, 62
Cysteine	+		13
Cytochalasin B	+		69, 95
Cytosine arabinosidase	+	+	52, 53, 91
Daunomycin (Rubidomycin)	+		126, 132
Diazepam	−		11, 108
Dichlorvos	+		14
Deoxycytidine		−	25
Diepoxybutane		+	135
Diethylhexylphthalate	+		124
Diethylnitrosamine		−	12
Dimethyldithiocarbamix acid (Ziram)	+		86
Dimethylnitrosamine		−	12
Diphenylhydantoin	+		112
Disulfiram (Antabuse)	+		58
DTPA		+	90
Epichlorhydrin	+		51
Ethanol	−		9
Ethyl methanesulfonate		+	4, 12, 35, 133
Ethyleneimines	+		76, 84, 92
2-3-5-tris-ethyleneiminobensoquinone	+		97

Table 8.1 Continued)

Compound	Chromosome aberrations	SCEs	References
Ergotoxin mesylate (dihydro-)	−	−	123
Ethyl nitroso urea		+	135
Hycanthone	+		74
Hydantoin	+		109
Frusemide	+		42
Hydroxyurea	+		77
Hydroxyquinoline sulfate	+		27
Lead acetate	+	−	3, 100
Lycurim		+	91
Lysergic diethylamide (LSD)	−		61, 70
	±		47
Melphelan	+		114
Methotrexate	+		113
8-Methoxypsoralen + UV	+	+	68, 130, 136
(Tri) methoxypsoralen	+	+	68, 130
Methylmezoxymethanol acetate		+	22
Methyl methanesulfonate		+	12, 92, 135
N-methyl-N-nitro-N-nitrosoguanidine		+	135
N-methyl-N-nitro-N-nitrosourea		+	129
Miracil D	+		73
Mitomycin C	+	+	41, 56, 57, 78, 91, 135
			12, 25, 104
Nitrogen mustards			
Mustine hydrochloride	+	+	84
3-bis(2-chloroethyl amino)-			
4-methylbenzoic acid	+		87
4-nitroquinoline oxide	+	+	37, 84, 135
Nitrosamides			
methylcapricamide	+		79
methylproprionamide	+		79
Organophosphorous compounds	+		106
(gasoline additives)			
Oxipurinol	−		115
Ozone	+		32
Phensuximide	+		109
Platinum (II) diamminedchloride	+	+	66
Quinacrine mustard		+	107
Saccharin (Na)		+	134
Selenite (Na)		+	92
Sodium acetate		−	3
Sodium fluoride	−		105
Styrene	+	+	59
Styrene oxide	+	+	59
TEPA (Thiophosphamide)	+	−	24, 52
Tetramethyl-uclphonyl-D-mannitol	+		54
Tobacco smoke condensate		+	39
Toluene	−	−	30
2-3-5 Triethyleneiminebenzoquinone 1-4	+		119
Trenimon	+	+	3, 34, 48, 105
Vincristine		+	91
		−	116
		−	52
Xylene	−	−	30

[a] In the data summarized in this Table, no attempt has been made to list the concentrations of chemicals used, the duration of exposure, or to indicate the degree of positive response noted. In some cases different authors have reported conflicting results and, for quantitative information, the reader is advised to consult the original quoted references.

hr. after culture initiation for detecting SCEs — and allow the substance to remain until the cells are harvested. At the same time, it is absolutely essential to run two parallel sets of control cultures: one with added saline and no mutagen (a negative control) and one with an added known active mutagen (a positive control). The substance to be tested is itself introduced over a range of concentrations, from a near toxic level which allows few or no cells to progress through to mitosis, to a series of dilutions below this level.

We have emphasized the fact that the chromosome aberrations induced by most chemical agents develop when the cells proceed through the S phase and arise as a consequence of misreplication. It is therefore important to ensure that, if a substance produces mutagenic lesions, these lesions are present during the S phase. Lymphocytes are in a G_0 or G_1 phase when introduced from peripheral blood into culture and the first cells to reach the S phase take at least 18 hr. to do so. Since the cells repair DNA lesions while they progress through the G_1 phase, the practice of introducing the test chemical into the culture at the time of its initiation may not be optimal unless the substance induces an activating system. In general, a better method would be to use two sets of cultures: one in which the test substance is added to the culture at its initiation and one 24 hr. later.

Finally, most agents that produce DNA lesions also produce mitotic delay and it is essential to use a number of fixation times to ensure that cells in their first posttreatment mitosis (in the case of scoring for chromosome aberrations) or second post-treatment mitosis (in the case of SCE scoring) are available for analysis. Employing the principles and methods outlined above, a large number of workers have utilized the human peripheral blood lymphocyte system to assay for the chromosome-damaging effects of a wide range of known and suspected mutagens. The two most favoured end-points utilized are chromosomal aberrations and sister chromatid exchanges observed in metaphase cells. The majority of studies on SCEs have been undertaken on Chinese hamster cells with their small number of large chromosomes, but the effects of many scores of compounds have been studies on human peripheral blood lymphocytes and some of the results obtained are summarized in Table 8.1. Some of these compounds are anticancer drugs that are administered to patients in vivo, e.g., Adriamycin, nitrogen mustard, etc; others are present in the environment, sometimes in the individual's working environment, e.g., Aflatoxin B_1, styrene oxide, and their effects can be detected in blood lymphocytes of exposed individuals. However, the relationship between in vivo and in vitro response of human blood lymphocyte chromosomes to a range of chemical mutagens is a somewhat uncharted area and further study in this field is certainly necessary.

Literature Cited

1. Allen, J. W.; Latt, S. A. In vivo BrdU-33258 Hoechst analysis of DNA replication kinetics and sister chromatid exchange formation in mouse somatic and meiotic cells. Chromosoma 58:325-40; 1976.

2. Ames, B. N.; Durston, W. E.; Yamosaki, E.; Lee, F. D. Carcinogens are mutagens: a simple test system combining liver homogenates for activation and bacteria for detection. Proc.Nat.Acad.Sci. USA 70:2281-85; 1973.

3. Beek, B.; Obe, G. The human leukocyte test sytem. VI. The use of sister chromatid exchanges as possible indicators for mutagenic activities. Humangenetik 29:127-34; 1975.

4. Bennet, S. P.; Bloom, A. D.; Nakamura, F. T.; Spence S. E.; Ohki, K. Mutation in human lymphoid lines (abstract). Amer.J.Hum.Genet. 29:22A 1977.

5. Brown, R. F.; Wu, Y. Induction of sister chromatid exchanges in Chinese hamster cells by chlorpropamide. Mutat.Res. 56:215-17; 1977.

6. Buckton, K. E. Identification with G and R banding of the postion of breakage points induced in human chromosomes by in vitro X-irradiation. Int.J.Radiat.Biol. 29:475-88; 1976.

7. Buckton, K. E.; Langlands, A. O.; Smith, P. G.; Woodcock, G. E.; Looby, P. C.; McLelland, J. Further studies on chromosome aberration production after whole body irradiation in man. Int.J.Radiat.Biol. 19:369-78; 1971.

8. Burgdorf, W.; Kurvink, K.; Cervenka, J. Elevated sister chromatid exchange rate in lymphocytes of subjects treated with arsenic. Hum.Genet. 36:69-72; 1977.

9. Cadotte, M.; Allard, S.; Verdy, M. Lack of effect of ethanol in vitro on human chromosomes. Ann.Genet. 16:55-56; 1973.

10. Cohen, M. M.; Shaw, M. W. Effects of mytomycin C on human chromosomes. J.Cell Biol. 23:386-95; 1964.

11. Cohen, M. M.; Hirschorn, K.; Frosch, W. A. Cytogenetic effects of tranquillizing drugs in vivo and in vitro. J.Am.Med.Assoc. 207:2425-26; 1969.

12. Craig-Holmes, A. P.; Shaw, M. W. Effects of six carcinogens on SCE frequency and cell kinetics in cultured human lymphocytes. Mutat.Res. 46:375-84; 1977.

13. Danielian, E. A.; Arutinnian, R. M.; Ispirian, S. M. Cytogenetic activity of radioprotectors in a culture of human lymphocytes. VI. Tsi.tol Genet. 12:437-40; 1978.

14. Dean, B. J. The effect of dichlorvos on cultured human lymphocytes. Arch. Toxicol 30:75-85; 1972.

15. Deknudt, G.; Deminatti, M. Chromosome studies in human lymphocytes after in vitro exposure to metal salts. Toxicology 10:67-75; 1978.

16. Dolimpio, D. A.; Jacobson, C.; Legator, M. Effect of aflatoxin on human leukocytes. Proc.Soc.Exp.Biol.Med. 127:559-62; 1968.

17. Dresp, J.; Schmid, E.; Bauchinger, M. The cytogenetic effect of bleomycin on human peripheral lymphocytes in vitro and in vivo. Mutat.Res. 56:341-53; 1978.

18. Evans, H. J. Population cytogenetics and environmental factors. *In* Jacobs; Price; Law eds. Human population cytogenetics, Pfizer Medical Monographs 5. Edinburgh University Press; 1970: 192-216.

19. Evans, H. J. Effects of ionizing radiation on mammalian chromosomes. *In* German, J., ed. Chromosomes and cancer. New York: John Wiley and Sons; 1974: 191-237.

20. Evans, H. J. What are sister chromatid exchanges? *In* de la Chapelle; Sorsan eds. Chromosomes today, 6. 1977: 315-26.

21. Evans, H. J.; Scott, D. Influence of DNA synthesis on the production of chromatid aberrations by X-rays and maleic hydrazide. Genetics 49:17-38; 1964.

22. Evans, L. A.; Kevin, M. J.; Jenkins, F. C. Human sister chromatid exchange caused by methylazoxymethanol acetate. Mutat.Res. 56:51-58; 1977.

23. Fukushima, T.; Zeldis, L. J. Chromosome aberrations induced with busulfan on human lymphocytes. Acta Haematol. Jap. 33:312-20; 1970.

24. Funes-Cravioto, F.; Yakovienko, K. N.; Kuleshov, N. P.; Zhurkov, V. S. Localization of chemically induced breaks in chromosomes of human leucocytes. Mutat.Res. 23:87-105; 1974.

25. Galloway, S. M. Ataxia Telangiectasia: The effects of chemical mutagens and X-rays on sister chromatid exchanges in blood lymphocytes. Mutat.Res. 45:343-49; 1977.

26. Galloway, S. M. What are sister chromatid exchanges? *In* Nichols, W. W.; Murphy, D. G. eds. DNA repair processes. Symposia Specialists, Miami, Florida, USA. 1977; 191-201.

27. Gebhart, E. Comparative studies on the influence of radioprotectors and amino acids on the chromosome-damaging activity of 8-hydroxyquinoline sulfate in human lymphocytes in vitro. Mutat.Res. 18:353-61; 1973.

28. Gebhart, E.; Kuppauf, H. Bleomycin and sister chromatid exchange in human lymphocyte chromosomes. Mutat.Res. 58:121-24; 1978.

29. German, J. Cytological evidence for crossing-over in vitro in human lymphoid cells. Science 144:298-301; 1964.

30. Gerner-Smidt, P.; Friedrich, U. The mutagenic effect of benzene, toluene and xylene studied by the SCE technique. Mutat.Res. 58:313-16; 1978.

31. Gibas, Z.; Lemoin, J. The induction of sister chromatid exchanges by 9-aminoacridine derivatives. Mutat.Res. 67:93-96; 1979.

32. Gooch, P. C.; Creasia, D. A.; Brewen, J. G. The cytogenetic effects of ozone: inhalation and in vitro exposures. Environ.Res. 12:188-95; 1976.
33. Hampel, K. E.; Lackner, A.; Schulz, G.; Busse, V. Chromosome mutation caused by azathioprine in human leukocytes in vitro. Z. Gastroenterol. 9:47-51; 1971.
34. Hayashi, K.; Schmid, W. The rate of sister chromatid exchange parallel to spontaneous chromosome breakage in Fanconi's anemia and to Trenimon-induced aberrations in human lymphocytes and fibroblasts. Humangenetik 29:201-6; 1975.
35. Hayez-Delatte, F.; Feremans, W. Chromosome aberrations induced by bleomycin in human lymphocytes in vitro. Bull.Cancer (Paris) 62:29-36; 1975.
36. Heddle, J. A.; Benz, R. D.; Countryman, P. I. Measurement of chromosomal breakage in cultured cells by the micronucleus technique. *In* Evans; Lloyd eds. Mutagen-induced chromosome damage in man. Edinburgh University Press 1978:191-200.
37. Hiragun, A.; Nishimoto, Y. Induction of chromosome aberrations in cultured human lymphocytes by 4-nitroquinoline 1-oxide. Gann. 64:183-87; 1973.
38. Honeycombe, J. R. The effect of busulphan on the chromosomes of normal human lymphocytes. Mutat.Res. 57:35-49; 1978.
39. Hopkin, J. M.; Evans, H. J. Cigarette smoke condensates damage DNA in lymphocytes. Nature 279:241-42; 1979.
40. Ilinskikh, N. N.; Ilinskikh, I. N.; Makarov, L. N.; Chernoskutova, S. A. Effect of ethanol and its metabolite acetaldehyde on the chromosomal apparatus of rat and human cells. Tsitologiia 20:421-25; 1978.
41. Ishii, Y.; Bender, M. A. Factors influencing the frequency of mitomycin C induced sister chromatid exchanges in 5-bromodeoxyuridine substituted human lymphocytes in culture. Mutat.Res. 51:411-18; 1978.
42. Jameela; Subramanyam, S.; Sadasivan, G. Clastogenic effects of frusemide on human leukocytes in culture. Mutat.Res. 66:69-74; 1979.
43. Jensen, E. M.; LaPolla, R. J.; Kirby, P. E.; Haworth, S. R. In vitro studies of chemical mutagens and carcinogens. I. Stability studies in cell culture medium. J.Nat.Cancer Inst. 59:941-44; 1977.
44. Kaneko, T. Chromosome damage in cultured human leukocytes induced by chromium chloride and chromium trioxide (author's transl.). Jpn.J.Ind.Health 18:136-37; 1976.
45. Kato, H. Spontaneous sister chromatid exchanges detected by a BUdr labelling method. Nature 251:70-72; 1974.
46. Kato, R. Chromosome breakage induced by a carcinogenic hydrocarbon in Chinese hamster cells and human leukocytes in vitro. Hereditas 59:120-41; 1968.
47. Kato, T.; Jarvik, L. F. LSD-25 and genetic damage. Dis.Nerv.Syst. 30:42; 1969.
48. Kaufmann, W.; Gebhart, E.; Horback, L. Determination of the threshold value of the mutagenic activity of Trenimon on human lymphocytes "in vitro". Humangenetik 20:1-8; 1973.
49. Kihlman, B. A. Deoxyribonucleotide synthesis and chromosome breakage. Chromosomes Today I:108-17; 1966.
50. Knuutila, S.; Helminen, E.; Knuutila, L.; Leisti, S.; Simmes, M.; Tammisto, P.; Westermarck, T. Role of clozapine in the occurrence of chromosomal abnormalities in human bone marrow cells in vivo and in cultured lymphocytes in vitro. Hum.Genet. 38:77-89; 1977.
51. Kucerova, M.; Polivkova, Z.; Sram, R.; Matousek, V. Mutagenic effect of epichlorohydrin. I. Testing on human lymphocytes in vitro in comparison with TEPA. Mutat.Res. 34:271-78; 1976.
52. Kucerova, M.; Polivkova, Z. *In* Scott, D.; Bridges, B. A.; Sobels, F. H. eds. Progress in genetic toxicology. Elsevier/North-Holland Biomedical Press; 1977: 319-25.
53. Kuleshov, N. P. Age sensitivity of the chromosomes of human lymphocytes to the action of cytosine arabinoside at stage G_2 of the cell cycle. Tsitologiia 14:1368-73; 1972.
54. Lapis, K. and Schuler, D. Effect of tetra-o-methyl-sulphonyl-D-mannitol (R-52) on the ultrastructure of lymphoid cells in tissue culture. Haematologia 5:5-23; 1971.
55. Latt, S. A. Microfluorometric detection of deoxyribonucleic acid replication in human metaphase chromosomes. Proc.Nat.Acad.Sci. USA 70:3395-99; 1973.
56. Latt, S. A. Sister chromatid exchanges, indices of human chromosome damage and repair: Detection by fluorescence and induction by mytomycin C. Proc.Nat.Acad.Sci. USA 71:3162-66; 1974.

57. Latt, S. A.; Stetten, G.; Juergens, L. A.; Buchanan, G. R.; Gerald, P. S. Induction by alkylating agents of sister chromatid exchanges and chromatid breaks in Fanconi's anemia. Proc.Nat.Acad.Sci. USA 72:4066-70; 1975.
58. Lilly, L. J. Investigations in vitro and in vivo, of the effects of disulfiram (Antabuse) on human lymphocyte chromosomes. Toxicology 4:331-40; 1975.
59. Linnainmaa, K.; Meretoja, T.; Sorsa, M.; Vainio, H. Cytogenetic effects of styrene and styrene oxide. Mutat.Res. 58:277-86; 1978.
60. Lloyd, D. C. The problems of interpreting aberration yields induced by in vivo irradiation of lymphocytes. In Evans; Lloyd, eds. Mutagen-induced chromosome damage in man. Edinburgh University Press; 1978:77-88.
61. Loughman, W. D.; Sarjent, T. W.; Israelstau, D. M. Leukocytes of humans exposed to lysergic acid diethylamide: lack of chromosomal damage. Science 158:508-10; 1967.
62. Madle, S.; Westphal, D.; Hilbig, V.; Obe, G. Testing in vitro of an indirect mutagen (cyclophosphamide) with human leukocyte cultures. Activation by liver perfusion and by incubation with crude liver homogenate. Mutat.Res. 54:95-99; 1978.
63. Massimo, L.; Dagna-Bricarelli, F.; Fossati-Guglielmoni, A. Effects of adriamycin on human lymphocytes stimulated with PHA "in vitro". Rev.Eur.Etud.Clin.Biol. 15:793-99; 1970.
64. Mauer, I.; Weinstein, D.; Solomon, H. M. Acetylsalic acid: no chromosome damage in human leukocytes. Science 169:198-201; 1970.
65. Meretoja, T.; Vainio, H.; Sorsa, M.; Harkönen, H. Occupational styrene exposure and chromosomal aberrations. Mutat.Res. 56:193-97; 1977.
66. Meyne, J.; Lockhart, L. H. Cytogenetic effects of cis-Platinum (II) Diamminedchloride on human lymphocyte cultures. Mutat.Res. 58:87-97; 1978.
67. Morimoto, K. Analysis of combined effects of benzene with radiation on chromosomes in cultured human leukocytes. Jpn.J.Ind.Health 18:23-34; 1976.
68. Mourelatos, D.; Faed, M. J. W.; Johnson, B. E. Sister chromatid exchanges in human lymphocytes exposed to 8-methoxypsoralen and long-wave UV radiation prior to incorporation of bromodeoxyuridine. Experientia 33:1091-93; 1977.
69. Mulherkar, L.; Navlurhar, P.; Joshi, M. V. Effects of cytochalasin B on human lymphocyte cultures. Indian J.Exp.Biol. 16:430-34; 1978.
70. Muneer, R. S. Effects of LSD on human chromosomes. Mutat.Res. 51:403-10; 1978.
71. Nevstad, N.P. Sister chromatid exchanges and chromosomal aberrations induced in human lymphocytes by the cytostatic adriamycin in vivo and in vitro. Mutat.Res. 57:253-58; 1978.
72. Newsom, Y. L.; Singh, D. N. Cytologic effects of adriamycin on human peripheral lymphocytes. Acta Cytol. 21:137-40; 1977.
73. Obe, G. Distribution of miracil D-induced achromatic lesions and chromatid breaks on the chromosomes of human leukocytes. Z.Naturforsch 25:115-16; 1970.
74. Obe, G. Action of hycanthone on human chromosomes in leukocyte cultures. Mutat.Res. 21:287-88; 1973.
75. Obe, G.; Beek, B; Vaidya, V. G. The human leukocyte test system. III. Premature chromosome condensation from chemically and X-ray-induced micronuclei. Mutat.Res. 27:89-101; 1975.
76. Obe, G.; Sperling, K.; Belitz, H. J. Comparative study on the effect of ethyleneimines on human leukocyte chromosomes in vitro. Arzneim Forsch 22:982-84; 1972.
77. Oppenheim, J. J.; Fishbein, W. N. Induction of chromosome breaks in cultured normal human leukocytes by potassium arsenite, hydroxyurea and related compounds. Cancer Res. 25:980-85; 1965.
78. Pant, G. S.; Makada, N.; Tanaka, R. Sister chromatid exchanges in peripheral lymphocytes of atomic bomb survivors and of normal individuals exposed to radiation and chemical agents. Hiroshima J.Med.Sci. 25:99-106; 1976.
79. Paton, G. R.; Allison, A. C. Chromosome damage in human cell cultures induced by nitrosamides. Mutat.Res. 21:289-91; 1973.
80. Pero, R. W.; Bryngelsson, C.; Mitelman, F.; Kornfältr, Thulin, T.; Norden, A. Interindividual variation in the responses of cultured human lymphocytes to exposure from DNA-damaging chemical agents: interindividual variation to carcinogen exposure. Mutat.Res. 53:327-341; 1978.

81. Perry, P. Chemical mutagens and sister chromatid exchange. In de Serres F. J.; Hollaender A. eds. Chemical mutagens. Vol 6. New York: Plenum Press; 1980:1-39.
82. Perry, P.; Wolff, S. New Giemsa method for the differential staining of sister chromatids. Nature 251:156-58; 1974.
83. Perry, P.; Evans, H. J. Cytological detection of mutagen-carcinogen exposure by sister chromatid exchange. Nature 258:121-25; 1975.
84. Perry, P. E.; Jager, M.; Evans, H. J. In Evans, H. J.; Lloyd, D. C. eds. Mutagen-induced chromosome damage in man. Edinburgh: Edinburgh University Press; 1978:201-7.
85. Petres, J.; Berger, A. The effect of inorganic arsenic on DNA-synthesis of human lymphocytes in vitro. Arch.Dermatol.Forsch. 242:343-52; 1972.
86. Pilinskaya, M. A. Cytogenetic action of the fungicide ziram in a culture of human lymphocytes in vitro. Sov.Genet. 7:805-9; 1971.
87. Popescu, N. C.; Costachel, O.; Liciu, F.; Cioloca, L. In vitro and in vivo cytogenetic effects of 3-(bis(2-chloreothyl)amino)-4-methylbenzoic acid (NSC-146171; 1OE 82) in human peripheral leukocytes. Cancer Chemother.Rep. 57:29-32; 1973.
88. Promchainant, C. Cytogenetic effect of bleomycin on human leukocytes in vitro. Mutat.Res. 28:107-12; 1975.
89. Promchainant, C.; Baimai, V.; Nondasuta, A. The cytogenetic effects of aflatoxin and gamma-rays on human leukocytes in vitro. Mutat.Res. 16:373-380; 1972.
90. Prosser, J.S. Induction of sister chromatid exchanges in human lymphocytes by DTPA. Br.J.Ind.Med. 35:174-76; 1978.
91. Raposa, T. Sister chromatid exchange studies for monitoring DNA damage and repair capacity after cytostatics in vitro and in lymphocytes of leukaemic patients in vivo. Mutat.Res. 57:241-51; 1978.
92. Ray, J. H.; Altenburg, L. C. Sister chromatid exchange induction by sodium selenite: dependence on the presence of red blood cells or red blood cell lysate. Mutat.Res. 54:343-54; 1978.
93. Reeves, B. R.; Margoles, C. Preferential location of chlorambucil-induced breakage in the chromosomes of normal human lymphocytes. Mutat.Res. 26:205-8; 1974.
94. Richmond, J. Y.; Kaufman, B. N. Studies on busulfan (myleran) treated leukocyte cultures. I. Cytological observations. Exp.Cell Res. 54:377-80; 1969.
95. Ridler, M. A.; Smith, G. F. The response of human cultured lymphocytes to cytochalasin B. J. Cell Sci. 3:595-602; 1968.
96. Ristow, H.; Obe, G. Acetaldehyde induces cross-links in DNA and causes sister chromatid exchange in human cells. Mutat.Res. 58:115-19; 1978.
97. Robustelli, della Cuna G.; Marini, G.; Maggi, G. Cytologic and cytogenetic aspects of the anticellular effect of 2-3-5-tris-ethyleneiminiobenzoquinone (TEIB) in vivo and in vitro. Haematologica (Pavia) 53:609-26; 1968.
98. Rüdiger, H. W.; Kohl, P.; Mangels, W.; von Wichert, P; Bartram, C. R.; Wöhler, W.; Passarge, E. Benzpyrene induces sister chromatid exchanges in cultured human lymphocytes. Nature 262:290-92; 1976.
99. Sander, C.; Nilan, R. A.; Kleinhofs, A.; Vig, B. K. Mutagenic and chromosome-breaking effects of azide in barley and human leukocytes. Mutat.Res. 50:67-75; 1978.
100. Schmid, E.; Bauchinger, M.; Pietruck, S.; Hall G. Cytogenetic action of lead in human peripheral lymphocytes in vitro and in vivo. Mutat.Res. 16:401-6; 1972.
101. Schmid, E.; Bauchinger, M. Comparison of the chromosome damage induced by radiation and Cytoxan therapy in lymphocytes of patients with gynaecological tumours. Mutat.Res. 21:271-74; 1973.
102. Sciorra, L. J.; Kaufmann, B. N.; Maier, R. The effects of butylated hydroxytoluene on the cell cycle of chromosome morphology of phytohaemagglutinin-stimulated leucocyte cultures. Food Cosmet. Toxicol. 12:33-44; 1974.
103. Sehested, J. A simple method for R-banding of human chromosomes, showing a pH-dependent connection between R and G bands. Humangenetik 21:55-58; 1974.
104. Shiraishi, Y.; Sandberg, A. A. Effects of mitomycin C on sister chromatid exchange in normal and Bloom's syndrome cells. Mutat.Res. 49:233-38; 1978.
105. Slacik-Erbn, R.; Obe, G. The effect of sodium fluoride on DNA synthesis, mitotic indices and chromosomal aberrations in human leukocytes treated with trenimon in vitro. Mutat.Res. 37:253-66; 1976.

106. Söderman, G. Chromosome-breaking effect of a gasoline additive in cultured human lymphocytes. Hereditas 71:335-38; 1972.
107. Solomon, E.; Bobrow, M. Sister chromatid exchange: A sensitive assay of agents damaging human chromosomes. Mutat.Res. 30:273-78; 1975.
108. Staiger, G. R. Studies on the chromosomes of human lymphocytes treated with diazepam in vitro. Mutat.Res. 10:635-44; 1970.
109. Stenchever, M. A.; Allen, M. The effect of selected antiepileptic drugs on the chromosomes of human lymphocytes in vitro. Am.J.Obstet.Gynaecol. 116:867-70; 1973.
110. Stenchever, M. A.; Allen, M. A.; Jerominski, L.; Petersen, R. V. Effects of bis(2-ethylhexyl)phthalate on chromosomes of human leukocytes and human fetal lung cells. J.Pharm.Sci. 65:1648-51; 1976.
111. Stenchever, M. A.; Jarvis, J. A. Effect of barbiturates on the chromosomes of human cells in vitro – a negative report. J.Reprod.Med. 5:215-17; 1970.
112. Stenchever, M. A.; Jarvis, J. A. Diphenylhydantoin: effect on the chromosomes of human leukocytes. Am.J.Obstet.Gynecol. 109:961-62; 1971.
113. Stevenson, A. C. Effects of twelve folate analogues on human lymphocytes in vitro. *In* Evans; Lloyd, eds. "Mutagen-induced chromosome damage in man. 1978:227-38.
114. Stevenson, A. C.; Roman, C. S.; Patel, C. R. Effects of amylobarbitone on the frequency of chromosomal aberrations in human lymphocytes determined by chlorambucil and melphalan in vitro. Mutat.Res. 19:225-29; 1973.
115. Stevenson, A. C.; Silcock, S. R.; Scott, J. T. Absence of chromosome damage in human lymphocytes exposed to allopurinol and oxipurinol. Ann.Rheum.Dis. 35:143-47; 1976.
116. Stoll, C.; Brogaonkar, D. S.; Levy, J. M. Effect of vincristine on sister chromatid exchanges of normal human lymphocytes. Cancer Res. 36:2710-13; 1976.
117. Stoltz, D. R.; Khera, K. S.; Bendall, R.; Gunner, S. W. Cytogenetic studies with cyclamate and related compounds. Science 167:1501-2; 1970.
118. Stone, D.; Lamson, E.; Chang, Y. S.; Pickering, K. W. Cytogenetic effects of cyclamates on human cells in vitro. Science 164:568-69; 1969.
119. Stosiek, M.; Gebhart, E. Protective effect of Reducdyn against the chromosome damaging activity of 2,3,5-triethyleneimine-benzoquinone-1,4 on human lymphocytes in vitro. Humangenetik 25:209-16; 1974.
120. Sumner, A. T.; Evans, H. J.; Buckland, R. A. A new technique for distinguishing between human chromosomes. Nature, New Biol. 232:31-32; 1971.
121. Tamura, H.; Sugiyama, Y.; Sugahara, T. Effect of bleomycin on the chromosomes of human lymphocytes at various cell phases. Gann. 65:103-7; 1974.
122. Taylor, J. H. Sister chromatid exchanges in tritium-labelled chromosomes. Genetics 43:515-29; 1958.
123. Tsuchimoto, T.; Malter, B. E.; Deyssenroth, H. Analysis of chromosome aberration and sister chromatid exchanges in human lymphocytes exposed in vitro to Hydergine. Mutat.Res. 67:39-45; 1979.
124. Turner, J. H.; Petricciani, J. C.; Crouch, M. L.; Wenger, S. An evaluation of the effects of diethylhexyl phthalate (DEHP) on mitotically capable cells in blood packs. Transfusion 14:560-66; 1974.
125. Vance, J. C.; Cervenka, J.; Ullman, S.; Kersey, J. H.; Sabad, A.; Green, N. Immunological and chromosomal studies in a patient with sezary syndrome. Arch. Dermatol. 113:1417-23; 1977.
126. Vig, B. K. Nature of chromosome aberrations induced in pre-DNA-synthesis period in human leukocytes by daunomycin. Mutat.Res. 12:411-52; 1971.
127. Vig, B. K. Effect of hypothermia and hyperthermia on the induction of chromosome aberrations by adriamycin in human leukocytes. Cancer Res. 38:550-55; 1978.
128. Vig, B. K.; Natarajan, A. T.; Zimmerman, F. K. Study on cytological effects of carofur – a new mutagen. Mutat.Res. 42:109-15; 1977.
129. Vijayalaxmi; Thomson, E.; Evans, H. J. Induction of sister chromatid exchanges in human lymphocytes and Chinese hamster cells exposed to aflatoxin B_1 and N-methyl-N-nitrosourea. Mutat.Res. 67:47-53; 1979.
130. Waksvik, H.; Brøgger, A.; Stene, J. Psoralen/UVA treatment and chromosomes. I. Aberrations and sister chromatid exchange in human lymphocytes in vitro and synergism with caffeine. Hum.Genet. 38:195-207; 1977.

131. Weinstein, D.; Mauer, I.; Solomon, H. M. The effects of caffeine on chromosomes of human lymphocytes. In vivo and in vitro studies. Mutat.Res. 16:391-99; 1972.

132. Whang-Peng, J.; Leventhal, E. G.; Adamson, J. W.; Perry, S. The effect of daunomycin on human cell in vivo and in vitro. Cancer 23:113-21; 1969.

133. Wissmüller, H. F. The cytogenetic action of Butazolidin in human lymphocyte cultures dependent on the protein concentration of the culture medium. Mutat.Res. 14:83-94; 1972.

134. Wolff, S.; Rodin, B. Saccharin-induced sister chromatid exchanges in Chinese hamster and human cells. Science 200:543-45; 1978.

135. Wolff, S.; Rodin, B.; Cleaver, J. E. Sister chromatid exchanges induced by mutagenic carcinogens in normal and xeroderma pigmentosum cells. Nature 265:347-49; 1977.

136. Wolff-Schreiner, E. C.; Carter, D. M.; Schwarzacher, H. G.; Wolff, K. Sister chromatid exchange in photochemotherapy. J.Invest.Dermatol. 69:387-91; 1977.

9
Assays for Chromosome Aberrations Using Mammalian Cells in Culture

by WILLIAM AU AND T. C. HSU*

Abstract

The protocol of chromosome aberration assays can be simplified such that it can be used as a routine short-term test for identification of mutagens and carcinogens. It involves the use of the near-diploid Chinese Hamster Ovary cell lines (CHO) with cell generation time of 12–14 hr. Our protocol calls for an initial assay of a 5 hr. continuous incubation with 3 concentrations of a test compound. If the test compound is observed to induce chromosome aberrations, it can be considered a potentially hazardous compound. If the finding is negative, additional protocols must be conducted to reduce the possibility of a false negative test. These protocols consisted of a pulse treatment and a continuous treatment of 24 hr. In addition, compounds which are capable of inducing mitotic abnormalities may be identified by the anaphase anomalies test. Other criteria including the use of a metabolic activating system (S9), the recognition and recording of chromosome aberrations, and anaphase anomalies are presented.

Introduction

For identification of carcinogens, animal experimentation has always been the system of choice. However, the shortcomings of this assay method are that it is costly and time consuming. It has been estimated that at its upper limit this assay can test 100–200 compounds in one year on a world-wide basis (Clayson 1977). This would be an insignificant fraction of the 24,000 to 30,000 man-made chemicals that enter the environment in substantial quantities or of the 500 to 800 new compounds introduced commercially each year (Clayson 1977). Thus, there is an urgent need for

*Section of Cell Biology, The University of Texas System Cancer Center M. D. Anderson Hospital and Tumor Institute, Houston, Texas 77030.

economical and effective short-term tests as preliminary assays. At present, the most widely used assay is the Ames' bacterial mutation system. In a survey of 300 compounds, McCann et al. (1975) found that the Ames' assay has an accuracy of 90% to correlate carcinogenicity with mutagenicity. Other investigators found the accuracy of the Ames' assay also in the 90% range (Coombs et al. 1976; Anderson and Styles 1978), while Andrews et al. (1978 a,b) found it to be in the 50–80% range. Other short-term assays are still undergoing extensive evaluation in an international study (de Serres 1978). Chromosome breakage assay with cells in vitro is one of them.

Chromosome damage in mammalian cells in culture has been used extensively in studies of environmental mutagens and carcinogens. Most cytogeneticists classify, record, and analyze all types of chromosome aberrations and interpret the mechanisms of action. Such investigations have contributed significantly to our understanding of mutagen effects on the structure and physiology of chromosomes which form the basis for assay systems, but for routine screening, the elaborate protocol, time, and manpower involved would be inefficient. Assay procedures should be simple, rapid, and economical, yet should provide sufficient information to cover as many aspects as practicable. Thus, a compromise between detail and rapidity is needed. A preliminary protocol for such an assay has been described by Hsu et al. (1977). In another study which employed a slightly different assay system, Ishidate and Odashima (1977) surveyed the cytogenetic effects of 134 chemicals, 34 of which have been shown to be carcinogenic. The chromosome aberration assay correctly identified 25 out of these 34 (73.2%) and incorrectly identified 9 (26.5%). The Ames' bacterial mutation system correctly identified 28 (82.4%) and incorrectly identified 6 (17.6%).

There are two important points to be made from this study. First, among the 9 carcinogens incorrectly identified in the Ishidate report, some of them require metabolical activation (e.g., dimethylnitrosamine, quinoline, etc.). Second, the chromosome assay identified 2 carcinogens (diethylstilbestrol and urethane) that were not detected by the bacterial test system.

Certain compounds (e.g., aromatic amines) are not active in vitro unless they are metabolically activated. Thus investigators began incorporating extracts of animal (usually rat) livers into their in vitro cytogenetic assays. Such a modification is similar to that developed by Ames et al. (1973) for their bacterial mutation assay. Several investigators (Natarajan et al. 1976; Weinstein et al. 1977) have found that the liver extract was able to metabolize chemical compounds in mammalian cell cultures. Au et al. (1978; 1979a) found that the clastogenic activity of gentian violet was nullified almost completely when a rat liver extract was introduced to the test cultures. Therefore, when a metabolic activation system is introduced to cell cultures, the false-positive and false-negative information of an agent identified by the conventional chromosome breakage analysis would be reduced.

Most clastogens presumably induce DNA damage in living cells. Data on chromosome breakage, therefore, can be compared with those from the bacterial mutation assays. However, there is a group of chemicals, collectively called mitotic poisons, which may not be detected by mutation assays. These agents allegedly interfere with the mitotic apparatus (spindle proteins, centrioles, etc.), but do not necessarily react with DNA. Colchicine and vincristine are excellent examples because they prevent the formation of the spindle thereby arresting mitosis at metaphase. Some agents, e.g., halothane, do not inhibit spindle formation but induce defective spindles, giving rise to abnormal karyokinesis and aneuploid daughter cells (Kusyk and Hsu 1976). The end result of such mitotic division is a gross genetic imbalance, which may produce a more devastating effect on the cells than point mutations or mild chromosome disjunction, and since these agents are not true "mutagens," mitotic poisons will not be detected by bacterial assays or other assays based on the mechanism of point mutation, but can be detected and estimated by cell culture assays.

In this chapter, we present, in some detail, chromosome breakage assay systems using mammalian cells in vitro. The protocol is based mainly on our own opinions, which may or may not be shared by other cytogeneticists. Moreover, as new knowledge becomes available, the protocol can be improved in order to increase its accuracy in detecting mutagens and carcinogens.

Test Materials

The test material should possess the following qualities:

1. Good chromosome characteristics, including low number, distinguishable morphology, easy banding, etc.
2. Relatively stable karyotype.
3. The cells should be able to propagate quickly for obtaining large quantities without maintaining large stock cultures.
4. Short generation time.
5. Easy synchronization if experiments so require. The best system is the mitotic shake-off. Other synchronization systems, such as double thymidine block, isoleucine starvation, etc., usually require working at odd hours which is inconvenient.
6. High cloning efficiency.
7. Easy hybridization with other somatic cells for a high frequency of heterokaryons.

If not for the senescence phenomenon (Hayflick and Moorhead 1962; Hayflick 1964), diploid human fibroblast lines should be highly acceptable as candidates. However, these lines cannot be propagated with vigorous growth for more than 20 to 25 passages. In the meantime, a relatively large stock supply should be maintained and stored in liquid nitrogen freezer. An alternative is to use short-term human lymphocyte cultures or human lymphoid cell lines (see Chapter 8). Unfortunately, each type of cell culture has its own shortcomings. In short-term lymphocyte cultures, the rate of spontaneous chromosome aberrations and that of responses to chromosome-damage agents may very from donor to donor (Littlefield and Goh 1973; Pero et al. 1978). Human lymphoid lines do not senesce and maintain a near-diploid constitution, but these lines are always suspect because of viral contamination, especially Epstein-Barr virus. Moreover, studies on clastogen effects comparing lymphoid lines with peripheral lymphocyte cultures from patients with Down's syndrome (Sasaki and Tonomura 1969; Lambert et al. 1976; Huang et al. 1977; Countryman et al. 1978) and ataxia telangiectasia (Cohen et al. 1979) showed that lymphoid cells are more resistant than lymphocytes. Thus, the genetic composition and physiology of lymphoid cells may have been changed.

Practically all cell lines derived from Chinese hamster tissues meet those described requirements. Under favorable growth conditions, the cell cycle time is approximately 12 hours and the diploid number is low (2n = 22). However, it is rather difficult to keep the Chinese hamster cell lines in a predominantly diploid constitution. Most of the cell lines would become aneuploid with chromosomes in the diploid range. In our laboratory, we chose the CHO line (Tjio and Puck 1958) as principal test material. The CHO cells maintained a modal chromosome number of 21. The same line was analyzed by Siciliano et al. (1978) for electrophoretic shift variants of enzymes coded by approximately 40 genetic loci. They concluded that the CHO cells are functionally diploid. Since assays for clastogen effects do not require analyses of chromosome numbers, deviation from diploidy is not a serious problem as long as its karyological composition is stable.

In cases where the experimental result is not clear, it is a good practice to use another cell type for confirmation. Short-term lymphocyte cultures or long-term cell lines of human origin are desirable. But, as a routine material, we recommend Chinese hamster cells.

Methods

Routine Cell Stock and Setting Up Cultures for Experiments

Stock CHO cultures can be maintained in Corning T-75 flasks in McCoy's 5A medium supplemented with 10–20% fetal bovine serum. They can be incubated at 37°C in a dry incubator with tightened culture flask cap or in moistened, 5–10% CO_2 incubator with loosened cap.

Confluent cultures (2–5 × 10^7 cells/flask) are subcultured by trypsinization. The trypsinized cells are diluted with medium to obtain a 100 ml suspension and 9 ml of it is dispersed into each T-25 culture flask. This innoculum size (approximatley 2–4 × 10^6 cells/flask) would yield exponentially growing cultures from 12–24 hr. for short-term experimentation. For long-term experiments (e.g., 24–48 hr.), the trypsinized cells should be diluted to obtain a 300–400 ml suspension. The cultures are always set up at approximately 16–18 hr. before the initiation of an experiment.

Test Compounds

The test compound (hereafter referred to as "drug" as a generic name) should be dissolved in sterile distilled water or saline in concentrate, preferably prepared just prior to experiments. If the drug is water insoluble, another solvent, e.g., acetone, alcohol, or dimethylsulfoxide, may be used. If the molecular weight of the compound is known, the solutions should be prepared in molar concentrations. For assay of each drug, we suggest the use of three final concentrations 0.1, 1.0 and 10 μM, or 0.1, 1.0 and 10 μg/ml if the molecular weight is not known. If a clinical or human consumption dose is known, this concentration should be used as the lowest experimental dose.

The innoculum size of the test compound should be adjusted such that 0.5 ml of the drug solution and 0.5 ml of the S9 mixture (to be described in the next section) are added simultaneously to 9 ml of culture. The cultures with and without the solvent and S9 are made up to 10 ml with culture medium.

Metabolic Activation System

In bacterial assay systems for mutagenicity, primarily the Ames' test, and in mammalian cell culture tests, procarcinogens (e.g., aromatic amines) show no mutagenic or clastogenic activities, thus producing false negative data. Ames et al. (1973) began the use of the microsomal fractions of the rat liver extract to metabolize chemicals in their bacterial mutation assay. Their results indicate that metabolic conversions can be achieved by using a postmitochondrial liver homogenate extract known as S9.

In clastogen assays using cells in vitro, employment of the S9 fraction in conjunction with drugs has not been extensive. Natarajan et al. (1976) initiated the use of S9 in cell cultures. They observed that two indirect-acting nitrosamines were activated in the presence of liver extract to become potent clastogens. Other investigators (Ishidate and Odashima 1977; Weinstein et al. 1977; Matsuoka et al. 1979) confirmed the usefulness of S9 in in vitro assays. In the reported cases, most investigators emphasized the activating capability of the S9.

In our study of clastogenic agents, we found that the potent clastogenic activity of gentian violet is nullified in the presence of S9 (Au et al. 1979a). In a study of anticancer drugs (unpublished data), we observed that S9 could modify the clastogenic activity of a number of agents, including activation (cyclophosphamide), potentiation (bleomycin), reduction (adriamycin and neocarcinostatin), and no effect (actinomycin D, cytosine arabinoside, mitomycin C, methotrexate and vincristine). Thus, to avoid false-positive or false-negative data, it is safe to run all experiments with and without S9.

The procedure for preparing S9 and cofactors is briefly described as follows:

The liver homogenate (S9) is prepared similar to the method described by Ames et al. (1973). We use young male Long-Evans rats (200–500 gm) obtained from Blue Spruce Farms, Altamont, N.Y. Animals are injected IP with 500 mg/Kg Aroclor 1254 in corn oil (200 mg/ml stock concentration). Five days after injection, the animals are sacrificed by cervical dislocation and the livers removed aseptically. From this stage on, all steps should be carried out at 0–4° with sterile instruments. Liver tissues are rinsed twice in cold 0.15 M KCl cut into small pieces, and homogenized in 3x volume of KCl with a Fisher apparatus equipped with a Teflon pestle (normally 50 gm of liver tissues can be obtained from 4 rats; therefore, use 150 ml of KCl solution). The homogenate is centrifuged under refrigeration for 10 min. at 9000x g, and the supernatant (S9) is dispensed in 2 ml aliquots into screw-capped tubes, quickly frozen and stored at $-70°$. For each experiment, the S9 aliquot is thawed at room temperature and used immediately. In all studies, the S9 fraction is used together with cofactors and is referred to as the S9 mixture. The mixture contains 0.3 ml of S9 fraction and 0.7 ml cofactor supplement (11.4 mM $MgCl_2$, 47 mM KCl, 7.1 mM glucose-6-phosphate, 5.7 mM NADP, and 140 mM potassium phosphate buffer at pH 7.4). The cofactor solution can be prepared and frozen in small batches at $-20°$. Preferably, S9 fraction and cofactors are not prepared to last more than a 6-month period.

Each batch of S9 should be tested for its toxicity as well as activity. However, there is no standard assay for S9. In our laboratory, we have tested different batches of S9 on their activities on chromosome breakage. Cultures were treated with S9 according to our experimental protocol: 5 hr., 1–5 hr. pulse and 23–19 hr. release, and 24 hr. before cultures were harvested. When cells were incubated with 0.5 ml S9 in 10 ml medium for 24 hr. continuously, the mitotic rate was much reduced. Thus, for the 24 hr. experiment, 0.25 ml of S9 was used. Fifty metaphases were scored for each treatment and the average number of breaks per metaphase was calculated. As shown in Table 9.1, S9 induced slight increase in breakage rate in the 5 hr. and 24 hr. continuous exposure but not in the pulse and release experiment. The toxicity of S9 in mammalian cell cultures has also been observed by other investigators (Madle and Obe 1977) and the use of dialysis bag has been suggested for the retention of S9. In the Ames' bacterial assay, S9 did not change the plating efficiency of the 4 strains of *E. coli* (M. A. Butler, personal communication). It is recommended that the ability of S9 to cause 7,12-dimethylbenz(a)anthracene (DMBA) to become mutagenic in bacterial assay should be used as a guideline for optimal metabolic activity (de Serres and Shelby 1979). In mammalian cells, we found the anticancer drug, cyclophosphamide, consistently activated by different batches of S9. Others (Stetka and Wolff 1976; Benedict et al. 1978) found that S9 was effective in converting cyclophosphamide into a potent chromosome-damaging (breakage and sister chromatid exchange) agent. Thus, we suggest the use of cyclophosphamide as an agent to assay the activity of S9. Significant dose-dependent increase in chromosome damage should be observed with 25, 50 and 100 μg/ml of the drug in the presence of S9 (Table 9.2).

Harvesting Cell Cultures

For observation of chromosome aberrations. For estimation of clastogen effects, the cultures are treated with the test compound for the desired duration and a mitotic arrestant such as Colcemid (0.02 μg/ml) or colchicine (0.5 μg/ml) is added during the last hour to accumulate mitosis and to condense the chromosomes.

The cells at the designated harvest time are trypsinized and centrifuged. After decanting the medium, the cells are suspended in 8 ml of hypotonic solution for approximately 10 min. Several kinds of hypotonic solution, 0.085 M KCl, 0.7% Na-citrate, or cell culture medium diluted with distilled water 1:3 or 1:4, can be used with equal effectiveness. The cells are recentrifuged, and the hypotonic solution decanted until approximately 0.5 ml is left. The cells are suspended in this small amount of hypotonic solution and the fixative is drained slowly along the side of the tube until 7–8 ml of the fixative is added. The fixative is prepared fresh by mixing

Table 9.1 Effect of S9 on Chromosome Breakage of CHO Cultures

| | Number of lesions per metaphase | | | | | |
| | 5 hr. continuous | | 5 hr. pulse | | 24 hr. continuous | |
Experiment	–S9	+S9	–S9	+S9	–S9	+S9
1	0.12	0.24	0.12	0.10	0.18	0.20
2	0.04	0.06	0.20	0.22	0.08	0.12
3	0.02	0.08	0.08	0.08	0.14	0.16
4	0.08	0.12			0.06	0.08
5	0.12	0.16				

Table 9.2 Effects of Cyclophosphamide on Chromosome Breakage With or Without S9

Concentration (µg/ml)	Number of lesions per cell			
	−S9		+S9	
	1	2	1	2
0	0.08	0.12	0.10	0.12
25	0.08	0.08	0.46	0.18
50	0.16	0.06	0.66	0.42
100	0.10	0.08	a	a

a Denotes extensive chromosome breakage.

glacial acetic acid (1 part) with methanol (3 parts). The cells are fixed in this mixture for 5 min., centrifuged, and then washed two more times at 5–10 min. each with the fixative. Final cell suspensions are dropped onto clean, wetted slides for air-dried preparations. It is always a good practice to inspect the pilot preparations with low power phase-contrast optics to determine the cell density, the degree of chromosome spreading, and the mitotic frequency. When mitotic figures are plentiful and the spreads are acceptable, no more than four slides should be needed; if not, a few more slides should be made. Many drugs cause tangling of the chromosomes (stickiness), hence poor spreading. Spreading may be slightly improved by the flame-drying technique (igniting the fixative immediately after the cell suspension is dropped onto the slides). Conventional Giemsa staining (4% in 0.01 M phosphate buffer for approximately 5–7 min.) may be applied to all slides. For routine assays, chromosome banding will not be necessary.

For observation of anaphase anomalies. For the observation of possible mitotic poisons, both the experimental and the control cultures should receive no mitotic arrestant and the cells are not treated with a hypotonic solution. The harvested cells are fixed directly with the methanol-acetic acid mixture and air-dried preparations can be made with the procedure described above. For critical observations, the classic squash technique may be applied. Trypsinized cells are centrifuged and the cell button is fixed in 45% acetic acid for 20 min. Decant the fixative, and add approximately 0.5 ml of 1–2% acetic orcein and suspend the cells in the stain. A drop of the cell suspension is then placed on a clean slide and covered with a 22 × 22 mm or 22 × 30 mm coverslip. Press the coverslip with thumb or with a squashing apparatus, absorbing excess stain with a filter paper, and seal the edges of the coverslip with Kronig's wax.

Recognition of Chromosome Damage

In order to identify agents with different effects on chromosomes, we shall use the term "clastogens" (Shaw 1970) for agents capable of causing chromosome breakage and "mitotic poisons" for agents capable of inducing mitotic anomalies (spindle, centriole, and cytokinetic abnormalities).

An understanding of the damage induced at various stages of the cell cycle would

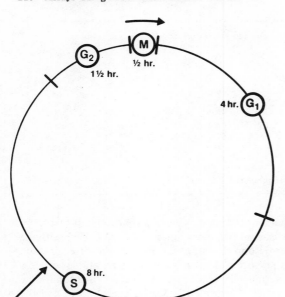

Figure 9.1 Diagrammatic representation of the cell cycle of CHO cells based on a 14 hr. cell generation time. Arrow on top indicates the direction of progression of the cycle. Arrow at the lower left indicates the position of the cycle when drug is administered for a 5 hr. experiment.

be important in identifying different types of stage-specific agents. Figure 9.1 is a diagramatic representation of a 14 hr. cell generation time for CHO. Chromosome breakage induced at different stages of the cycle would be observed as different chromosome aberrations at the ensuing mitotic (M_1) stage. Specifically, aberrations induced in G_1 and very early S phases are of the chromosome-type, whereas those induced in the remainder of S and G_2 phases are chromatid-type, (Fig. 9.2). Chromosome damage received during mitotic stage is not immediately detectable, but is detectable as chromosome-type aberrations in the mitosis of the next cell generation (M_2) similar to those of G_1 phase (Fig. 9.2a,b).

In the 5 hr. sample, only cells originally in the mid- or late S phase move to mitosis. Therefore, no chromosome-type aberrations are expected (Fig. 9.2c,d). The few observed are isochromatid breaks (breaks at identical or nearly identical loci of sister chromatids). In the 24 hr. sample and in the recovery sample, both chromatid- and chromosome-types of lesions may be present.

The classification of chromatid and chromosome aberrations and the method for converting each aberration into numbers of breaks have been discussed in Chapter 1. In raw data, it is advisable to record the number of various types of aberrations in a simplified form. We use the following abbreviations to record the aberrations: b (breaks), g (gaps), ib (isochromatid break), t (exchange), d (interstitial deletion), D (dicentric), R (ring) and F (acentric fragment). The capitalized abbreviations represent chromosome-type aberrations whereas the others are chromatid-type. Multiple exchanges are counted individually without attempt to decipher which chromosomes are involved, e.g., 6t or tttttt means six exchanges noted in that metaphase. Quadriradials and triradials are terms for describing certain chromatid aberrations. In essence, a quadriradial represents a chromatid exchange between two metacentric chromosomes, and triradial, the same between a metacentric and acrocentric.

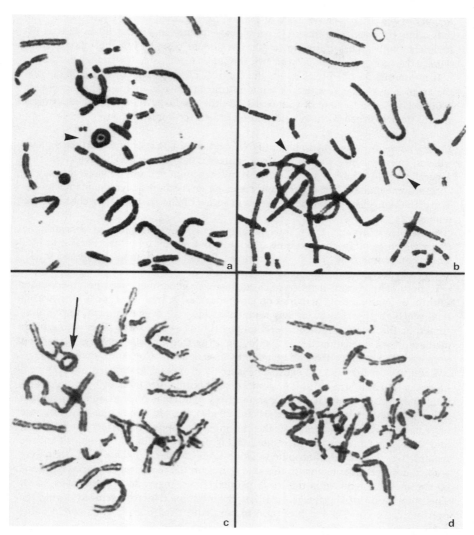

Figure 9.2 Metaphases of CHO cells showing chromosome types (a,b) and chromatid types (c,d) of damage. Note the difference between the chromosome (arrow) and chromatid rings (arrowhead).

Experimental Procedure

Choosing the Treatment Protocol

Continuous treatment. We routinely run a 5 hr. treatment with three doses of the drug. Before we actually code the slides and score for chromosome aberrations, we normally scan the unstained or stained slides for abnormalities related to the general toxicity of the drug. The presence of breaks, micronuclei, and pycnotic nuclei is an indication of positive finding. If no damage occurs in all the doses tested, experiments with higher doses up to 100 μM should be run. Whether chromosome damage is recorded or not, a 24 hr. run should be made using lower concentrations.

If accumulation of mitosis above the control level is observed, it is possible that the agent is a mitotic arrestant. Then a mitotic count (percentage of metaphases in 2000 cells) should be made to construct a mitotic index. An additional experiment should be performed without colcemid treatment for confirmation.

Pulse treatment. Many drugs inhibit DNA synthesis. The cells in the earlier S phase will fail to enter mitosis when the drug is present continuously. The treated culture generally would exhibit a greatly reduced mitotic count without a high frequency of chromosome damage because the cells receiving severe damage remain in interphase. Thus, the data may give a false picture in terms of the frequency of chromosome damage.

Some drugs, notably DNA intercalating agents, cause tangling of chromosome fibers which result in chromosome stickiness, but not chromosome breaks, in M_1. The chromosome fibers presumably rupture during anaphase movement and the lesions will express as chromosome aberrations in M_2.

In both of the above-mentioned cases, recovery experiments will provide a more faithful estimate of drug activities on chromosomes. A treatment for 3-4 hr. with a relatively high dose may be used as a standard. After the drug treatment, the cultures should be thoroughly washed, preferably by rinsing two or three times with growth medium, then reincubated for 24 hr. in drug-free growth medium. Air-dried preparations are made from these cultures after 1 hr. of Colcemid treatment.

Recently, we analyzed the clastogenic activity of biological stains in our routine 5 hr. assay (Au and Hsu 1979b). Chromosome abnormalities induced by azure A and toluidine blue was unusual. Neither showed any significant chromosome damage at a low dose (1 μM), but at high doses (10 and 20 μM), instead of having a dose response increase in chromosome damage, the chromosomes appeared peppery and puffed, and some of them even appeared banded (Fig. 9.3). In addition, the majority of the metaphases showed chromosome stickiness. Subsequently, these two dyes were studied in pulse experiments. In this study, most of the recovering metaphases showed chromosome damage in the form of breaks, exchanges, and rings (Fig. 9.2a,b). The observation of rings suggested that chromosome damage occurred during the previous cell cycle. The result of the 5 hr. and pulse experiments are shown in Table 9.3.

Recording of Abnormalities

All preparations should be coded and preferably scored by two individuals to eliminate bias.

Mitotic index. Either the Colcemid-treated series or the direct fixation series can be used for the estimate of mitotic index. Theoretically, the latter is more representative because the effects of mitotic poisons can be confidently detected without the influence of another mitotic arrestant. In untreated cells, all stages of mitosis are present. In our opinion, to record all stages is unnecessary for the following reasons: 1)

Assays Using Mammalian Cells in Culture 213

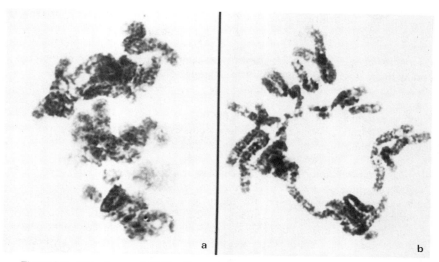

Figure 9.3 Metaphases of CHO cells treated with 10 µM toluidine blue for 5 hr., showing bloated (a) and banded (b) chromosomes.

The data represent only an approximation. To record all stages of mitosis requires more time than it is worth. 2) It is always a problem, even for experienced investigators, to recognize early prophase and late telophase stages, especially the latter. Telophase is followed by cell cleavage, and it is always tempting to record early daughter cells also when telophase is already completed. Thus, reading only metaphases in each sample is probably a better procedure than reading all stages of mitosis. As long as the experimental data are compared with those of control samples, a 1 hr. Colcemid arrest gives more discrete metaphase figures. A mitotic

Table 9.3 Effect of Azure A and Toluidine Blue on Chromosome Damage

Compound	Dosage (µM)	Percentage of metaphase with breaks	Percentage of metaphase with exchanges	Number of lesions per metaphase
Azure A	1	20	0	0.24
	10	—	—	—[a]
	10[b]	40	8	0.96
	20[b]	52	18	1.68
Toluidine blue	1	24	0	0.28
	10	—	—	—[a]
	10[b]	40	32	1.56
	20[b]	86	26	3.04
Control 1	—	14	0	0.16
Control 2[b]	—	16	0	0.20

[a] Chromosomes were puffed, tangled, and banded, therefore breaks were impossible to observe.
[b] Cultures were pulsed with compound for 4 hr. and recovered for 25 hr. The others were treated continuously for 5 hr.

poison would still drastically increase the frequency of metaphases in a 5 hr. run. A sample of 2000 cells should be sufficient for an estimate.

Anaphase/metaphase ratio. In a growing cell sample, the frequency of metaphases is usually higher than that of anaphase because metaphase lasts longer than anaphase. Under the influence of a mitotic arrestant, mitotis is arrested at metaphase and anaphase movement does not take place. Thus the anaphase/metaphase ratio becomes extremely low. This ratio and the metaphase index should be effective indicators of mitotic poisons. For recording, cultures are harvested without Colcemid and 200 anaphases and metaphases are scored.

Anaphase abnormalities. In order to score abnormal anaphases, slides prepared by the squash technique are preferred. Anaphase abnormality is one of the most difficult to record because false positive identification of chromosome stickiness may be made even by experienced investigators. When the nucleoli break down at the end of prophase, some of the nucleolar material adheres to the chromosomes. During anaphase, as sister chromatids separate, the nucleolar material may still be present at the terminal ends of the chromosomes, giving a sticky appearance. This type of stickiness is different from the true sticky chromosomes in which the chromatin fibers tangle, either between sister chromatids or between unrelated chromosomes (McGill et al 1974).

One crude way of differentiating normal anaphase and sticky anaphase is to examine late anaphase with special attention to the unseparated portion. In normal anaphase, the sister chromatids lie parallel in the equator. Sticky chromatids or chromosomes frequently attach to one another at various angles, and in severe cases the chromosomes fail to separate. However, even with such criteria, ambiguity may still exist. Our advice is that reading should be as conservative as possible. In other words, any anaphase with possible sticky chromosomes that is also possibly normal should be recorded as normal.

Another abnormality that may be confused with normal anaphase is the sidearm bridge formation (Humphrey and Brinkley 1969). Exchanges between sister chromatids occurring in late G_2 and prophase result in sidearm bridges in anaphase. In normal anaphase, the sister chromatids separate orderly so that the width of the chromatids is uniform (Fig. 9.4a). With an exchange between the sister chromatids, the point of exchange prevents the normal separation of the sisters as if they were locked at this point. As the chromosomes move toward the poles, the dividing chromatids would become progressively thinner at the "locked" point. Therefore, the distinction between a sidearm bridge and normal sister chromatid separation can be detected only in late anaphase.

Other categories of anaphase anomalies are relatively easy to identify, especially at late anaphase and early telophase. The following categories are suggested in recording: normal, sticky chromosomes, chromatin bridges, lagging chromosomes, and multipolarity. Several examples are shown in Figure 9.4.

Chromosome breakage. In scoring chromosome aberrations, metaphases in the diploid range with a reasonable degree of spread are randomly selected for detail observation. At least 100 metaphases should be scored for each experimental (dose-time) point.

For convenience, metaphases with 10 or more breaks are recorded as 10 breaks per

Assays Using Mammalian Cells in Culture 215

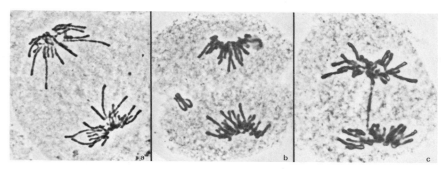

Figure 9.4 Some examples of anaphases showing (a) normal morphology, (b) lagging chromosome, and (c) sticky chromosome.

cell. From the recorded breaks and exchanges, etc., one can calculate the total number of breaks (or lesions) observed and the average number of lesions per cell. One can also calculate the percentage of cells with chromosome aberrations and the percentage of cells with exchanges. When reporting the results, all the data should be made available in descriptive and/or in tabular forms, emphasizing the number of lesions per cell, number of cells with aberrations, and number of cells with exchanges. With these indexes, the potency of different clastogens can be estimated on a molar solution or on μg/ml basis.

Other abnormalities. Other abnormalities include: c-mitosis, endoreduplication, and polyploidy. These can be recorded in percentages with respect to normal metaphases.

Analysis of Data

One of the criticisms of the assay using chromosome breakage is that if a high enough dose is used, breakage would always be observed. Most clastogens induce chromosome lesion at concentrations between 0.1–20 μM. If activity is observed only at 100 μM, it is relatively safe to conclude that the drug is not a potent clastogen.

One must make a decision as to which agent induces significant increase in chromosome breakage and which does not. In laboratories where routine surveys are carried out, the background spontaneous breakage rate of a cell line scored by a particular individual can be standardized statistically. Thus, the degree of significance of damage induced by an agent can be analyzed. On the other hand, statistical analysis (e.g., t-test) of an individual study can be performed. Many studies involve small sample size and it would be difficult to justify the use of any particular statistical method. Because of this, a baseline value should be calculated using two times the upper range of the control value as a rule of thumb. That is to say, if the damage inflicted by an agent is more than twice the baseline value, it would be considered significant (Au and Hsu 1979b).

Time-saving Steps

We suggest that each drug be treated first with the 5 hr. assay system, both for clastogens and for mitotic poisons. If definite evidence for clastogenic activity (e.g., significant levels of chromatid breaks and exchanges) is present, it can be concluded that the agent is a clastogen. There is really no need for further experimentation unless one wishes to delve into more details. In other words, the simple 5 hr. preliminary test can identify a large number of frank clastogens without additional experiments. On the other hand, if no significant clastogenic activity is recorded, then a second experiment should be set to increase the concentration up to 100 μM for both 5 hr. and 24 hr. harvests. If the 5 hr. sample suggests chromosome stickiness, or if the mitotic rate is exceedingly low, then a recovery experiment should be run. This sequence of experimentation saves unnecessary time, labor, and expenses, even though some chemicals must be tested more than once.

Time Required for Obtaining Preliminary Data

In one day, an experienced technician can handle four drugs for the 5 hr. experiments (three concentration samples for each drug) plus various controls. This includes the slide preparation also. Many factors contribute to the time required for microscopic evaluation, but an experienced investigator should have no problem reading slides for all three concentrations of a drug in one to two days, including the S9 samples.

Discussion

In order that an assay system for carcinogens and mutagens can be used with confidence, the system should be able to: (1) accurately identify a known carcinogen; (2) identify a noncarcinogen; and (3) predict the carcinogenicity of an unknown compound. Therefore, the accuracy of such assay systems is customarily compared with data obtained in carcinogenesis. The Ames' assay system is known to be approximately 90% confident. Purchase et al. (1978) concluded, from a survey of six short-term tests, that cell transformation and bacterial mutation tests are among the best. However, clastogen assay was not evaluated in their study. Tsutsui et al. (1977) reported that the chromosome aberration test was as confident as the mutation assay. It is obvious that no one system can correctly identify all cancer-causing agents. Thus, it is necessary to use a battery of selected tests such that the probability of detecting all carcinogens can be increased. A cytogenetic assay can definitely complement the bacterial mutation system.

Most neoplasms exhibit chromosome changes, in number and/or in rearrangements (Atkin 1976; Rowley 1977; Mitelman and Levan 1978). In several chemical-induced tumors, chromosome changes were shown to be an initial observable event (Kurita et al. 1969; Rees et al. 1970) or as a continuous process (Nowell 1976; Connell and Ockey 1977). Thus, cytogenetic analysis should be an important assay method for environmental mutagens.

One group of genetically hazardous compounds is known as procarcinogens (e.g., aromatic amines). Procarcinogens require metabolic activation to become ultimate carcinogens. In order to identify these agents by the chromosome breakage assays, liver S9 homogenate should be incorporated into the assay systems. A recent survey study by Matsuoka et al. (1979) showed that 10 out of 16 carcinogens were positive in their assays in the presence of S9. However, they use only one treatment protocol of 3 hr. incubation and 24 hr. recovery. With multiple incubation time, the confidence of their assay should be increased.

Another group of possible carcinogens, the mitotic poisons, probably will be negative in bacterial mutation or other short-term test systems. Since cancer often exhibits heteroploidy, and since congenital aneuploidy is not uncommon in human and experimental animals, cytogenetic assay systems should be of special value for this neglected group of compounds. Diethylstilbestrol, a human carcinogen, induced aneuploidy and polyploidy (Sawada and Ishidate 1978) without chromosome breakage (Bishun et al. 1978).

In order to maximize an assay system for identification of chemical mutagens with diversified mechanisms of actions, our proposed 5 hr. protocol (three concentrations with and without S9) is strongly recommended. Positive clastogens can be identified with a minimum amount of effort and expense. When negative data are obtained in the 5 hr. samples, additional experiments (such as 24 hr. harvest and recovery experiments) should be run to substantiate the conclusion.

We consider the assay system proposed here as a reasonably good start. As more and more information and new methodologies become available, the system should be modified further to improve its accuracy and confidence.

Literature Cited

1. Ames, B. N.; Durston, W. E.; Yamasaki, E.; Lee, F. D. Carcinogens are mutagens: a simple test system combining liver homogenates for activation and bacteria for detection. Proc. Nat. Acad. Sci. USA 70:2281-85; 1973.
2. Anderson, D.; Styles, J. A. The bacterial mutation test. Br. J. Cancer 37:924-30; 1978.
3. Andrews, A. W.; Thibault, L. H.; Lijinsky, W. The relationship between carcinogenicity and mutagenicity of some polynuclear hydrocarbons. Mutat. Res. 51:311-18; 1978a.
4. Andrews, A. W.; Thibault, L. H.; Lijinsky, W. The relationship between mutagenicity and carcinogenicity of some nitrosamines. Mutat. Res. 51:319-26; 1978b.
5. Atkin, N. B. Cytogenetic aspects of malignant transformation. S. Karger; Basel, Switzerland: 1976.
6. Au, W.; Pathak, S.; Collie, C. J.; Hsu, T. C. Cytogenetic toxicity of gentian violet and crystal violet on mammalian cells in vitro. Mutat. Res. 58: 269-76; 1978.
7. Au, W.; Butler, M. A.; Bloom, S. E.; Matney, T. S. Further study of the genetic toxicity of gentian violet. Mutat. Res. 66:103-12; 1979a.
8. Au, W.; Hsu, T. C. Studies on the clastogenic effects of biological stains and dyes. Env. Mut. 1:27-35; 1979b.
9. Benedict, W. F.; Banerjee, A.; Venkatesan, N. Cyclophosphamide-induced oncogenic transformation, chromosomal breakage, and sister chromatid exchange following microsomal activation. Cancer Res. 38:2922-24; 1978.
10. Bishun, N. P. Cytogenetic studies with diethylstilbestrol. Clin. Oncol. 4:159-65; 1978.
11. Clayson, D. B. Principles underlying testing for carcinogenicity. Cancer Bulletin 29:161-66; 1977.
12. Cohen, M. M.; Sagi, M.; Ben-Zur, Z.; Schaap, T.; Voss, R.; Kohn, G.; Ben-Bassat, H. Ataxia telangiectasia: chromosomal stability in continuous lymphoblastoid cell lines. Cytogenet. Cell Genet. 23:44-52; 1979.
13. Comings, D. E. What is a chromosome break? In German, J. ed. Chromosomes and cancer. New York: John Wiley and Sons; 1974: 95-135.
14. Connell, J. R.; Ockey, C. H. Analysis of karyotype variation following carcinogen treatment of Chinese hamster primary cell lines. Int. J. Cancer 20:768-79; 1977.
15. Coombs, M. M.; Dixon, C.; Kissonerghis, A. M. Evaluation of the mutagenicity of compounds of known carcinogenicity, belonging to the benz(a)anthracene, chrysene, and cyclopenta(a)phenanthrene series, using Ames' test. Cancer Res. 36:4525-29; 1976.
16. Countryman, P. I.; Heddle, J. A.; Crawford, E. The repair of X-ray induced chromosomal damage in trisomy 21 and normal diploid lymphocytes. Cancer Res. 37:52-58; 1978.

17. deSerres, F. J. International study of short-term carcinogenicity tests. Nature 274:740-41; 1978.
18. deSerres, F. J.; Shelby, M. D. The Salmonella mutagenicity assay: recommendation. Science 203:563-65; 1979.
19. Hayflick, L.; Moorhead, P. S. Handbook on growth. Washington, D.C.: Fed. Am. Assoc. Exper. Biol.; 1962.
20. Hayflick, L. The limited in vitro lifetime of human diploid cell strain. Exp. Cell Res. 37:614-36; 1964.
21. Hsu, T. C.; Collie, C. J.; Lusby, A. F.; Johnston, D. A. Cytogenetic assays of chemical clastogens using mammalian cells in culture. Mutat. Res. 45:233-47; 1977.
22. Huang, C. C.; Banerjee, A.; Tan, J. C.; Hou, Y. Comparison of radiosensitivity between human hematopoietic cell lines derived from patients with Down's syndrome and from normal individuals. J. Nat. Cancer Inst. 59:33-36; 1977.
23. Humphrey, R.; Brinkley, B. Ultrastructural studies of radiation induced chromosome damage. J. Cell Biol. 42:745-53; 1969.
24. Ishidate, M., Jr.; Odashima, S. Chromosome test with 134 compounds on Chinese hamster cells in vitro—a screening for chemical carcinogens. Mutat. Res. 48:337-54; 1977.
25. Kurita, Y.; Sugiyama, T.; Nishiyuka, Y. Cytogenetic analysis of cell population in rat leukemia induced by pulse doses of DMBA. Gann. 60:527-35; 1969.
26. Kusyk, C. J.; Hsu, T. C. Mitotic anomalies induced by three inhalation halogenated anesthetics. Env. Res. 12:366-70; 1976.
27. Lambert, B.; Hansson, K.; Bui, T. H.; Funes-Cravioto, F.; Lindsten, J. DNA repair and frequency of X-ray- and UV light-induced chromosome aberrations in leukocytes from patients with Down's syndrome. Ann. Hum. Genet. 39:293-305; 1976.
28. Littlefield, L. G.; Goh, K. Cytogenetic studies in control men and women. I. Variations in aberration frequencies in 29,709 metaphases from 305 cultures obtained over a three-year period. Cytogenet. Cell Genet. 12:17-34; 1973.
29. Madle, S.; Obe, G. In vitro testing of an indirect, mutagen (CTN) with human leukocyte cultures: activation with liver microsomes and use of a dialysis bag. Mutat. Res. 56:101-3; 1977.
30. Matsuoka, A.; Hayashi, M.; Ishidate, M., Jr. Chromosome aberration tests on 29 chemicals combined with S9 mix in vitro. Mutat. Res. 66:277-90; 1979.
31. McCann, J.; Choi, E.; Yamasaki, E.; Ames, B. N. Detection of carcinogens as mutagens in the Salmonella/microsome test: assay of 300 chemicals. Proc. Nat. Acad. Sci. USA 72:5135-39; 1975.
32. McGill, M.; Pathak, S.; Hsu, T. C. Effects of ethidium bromide on mitosis and chromosomes: a possible material basis for chromosome stickiness. Chromosoma 47:157-67; 1974.
33. Mitelman, F.; Levan, G. Clustering of aberrations to specific chromosomes in human neoplasms. III. Incidence and geographic distribution of chromosome aberrations in 856 cases. Hereditas 89:207-32; 1978.
34. Natarajan, A. T.; Tates, A. D.; Van Buul, P. P. W.; Meijers, M.; de Vogel, N. Cytogenetic effects of mutagens/carcinogens after activation in a microsomal system in vitro. I. Induction of chromosome aberrations and sister chromatid exchanges by diethylnitrosamine (DEN) and dimethylnitrosamine (DMN) in CHO cells in the presence of rat liver microsomes. Mutat. Res. 37:83-90; 1976.
35. Nowell, P. C. The clonal evolution of tumor cell population. Science 194:23-28; 1976.
36. Pero, R. W.; Bryngelsson, C.; Mitelman, F.; Kornfält, R.; Thulin, T.; Norden, A. Interindividual variation in the responses of cultured human lymphocytes to exposure from DNA damaging chemical agents. Mutat. Res. 53:327-41; 1978.
37. Purchase, I. F. H.; Longstaff, E.; Ashby, J.; Styles, J. A.; Anderson, D.; Lafevre, P. A.; Westwood, F. R. An evaluation of six short-term tests for detecting organic chemical carcinogens. Br. J. Cancer 37:873-903; 1978.
38. Rees, E. D.; Majumdar, S. K.; Shuck, A. Changes in chromosomes of bone marrow after intravenous injections of 7,12-dimethylbenz (a)anthracene and related compounds. Proc. Nat. Acad. Sci. USA 66:1228-35; 1970.
39. Rowley, J. D. Mapping of human chromosomal regions related to neoplasia: evidence from chromosomes 1 and 17. Proc. Nat. Acad. Sci. USA 74:5729-33; 1977.

40. Sawada, M.; Ishidate, M., Jr. Colchicine-like effect of diethylstilbestrol (DES) on mammalian cells in vitro. Mutat. Res. 57:175-82; 1978.

41. Sasaki, M. S.; Tonomura, A. Chromosomal radiosensitivity in Down's syndrome. Jpn. J. Human Genet. 14:81-91; 1969.

42. Shaw, M. W. Human chromosome damage by chemical agents. Ann. Rev. Med. 21:409-32; 1970.

43. Siciliano, M. J.; Siciliano, J.; Humphrey, R. M. Electrophoretic shift mutants in Chinese hamster ovary cells: evidence for genetic diploidy. Proc. Nat. Acad. Sci. USA 75:1919-23; 1978.

44. Stetka, D. G.; Wolff, S. Sister chromatid exchange as an assay for genetic damage induced by mutagen-carcinogens. II. In vitro test for compounds requiring metabolic activation. Mutat. Res. 41:343-50; 1976.

45. Tjio, J. H.; Puck, T. T. Genetics of somatic mammalian cells. II. Chromosomal constitution of cells in tissue culture. J. Exp. Med. 108:259-68; 1958.

46. Tsutsui, T.; Umeda, M.; Maizumi, H.; Saito, M. Comparison of mutagenicity and inducibility of DNA single-strand breaks and chromosome aberrations in cultured mouse cells by potent mutagens. Gann. 68:609-17; 1977.

47. Weinstein, D.; Katz, M. L.; Kazmer, S. Chromosomal effects of carcinogens and noncarcinogens on WI-38 after short-term exposure with and without metabolic activation. Mutat. Res. 46:297-304; 1977.

10
The Micronucleus Test: An In Vivo Bone Marrow Method

by W. SCHMID*

Abstract

The micronucleus test is an in vivo cytogenetic screening procedure used for the detection of freshly induced structural chromosome aberrations in bone marrow of experimental animals. Furthermore, it is able to reveal chromosome loss due to impairment of the spindle apparatus. Micronuclei originate from chromatin which has been lagging in anaphase. The tested cell population consists of erythroblasts undergoing their last chromosome replication and mitosis. The effect is scored after expulsion of the main nucleus in young polychromatic erythrocytes. For many routine purposes this test is a substitute for the tedious and often difficult core of bone marrow metaphase analysis for chromosome breaks.

The micronucleus test (m.t.) is an in vivo cytogenetic screening procedure for the detection of freshly induced structural chromosome aberrations and for revealing chromosome loss due to partial impairment of the spindle apparatus. Micronuclei, i.e., small, incomplete nuclei, originate from chromatin which has been lagging in anaphase. In the course of telophase, this material is included into one or the other daughter cell where it can either fuse with the main nucleus or form one or several secondary nuclei (Fig. 10.1 a-f). Thus, the lagging elements—and later micronuclei—represent acentric chromosome fragments and di- or multicentrics connected by bridges or, in the case of spindle malfunction, they consist of entire chromosomes.

The process of micronuclei formation occurs in all dividing cell types, but not all tissues are suitable for quantitative scoring of induced micronuclei. Indeed, the

*Institute of Medical Genetics, University of Zurich, Zurich, Switzerland.

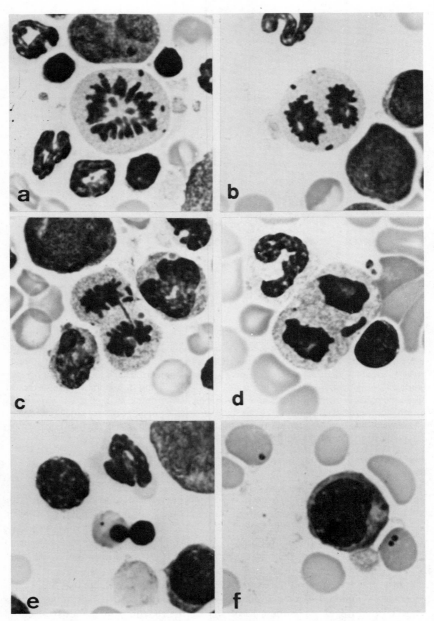

Figure 10.1 The events leading to the formation of micronucleated polychromatic erythrocytes. In (a), chromosome fragments in a metaphase; in (b), fragments outside of the spindle in anaphase stage; in (c), beside fragments an anaphase bridge is visible; in (d), the transition from telophase to G_1 nuclei; (e), the main nucleus is slipping out of an erythroblast, leaving behind a micronucleus; in (f), micronuclei are visible in two erythrocytes and a nucleated marrow cell. (Reproduced from Chemical Mutagens, Vol. 4, Plenum Publishers, with permission.)

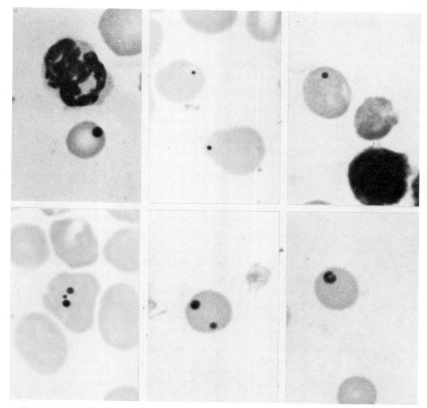

Figure 10.2 The morphology of micronuclei. Normally they are round and have a sharp contour. The majority measures from 1/20 to 1/5 of the diameter of an erythrocyte. Cells with more than one micronucleus are in minority and so are cells with irregularly shaped micronuclei. Even after treatments with most effective doses of clastogens, which produce visible breaks in almost every metaphase, it is rare that more than 8% of the polychromatic erythrocytes are micronucleated. (Reproduced from Chemical Mutagens, Vol. 4, Plenum Publishers, with permission.)

feature that renders the m.t. practicable for screening possible mutagens resides entirely in the fact that in bone marrow the great majority of micronuclei occurs in a cell type which expels its main nucleus, i.e., the anucleate young erythrocyte. A few hours after completion of the last mitosis, erythroblasts lose their nucleus (Fig. 10.1 e). For unknown reasons, micronuclei remain behind and become very obtrusive structures which, in hematological routine, have long been known as Howell-Jolly bodies (Fig. 10.2). A great advantage for the m.t. lies in the fact that the very young erythrocytes stain differently from the older forms, i.e., the mature red erythrocytes (normocytes). For approximately 24 hr., the young erythrocytes stain not red but bluish and are called polychromatic erythrocytes. Micronuclei present in this cell type are bound to have arisen mostly, if not exclusively, during the course of the immediately preceding cell cycle and under the influence of the temporally limited action of the mutagen. The morphology of all other types of bone marrow cells does not reveal whether they were mitotically active during the test period or not.

The m.t. is a substitute for the difficult chore of bone marrow metaphase analysis for chromosome breaks. Technical difficulties with this unfavorable material and the scarcity of mitoses often turn out to be a serious handicap. In the m.t., much higher numbers of cells can be studied; as a matter of fact, there is almost no limit to the number of scorable cells. The spontaneous background rate is low and consistent, a feature, within technical limits, shared by the metaphase method. Furthermore, the m.t. is independent of a favorable species karyotype. This permits a fast comparison of test results in different mammalian species.

For in vivo scoring of chromosome loss due to impairment of the spindle apparatus, metaphase studies cannot replace the m.t. The metaphase method is inadequate for this task for two reasons. Firstly, it can register loss or gain of chromosomes only in the metaphase following the one during which spindle impairment took place, a situation fundamentally different from the one in which chromatid breaks are scored. Aneuploid cells, however, often do not enter another mitosis and, therefore, are lost to scoring. Secondly, the metaphase method has a serious technical limitation when it comes to chromosome counting. In order to obtain analyzable metaphase spreads from bone marrow cells, the treatment of mitotic cells has to be so harsh that technical chromosome losses on the slides are unavoidable. Another alternative, anaphase scoring in bone marrow, is unsuitable for quantitative purposes, not only because of the shortage of anaphase figures, but for technical reasons as well: contrary to the convenient horizontal alignment of anaphase cells grown on glass surfaces, such an orientation is lacking in a smear.

The Technical Procedures

Premises

There are a number of principles which must be clearly borne in mind in order to appreciate the various technical steps in preparation and scoring.

1. The tested cell population consists of erythroblasts undergoing their last chromosome replication and mitosis. The effect is scored after expulsion of the main nucleus in young polychromatic erythrocytes.
2. The bone marrow preparation in the m.t. comprises a representative sample prepared from the full content of one or several marrow cavities. It is not a smear from a haphazard part of the marrow. Shifts in the proportions of different cell types can thus be recognized, as e.g., lack of polychromatic erythrocytes or an increase in normocytes.
3. Morphology and distribution of micronuclei are well-defined features.
4. Preparation and scoring are performed in a manner providing safeguards against scoring artifacts.
5. The normal application scheme with an exposure time of 30 hr. permits some conclusions regarding the cell-cycle-specific period of action of the tested agent.

Test Animals

For routine purposes the laboratory mouse is of perfect size for the m.t. If larger animals are utilized, a smaller bone than the femur should be chosen in order to be able to process the complete content of an entire marrow cavity. Fat in the bone marrow of adult or old animals has a deleterious effect on the quality of the smears. Therefore, adolescent animals, e.g., mice 7–12 weeks old, ought to be used.

Treatment of Animals with Test Substances

It is clear that problems of solubility of the test substance, mode of application, and dosage must be clarified before a test is started. If necessary, the aid of experienced toxicologists should be sought. Since all in vivo cytogenetic methods are not very sensitive, test series always should include the dosage range from usage level up to the highest tolerable dose.

The Preparation of Bone Marrow

Material. Scissors and forceps for dissection; gauze; disposable syringes, 1 ml; disposable needles (25 g, or approximately 0.5 × 16 mm); siliconized, conical 5 ml centrifuge tubes; siliconized Pasteur pipettes (long type) fitted with rubber bulbs; fetal calf serum, approximately 7 ml per animal; precleaned slides; polished cover glasses (the type used for hematological counting chambers); laboratory centrifuge (required speed 1000 rpm) are needed.

Dissection and handling of the bone marrow.

1. Immediately after sacrificing the animal, both femora are removed in toto, which means cutting through the pelvic bones and below the knee. The bones are freed from muscle by the use of gauze and fingers. Cleaning the distal end is quite simple since, by gentle traction, the knee with all surrounding tissue will separate from the shaft in the epiphyseal line, leaving the marrow cavity closed. The proximal end of the femur is carefully shortened with scissors until a small opening to the marrow canal becomes visible.

2. A centrifuge tube is filled with serum up to the rim. With the needle mounted, about 0.2 ml serum is aspirated into the syringe. Subsequently, the needle is inserted a few millimeters deep into the proximal part of the marrow canal which is still closed at the distal end. Then the femur is submerged in the serum and squeezed against the wall to prevent it from slipping off the needle. By gentle aspirations and flushings, the marrow is forced out through the opening around the needle. In this way, the cells get into the serum in the form of the desired fine suspension and not as gross particles. Only in the case where the needle becomes obstructed at the beginning does one flush out the serum in the syringe; otherwise the flushing begins with aspiration.

3. The tube is centrifuged at 1000 rpm for 5 min. The supernatant is drawn off completely with a Pasteur pipette. The cells in the sediment are carefully mixed by repeated aspiration into the capillary part of a new, dry, siliconized pipette.

4. A small drop of the viscous suspension is put on the end of a slide and spread by pulling the material behind a polished cover glass held at an angle of 45°. The size of the droplet is chosen so that all material is used up at a distance of 2–3 cm. The preparations are then air-dried.

Staining. The best results are obtained if staining is performed the day after preparation of the slides. If it is done immediately, the slides should be flamed for a short time.

The materials needed are as follows: ordinary vertical staining jars used for all steps; May-Gruenwald solution (e.g., Merck, Art. 1424); Giemsa's solution (e.g., Merck, Art. 9204); Aqua dest.; methanol; xylene; mounting medium (e.g., Euparal); cover glasses; and filter paper.

Stepwise procedure:

1. Undiluted May-Gruenwald solution, 3 min.
2. May-Gruenwald, diluted with Aqua dest. 1 : 1, for 2 min.
3. Rinse shortly in two changes of Aqua dest.
4. Diluted Giemsa solution for 10 min. (1 part Giemsa, 6 parts Aqua dest.).
5. Rinse well in Aqua dest.
6. Blot gently between two layers of filter paper.
7. Clean backside of preparation with methanol.
8. Xylene for 5 min.; embed.

Contrary to hematological routine, these preparations have to be mounted with cover glasses so that they can be analyzed at high as well as medium magnifications.

Scoring of slides and aberrations. As a first step, the slides are screened, at a medium magnification, for regions of suitable technical quality. In such regions, which are often located in a zone close to the end of the smear, the cells must be well-spread and perfectly stained. An excellent morphology of the nucleated cells serves as a criterion for good quality, even though the nucleated cells are not evaluated in the test. The erythrocytes must be separated from each other and well spread, neither globular nor having slurred contours. Their staining must be vigorous—red in mature erythrocytes (anulocytes or normocytes) and with a strong bluish tint in the young, immature forms (polychromatic erythrocytes). The routine timing scheme for the application of test substances comprises two doses given 30 hr. and 6 hr. before sacrificing the animals. This schedule is not absolutely optimal, but it is compatible with normal working hours in the laboratory. In respect to chromosome breaking substances, the application 6 hr. prior to preparation certainly is too late to increase the yield of the first dose; it should, however, increase the yield of spindle poisons.

There is no reason not to increase the number of doses given in several intervals over the period of 30 hr. The reasons not to go beyond the 30 hr. limit is that after an average of 24 hr., polychromatic erythrocytes mature into red normocytes. The time interval from the last S phase over mitosis, expulsion of the main nucleus, and maturation into a normocyte practically always exceeds 30 hr. Chromosome damage caused during the last S period or earlier, therefore, manifests itself only in micronucleated *polychromatic* erythrocytes.

Although it is possible to score thousands of cells in the preparations obtained from a single mouse, it is clear that for many reasons a sufficient number of test animals should be investigated. The minimum number for a given dose should not be less than ten animals.

In preparations from untreated, adolescent mice the proportion of total erythrocytes is on the order of 8%, and roughly half of these are polychromatic. If the test agent inhibits the proliferation of the erythroblasts, the proportion of polychromatic erythrocytes is reduced. If chromosome-breaking agents or spindle poisons destroy a sizable proportion of the nucleated cells, the lumen in the marrow cavity is replenished with peripheral blood, and therefore, in the preparations, the proportion of blood, i.e., red normocytes, increases. Thus, it is mandatory to begin scoring with a knowledge of a) the proportion of nucleated cells and b) the proportion of polychromatic forms within the erythrocyte population.

In the next step, at medium to high magnification, 1000 polychromatic erythrocytes are screened for the presence of micronuclei. The scored elements are micronucleated cells and not the number of micronuclei.

For two important reasons, it is necessary to register separately the number of micronucleated normocytes in the optical fields containing the 1000 polychromatic erythrocytes. The first reason is to safeguard against counting artifacts. Difficulties sometimes can occur with residues of stain, but this is easy to remedy. Another source of artifacts is noticed sometimes, especially in bone marrow from rats: basophilic granula of damaged precursors of basophilic leucocytes can resemble micronuclei, but in such cases the particles occur indiscriminately over all types of cells as well as in between them. When learning the m.t., it is absolutely mandatory that the experimenter take pains to look at positive and negative control preparations from an experienced laboratory. Otherwise, there is a considerable risk that artifacts enter into the scores.

The second reason for counting micronuclei over the normocytes as well is the following: agents or doses of agents acting predominantly during the S phase of the cell cycle do not produce an appreciable amount of micronucleated normocytes, since 30 hr. is too little time for completing part of the S phase, G_2 mitosis, expulsion of the nucleus, and maturation. The proportion of micronucleated normocytes increases, however, if the agents act on the spindle apparatus. Such an action may, however, only be suspected if 80% or more of the total micronuclei are present in polychromatic erythrocytes.

When working with clastogens, artifacts usually must be suspected if more than five micronucleated normocytes are registered in the fields containing the counted 1000 polychromatic cells.

This protocol of analysis may be referred to as the *routine procedure*. One needs to be aware that this protocol, for the purpose of keeping screening as simple as possible, does not provide strictly quantitative results and that the score is slightly influenced by two variables: (1) the transition from the polychromatic to the mature erythrocyte is a short but nevertheless continuous process. Whether an erythrocyte in transition should be scored as polychromatic or as red will always be a more or less subjective decision made by the investigator. (2) For determining the incidence of micronucleated normocytes, the total number of red cells is not counted, the reference being given by the optical fields containing the scored 1000 polychromatic erythrocytes. The author recommends the use of the routine procedure in the following situations: (1) If the m.t. is carried out only occasionally, and if it is performed on a wide spectrum of substances. (2) If it is known that the agents can affect the normal proliferation of bone marrow cells, as is the case, for example, with cytostatics and antimetabolites. (3) If dose-effect relationships of cytogenetically active substances are studied.

In the screening of numerous other compounds, a simplified protocol of analysis can be applied and is based on the following considerations. In controls, as well as after treatments not markedly affecting the proliferation of the bone-marrow cells, the relation of polychromatic to mature erythrocytes is fairly consistent, within the region of 1 : 1. Analysis of 2000 total erythrocytes in place of 1000 polychromatic erythrocytes reduces the number of subjective decisions; furthermore, the results of different investigators are easier to compare. Negative results obtained in this way do not pose further problems with respect to cytogenetic activity in bone marrow. Positive results, however, may require the establishment of a dose-effect curve by applying the routine procedure.

Comments

A main advantage of the test is its simplicity, which permits the thorough testing of a substance over a wide dose range, after different modes of application and in different species. On the other hand, this simplicity should not be overrated; the method needs to be learned carefully and the quality of the preparations must be compared to those of experienced laboratories. Other advantages are the independence of a favorable species karyotype and the suitability to detect chromosome loss due to impairment of the spindle apparatus. Furthermore the m.t. permits a quick orientation about proliferation in the bone marrow. Inhibition of proliferation due, e.g., to the antimetabolic action of a compound is readily recognized by a deficit of polychromatic erythrocytes. Any severe cell destruction occuring, e.g., after the action of a strong clastogen is recognizable by an influx of peripheral blood into the marrow.

The m.t. was introduced in the early 1970s (1;2;3;4;5) and since was applied to a great number of chemicals; for reviews, see e.g., Schmid (6), Maier and Schmid (7) and Matter and Grauwiler (8). Compared to metaphase scoring, it was found that the micronucleus test is at least as sensitive as the former method.

Despite its many advantages, the m.t. is not intended to be more than a screening procedure. It is mandatory that an unexpected, newly discovered effect must be characterized by suitable cytogenetic techniques. This is the place where in vitro cytogenetic procedures should be applied. For confirming clastogenicity, one should use metaphases of cell cultures, and for confirming a suspected effect on the spindle apparatus, anaphases of cell cultures grown on cover glasses should be studied.

In respect to mutagens acting on the spindle apparatus, and therefore able to cause genome mutations, the following comment seems to be in place: it is probably not justified to assume that spindle poisons are responsible for the bulk of numerical aneuploidies in man, such as autosomal trisomies and sex chromosome anomalies. The maternal age effect suggests that their origin must usually be more complicated and, so far, is not understood. It is, however, reasonable to admit the possibility that spindle poisons could play a partial role in this quantitatively important problem of human pathology. One step to confirm or refute a suspicion of this kind would be an extension of testing by the use of the dominant lethal assay.

Another aspect of the induction of genome mutations which should not be neglected is its possible role in carcinogenesis, at least in the sense of a cocarcinogenic factor. An investigation into that problem actually seems to be overdue.

Literature Cited

1. Boller, K.; Schmid, W. Chemische Mutagenese beim Säuger. Das Knochenmark des Chinesischen Hamsters als in vivo-Testsystem. Hämatologische Befunde nach Behandlung mit Trenimon. Humangenetik 11:35–54; 1970.
2. Matter, B.; Schmid, W. Trenimon-induced chromosomal damage in bone marrow cells of six mammalian species, evaluated by the micronucleus test. Mutat. Res. 12:417–25; 1971.
3. von Ledebur, M.; Schmid, W. The micronucleus test. Methodological aspects. Mutat. Res. 19:109–17; 1973.
4. Schmid, W. Chemical mutagen testing on in vivo somatic mammalian cells. Agents Actions 3:77–85; 1973.
5. Schmid, W. The micronucleus test. Mutat. Res. 31:9–15; 1975.
6. Schmid, W. The micronucleus test for cytogenetic analysis. In Hollaender, A. ed. Chemical mutagens. Vol. 4. New York: Plenum Publishing Co.; 1976: 31–53.
7. Maier, P.; Schmid, W. Ten model mutagens evaluated by the micronucleus test. Mutat. Res. 40:325–38; 1976.
8. Matter, B. E.; Grauwiler, J. Micronuclei in mouse bone-marrow cells. A simple in vivo model for the evaluation of drug-induced chromosomal aberrations. Mutat. Res. 23:239–49; 1974.

11

The Use of the Host-Mediated Assay for Cytogenetic Studies

by R. J. PRESTON* AND J. G. BREWEN†

Abstract

The host-mediated assay enables cells to be exposed to a chemical in an essentially in vivo situation. There are several advantages to this when compared to an in vitro exposure. The requirement for an added metabolic activation system is removed, since the host animal metabolises the test compound. It also allows for the fact that the actual dose exposure to cells in vivo can be very different from that in vitro. It also makes it possible to study cells in an in vivo environment that would otherwise not be amenable to such exposure (e.g., human cells). Furthermore, comparisons can be made between the test cell type and various cell types of the host animal (e.g., bone marrow, lymphocytes, and spermatogonia).

The method most commonly used for cytogenetic studies is described, together with the results from the relatively small number of reported uses of the assay. The technique has not been widely used, despite its advantages over in vitro assays, because of the intricacy of the technique and the rather lengthy time it takes to perform. However, it still remains as a most valuable assay for studying human cells, in particular, "in vivo".

Introduction

In order to test the potential of a particular compound to produce an effect which would present a genetic hazard to man, it has generally been assumed that a cytogenetic assay is important. Such an assay can be performed on many different cells from different species, but, in order that the results can be more readily extrapolated to the effectiveness in man, it is desirable to use human cells as some part of the assay.

*Biology Division, Oak Ridge National Laboratory Oak Ridge, Tennessee 37830.
†Medical Affairs, Allied Chemical Corporation, Morristown, N. J. 07960.

The most common system employed which meets these requirements is the treatment of human peripheral lymphocytes in vitro. There are many advantages to the lymphocyte system. The population is relatively synchronous even after mitogenic stimulation, and thus cells can be treated in specific stages of the cell cycle. Because of this synchrony, it is possible to analyze cells which are in their first metaphase after treatment, which is a prerequisite for an in vitro cytogenetic assay. However, there are drawbacks to this in vitro assay. Some compounds require activation in order to be mutagenic, carcinogenic, or clastogenic, and this activation in many cases is not performed by lymphocytes. Also, the actual dose exposure to cells in vitro versus those in vivo can be very different.

In order to use the advantage of the lymphocyte system and to overcome the disadvantages, the host-mediated assay system has been employed for cytogenetic studies. In such an assay, the human cells themselves are in an essentially in vivo situation, with the host animal providing the activation and metabolism of compounds. It is also possible to compare the results for the human cells with those obtained from analysis of the lymphocytes, bone marrow, or germ cells of the host animal itself. Furthermore, not only can cells from other mammalian species be treated within the same host for comparison with human cells, but the host-mediated assay can utilize bacterial systems (1), yeast (2), or *Neurospora* (3), allowing a comparison between mutagenic effects on many cell types within the same environment. The method described here uses peripheral lymphocytes as the human cell type, but of course other human cell types can be used equally well. This method is based on the procedure described originally by Buckton and Nettesheim (4) and modified by Brewen et al. (5; 6).

Methods

Preparation of Diffusion Chambers

The diffusion chambers consist of Lucite rings with an outside diameter of 22 mm, an inside diameter of 14 mm, a height of 6 mm, and a 1 mm hole drilled through the ring to enable the chamber to be filled with the cell suspension. (The rings can be obtained from Piedmont Plastics Inc., 4927 Selabert Ave., Charlotte, N.C. 28261. Soap lubricant must be used instead of cutting oil in the manufacture.) Millipore filters with a pore size of 0.45 μ are glued to the sides of this ring with DEK-adhese cement (which can be obtained from Donald Tulloch, Jr., P.O. Box 146, Chadds Ford, Pa. 19217). The chamber is dry-heat sterilized at 180°C for 48 hr. The filters still contain toxic substances, in particular, detergents. Therefore, 24 to 36 hr. before they are to be used, they are filled with a sterile 1% solution of mouse serum in Hanks' balanced salt solution (HBSS), and then placed in a container of this same solution at 4°C. The HBSS/serum solution is removed with a 22-gauge needle through the hole in the Lucite ring immediately prior to using the chamber.

Cell Preparation

The technique for lymphocytes is described because some details of the procedure differ from those for normal leukocyte cultures. For other cell types, e.g., fibroblast or lymphoid cell lines, suspensions of cells in fresh growth medium are loaded into the chamber at concentrations which will vary from cell type to cell type.

Freshly drawn heparinized blood samples are divided into 4 ml aliquots and centrifuged at 350 g (half maximum speed) in a clinical centrifuge for 8 min. The buffy coat layer is removed with a Pasteur pipette and put into a 1:1 mixture of HBSS and phytohemagglutinin (PHA) with a pH of about 7.0. The cell concentration is 1 buffy coat in 0.2 ml of the HBSS/PHA mixture. The buffy coat samples from one donor are pooled at this time.

The leukocyte suspension is injected into the diffusion chamber through the 1 mm hole with a 22-gauge needle. The hole is then sealed with melted paraffin to give a completely sealed diffusion chamber.

Implantation of Diffusion Chamber in Host Animal

It is possible to use any host animal, but normally the most suitable are mice, rats, guinea pigs, or Chinese hamsters. The animal is anesthetized with ether and the hair shaved from the abdomen. An incision is made slightly to one side of the midline and the diffusion chamber is inserted through this into the peritoneal cavity. The incision is then closed with metal suture clips.

Treatment of Host Animal and Cells in Diffusion Chamber

The host animal can be treated with a test compound at different times after implantation of the diffusion chamber. At the time the leukocytes are placed in the chamber, they are all in G_0. About 24 to 36 hr. later, almost all transformed cells will be in the S phase. If cells are treated 40 hr. or more after stimulation and fixed 2 to 3 hr. later, the cells analyzed in metaphase will have been in G_2 at the time of treatment. It is possible, therefore, to treat cells in all stages of the cell cycle. The cells are usually fixed between 56 to 72 hr. after implantation when there is a high mitotic index.

Preparation of Cells for Analysis

The host animal is killed and the chamber removed. The Millipore filters are cut from the Lucite ring with a razor blade and are placed with the contents of the chamber in 3 ml of tissue culture medium (in our case Medium 199 from GIBCO) which contains 1×10^{-6} M colchicine. This is then incubated at 37°C for 2 to 4 hr. It is usual, but not always necessary, to add 1 ml of 1% pronase to the medium and then gently agitate the mixture with a magnetic stirrer for 15 to 30 min. This breaks up blood clots which often form in the culture during its incubation in the diffusion chamber. The cells are then pelleted out of the medium by centrifugation.

The method of slide preparation described below is the one we use routinely for blood cultures in this laboratory, but of course there are many other equally satisfactory procedures.

Following incubation with colchicine and pronase, the cells are washed with HBSS, centrifuged, and resuspended in 3 parts of distilled water to 1 part of HBSS. The cells are incubated in this hypotonic solution for 15 min. at 37°C and then centrifuged. The cell pellet is slowly fixed with 3 parts of methanol to 1 part of glacial acetic acid. The undisturbed pellet of cells is allowed to fix for 15 to 30 min. The cells are washed three times with fixative, and then a drop of cells at the desired concentration is dropped onto a clean wet slide. The slide is allowed to air-dry. The dried slide is stained with 2% Giemsa or 1% aceto-orcein and mounted in Euparal.

This technique has been modified by Huang (7) for the study of human lymphoid cell lines. The only differences between the two techniques are that the host animal itself is injected with 0.5 ml of 0.04% colchicine to arrest cells in the diffusion chamber in metaphase, and the cell clots still within the chamber are disassociated with 0.5% pronase and 5% Ficoll for 1 hr.

Discussion of Results

Despite the obvious advantages of this technique for studying human cells in an essentially in vivo environment, the somewhat complicated experimental design has limited its use as a test system. There are very few published reports of cytogenetic data obtained from the host-mediated assay.

Brewen et al. have studied the effects of methyl methanesulfonate (5) and cyclohexylamine (6) on human leukocytes cultured in diffusion chambers using mice and Chinese hamsters, respectively, as host animals. The results showed that intravenous or intraperitoneal injections of methyl methanesulfonate given 12 or 24 hr. before the chambers were removed for cell fixation produced a significant level of chromatid aberrations compared to the control at doses of 65 mg/kg and higher. The cyclohexylamine was administered as daily intraperitoneal doses of 50, 150, or 450 mg/kg. No difference in yield of chromosome aberrations was found between treated and control.

Huang (7) studied the growth characteristics of human lymphoid cell cultures in diffusion chambers implanted in C_3H/St mice, and also the cytogenetic effects of cyclophosphamide on human cells and bone marrow cells of the host mice. He found that the maximum cell number was obtained 3 to 5 days after implantation, at that time the cell concentration was 4 to 15 times that of the initial concentration of about 5×10^5 cells per chamber. It was possible to show that cells could be recultured even after two weeks in the diffusion chamber. There was a significant increase in chromatid aberrations in the human cells following cyclophosphamide treatment of the host with doses of 40 mg/kg and above. There was also a significant increase in aberrations in the host bone marrow at a dose of 80 mg/kg. No direct comparison of the yields in the bone marrow and human lymphoid cells was made.

There is also a preliminary report by Furukawa and Huang (8) on the induction of sister chromatid exchanges in human cells cultured in diffusion chambers in mice. The cells used were two human hematopoietic cell lines, and the technique was basically the same as that for aberration studies described above, except that bromodeoxyuridine (BrdU) had to be incorporated to allow for differential chromosome staining for detection of sister chromatid exchanges. This was achieved by giving the host animals (C3H mice) 12 hourly injections of 0.2 ml of a solution containing 3 parts of 5 BrdU (10^{-2} M) and 2 part deoxycytidine (5×10^{-2} M). Cyclophosphamide was given as a single injection 12 to 14 hr. after the first injection of BrdU. The cells were analyzed 36 to 40 hr. after the first injection of BrdU. There was an increase in sister chromatid exchanges at all cyclophosphamide doses (7.5 to 30 μg/gm). Thus, the host-mediated assay also seems to be adaptable to sister chromatid exchange studies.

Evaluation of the Assay

The greatest drawback to the host-mediated assay for cytogenetic studies is the somewhat time-consuming and intricate nature of the technique. However, in many ways it is a desirable addition to a testing protocol because it provides additional information to other in vivo and in vitro tests. It is possible to study human cells in an in vivo environment, thereby taking into account the requirement of metabolic activation of many compounds and also the distribution of a particular compound within an organism. It is also feasible to make a direct comparison between aberration frequencies in human cells and various dividing cells of the host animal. Clearly these advantages have not been exploited as there are very few reports of the use of the host-mediated assay for cytogenetic studies.

Literature Cited

1. Gabridge, M. G.; Legator, M. S. A host-mediated microbial assay for the detection of mutagenic compounds. Proc. Soc. Exp. Biol. Med. 130:831-34; 1969.
2. Fahrig, R. Metabolic activation of aryldialkyltriazenes in the mouse: induction of mitotic gene conversion in *Saccharomyces cerevisiae* in host-mediated assay. Mutat. Res. 13:436-39; 1971.
3. Legator, M. S.; Malling, H. V. The host-mediated assay: a practical procedure for evaluating potential mutagenic agents in mammals. *In* Hollaender, A. ed. Chemical mutagens. Vol. 2. 11. New York: Plenum Press; 1971: 569-90.
4. Buckton, K. E.; Nettesheim, P. In vitro and in vivo culture of mouse peripheral blood for chromosome preparations. Proc. Soc. Exp. Biol. Med. 128:1106-10; 1968.
5. Brewen, J. G.; Nettesheim, P.; Jones, K. P. A host-mediated assay for cytogenetic mutagenesis: preliminary data on the effects of methyl methanesulfonate. Mutat. Res. 10:645-49; 1970.
6. Brewen, J. G.; Pearson, F. G.; Jones, K. P.; Luippold, H. E. Cytogenetic effects of cyclohexylamine and N-OH-cyclohexylamine on human leukocytes and Chinese hamster bone marrow. Nature (London) New Biol. 230:15-16; 1971.
7. Huang, C. C. A modified host-mediated assay using cultured human lymphoid cells in diffusion chambers in mice and cytogenetic effects of cyclophosphamide. Environ. Res. 13:267-77; 1977.
8. Furukawa, M.; Huang, C. C. Increased frequencies of sister chromatid exchanges in human cells cultured in diffusion chambers in mice treated with cyclophosphamide. Mamm. Chromosome Newsl. 18:22; 1977.

12
Measurement of DNA Repair Synthesis in Cultured Human Fibroblasts as a Short-term Bioassay for Chemical Carcinogens and Carcinogenic Mixtures

by R. H. C. SAN AND H. F. STICH*

Abstract

A good correlation exists betwen the carcinogenicity of a chemical compound and its capacity to induce non-semiconservative DNA synthesis in cultured mammalian cells. Monitoring this unscheduled DNA synthesis (UDS) thus provides a means of identifying chemical carcinogens. Cultured fibroblasts of human or rodent origin are prevented from undergoing DNA replicative synthesis by deprivation of an essential amino acid (e.g., arginine) or by addition of a chemical inhibitor (e.g., hydroxyurea) and then exposed to the test chemical concomitantly with tritiated thymidine (^{3}HTdR). The unscheduled incorporation of the radioisotope is detected by autoradiography or liquid scintillation counting.

One advantage of the UDS assay is that the endpoint (unscheduled ^{3}HTdR incorporation) is a general phenomenon not limited to one specific type of DNA damage. The relevance of using human cells as test subjects is another feature that cannot be ignored.

The limited capability of cultured fibroblasts for the metabolic activation of precarcinogens can be alleviated by the inclusion of a liver microsomal preparation in the incubation mixture. Alternatively, rat hepatocyte primary cell cultures provide a test system with a built-in carcinogen metabolizing capability.

Except for a group of structurally related compounds (4NQO derivatives), it remains to be shown that the level of DNA repair triggered by a carcinogen provides a quantitative estimation of the degree of its carcinogenic potential.

In view of the growing concern over the activation and the deactivation of carcinogens which cannot be studied in most in vitro test systems, a procedure involving in vivo administration of

*Environmental Carcinogenesis Unit, British Columbia Cancer Research Centre, Vancouver, B.C., Canada

a test compound followed by in vitro monitoring of DNA repair in various tissues may turn out to be a technique of choice.

Introduction

In view of the increasing need for the carcinogenicity testing of the plethora of both man-made and naturally occurring chemicals in man's environment, the last decade has witnessed the introduction of numerous fast and economical prescreening procedures. Most of these newly developed methods that appear suitable for a large scale prescreening programme depend on the capacity of carcinogens to induce mutations or to affect the DNA of indicator organisms. Currently, a great deal of attention is focused on the use of various bacteria, yeast, fungi, plants, insects, and mammalian cell cultures in such short-term bioassays (see review by Stich et al. 1975).

The monitoring of DNA repair synthesis in cultured human fibroblasts provides a means to detect the DNA damaging action of chemical carcinogens. Several methods are available to estimate DNA repair in eukaryotic cells, e.g., (1) by chromatographic detection of the rate of the disappearance of altered nucleotides (Setlow and Carrier 1964), or (2) by sedimentation of DNA through alkaline sucrose gradients for detection of DNA fragmentation and rejoining (McGrath and Williams 1966). The most widely used methods of monitoring DNA repair synthesis involve the "resynthesis" of short sections of the DNA molecule which were eliminated by endo- and exonuclease enzymes following exposure to exogenous DNA damaging agents (Rasmussen and Painter 1966). Common techniques that utilize the phenomenon of "unscheduled DNA synthesis" or "DNA repair synthesis" track the incorporation of 5-bromodeoxyuridine or tritiated thymidine into nuclear DNA of repairing cells, although other purine or pyrimidine precursors can also be successfully employed (Cleaver 1972).

Examination of over 100 chemical compounds representing the major groups of carcinogenic substances revealed a good correlation between the carcinogenic activity and the capacity to elicit DNA repair synthesis in cultured mammalian cells (Stich et al. 1971; San and Stich 1975; Williams 1976, 1977, 1978; Martin et al. 1978). The general procedure for the autoradiographic detection of carcinogen-induced DNA repair in tissue culture systems is described here. Included in the discussion are the advantages, limitations, and applications of the DNA repair test.

Materials and Methods

Cell Cultures

Primary cultures of human skin fibroblasts are initiated from small punch skin biopsies about 3 mm in diameter taken from the forearm of male or female donors. Each tissue sample is torn into about 50 tiny pieces which are then sandwiched between two 22 × 22 mm coverslips, placed into a 35 mm petri dish and maintained in Eagle's minimal essential medium supplemented with antibiotics and 15% fetal calf serum. The cultures are kept in a humidified incubator (37°C) staffed with 5% CO_2 and 95% air. After 14 days, the outgrowing fibroblasts are removed from the coverslips and cultured in monolayer in 100 mm petri dishes (Falcon).

Culture Media

For routine maintenance of cell cultures, Eagle's minimal essential medium (MEM) is used. Arginine deficient medium (ADM) is prepared according to the standard formulation for Eagle's minimal essential medium (Eagle 1959; Merchant et al. 1964). The various essential amino acids, with the exception of arginine, are weighed out and dissolved in the manner described in the Appendix. The nonessential amino acids (in the form of 100x concentrated mixture, Flow Laboratories), as well as the vitamins, are added and the culture medium sterilized by passage through a 0.2 μ millipore filter.

The following antibiotics are added to all culture media: 100 units/ml penicillin (General Biochemicals), 100 μg/ml streptomycin sulfate (General Biochemicals), 100 μg/ml kanamycin (Grand Island Biological Co.), and 2.5 μg/ml fungizone (Grand Island Biological Co.).

MEM is supplemented with 15% fetal calf serum (Grand Island Biological Co., heat inactivated at 56°C for 30 min.), whereas ADM is supplemented with 5% FCS.

Procedure for Autoradiographic Detection of DNA Repair Synthesis

Preparation of Cell Cultures

Cells from the third and fourth passage of human skin fibroblast cultures are seeded onto 22 mm^2 cover glasses (Clay Adams) in 35 mm petri dishes (Falcon Plastic) at 5 × 10^4 cells per dish and covered with MEM + 15% FCS. The cell cultures are always used for experimentation before the monolayer has reached confluency. This permits better cytologic preparation, uniform exposure to the test compound, and lower background count when processed for autoradiography.

To distinguish between DNA replication at S phase and DNA repair synthesis, a modification of the procedure developed by Freed and Schatz (1969) is employed. By depriving the cell culture of the essential amino acid arginine, DNA synthesis associated with chromosome replication is drastically reduced (Stich and San 1970). Upon reaching 80% confluency (four days after seeding), the cells are placed into ADM. This is done by dipping the coverslips about five times into each of two beakers of ADM, and then transferring them to new 35 mm petri dishes containing ADM (5% FCS). After three days in ADM, more than 90% of the cells would be arrested at G_1.

Solutions of the Test Compound

Test compounds that are readily soluble in water are dissolved directly in ADM. Unless higher concentrations are required, a 10^{-3}M stock solution in ADM (2.5% FCS) is usually prepared immediately before use. Serial dilutions are then made to obtain the desired final concentrations.

For agents that are not readily soluble in water, a stock solution is made in DMSO and appropriate serial dilutions made in the same solvent. Final dilution with culture medium to give the desired concentration are made just prior to addition to the cell cultures. The final concentration of DMSO does not exceed 1%.

Since many chemicals readily bind to serum proteins, a standard concentration

(2.5%) of FCS is used. In some laboratories, solutions of the test compounds are prepared in serum-free medium (Williams 1976, 1977).

Exposure to Test Compound and Radioactive-Label Incorporation

Previously, radioactive labelling with ^{3}HTdR was done immediately following exposure to the test compound (Stich and San 1970; San and Stich 1975). In view of the fact that some test compounds have very labile reactive metabolites and that DNA repair synthesis is initiated shortly after the DNA damage is inflicted, the procedure is modified to permit simultaneous administration of the test compound and the radioactive label.

Tritiated thymidine (specific activity 20 Ci/mmol) is obtained as thymidine (methyl-^{3H}) in aqueous solution from New England Nuclear and diluted to a concentration of 20 μCi/ml with ADM (2.5% FCS). Upon mixing with an equal volume of a solution of the test compound, the final concentration of ^{3}HTdR is 10 μCi/ml.

Exposure of the cell cultures to ^{3}HTdR and the test compound is carried out by replacing the medium in the petri dish with 0.5 ml of ^{3}HTdR at 20 μCi/ml and 0.5 ml of the test compound at twice the desired final concentration.

Preparation of Activation Mixture

For test compounds which require metabolic activation, the postmitochondrial fraction (S9) from adult rat liver is included in the incubation mixture.

Adult rat liver microsomes are available commercially from Lytton Bionetics. Alternatively, this can be prepared by the following procedure. Adult Fischer rats (males with average body weight of 200–300 gm) are killed by cervical dislocation. The livers are removed immediately, trimmed of adhering connective tissue and placed into a beaker containing PBS/sucrose buffer at 4°C (PBS, with calcium and magnesium ions, containing 0.25 M sucrose, pH 7.5). All subsequent steps are conducted with solutions and containers maintained at 4°C. Liver from several animals are pooled, washed with PBS/sucrose buffer, dabbed dry on absorbent tissue, weighed, and quickly transferred to separate beakers of fresh PBS/sucrose buffer such that 6 gm of liver are contained in 10 ml buffer. The livers are minced with scissors and the contents of each beaker are homogenized by 10 up-and-down strokes in a loose-fitting Potter-Elvejhem homogenizer operating at 1,000 rev./min. The homogenate is centrifuged at 9000x g for 10 min. (at 4°C) in a precooled rotor (Beckman, Type 40) and nitrocellulose centrifuge tubes. The resulting postmitochondrial supernatant fractions (commonly referred to as S9 fractions) are pooled, mixed thoroughly to ensure homogenicity of each batch, and then distributed in measured aliquots to polypropylene tubes, capped, immediately frozen in liquid nitrogen, and stored at -70°C (Stich and Laishes 1975). To optimize the activating capability of the S9 fraction, liver microsomes from animals induced with Arochlor 1254 for five days and prepared as reported by Ames et al. (1973) may be employed.

The activation system is prepared by adding, for each cell culture, 4.0 μmoles NADPH or NADP, 25 μmoles MgCl$_2$, and 20 μmoles G6P in 0.4 ml of S9 fraction (thawed immediately before use) and adjusting the pH to 7.2 with 0.1N NAOH.

Appropriate volumes of a solution of the test compound, ^{3}HTdR and activation system are mixed, vortexed quickly, and the resulting activation mixture is added immediately to the cell cultures.

Fixation

Following exposure of the cell cultures concomitantly to ^{3}HTdR and the test compound for the appropriate period of time (usually 2 to 3 hrs.), the coverslips are rinsed by dipping two or three times into Hanks' balanced salt solution, immersed in 1% sodium citrate for 15 min., followed by 10 min. fixation in acetic acid/ethanol (1:3, v/v), and allowed to air-dry.

To facilitate manipulation of the fragile coverslips, they are mounted with the cell monolayer facing upwards on glass slides using a small quantity of molten paraffin wax.

Coating with Photographic Emulsion

Acetic acid has been reported to react with photographic emulsions, resulting in an increase in background grain count (Stocker and Muller 1967). As a precaution to remove final traces of acetic acid from the fixation procedure, the slides are passed through a graded alcohol series, 95% ETOH, 70% ETOH, 20% ETOH (10 min. each), two changes of distilled water, one change of PBS, two more changes of distilled water (10 min. each) and then left to air-dry.

The slides are coated with Kodak NTB-3 nuclear-track emulsion in a dark room equipped with a Kodak Wratten Series 2 red filter. The emulsion is thawed at 43°C in a water-bath in the dark, diluted with an equal volume of distilled water, and checked (using a blank slide) for the absence of air bubbles. The glass slides bearing the fixed cell cultures are then individually dipped in the emulsion, air-dried in a vertical position, and then stored at 4°C in light-tight boxes for 14 days.

Processing and Staining of Autoradiograms

Autoradiograms are brought back to room temperature after 14 days of exposure. Processing is done at 18°C in Kodak D-19 developer (3 min.), stop bath (30 sec.), Kodak fixer (10 min.), and hypoclearing agent (1 min.). After a 30-min. rinse in running water (18°C), the slides are stained with 2% aceto-orcein for 10 min., dehydrated through successive immersion (1.5 min. each) in 95% ethanol, butanol, butanol/xylol, two changes of xylol, and mounted in Permount (Fisher Scientific Co.) by superimposing another converslip over that bearing the cell monolayer.

Analysis of Autoradiograms

The amount of DNA repair synthesis is estimated by counting the number of grains over each nucleus. This can be done manually using a light microscope or with one of the automatic grain counters available on the market (e.g., Artek Model 880). Care is taken that nuclei of comparable size are selected so that only cells of the same ploidy are used. Routinely, grain counts are made on small interphase nuclei.

Background count is taken into consideration by reckoning the number of grains over an area equal in size to that of the nucleus. At least 30 nuclei, at random locations throughout the entire coverslip culture, are analyzed for grain number. When the grain number is below 10 per nucleus, at least 100 nuclei are scored. Based on statistical calculations, Rogers and England (1973) have demonstrated that the accuracy of estimating the radioactivity per nucleus will depend, not on the number of nuclei counted, nor on the total area of emulsion scanned, but on the total number of silver grains counted in one sampling of the cell population. Thus, for a certain ratio of the nuclear to background grain counts, there is an optimum number of nuclei to be scored.

Liquid Scintillation Counting Technique

The analysis of autoradiograms with a light microscope is a time consuming process, and this may constitute a limiting factor in a large-scale screening programme. Even with the aid of an automatic grain counter, the scoring still has to proceed on an individual cell basis. Using liquid scintillation counting to monitor the unscheduled incorporation of ^{3}HTdR, Martin et al. (1978) recently demonstrated a correlation between carcinogenicity and increase in ^{3}HTdR uptake.

Advantage of DNA Repair Test

DNA repair synthesis can be detected in human fibroblasts following exposure to carcinogens of various chemical structures. Unlike some of the mutagenicity tests, the DNA repair assay has the advantage that the endpoint (unscheduled ^{3}HTdR incorporation) is a general phenomenon not limited to one specific type of DNA damage but encompasses a variety of alterations or lesions in the DNA molecules.

Considering that the aim of the DNA repair assay is to identify environmental carcinogens and mutagens that may pose a hazard to man, the relevance of using human cells as test subjects is an advantage that cannot be ignored.

Human fibroblasts can be readily obtained from skin biopsies of "normal" and cancer-predisposed persons. Since the unscheduled DNA synthesis triggered by a carcinogen occurs in more than 99% of the treated cells, only a small number of cells are required.

In addition, the autoradiographic preparations provide a number of features which are not available from the liquid scintillation counting procedure. These include a record of the cytological location of the incorporated thymidine, the identification of the cell type engaged in DNA repair synthesis, the proportion of cells involved in DNA repair, and the variation of DNA repair levels within a cell population.

The technique can also be adapted for peripheral blood lymphocytes. Minute quantities of human blood are adequate to provide sufficient numbers of lymphocytes for short-term cultures (48–72 hr.), during which time a DNA repair assay can be performed (Burk et al., 1971; Lieberman et al. 1971; Jacobs et al. 1972).

Limitations

In spite of its many advantages and its relevance to man, the DNA repair bioassay is not without its limitations. The technique suffers from the fact that it does not throw any light on the precise DNA-carcinogen interaction that triggers DNA repair synthesis. For example, alkylation of DNA may occur at N1, N3, or N7 of adenine; N3, N7, O6 of guanine; N3 of cytosine; and N3 and O4 of thymine (Lawley and Brookes 1963; O'Conner et al. 1972). Which of these events initiates repair and whether all alkylation products are removed by excision repair and at what rate remains to be elucidated.

There are a few other aspects worthy of some discussion which are presented in the following sections.

Use of Rat Liver Microsomal Activation Mixture (S9)

As in most in vitro bioassays for carcinogens and mutagens, cultured fibroblasts have a limited capability for metabolic activation of precarcinogens. Thus, the DNA repair test will readily identify agents that are administered in their reactive forms (e.g., alkylating agents, N-acetoxy-2-acetylaminofluorene) or ones that are readily converted into DNA damaging species without the involvement of mixed function oxidases (e.g., 4-nitroquinoline 1-oxide; N-hydroxy-2-acetylaminofluorene). Other precarcinogens may yield "false" negative results. This difficulty may be circumvented by the inclusion of a 9000x g supernatant (S9) from rat liver in the incubation mixture (Garner and Hanson 1971; Ames et al. 1973). Using this exogenous source of microsomal enzymes, it has been possible to activate several precarcinogens in vitro, e.g., dimethylnitrosamine (Malling 1971; Laishes and Stich 1973), aflatoxin B_1 (Garner and Hanson 1971; Stich and Laishes 1975), and sterigmatocystin (Stich and Laishes 1975).

Use of Rat Hepatocyte DNA Repair Test

Despite the popularity of using an exogenous liver microsomal preparation to activate precarcinogens to their reactive forms, there is growing evidence that such a procedure does not accurately reflect the reactions occurring in vivo. For example, mouse liver microsomes can activate aflatoxin B_1 to DNA damaging species (Stich and Laishes 1975), although this precarcinogen has not been shown to induce liver tumors in the mouse (Wogan 1966, 1973). Microsomes prepared from rats pretreated with phenobarbitone are more effective than those obtained from control animals in activating aflatoxin B_1 in vitro to a reactive derivative which is toxic to *S. typhimurium* (Garner et al. 1972) and which inhibits RNA polymerase (Neal and Godoy 1976). However, pretreatment with phenobarbitone renders the rats less susceptible to the carcinogenic action of aflatoxin B_1 (McLean and Marshall 1971).

The possibility of initiating rat hepatocyte primary cell (HPC) cultures (Williams et al. 1977) answers the need for a test system with a built-in carcinogen metabolizing capability and one in which activated carcinogenic species can be readily delivered to the target (indicator) cells. Such HPC cultures are highly viable and composed almost exclusively of hepatocytes. In addition, they have been shown to retain a high degree of the functional activity of the liver within the first 24 hours after initiation (Laishes and Williams 1976; Bissell et al. 1973; Bonney 1974; Lin and Snodgrass 1975; Michalopoulos and Pitot 1975). Use of HPC cultures in carcinogen-induced DNA repair studies revealed that they can activate a wide spectrum of precarcinogens requiring different pathways of metabolism, including polycyclic aromatic hydrocarbons, aromatic amines, azodyes, nitrosamines, and mycotoxins (Williams 1976, 1977, 1978).

Carcinogenic Potential and Detectable DNA Repair

It remains to be shown that the level of DNA repair triggered by a carcinogen provides a quantitative estimation of the degree of its carcinogenic potential. A good correlation between these two factors was demonstrable within a group of structurally related compounds, e.g., strong and weak carcinogenic 4NQO isomers and derivatives (Stich et al. 1971, 1974). However, no such correlation was obvious when carcinogens of different molecular structures were compared (San and Stich 1975). The absence of a good correlation may be attributable to several factors: (1) the difficulty in placing a quantitative value on carcinogenicity from in vivo rodent assays; (2) the lack of uniformity in the exposure times required for different chemical carcinogens to elicit a detectable level of unscheduled ^{3}HTdR incorporation. In the case of precarcinogens, this may be linked to the time period within which active intermediates, or metabolites are produced or (in the case of both precarcinogens and ultimate carcinogens) to the time required for the formation of DNA-carcinogen complexes to bring about an alteration in the DNA structure.

DNA Repair Inhibition

The absence of a DNA repair synthesis following the administration of a test compound could be due to the lack of its interaction with DNA (e.g., noncarcinogenic agents) or due to a blockage of repair processes by a chemical with a strong inhibitory property (e.g., acriflavine, daunomycin, mitomycin C: Stich et al. 1974). Studies by Gaudin et al. (1972a,b) demonstrated that all cocarcinogens examined inhibit DNA repair synthesis. Combined application of UV-irradiation and a chemical carcinogen revealed that various carcinogenic compounds differ greatly in their capacity to inhibit repair (Stich et al. 1974; Lo and Stich 1978). An unanswered question remains as to whether a chemical could have a carcinogenic effect by merely inhibiting repair of "spontaneously" occurring DNA lesions and in this manner contribute to genetic anomalies and neoplastic transformation.

Applications

The use of the DNA repair bioassay is by no means restricted to the simple detection of individual carcinogens and mutagens. It has been successfully applied in studying the formation of carcinogens from noncarcinogenic precursors (Lo and Stich 1975, 1978). It can be used to unravel the complex interactions between carcinogens or cocarcinogens and modulating factors (Stich et al. 1977; Lo and Stich 1978). It can provide information on variations within the human population in sensitivity to carcinogenic agents. It provides a convenient endpoint to correlate in vitro and in vivo responses to chemical carcinogens and to identify the target tissue (Stich and Kieser 1974). In view of the growing concern over the activation and deactivation of carcinogens which cannot be studied in most in vitro test systems, a procedure involving in vivo administration of a test compound followed by in vitro monitoring of DNA repair in various tissues may turn out to be a technique of choice.

Literature Cited

Ames, B. N.; Durston, W. E.; Yamasaki, E.; Lee, F. D. 1973. Carcinogens are mutagens: a simple test system combining liver homogenates for activation and bacteria for detection. Proc. Nat. Acad. Sci. USA 70:2281-83.

Bissell, D. M.; Hammaker, L. E.; Meyer, U. A. 1973. Parenchymal cells from adult rat liver in nonproliferating monolayer culture. 1. Functional studies. J. Cell Biol. 59:722-34.

Bonney, R. J. 1974. Adult rat liver parenchymal cells in primary culture: characteristics and cell recognition standards. In vitro 10:130-42.

Burk, P. G.; Lutzner, M. A.; Clarke, D. D.; Robbins, J. H. 1971. Ultraviolet-stimulated thymidine incorporation in Xeroderma pigmentosum lymphocytes. J. Lab. Clin. Med. 77:759-67.

Cleaver, J. E. 1972. DNA repair with purines and pyrimidines in radiation- and carcinogen-damaged normal and Xeroderma pigmentosum human cells. Cancer Res. 33:362-69.

Eagle, H. 1959. Amino acid metabolism in mammalian cell cultures. Science 130:432-37.

Freed, J. J.; Schatz, S. A. 1969. Chromosome aberrations in cultured cells deprived of a single essential amino acid. Exp. Cell Res. 55:393-409.

Garner, J. V.; Hanson, R. S. 1971. Formation of a factor lethal for *S. typhimurium* TA1530 and TA1531 on incubation of aflatoxin B_1 with rat liver microsomes. Biochem. Biophys. Res. Commun. 45:774-79.

Garner, R. C.; Miller, E. C.; Miller, J. A. 1972. Liver microsomal metabolism of aflatoxin B_1 to a reactive derivative toxic to *Salmonella typhimurium* TA1530. Cancer Res. 32:2058-66.

Gaudin, D.; Gregg, R. S.; Yielding, K. L. 1972a. DNA repair inhibition: a possible mechanism of action of cocarcinogens. Biochem. Biophys. Res. Commun. 45:630-36.

Gaudin, D.; Gregg, R. S.; Yielding, K. L. 1972b. Inhibition of DNA repair by cocarcinogens. Biochem. Biophys. Res. Commun. 48:945-49.

Jacobs, A. J.; O'Brien, R. L.; Parker, J. W.; Paolilli, P. 1972. Abnormal DNA repair of 4-nitroquinoline-1-oxide-induced damage by lymphocytes in Xeroderma pigmentosum. Mutat. Res. 16:420-24.

Laishes, B. A.; Stich, H. F. 1973. Repair synthesis and sedimentation analysis of DNA of human cells exposed to dimethylnitrosamine and activated dimethylnitrosamine. Biochem. Biophys. Res. Commun. 52:827-33.

Laishes, B. A.; Williams, G. M. 1976. Conditions affecting primary cell cultures of functional adult rat hepatocytes. I. The effect of insulin. In Vitro 12:521-32.

Lawley, P. D.; Brookes, P. 1963. Further studies on the alkylation of nucleic acids and their constituent nucleotides. Biochem. J. 89:127.

Lieberman, M. W.; Baney, R. N.; Lee, R. E.; Sell, S.; Farber, E. 1971. Studies on DNA repair in human lymphocytes treated with proximate carcinogens and alkylating agents. Cancer Res. 31:1297-1306.

Lin, R. C.; Snodgrass, P. J. 1975. Primary culture of normal adult rat liver cells which maintain stable urea cycle enzymes. Biochem. Biophys. Res. Commun. 64:725-34.

Lo, L. W.; Stich, H. F. 1975. DNA damage, DNA repair and chromosome aberrations of Xeroderma pigmentosum cells and controls following exposure to nitrosation products of methylguanidine. Mutat. Res. 30:397-406.

Lo, L. W.; Stich, H. F. 1978. The use of short-term tests to measure the preventive action of reducing agents on the formation and activation of carcenogenic nitroso compounds. Mutat. Res. 57:57-67.

McGrath, R.; Williams, R. W. 1966. Reconstruction in vivo of irradiated *Escherichia coli* deoxyribonucleic acid: the rejoining of broken pieces. Nature 212:534-35.

McLean, A. E. M.; Marshall, A. 1971. Reduced carcinogenic effects of aflatoxin in rats given phenobarbitone. Brit. J. Exp. Path. 52:322-29.

Malling, H. V. 1971. Dimethylnitrosamine: formation of mutagenic compounds by interaction with mouse liver microsomes. Mutat. Res. 13:425-29.

Martin, C. N.; McDermid, A. C.; Garner, R. C. 1978. Testing of known carcinogens and noncarcinogens for their ability to induce unscheduled DNA synthesis in HeLa cells. Cancer Res. 38:2621-27.

Merchant, D. J.; Kahn, R. H.; Murphy, W. H. 1964. Handbook of cell and organ culture. Minneapolis: Burgess Publishing Company.

Michalopoulos, G; Pitot, H. C. 1975. Primary culture of parenchymal liver cells on collagen membranes. Exp. Cell Res. 94:70-78.

Neal, G. E.; Godoy, H. M. 1976. The effect of pretreatment with phenobarbitone on the activation of aflatoxin B_1 by rat liver. Chem. Biol. Interactions 14:279-89.

O'Conner, P. J.; Capps, M. J.; Craig, A. W.; Lawley, P. D.; Shah, S. A. 1972. Differences in the patterns of methylation in rat liver ribosomal ribonucleic acid after reaction in vivo with methylmethansulfonate and N-N-dimethylnitrosamine. Biochem. J. 129:519.

Rasmussen, R. E.; Painter, R. B. 1966. Radiation-stimulated DNA synthesis in cultured mammalian cells. J. Cell Biol. 29:11-19.

Rogers, A. W.; England, J. M. 1973. The statistical analysis of autoradiographs. *In* Rogers, A. W., ed. Techniques of autoradiography. Amsterdam: Elsevier Scientific Publishing Co.; 216-25.

San, R. H. C.; Stich, H. F. 1975. DNA repair synthesis of cultured human cells as a rapid bioassay for chemical carcinogens. Int. J. Cancer 16:284-91.

Setlow, R. B.; Carrier, W. L. 1964. The disappearance of thymine dimers from DNA: an error-correcting mechanism. Proc. Nat. Acad. Sci. USA 51:226-31.

Stich, H. F.; Kieser, D. 1974. Use of DNA repair synthesis in detecting organotropic actions of chemical carcinogens. Proc. Soc. Exp. Biol. Med. 145:1339-42.

Stich, H. F.; Laishes, B. A. 1975. The response of Xeroderma pigmentosum cells and controls to the activated mycotoxins, aflatoxins, and sterigmatocystin. Int. J. Cancer 16:266-74.

Stich, H. F.; San, R. H. C. 1970. DNA repair and chromatid anomalies in mammalian cells exposed to 4-nitroquinoline 1-oxide. Mutat. Res. 10:389-404.

Stich, H. F.; San, R. H. C.; Kawazoe, Y. 1971. DNA repair synthesis in mammalian cells exposed to a series of oncogenic and nononcogenic derivatives of 4-nitroquinoline 1-oxide. Nature (London) 229:416-19.

Stich, H. F.; Kieser, D.; Laishes, B. A.; San, R. H. C. 1974. The use of DNA repair in the identification of carcinogens, precarcinogens, and target tissue. *In* Scholefield, P. G., ed. Canadian cancer conference: proceedings of the Tenth Canadian cancer conference. Toronto: University of Toronto Press; 83-110.

Stich, H. F.; Lam, P.; Lo, L. W.; Koropatnick, D. J.; San, R. H. C. 1975. The search for relevant short-term bioassays for chemical carcinogens: the tribulation of a modern Sisyphus (Invitation Paper). Can. J. Genet. Cytol. 17:471-92.

Stich, H. F.; San, R. H. C.; Lam, P.; Koropatnick, J.; Lo, L. 1977. Unscheduled DNA synthesis of human cells as a short-term assay for chemical carcinogens. *In* Hiatt, H. H.; Watson, J. D.; Winsten, J. A., eds. Origins of human cancer. New York: Cold Spring Harbor Laboratory; 1499-1512.

Stocker, E.; Muller, H. A. 1967. Zur chemischen Induktion von Silberkornen im stripping Film bei der Orcein-Quetsch-Technik. Histochemie 11:167-70.

Williams, G. M. 1976. Carcinogen-induced DNA repair in primary rat liver cell cultures: a possible screen for chemical carcinogens. Cancer Letters 1:231-36.

Williams, G. M. 1977. Detection of chemical carcinogens by unscheduled DNA synthesis in rat liver primary cell cultures. Cancer Res. 37:1845-51.

Williams, G. M. 1978. Further improvements in the hepatocyte primary culture DNA repair test for carcinogens: detection of carcinogenic biphenyl derivatives. Cancer Letters 4:69-75.

Williams, G. M.; Bermudez, E.; Scaramuzzino, D. 1977. Rat hepatocyte primary cell cultures. III. Improved dissociation and attachment techniques and the enhancement of survival by culture medium. In Vitro 13:809-17.

Wogan, G. N. 1966. Chemical nature and the biological effects of the aflatoxins. Bacteriol. Rev. 30:460-70.

Wogan, G. N. 1973. Aflatoxin carcinogenesis. Methods in Cancer Res. 7:309-433.

APPENDIX
Arginine Deficient Medium (ADM)

To prepare 10 litres of ADM, the procedure is as follows:
1. Hank's Balanced Salt Solution (BSS)
   To prepare 1 litre of 10x stock solution:

   A. Sodium Chloride (NaCl) — 80 gm
      Potassium Chloride (KCl) — 4 gm
      Magnesium Sulphate (MgSO$_4$•7H$_2$O) — 1 gm
      Sodium Phosphate, dibasic (Na$_2$HPO$_4$) — 0.48 gm
      Potassium Phosphate, monobasic (KH$_2$PO$_4$) — 0.6 gm
      Glucose — 10.0 gm
      (Dissolved in 800 ml of distilled water)
   B. Calcium Chloride (CaCl$_2$) — 1.4 gm
      (Dissolved in 100 ml of distilled water)
   C. Phenol Red — 0.1 gm
      (Dissolved in distilled water. Before making up to a final volume of 100 ml, the pH has to be adjusted to 7.0 with 0.05 N NaOH)
   D. Mixing the above three solutions gives 1 litre of 10x stock Hank's BSS.

2. Essential Amino Acids

L-Histidine	310 mg
L-Leucine	520 mg
L-Lysine	580 mg
L-Isoleucine	520 mg
L-Methionine	150 mg
L-Phenylalanine	320 mg
L-Threonine	480 mg
L-Tryptophan	100 mg
L-Valine	460 mg

   (Dissolved in 100 ml 1x Hank's BSS)

L-Tyrosine	360 mg (Dissolved in 100 ml 0.1 N HCl)
L-Cystine	240 mg (Dissolved in 100 ml 0.1 N HCl)
L-Glutamine	2.92 gm (Dissolved in 100 ml 1x Hank's BSS)

3. Nonessential Amino Acids

L-Alanine	89 mg
L-Asparagine	150 mg
L-Aspartic Acid	133 mg
L-Glutamic Acid	147 mg
L-Proline	115 mg
L-Serine	105 mg
Glycine	75 mg

   (Dissolved in 100 ml 1x Hank's BSS)

4. Vitamins

Choline Chloride	100 mg
Nicotinamide	100 mg
i-Inositol	200 mg
Pyridoxal	100 mg
Riboflavin	10 mg
D-Ca-Pantothenate	100 mg
Thiamine HCl	100 mg
(Dissolved in 100 ml 1x BSS)	
Folic Acid	10 mg
(Dissolved in 100 ml 1x BSS)	

The solutions from steps 2, 3, 4, and the amount left from step 1 are thoroughly mixed. Distilled water is added to bring the final volume to 10 litres.

The culture medium can be sterilized by passage through a millipore filter (pore size: 0.22 microns; Millipore Filter Corporation, Mass., USA).

Antibiotics, fetal calf serum, and sodium bicarbonate are added to the culture just prior to use.

In lieu of weighing out the individual items, the vitamins and nonessential amino acids are obtainable in the form of 100x concentrated mixture from Flow Laboratories, Inc., Inglewood, California.

13
Male Germ Cell Cytogenetics

by I. -D. ADLER*

Abstract

Preparation techniques for testicular material of mammals are compared and a simple procedure for mouse testes is described in detail.

The literature is reviewed for cytogenetic effects of ionizing radiation and chemical mutagens on male germ cells. Additionally, experimental results with gamma-irradiation and two chemical mutagens, procarbazine and TEM, are presented.

The differences in spermatogonial response to ionizing radiation and chemical mutagens are discussed and explained. Furthermore, the cytogenetic characteristics of ionizing radiation and chemical mutagens are demonstrated by describing the effects on prophase stages of male meiosis.

It is concluded that male germ cell cytogenetics are not well suited for mutagenicity screening. However, they are a valuable tool to characterize the cytogenetic potentials of mutagens in dividing germ cells.

Introduction

Chromosome analysis in male germ cells is not just one test procedure in the field of mutation reseach but a complex system with which detailed questions about germ cell sensitivity can be answered. For cytogenetic analysis, spermatogonial mitoses as well as first and second meiotic divisions are available in the male germ line. The mouse is extensively used in mutation experiments, and, therefore, the development of male and female germ cells was studied in detail. The pattern of spermatogonial stem cell renewal and differentiation in the mouse was described by Oakberg (1956,

*Department of Genetics, Gesellschaft f. Strahlen- u. Umweltforschung, D-8042 Neuherberg, Germany. The author wants to express her gratitude to all coworkers who participated in the various experiments: Kerry Fagan, Ruth Schmöller, Elfriede Seebacher and Werner Wittman.

1971) who also determined the length of the different developmental stages (Oakberg 1960). The diagram in Figure 13.1 illustrates the different stages of spermatogenesis in mice and their duration.

The present survey will describe a simple preparation technique for testicular material and will review the literature on three aspects. The first aspect is the direct observation of cytogenetic effects by chemical mutagens or ionizing radiation on spermatogonia in analysing mitoses of differentiated spermatogonia. The second aspect is to assess indirectly the effect on spermatogonial stem cells by analysing diakineses-metaphases I of primary spermatocytes for reciprocal translocations after an appropriate interval between treatment and analysis. The third aspect is the assessment of effects on primary spermatocytes themselves by analysing chromatid type aberrations at diakinesis-metaphase I.

The majority of data have been accumulated from the translocation test, which is the chromosome analysis of diakinesis-metaphase I after treatment of spermatogonia at the stem cell stage. The effect of ionizing radiation on spermatogonia has been studied by that procedure in great detail. With a variation of the preparation technique of testicular material, it was possible to include spermatogonial mitoses in the cytogenetic analysis. Especially for several chemical mutagens, the effect on spermatogonia has been evaluated by direct observation. The possibility of comparing directly observed effects with those assessable in derived spermatocytes offers a tool to study the phenomenon of germinal selection and elucidate some of the problems that arise from the fact that chemical mutagens, unlike radiation, do not produce heritable chromosomal changes in mammalian spermatogonia.

Preparation Techniques

Most of the cytogenetic studies on male germ cells were performed with mice. Even though the mitotic chromosome complement of the mouse, which consists of 40 acrocentric chromosomes of quite similar size, does not favor the species, it is just that uniformity that makes meiotic chromosome analysis fairly easy.

Since a series of preparation techniques had been published by Welshons et al. (1962), Evans et al. (1964), and Schleiermacher (1966) for meiotic chromosomes, Hoo and Bowles (1971) combined these methods to achieve a better yield of spermatogonial cells. Like any other cytogenetic preparation procedure, the processing of the cell material consists of five major steps: isolation of cells, hypotonic treatment, fixation, slide making, and staining. The different preparation techniques have recently been compared (Adler 1978), and a general description was given of the laboratory variation used in our group. Although the described procedure is recommended for meiotic chromosome preparations, it can be applied successfully to spermatogonial mitoses by introducing a colchicine treatment of the animals 3.5 to 5 hr. prior to sampling of the testes. Colchicine is intraperitoneally (i.p.) injected at a dose of 4 mg/kg for mice and rats or 10 mg/kg for Chinese hamsters if spermatogonia are to be analysed. For spermatocyte analysis, the colchicine pretreatment has a certain disadvantage in contracting the bivalents to an extent that makes recognition of the chromatid structure rather difficult.

Spermatogonia are the bottom layer of the tubules (Schleiermacher 1970) and cannot, therefore, be freed as easily as spermatocytes from the tubules by mechanical forces, i.e., forceps or rubber rollers. Hoo and Bowles (1971) suggested that the fix-

SPERMATOGENESIS OF MICE

Days	Stage			DNA S.	RNA S.	Morphology
52			A_s			
51						
50						
49						
48	STEM					
47						
46			A -pairs			
45		SPERMATOGONIA				
44			A -chains			
43						
42			A_1			
41	DIFFERENTIATED		A_2			
40			A_3			
39						
38			A_4			
37			I_n			
36						
35			B			
34		PRELEPTOTENE				
33		LEPTOTENE				
32		ZYGOTENE				
31						
30						
29	SPERMATOCYTES					
28						
27		PACHYTENE				
26						
25						
24						
23		DIPLOTENE				
22		DIAKINESIS/MI-MII				
21		I				
20		II				
19		III				
18		IV				
17		V				
		VI				
16		VII				
15		VIII				
14	SPERMATIDS	IX				
13		X				
12		XI				
11		XII				
10		XIII				
		XIV				
9		XV				
8		XVI				
7						
6						
5	SPERMATOZOA	IN EPIDIDYMIS				
4						
3						
2						
1						

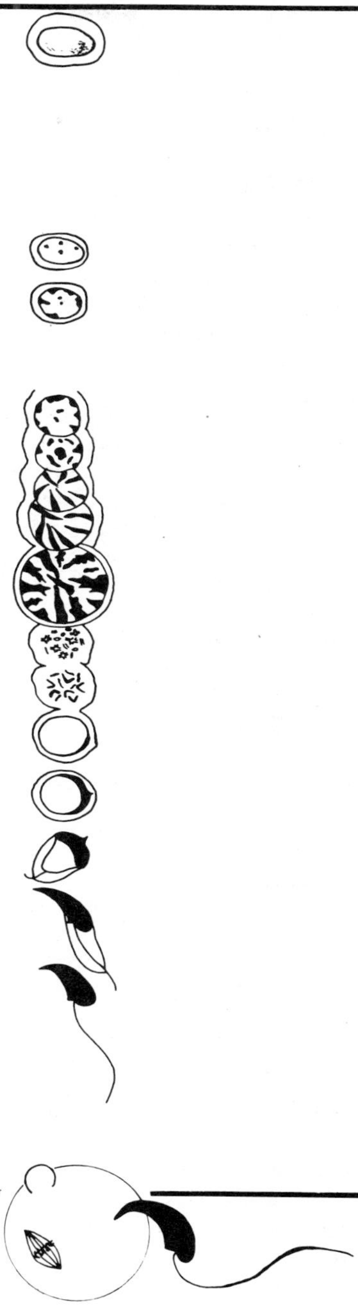

Figure 13.1 Timetable of the different stages of mouse spermatogenesis assembled by J. Kratochvilova (unpublished) according to the data of Oakberg (1960).

ed tubules be immersed in 50–60% acetic acid. Thereby one obtains a cell suspension that contains a large number of spermatogonia. Slides can be made by various methods. It is important that the glass is clean and especially grease-free and that the fluid of the cell suspension evaporates quickly.

For diakinesis analysis, the Orcein staining and phase contrast microscopy is highly recommended since more details of finer structures can be recognized in the meiotic chromosome. Giemsa staining and normal light microscopy is adequate for spermatogonial mitoses.

The preparation procedure can be summarized as follows:

1. Testes are removed from the animal and placed in a watchglass with 2% sodium citrate until sampling of testes from a number of animals is completed.

2. The tunica is removed and the tubules are transferred to another watchglass with 1% sodium citrate and left for about 20 min. at room temperature.

3. The hypotonic solution is readily removed by a Pasteur pipette and the tubules are quickly covered with cold fixative (1 part glacial acetic acid and 3 parts methanol). The fixative is changed after 10 min. at room temperature. The fixed tubules can be stored in the refrigerator for hours or even days.

4. The fixed tubules are transferred to a centrifuge tube containing 5 ml of 50% acetic acid and pipetted rigorously until the tubules become opaque. The cells are then centrifuged at 1500 rpm for 10 min. the supernatant is discarded, and the fixative is added to obtain a cell suspension for slide making. Depending on the size of the sediment the amount of fixative can vary between 0.5 and 3 ml.

5. Clean slides are kept in 70% alcohol for 24 hr. Three to four drops of the cell suspension are placed on an alcohol-wet slide and flame-dried.

6. Spermatogonia preparations are stained in 5% Giemsa solution (Merck) at pH 6.8 for 10 min., rinsed in acetone, acetone-xylene 1:1, and xylene before mounting.

7. Spermatocyte preparations are stained in 2% Orcein (in 50% acetic acid) (Merck) for 30 min., differentiated in 70% alcohol for 30 sec., dried via 90% and 98% alcohol, and washed in xylene before mounting.

For analysis of chromatid aberrations at diakinesis-metaphase I of meiosis the described method is not optimal. Treatment of the cells with 50% acetic acid disturbs some of the fine structures of the chromatids in meiotic chromosomes. The "milder" preparation technique described by Evans et al. (1964) allows a more detailed analysis for chromatid interchanges. Figures 13.2, 13.3, and 13.4 show examples of aberrations scored in spermatogonia and spermatocytes of mice.

Effects of Radiation and Chemical Mutagens on Spermatogonia

Analysis of Spermatogonial Mitoses

Chemical mutagens. The majority of experiments using spermatogonial chromosome analysis were performed to study the effect of chemicals. Among the first reports was an experiment with lysergic acid diethylamide (LSD-25) (Cohen and Mukherjee 1968). The authors stated a positive result in mouse spermatogonia over a period of 21 days. The effect reported as breaks must have been of the chromatid type, and its seems doubtful that chromatid aberrations can be scored 7 to 21 days after a single treatment in both bone marrow and long-cycling spermatogonia. The experiments were repeated by Egozcue and Irwin (1969). They obtained a negative

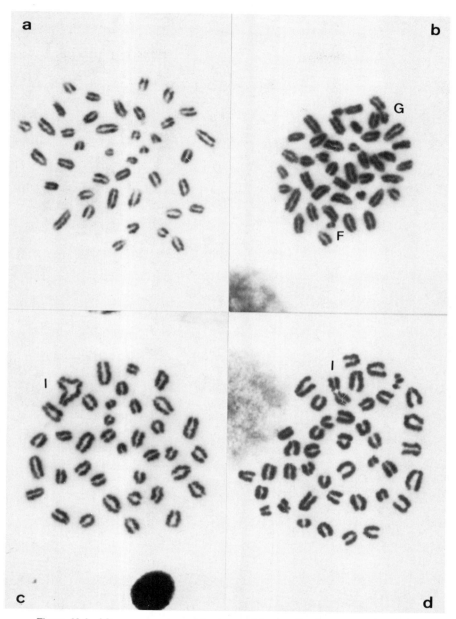

Figure 13.2 Mouse spermatogonial mitoses: (a) normal cell with 40 acrocentric chromosomes; (b) cell with a gap (G) and a double fragment (F); (c) cell with an asymmetrical interchange (I); (d) cell with a symmetrical interchange.

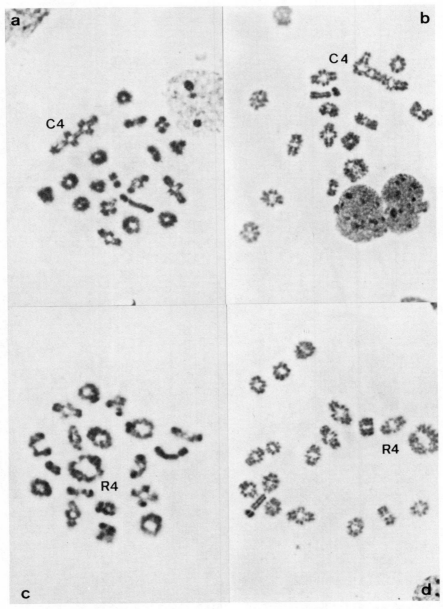

Figure 13.3 Mouse diakinesis-metaphase I: (a) and (b): cells with 18 bivalents and 1 chain quadrivalent (C4); (c) and (d): cells with 18 bivalents and 1 ring quadrivalent (R4).

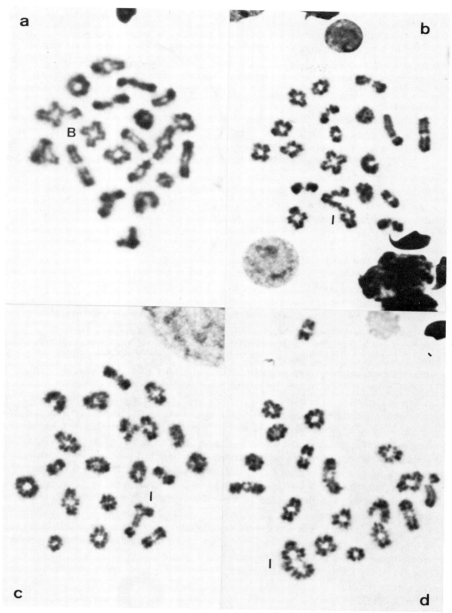

Figure 13.4 Mouse diakinesis-metaphase I: (a) cell with a chromatid break (B); (b) cell with an incomplete chromatid exchange (I); (c) cell with a chromatid exchange between X and an autosome (I); (d) cell with a symmetrical chromatid exchange (I).

response in mouse spermatogonia in an acute and in a chronic study. At about the same time, Jagiello and Polani (1969) reported on an acute and on a chronic LSD experiment where spermatogonial mitoses meiotic metaphase I and II of male mice, and meiotic metaphase I and II of female mice were analysed; the outcome was negative. These experiments and the way the data were reported are quite inconsistent, demonstrating the necessity for standardized and well-defined experimental conditions.

Another example of contradicting results on spermatogonial analysis is the artificial sweetner cyclamate or its metabolite, cyclohexylamine. In an experimental protocol using five daily injections of cyclohexylamine at doses between 1 and 50 mg/kg/day, Legator et al. (1969) found a significant increase in breaks 24 hr. after the last injection. On the other hand, Machemer and Lorke (1975) found no increase in the aberration yield above control level after five daily oral applications of 2000 mg/kg of sodium cyclamate to Chinese hamsters when they analysed spermatogonial mitoses 24 hr. after the end of the treatment. They also obtained a negative result when they gave five daily oral applications of 150 mg/kg of cyclohexylaminesulfate and analysed the spermatogonia of Chinese hamsters 24 hr. after the last treatment (Machemer and Lorke 1976).

In the same paper, cyclophosphamide and trimethylphosphate were reported positive, while sodium saccharin gave a negative response. In the cyclophosphamide experiment, Machemer and Lorke studied the effect of two daily oral treatments and analysis of spermatogonia 24 and 48 hr. later, as compared to five oral treatments and chromosome analysis 24 and 72 hr. after the last dose. The aberration yields at 48 hr. were higher than at 24 hr. after two treatments. After five treatments, the scores were higher at 24 hr. than at 72 hr. These findings are somewhat difficult to interpret and require some general discussion of the so-called subacute treatment schedule, by which toxicologists mean to optimise the effect.

The majority of spermatogonial mitoses analysed in a preparation 24 hr. after treatment are differentiating spermatogonia. According to Monesi (1962) the cell cycle in the mouse varies between 21 and 30.5 hr. with A_1-spermatogonia having the longest and B-spermatogonia the shortest average life span. If chromatid aberrations are evaluated, one can assume that the cells analysed at mitoses have been in the same round of the cell cycle during the time of treatment. Thus, if a chromosome study is performed 72 hr. after the last of five treatments, one would not expect to obtain a maximum in the chromatid aberration yield even if the cell cycle was prolonged by the treatment. Even 48 hr. after the second treatment, which would be 72 hr. after the first one, it is quite surprising to find higher aberration yields than after 24 hr. But obviously the cell cycle of spermatogonia in Chinese hamsters differs considerably from that in mice. It seems to be far longer since G_2 phase of the cell cycle is longer than in somatic cells (Utakoji and Hsu 1965), but detailed information comparable to that for mice and rats (Oakberg 1956; Hilscher et al. 1966; Hilscher 1967) is not available.

There is another factor to be included in the discussion of subacute versus acute treatment and that is the cell-killing effect. Spermatogonia have been demonstrated to be extremely sensitive to cell-killing by hycanthone and by ionizing radiation (Oakberg 1974). The more the spermatogonia progress in differentiation from type A_1 to type B, the higher the sensitivity to cell-killing. Therefore, one should be aware of the possibility of decreasing the aberraton yield by repeated treatments spaced 24

Table 13.1 Distribution of Types of Exchanges among the Total Number of Exchanges Observed in Mouse Spermatogonia after Treatment with Procarbazine or Triethylenemelamine (TEM)

Compound	Total number of exchanges	Symmetrical interchanges		Asymmetrical interchanges		Intrachanges		Interchanges between 3 chromosomes	
		no.	%	no.	%	no.	%	no.	%
Mitomycin C[a]	806	733	90.0	31	3.9	0	0	42	5.2
Procarbazine	21	14	66.7	7	33.3	0	0	0	0
TEM	63	46	73.0	16	25.4	1	1.6	0	0
Total	84	60	71.4	23	27.4	1	1.2		

[a] From Adler 1974.

hr. apart. The first treatment might cause considerable cell death. Furthermore, the chromatid aberrations caused by the first treatment will not be recognized in cells that have progressed into the next cell cycle unless derived chromosome aberrations are analysed by banding techniques. An example for this argument is the report by Röhrborn and Basler (1977) where five daily i.p. injections of 40 mg/kg of cyclophosphamide did not yield an increase in aberrations when spermatogonia of Chinese hamsters were scored 24 hr. after the last injection. In mice, however, a single i.p. injection of 200 mg/kg of cyclophosphamide resulted in 8% of the cells with aberration 24 hr. after treatment (Rathenberg 1975). Also, the yield of chromatid aberrations in mouse spermatogonial mitoses with 2-chloroethyl-1-nitrosoureido-ethol (CNU-ethanol) after a single injection with the lowest dose was higher at 24 hr. than at 48 hr. with the highest dose, cells could only be scored from slides of the 24 hr. interval (Tates and Natarajan 1976).

One of the more extensively studied compounds in the spermatogonial assay is mitomycin C (MC). Independently, Manyak and Schleiermacher (1973) and Adler (1974) found that a characteristic type of chromatid interchange was induced in mouse spermatogonia. Using the C-banding technique (Gagné et al. 1971) to reveal centromeric heterochromatin, Adler demonstrated that the exchange configurations involved the centromeric ends of the mouse chromosomes. More than 90% of these interchanges were symmetrical (Table 13.1). The table shows that with other chemical mutagens, procarbazine and triethylene melamine (TEM), the induced symmetrical interchanges also outnumber the asymmetrical ones. However, the total yield of interchanges is lower than the yield of chromatid breaks for procarbazine as well as for TEM (Tables 13.2 and 13.3) (Adler, unpublished). When procarbazine was given in single repeated doses spaced 2 hr. intervals, the aberration yields 12 and 18 hr. after the last treatment corresponded to the ones after single treatment (Table 13.4). Therefore, a cell-killing effect could be debated for procarbazine. However, Table 13.5 shows that the relative frequencies of mitoses in relation to first and second meiotic divisions were reduced 24 hr. after single, as well as after repeated, treatments. The TEM results are almost identical with the ones published by Luippold et al. (1978). Both sets of data demonstrate the cell-killing effect of TEM on differentiating spermatogonia by the lack of mitoses at sampling times later than 18 hr. after treatment with higher doses.

Table 13.2 Frequencies of Chromatid Aberrations in Mouse Spermatogonia after Treatment with Procarbazine

Dose (mg/kg)	Time[a] (hr.)	Number of animals	Total number of cells	Number of Gaps	Number of Breaks	Number of Exchanges	Aberrant cells (percent ± SE)
0	—	24	1600	9	2	0	0.13 ± 0.09
200	6	6	300	2	2	0	0.67 ± 0.47
	12	6	300	5	0	0	0
	24	6	600	4	4	4	1.33 ± 0.45[b]
400	6	6	300	2	0	0	0
	12	6	300	4	2	0	0.67 ± 0.47
	24	6	600	7	6	3	1.17 ± 0.44[b]
600	6	6	300	2	1	0	0.33 ± 0.33
	12	6	300	6	2	2	1.33 ± 0.55[b]
	24	6	600	10	12	2	2.17 ± 0.66[b]
800	6	6	300	12	6	0	2.00 ± 0.80[b]
	12	6	300	13	13	3	4.67 ± 1.12[b]
	24	6	600	10	4	7	1.67 ± 0.52

[a] Interval between treatment and preparation of testicular tubules.
[b] $P \leq 0.01$ (Mann-Whitney-Wilcoxon).

Table 13.3 Frequencies of Chromatid Aberrations in Mouse Spermatogonia after Treatment with Triethylenemelamine (TEM)

Dose (mg/kg)	Time[a] (hr.)	Number of animals	Total number of cells	Number of Gaps	Number of Breaks	Number of Exchanges	Aberrant cells (percent ± SE)
0	—	28	1400	11	2	0	0.14 ± 0.10
0.05	6	6	300	2	2	0	0.67 ± 0.47
	12	5	250	3	3	1	1.33 ± 0.66
	18	6	300	1	3	1	1.33 ± 0.66
	24	6	300	6	2	3	1.67 ± 0.74
0.10	6	6	300	5	1	0	0.33 ± 0.33
	12	6	300	5	1	4	1.67 ± 0.74
	18	6	300	1	2	1	1.00 ± 0.57
	24	6	300	12	8	13	6.33 ± 1.41[b]
0.20	6	6	300	2	3	0	0.67 ± 0.47
	12	6	300	5	7	9	5.33 ± 1.30[b]
	18	6	300	8	10	8	6.00 ± 1.37
	24	6	No divisions				
0.50	6	6	300	3	2	0	0.33 ± 0.33
	12	6	300	7	11	5	5.00 ± 1.26[b]
	18	6	300	7	14	18	9.33 ± 1.68[b]
	24	6	No divisions				

[a] Interval between treatment and preparation of testicular tubules.
[b] $P \leq 0.01$ (Mann-Whitney-Wilcoxon).

Table 13.4 Frequencies of Chromatid Aberrations in Mouse Spermatogonia after Fractionated Treatment with Procarbazine

Dose (mg/kg)	Time[a] (hr.)	Number of animals	Total number of cells	Number of Gaps	Number of Breaks	Number of Exchanges	Aberrant cells (percent ± SE)
0		10	20,000	125	28	1	0.15 ± 0.03
4 × 50	12	6	300	7	2	0	0.67 ± 0.47
	18	6	300	5	4	0	1.33 ± 0.55
4 × 100	12	6	300	15	10	2	4.00 ± 1.13[b]
	18	6	300	15	11	2	4.33 ± 1.18[b]
4 × 200	12	6	300	12	11	5	5.33 ± 1.30[b]
	18	6	300	7	2	5	2.33 ± .87[b]

[a]Interval between last treatment and preparation of testicular tubules.
[b]$p < 0.05$ (Mann-Whitney-Wilcoxon).

Radiation. Very little has been published on analyses of spermatogonial mitoses after ionizing radiation. Chromatid aberrations were analysed in Chinese hamster spermatogonia 6 hr. after gamma-irradiation with doses between 50 and 125R delivered at a dose rate of 16.6R/min (Brooks and Lengemann 1967). It was found that the number of exchanges increased more than linearly to the dose, whereas the chromatid and isochromatid deletions increased linearly. The aberration yield varied with the interval between the end of radiation delivery and preparation. Using a dose of 100R the maximum for deletions was obtained after a 3 hr. interval while the maximum for interchanges occured at the 6 hr. interval. The abberration yields decreased drastically and almost reached control levels 24 hr. after the end of the irradiation period. A similar experiment was performed comparing mouse and Chinese hamster. In the dose-response study 6 hr. after X-irradiation with doses from 25 to 125R, the aberration yield in mouse spermatogonia was slightly above the one in Chinese hamsters. When the results from both species were compared 12 and 24 hr. after irradiation with 100R, the mouse was four to five times as sensitive as the Chinese hamster (Rathenberg et al. 1976).

In recent experiments, five male mice per dose were irradiated with 50, 100, 200 or 300R gamma-rays (Adler and Wittmann, unpublished). Spermatogonial mitoses

Table 13.5 Ratios of Spermatogonial Mitoses, First, and Second Meiotic Divisions in Preparations of Testicular Tubules after Single or Repeated Treatment of Mice with Procarbazine

Time[a] hr.	200 mg/kg			4 × 200 mg/kg[b]		
	Mitoses (percent)	Diakinesis (percent)	Metaphases I (percent)	Mitoses (percent)	Diakinesis (percent)	Metaphases II (percent)
12	46.1	35.8	18.1	43.2	42.5	14.3
18	—	—	—	44.8	42.8	12.6
24	15.5	54.0	30.5	14.5	65.5	20.0
48	4.5	60.8	34.7	—	—	—

[a]Interval between treatment or last of four treatments.
[b]Interval between i.p. injections was two hr.

Table 13.6 Frequencies of Chromatid Aberrations in Spermatogonial Mitosis 10 hr. after Gamma-Irradiation of Male Mice

Dose R (60 R/min)	No. of animals	No. of cells scored	Number of						No. of cells with aberrations (excluding gaps)	Percent
			Gaps	%	Breaks	%	Exchanges	%		
Control	4	400	0	0	0	0	0	0	0	0
50	5	500	20	4.0	28	5.8	27	5.4	54	10.8 ± 1.38
100	5	500	54	10.8	51	10.2	70	14.0	118	23.6 ± 1.89
200	5	500	98	19.6	140	28.2	176	35.2	256	51.2 ± 2.23
300	5	500	161	32.2	379	75.8	367	73.4	376	75.2 ± 1.93

were analysed 10 hr. after irradiation. The aberrations observed were exclusively of the chromatid type which implies that the cells had been in G_2 of the preceding cell cycle at the time of irradiation. The data are presented in Table 13.6. The dose-effect curves for deletions and exchanges are both characterised by a quadratic component. Among the exchanges, symmetrical and asymmetrical exchanges were discriminated. The ratio varied with increasing dose towards a surplus of asymmetrical exchanges (Table 13.7). These data were the basis for the calculation of an expected frequency of translocation multivalents in primary spermatocytes derived from irradiated stem cell spermatogonia and will be discussed again in the next section.

In conclusion, there are three points for the experimental protocol of the test that should be emphasized:

1. Single treatments and repeated fixation times should be used rather than repeated treatments and a single fixation interval.

2. The analysis of chromatid aberrations in first mitotic divisions after treatment is more reliable than the analysis of chromosome type aberrations in second or third mitoses. An exception may be made with Chinese hamsters where the chromosome complement of 11 distinct chromosome pairs facilitates the recognition of reciprocal translocations.

3. However, until the cell cycle kinetics and the timing of the whole pattern of spermatogenesis in Chinese hamsters is well established, it seems more feasible to work with mice or rats.

Analysis of Translocation Multivalents at Diakinesis.

Analysis of translocation multivalents in primary spermatocytes after treatment of spermatogonia is also called "translocation test." It has been a classical method in radiation genetics and was applied to study spermatogonial radiosensitivity, the effects of dose fractionation and dose rate, the relative biological effectiveness (RBE) of different types of radiation, the differences in radiosensitivity among species, and the possible extrapolation of experimental data from mouse to man.

Radiation. The concept of Snell (1934; Snell and Picken, 1935) and Hertwig (1940) that X-ray–induced semisterility in male mice is caused by a reciprocal translocation was confirmed cytogenetically. Koller (1944) observed translocation multivalents—

Table 13.7 Ratios of Asymmetrical to Symmetrical Interchanges after Gamma-Irradiation of Spermatogonia in G_2 of the Cell Cycle

Dose (R)	Asymmetrical interchanges	Symmetrical interchanges	Ratio
50	7	20	0.35
100	35	35	1.00
200	124	52	2.38
300	251	116	2.16

associations of four chromosomes—in meiotic metaphases of semisterile mice. The improvement of the cytogenetic preparation technique by Evans and coworkers (1964) made it possible to study the induction of translocations in spermatogonia no longer by raising F_1 progeny. Instead, cytogenetic analysis of meiotic chromosomes in primary spermatocytes of treated animals at appropriate intervals after treatment became a widely used method to evaluate the radiosensitivity of spermatogonial stem cells to translocation induction.

Léonard and Deknudt (1967) reported a linear dose response for the translocation induction by X-rays in mouse spermatogonia in a dose range from 100 to 600R. The interval in these experiments between irradiation and analysis of meiotic metaphases was 10 weeks. In another experiment, the same authors analysed the persistence of translocations in spermatocytes over a period from 50 to 600 days after irradiation (Léonard and Deknudt 1970). The pattern of translocation frequencies showed an increase from 8 to 13% between 60 to 100 days after irradiation with 600R. A plateau was reached between 100 and 200 days, and the frequency of translocations then decreased to 5% between 200 and 450 days after treatment. The authors argued that the majority of translocations induced in spermatogonial stem cells by ionizing radiation are not selectively eliminated, but persist in the germ line.

Lyon and Morris (1969) extended the study of the dose-response curve in spermatogonia of mice for X-ray–induced translocations to 1000R. The obtained translocation yield was far below the one expected from linear extrapolation. The linear dose-response and the deviation from linearity above 600R in form of a decrease was confirmed by numerous authors and arrays of publications (Léonard and Deknudt 1968, 1969; Léonard 1971; Evans et al. 1970; Muramatsu et al. 1971; Savkovic and Lyon 1970; van Buul 1977). Linear shapes of the dose-response curve were also obtained for gamma-rays (Searle et al. 1971) and fast neutrons (Muramatsu et al. 1973; Searle et al. 1969). Léonard and coworkers claimed that, in mouse spermatogonia, translocation induction was a one-track event (Léonard and Deknudt 1969, 1971).

However, from classical radiation genetics (Sax 1940; Lea and Catcheside 1942), it is known that translocation production by low LET irradiation is a mainly two-track event, and the dose-response for their induction contains a quadratic component. Therefore, the linearity observed in the mouse translocation experiments was explained as not representing the actual rates of induction, but as resulting from distortion factors (Searle et al. 1971).

By varying the preparation time such that the first meiotic wave was obtained after the various X-ray doses, Preston and Brewen (1973) could show that the transloca-

Table 13.8 Translocation Frequencies in Meiotic Metaphase I after Treatment of Mouse Spermatogonia with Different Doses of Gamma Rays (60R/min)

Dose (R)	Cells with translocations					Percent cells with translocations	Percent translocations
	0	1	2	3	4		
0	900	—	—	—	—	0	0
50	493	4	2	1	—	1.4 ± 0.5	2.2 ± 0.7
100	489	10	1	—	—	2.2 ± 0.7	2.4 ± 0.7
200	478	17	5	—	—	4.8 ± 1.0	5.4 ± 1.0
300	460	29	10	1	—	9.2 ± 1.4	10.4 ± 1.4
400	440	42	15	2	1	15.0 ± 1.7	16.4 ± 1.7
600	4081	61	21	7	3	24.0 ± 1.9	27.2 ± 2.0

tion yields up to 500R X-rays actually followed a dose-response relationship with a significant two-track component. For acute gamma-irradiation and comparable preparation times, the dose-response of reciprocal translocations also showed a positive deviation from linearity in the dose-range from 50–600R (Table 13.8) (Adler and Wittman, unpublished).

From the data of irradiated differentiated spermatogonia (see above and Tables 13.6 and 13.7), one can calculate the expected frequency of reciprocal translocations in primary spermatocytes. Several assumptions, however, are necessary, although they surely oversimplify the system: (1) differentiated and stem cell spermatogonia are equally sensitive to aberration induction by ionizing radiation; (2) cells in G_2 and G_1 are equally sensitive; (3) half of the cells are in G_1 and half of them in G_2. In Table 13.9, the frequencies of symmetrical interchanges in spermatogonia are corrected for accompanying cell-lethal events. Only one-fourth of the chromatid interchanges induced in G_2 will give a translocation-carrying daughter cell, while all chromosome exchanges induced in G_1 will be transmitted to the next cell generation. The expected frequency of interchanges, therefore is 0.5×0.25 % symmetrical interchanges for G_2 plus 0.5×100% of symmetrical interchanges for G_1. The expected and the actually observed translocation frequencies are in rather good agreement. This indicates that in the dose range from 50 to 300R, there should be little selective disadvantage for spermatogonia carrying a reciprocal translocation.

Table 13.9 Frequencies of Expected and Observed Translocations in Spermatocytes after Gamma-Irradiation of Spermatogonia of Mice

Dose (R)	Percent symmetrical interchanges in spermatogonia[a]	Percent expected translocations in spermatocytes[b]	Percent observed translocations in spermatocytes
50	4.0	2.5	2.2
100	7.0	4.4	2.4
200	9.8	6.2	5.4
300	16.6	10.4	10.4

[a] Values corrected for accompanying deletions, but not corrected for cells with two interchanges.
[b] $0.5 \times 0.25 \times$ percent symmetrical interchanges + $0.5 \times$ percent symmetrical interchanges.

The hump in the dose-response curve at about 600R was interpreted as a result of a coincidence between cell death and translocations in parts of the spermatogonial stem cell population (Lyon and Morris 1969). The model of Oftedal (1968) describes mathematically that a humped dose-response curve is to be expected if the sensitivity of cell killing and mutation induction coincide in a heterogeneous cell population. Thus, it was argued by most of the authors that the spermatogonial stem cell population consisted of cells sensitive and resistant to both cell killing and translocation induction (Lyon and Morris 1969; Gerber and Léonard 1971; Evans et al. 1970; Léonard 1971).

A whole series of fractionation experiments by Long and coworkers supported the idea of heterogeneity in the stem cell population (Lyon et al. 1970a, 1970b, 1972, 1973). But the authors could not decide on the basis of their results whether there were two distinctly different cell populations present or whether cells in different stages of the cell cyle responded differently. Gerber and Léonard (1971) theoretically argued for two subpopulations with differing sensitivities and different kinetics of stem cell renewal. Experimental data on fractionation of fast neutrons into unequal doses, however, were interpreted as evidence against a radioresistant spermatogonial population by Searle and coworkers (1971). Dose-rate studies showed a distinct dose-rate effect for gamma-rays. It was less pronounced for X-rays (Searle et al. 1968, 1972). The authors found a marked divergency from a Poisson distribution of spermatocytes carrying one, two, three, or more translocations, and again interpreted this as a result of differential radiosensitivity, most likely during the spermatogonial cell cycle. But they still could not entirely rule out the possibility of independent subpopulations.

It was only recently, through experiments by Cattanach and coworkers (1976, 1977) and by Preston and Brewen (1976), that the understanding of spermatogonial stem cell sensitivity was completed. Fractionation of equal or unequal doses of X-rays with different fractionation intervals showed that different cell stages of stem cell spermatogonia that actually accounted for the different sensitivities. The varying responses at short intervals between 3 and 72 hr. were interpreted as being due to partial synchronization of the cells by the killing of sensitive cell stages. The remaining cells then pass through cell stages with differing sensitivities for aberration induction. The lower yields after fractionation intervals up to 4 weeks reflected a change in cell cycle length and thus relative duration of the various cell cycle phases in repopulating spermatogonia. Thus it was assumed that G_2 cells are about twice as sensitive as G_1 and S cells. The translocation recovery from G_1 cells in the following cell generation, however, is 1.0, while that from G_2 cells in only 0.25 of the originally induced, since they were of the chromatid type. Thus, short-cycling cells with relatively shorter G_1 phases, in relation to constant S, G_2, and M phases, yielded lower translocation rates. At 6 to 8 weeks after the initial X-ray dose, the stem cell population had returned to normal size and, therefore, also normal cell cycle length and responded in additive yields to the two radiation doses.

The sensitivity to X-irradiation of gonocytes in male fetuses at 13.5 days of pregnancy seemed to be much lower than that of adult spermatogonia (Ivanov et al. 1973). Also, for fast neutrons, embryonic germ cells were less sensitive than the adult ones (Searle and Phillips 1971). After birth, the immature spermatogonia resembled those fast-cycling, repopulating spermatogonia in the fractionation experiments. By 10 days of age, the spermatogonia are thought to have reached a condition similar to that in adult testes. The translocation yields came close to the values for adults (Cattanach et al. 1977).

The relative biological effectiveness (RBE) of different radiation qualities was studied mostly by Searle and his coworkers. For acute gamma-rays versus X-rays, they derived an RBE value of 0.6 (Searle et al. 1971). For acute irradiation at low doses, the fast neutrons to X-ray RBE was about 4 in the linear part of the curves (Muramatsu, Nakamura, and Eto 1973), while, for chronic irradiation, the fast neutron to gamma-ray RBE increased to 20–25 (Searle et al. 1969). Recently, the effects of alpha particles from plutonium 239 were also compared to gamma-rays from cobalt 60, and an RBE of 24 was estimated for translocation induction in spermatogonia (Beechey et al. 1975; Searle et al. 1976).

Transmission of reciprocal translocations from spermatocytes to the next generation was studied by comparing the results in the translocation test to the number of F_1 translocation carriers obtained in the heritable translocation test. Through meiotic chromosome segregation, a spermatocyte with a reciprocal translocation has roughly a one in four chance to produce a translocation carrying F_1 offspring. When the transmission of translocations from spermatocytes to the F_1 generation was studied at a relatively high dose, namely 1200R X-rays given in two fractions of 600R separated by 8 weeks, it was found that only one-half of the expected progeny with a reciprocal translocation could be recovered (Ford et al. 1969). Therefore, it was concluded that a selection process operates against reciprocal translocations in the postmeiotic diploid germ cells or in the diploid zygote after fertilization. However, for a single dose of 150R of X-rays, it was demonstrated by Brewen and coworkers (1974) that the ratio of F_1 translocation heterozygotes to the translocations scored in spermatocytes of treated males was the expected 0.25%. At single doses between 300 and 1200R, the ratio dropped to 0.12%, i.e., half of the expected. The authors suggested that higher doses of X-rays induced more events in the diploid spermatogonia that are lethal in the haploid germ cells, and they conclude that the recovery rate of balanced translocation carriers after low radiation doses is in the order of magnitude theoretically expected from cytogenetic analysis in the translocation test.

These results lead to another area of cytogenetic mutation research where the main work has only begun. That area is the extrapolation of cytogenetic experimental data from animals to man. For that purpose, species comparisons in terms of radiation sensitivity to translocation induction have been performed. Among the species studied, it was found that both Guinea pigs and rabbits gave a humped dose-response curve to acute X-irradiation similar to mice, but with a peak yield at around 200–300R. Golden hamsters gave a very low response at 200R, which was the only dose tested (Lyon and Smith 1971). Fractionation experiments with Guinea pigs and golden hamsters showed a result similar to mice, except that, at the 8 week fractionation interval, the yield had not quite reached additivity. Therefore, it was concluded that the repopulation kinetics of spermatogonia in these species are different from the mouse in that they are somewhat slower (Lyon and Cox 1975). A humped dose-response curve was also found for Rhesus monkeys, again with the highest yield at about 200R and a rather low overall yield (Lyon et al. 1976).

Similarly, limited data on man indicated a hump at about 200 rad, while, in the marmoset, the aberration yield reached a plateau at 100 rad which slightly bends down to 300 rad. The absolute yields were about the same in marmoset and man, i.e., 7.5% and 7.0% reciprocal translocations at 200 rad. In the mouse, however, the comparing value was 4.3% reciprocal translocations (Preston and Brewen 1973; Brewen and Preston 1975). These observations led to the conclusion that man was twice as sensitive as the mouse to the induction of translocations by low LET-irradiation. A similar observation of human cells being twice as sensitive as mouse cells had been made in peripheral leukocytes when asymmetrical interchanges (i.e., dicentrics) were analyzed (Brewen et al 1973). The yield of dicentrics at 300R of X-rays in the in vitro irradiated mouse blood was 31% (Brewen and Preston 1973a); the comparable value in human blood was 76.5% (Brewen and Luippold 1971). The ratio of somatic cell dicentrics to germ cell translocation was about 1:2 in man and 3:2 in the mouse at X-ray doses of 100R and below.

The doubling dose for reciprocal translocations in the first generation after X-ray exposure was calculated by Brewen and Preston (1975). Based on the limited data from analysis of spermatocytes at diakinesis-metaphase I from human material, a mean rate of 7.7×10^{-4} induced translocations per spermatogonia per rad was calculated for man (Brewen and Preston 1975). Introducing the expected ratio of transmission from spermatocytes to F_1 offspring of irradiated male mice of 1:4 for low doses and 1:8 for high X-ray doses, the rate of transmissable reciprocal translocations was calculated to range from $0.96-1.93 \times 10^{-4}$ translocations per gamete per rad ($7.7 \times 10^{-4}:4$ and $7.7 \times 10^{-4}:8$). In the human population, the newly occurring rate of viable translocations per gamete per generation was estimated to be 4×10^{-4} (Jacobs et al. 1972). With conventional staining techniques a range of about 0.25 to 1.00 of the true spontaneous translocation rate can be assumed to be detected by somatic cytogenetics in humans. Therefore, the spontaneous rate may range from $4 - 16 \times 10^{-4}$ translocations per gamete per generation. Based on these values (induced rate being roughly $1 - 2 \times 10^{-4}$ translocations/gamete/rad; spontaneous rate being $4 - 16 \times 10^{-4}$ translocations/gamete/generation), the doubling dose for acute X-rays was estimated to be on the order of 2 - 16 rad per generation. It was conservatively assumed that the doubling dose for chronic low-dose X-ray exposure might be in the same order of magnitude with a maximum of 4 - 30 rad (Brewen and Preston 1975).

Chemical mutagens. In contrast to the vast amount of data and information that has been accumulated in radiation genetics using the translocation test, the application of the system to the area of chemical mutagenesis has given mostly negative results. Even with compounds that were highly active in breaking spermatogonial chromosomes and producing symmetrical interchanges, like mitomycin C, no translocations were obtained in derived spermatocytes (Adler 1974). A comprehensive review was recently given by Léonard (1976). In Table 3 of Léonard's review, the frequencies of induced translocation multivalents reported for the different chemical mutagens with a positive response do not exceed the 1% margin. One exception in the table are the data of Sram and coworkers (1970). They reported as many as 50% translocations to have been induced in spermatogonia with 2.5 mg/kg of tris(1-aziridinyl)phosphine oxide (TEPA). The study, however, was not performed on spermatocytes of treated males but estimated the number of F_1 progeny with translocations from matings sampling spermatogonia. The same protocol was used in the thio tris(1-aziridiny l)phosphine oxide (thio-TEPA) experiments by Malashenko and Surkova (1974). They obtained 3% translocation carriers with 5 mg/kg of thio-TEPA. Both values are surprisingly high. In the first case, the presence of preexisting translocations in the stock of mice cannot be excluded. In another experiment, the effect of 1 mg/kg of TEPA on mouse spermatogonia was analyzed in derived spermatocytes, and the translocation frequency was 0.4% (Sram and Kocisova 1975).

Among the more recent reports not included in the review by Léonard (1976) are two on nitrosourea derivatives, the monofunctional CNU-ethanol and the bifunctional 1, 3-bis-(2-chloroethyl)-3-nitrosourea (BCNU) (Tates and Natarajan 1976; Tates et al. 1977). CNU-ethanol, which had given a positive response when mouse spermatogonial mitoses were analysed, also showed a few translocation multivalents (0.6%), but only in one of the three dose groups. The results with BCNU were negative. Similarly, a few translocations here and there were found in mouse spermatocytes after treatment of spermatogonia with TEM in the study by Luippold and coworkers (1978). The negative results in the translocation test with chemical mutagens led to the impression that spermatogonia were insensitive to the mutagenic effect of chemicals. One of the explanations for the lack of translocation configurations with mitomycin C was seen in the specific type of symmetrical exchange induced by the compound. The exchanges mostly involved whole chromosome arms with the breaks located in the centromeric heterochromatin. In mouse chromosomes, which are acrocentric, these translocations have only a very low chance to form multivalent configurations in spermatocytes (Adler 1974). If this was true, species with other chromosome complements including metacentric chromosomes should give a positive result. However, this was not found for Chinese hamsters (Table 13.10) when both mitomycin C and TEM were investigated. TEM also gave a negative response in golden hamsters and Guinea pigs (Cox and Lyon 1975).

Cattanach (1975) discussed two reasons for the obvious differences between the effects of ionizing radiation and chemical mutagens on spermatogonial stem cells. One was the specificity of action, the other the ability to produce delayed mutations. The specificity of chemicals was understood as cell-cycle-specificity, which means that only cells in a certain stage of the cell cycle would be affected by the chemical mutagen. Only mitotically dividing cells undergoing DNA synthesis might be affected since it is well established that aberration production by chemical mutagens depends on DNA synthesis (Bender et al. 1974). In spermatogonial stem cells with a

Table 13.10 Results of Chromosome Analyses in Primary Spermatocytes of Chinese Hamsters 106 to 133 Days after Treatment of Spermatogonia with Mitomycin C (MC) or Triethylenemelamine (TEM)

			Number of cells with					
			XY univalents		Autosomal univalents		Presumed translocation multivalents	
Compound	Dose (mg/kg)	Number of cells[a] analyses	Number	Percent	Number	Percent	Number	Percent
Control	0	800	2	0.25	2	0.25	0	0
MC	5.0	800	2	0.25	3	0.38	0	0
TEM	1.0	800	4	0.50	3	0.38	0	0

[a] 100 cells analysed per male.

cell cycle of several days, the relative duration of S phase is rather short, and thus the majority of the cells will be in a less sensitive stage. The specificity was also seen in the location of chemically induced chromosome breaks. In translocation heterozygotes induced with chemical mutagens, it is obvious that the multivalent frequency is lower than in translocation carriers produced by X-irradiation (Adler 1978). Thus, even though a translocation is present, multivalents are not formed as readily, which seems to be due to the size of the exchanged chromosome parts or the terminal location of the break points. With a crossing-over frequency of about 1.2 per chromosome, the chances of homologous chromosome parts on nonhomologous chromosomes to be linked by a chiasma become lower as their respective sizes decrease.

The simplest and, therefore most plausible, explanation, however, for the low translocation yield by chemical mutagens was given by Brewen and Preston (1973b). Aside from the factor already mentioned, which is the distinct cell cycle specificity in the action of chemical mutagens, there is another basic difference in the type of aberrations produced by radiation and chemicals. While in G_1 of the cell cycle, the radiation-induced aberration is of the chromosome type; the aberration induced by the majority of the chemical mutagens is of the chromatid type. Thus, as already mentioned, every induced symmetrical interchange will only have a chance of one in four to result in a reciprocal translocation after mitosis. With free segregation of the chromatids, there is a 50% chance that the chromosome complements of the two daughter cells will carry a duplication-deficiency which would be cell-lethal. Therefore, half of the induced symmetrical interchanges will be lost. From the remaining, only one of the two daughter cells will carry the translocation; the other one will be normal.

The conclusion has to be that none of the above-mentioned factors alone can explain the low translocation rate itself, but all of them act together in reducing the translocation yield from chemically treated spermatogonia. In view of this argument, one could assume that, for hazard evaluation, the action of chemical mutagens on spermatogonia is irrelevant. Unfortunately, such a safe declaration is not possible. The reasons are twofold. First, hardly any of the environmental chemicals exerts its effect on humans only at one single occasion, and so far we have no experimental data for this particular test on chronic or repeated treatments. It is conceivable that repeated or chronic treatments with low doses of a chemical mutagen will hit a larger part of the spermatogonial stem cell population as more cells pass through the sensitive stage of the cell cycle during the presence of the chemical mutagen. Thus, the translocation yield could be increased while the cell killing might be reduced. Repeated treatments with procarbazine at intervals of 2 hr. gave a first indication that this might be the case (Table 13.11). However, the treatment intervals will have to be studied and varied according to pharmacological information on activation, inactivation, and distribution of each particular chemical mutagen. The second reason is that evidence exists that point mutations induced in spermatogonia by chemical mutagens pose a quantifiable hazard (Ehling and Neuhäuser 1978).

Effects of Radiation and Chemical Mutagens on Mouse Spermatocytes

All of the work published so far on cytogenetic effects during meiotic prophase was performed with mice. When spermatocytes are treated during prophase, there are

Table 13.11 Results of Chromosome Analyses in Primary Spermatocytes 100-140 Days after Treatment of Mouse Spermatogonia with Procarbazine

Dose (mg/kg)	Total number of cells	Number of cells with univalents		Fragments	Presumed translocation multivalents	
		XY	autosomal			
0	2000	90	36	1	1	0.05
200	1000	81	32	2	0	0
400	600	32	10	0	1	0.17
600	1000	52	18	2	1	0.10
800	1000	45	15	1	1	0.10
4 × 50	1000	62	26	1	4	0.40
4 × 100	800	56	14	1	2	0.25
4 × 200	1000	83	25	4	4	0.40

several possibilities to analyze chromosomal damage during both meiotic divisions. In the earlier studies, anaphase I and II were scored for the presence of chromosome, chromatid, and subchromatid bridges and fragments (Oakberg and DiMinno 1960; Moutchen 1969). Recently it has been recommended to study metaphases II for numerical deviations (Pathak and Hsu 1977). The recommended experimental animal was the Chinese hamster with 22 chromosomes forming 11 bivalents. The majority of experiments, however, were performed analysing diakinesis-metaphase I chromosomes. The expected aberrations should be of the chromatid type and of the subchromatid type as the treatment approaches the end of prophase (Evans 1962). Another effect is sometimes described as chromosome stickiness (Wennström 1971; Clark 1974; Bulsiewicz et al. 1976).

Radiation

Contradicting results have been published in regard to the radiation sensitivity of the different prophase stages of meiosis. In the early experiments of Oakberg and DiMinno (1960), the sensitivities for cell killing and for aberration induction were exactly opposed to each other. While preleptotene and leptotene were the most sensitive stages to cell killing, diakinesis-metaphase I were the most resistant stages. For induction of chromosome breakage, preleptotene and leptotene were reported to be the most resistant and diakinesis-metaphase I the most sensitive stages. The authors had evaluated anaphase I and anaphase II for abnormalities in 10 μm sections of embedded tubules. Fragments, bridges, and sticky bridges were scored, but only classified as abnormal anaphases. The sensitivity decreased slightly from diakinesis to zygotene and showed a drastic drop to leptotene and preleptotene. Since, at these stages, the sensitivity to cell killing was highest, the authors discuss the possibility of aberration elimination in these stages as being extremely high.

Wennström (1971) analysed breaks, fragments, and chromatid exchanges in diakinesis-metaphase I at 2.5 hr., 1, 3, and 7 days after acute X-irradiation with 200R. Through the time intervals, he sampled cells from irradiated diakinesis, diplotene, and late and mid-pachytene. The highest frequencies of aberrations were obtained one day after irradiation and then subsequently decreased. The dose-responses studied on day 1 after irradiation with 50 to 400R showed a quadratic component for breaks, fragments, and chromatid exchanges.

In another study, the diakinesis-metaphases I were evaluated 2, 4, 8, 24, and 48 hr. after acute X-irradiation with 500R (McGaughey and Chang 1973). The frequencies of unrejoined breaks increased from 2 to 4 hr. and decreased from 4 to 8 hr. after irradiation. The frequencies of total breaks including those calculated from the observed rearrangements increased linearly from 0 to 8 hr. No significant change in the number of unrejoined or total breaks was found between 8 and 48 hr. after irradiation. In the calculation of the number of breaks per cell, isolated fragments or distinct gaps were counted as one unrejoined break. Rearrangements between two bivalents were counted as two, between three bivalents as three, and so on to five bivalents as five breaks. It is conceivable that the linearity of the dose-response is due to an underestimate in the number of breaks involved in the rearrangements since a rearrangement between three bivalents requires a minimum of four breaks.

In contrast to the data of Wennström (1971) Tsuchida and Uchida (1975) found a higher frequency of abnormal diakineses on day 5 corresponding to irradiated pachytene than on day 1 representing diplotene. They used gamma-rays at a dose rate of 29R/min. Fragments mainly contributed to the increase in aberration yield, while the frequencies of rearrangements were similar at both sampling times.

When rearrangements and fragments were analysed at days 1, 5, 9, and 11 after acute gamma-irradiation with 300R, the highest yield of rearrangements was obtained from day 9, which corresponded to irradiated zygotene (Adler 1977). The highest frequency of fragments was observed at day 1, similar to the other reports. A dose-response study at day 9 from 50 to 400R again showed a quadratic factor being involved in both the exchange and the fragment formation. As in the early report from Oakberg and DiMinno (1960), the leptotene stage was by far the least sensitive.

A comparison between acute X-irradiation and protracted gamma-irradiation was performed by Walker (1977). The sensitivity pattern after 300R of X-rays revealed the highest aberration frequencies for both fragments and rearrangements at diakinesis and the lowest at days 11 and 12.5. The overall aberration yield was considerably lower than in the experiments described above. The experiments with protracted gamma-irradiation were performed with two strains of mice, CBA and PT stock, and the results did not differ significantly, although the inbred CBA strain appeared to be more sensitive. An average value of structural chromosome aberrations from acute irradiation was compared to the value observed after protracted irradiation and was found not to differ significantly.

The obvious discrepancy is that one group of researchers (Oakberg and DiMinno 1960; Wennström 1971; Walker 1977) found diplotene to diakinesis to be the most sensitive stage of prophase, while another group (Tsuchida and Uchida 1975; Adler 1977) found pachytene to zygotene to be more sensitive to aberration induction by ionizing irradiation. Besides differences in the experimental protocol, such as preparation technique, strains of mice, irradiation sources, and preparation intervals, the main reason for the discrepancies could be seen in the interpretation of chromosomal damage observed. As pointed out by Evans (1962), chromatid aberrations are induced in meiotic prophase from S phase in preleptotene to diakinesis, and subchromatid aberrations simultaneously occur starting at mid-pachytene to metaphase I. Evans also discussed the phenomenon of chromosome stickiness that was observed in plant mitosis or meiosis when irradiated during or shortly before cell division. There were two types of stickiness reported, one of which generally affected all chromosomes and led to chromosomal clumping. The other type of stickiness was more localized and has been interpreted as exchange between half chromatids. It was found in meiotic cells that these subchromatid exchanges were only induced in prophase stages which contained visible chromosomes, i.e., at or after the time of chiasma formation. Scored at anaphase, typical subchromatid bridges can be seen. Scored at metaphase, however, it depends largely on the quality of the preparations whether a configuration is evaluated as stickiness or subchromatid exchange. None of the cited reports have tried to discriminate between chromatid and subchromatid aberrations, but most of them have avoided including chromosome stickiness phenomena in the score of true aberrations. A variation between authors in the discrimination of the latter type may have caused some of the discrepancy in the results.

Chemical Mutagens

In the area of chemical mutagenesis, the early work again involved analysis of anaphase I and II for chromosome and chromatid bridges, as well as fragments. The effects of TEM, diepoxybutane, and myleran were evaluated 20 hr., 2, 5, 10, and 20 days after single injections (Moutchen 1961), and it was found that the sensitivity of various stages is different for the three compounds. When analysed at anaphase I, myleran showed two peaks of sensitivity at days 1-2 and day 10. It induced mainly fragments while the number of bridges was not significantly different from the controls. TEM was found to produce a maximum for fragments at day 1 and their frequency decreased to day 10, while the number of bridges observed did not change from day 1 to day 7. Diepoxybutane also revealed the maxima for fragments at day 1 and for bridges at day 1 and 2. The dose-response study for all three chemicals at day 1 (i.e., 20 hr. after treatment) showed an increase with dose for both parameters. Comparing the fragment yields at anaphase I and anaphase II, elimination of aberrations was higher for diepoxybutane than for TEM and myleran. With myleran, the yields were not significantly different. The authors interpreted the result as the new genesis of fragments from subchromatid damage by myleran being higher than the elimination of fragments. Methyl methanesulfonate (MMS) was investigated in a similar way (Moutchen 1969). The end of pachytene (day 3) appeared to be the most sensitive stage and the elimination of fragments was high. No difference in aberration yield was found between the two doses of 30 and 60 mg/kg.

The effects of ethylnitrosourea (ENU) and hydroxylamine (HA) were analysed at diakinesis-metaphase I after treatment of dormant spermatocytes (preleptotene), at mid-pachytene, and at diplotene (Ramaiya 1969). The aberration frequency was about the same for all three stages after treatment with the lower dose of ENU, while, with the higher dose, the leptotene stage gave the highest aberration yield. HA treatment resulted in more cells with aberrations from leptotene than from the other two stages. The observed chromatid rearrangements were discriminated in symmetrical ones and asymmetrical ones, and the latter were more frequent. Fragments were relatively infrequent and univalents showed no consistent tendency. A concurrent experiment with X-irradiation at doses from 100 to 600R caused the authors to conclude that radiation induced relatively fewer translocations and correspondingly more univalents and especially fragments.

It is surprising that positive results were obtained both in the study of anaphase and of metaphase chromosomes after treatment of prophase stages later than leptotene. In view of the fact that alkylating chemicals require a round of DNA replication for translation of the damage into a chromosomal event, one would expect not to find any aberrations in cells treated after DNA synthesis had taken place. Several recent experiments gave a negative response during prophase stages later than leptotene. For methyl nitrosourea (MNU), the frequency of rearrangements induced at diplotene and pachytene has been reported to be below 1% (Goetz 1973). 6-Mercaptopurine was found to induce aberrations mainly on days 14 and 15 prior to diakinesis (Generoso et al. 1975). Since the majority of aberrations were chromatid deletions, the cells were most likely in preleptotene to leptotene at the time of treatment. When the effect of mitomycin C on primary spermatocytes was analysed, the aberration yield for both rearrangements and fragments was highest 11 to 12 days after treatment. Simultaneous labelling with ^{3}H-thymidine revealed a high coincidence in the cells of label and chromatid aberrations (Adler 1976). A similar experiment was performed with TEM (Luippold et al. 1978), and a coincidence of labelling and aberrations was found on days 12 and 13 after treatment. No aberrations could be observed after treatment of pachytene.

Likewise, Generoso and coworkers (1977) had found no increase in the aberration yield after TEM treatment when analysing diakinesis-metaphase I 2.5 to 6.5 days later, i.e., at pachytene. However, a sizable number of F_1 progeny from TEM-treated spermatocytes were carriers of reciprocal translocations. The authors postulated a repair process acting during spermatid stages that translated the lesions induced in spermatocytes into chromosomal aberrations. But to date, there is no evidence that the translation of the damage had occured prior to postfertilization DNA synthesis. On the other hand, in cytogenetic analyses at metaphase I and first cleavage of TEM-treated oocytes, it was found that the chromatid aberrations only appeared at first cleavage when the cells had gone through a round of DNA replication (Brewen and Payne 1978). Therefore, the recent data on aberration induction in spermatocytes support the view that chromosomal damage induced by alkylating chemicals is mediated through semiconservative DNA replication (Evans 1977).

Summary and Conclusions

The different applications of male germ cell cytogenetics have provided information on both the effects of radiation and of chemical mutagens on spermatogonia and primary spermatocytes. When spermatogonial mitoses were analysed, it was found that chemical mutagens produced chromosomal damage. However, no translocation multivalents were found in spermatocytes after treatment of spermatogonia. Explanations for this lack of reciprocal translocations were the cell cycle specificity of chemical mutagens, the nonrandom distribution of chemically induced breaks, and the fact that most chemical mutagens induce primarily chromatid type aberrations. It was suggested that repeated treatments with intervals according to the pharmacokinetics of the chemical might enlarge the fraction of the cell population that will be affected during the specifically sensitive stage of the cell cycle.

In the area of radiation genetics, the amount of information is such that a rather complete understanding of spermatogonial radiation sensitivity has emerged. It seems quite clear that the spermatogonial stem cells pass through cell cycle stages with differing sensitivities to both cell killing and translocation induction. The original findings that translocation induction followed a linear dose-response kinetic have been demonstrated to be due to distortion of the primary event by concurrent cell killing. Recent studies found a significant two-track component in the kinetics of aberration induction in differentiated spermatogonia, stem cell spermatogonia analysed at diakinesis-metaphase I, and in prophase of spermatocytes. Thus, the male germ cells respond to ionizing radiation in a fashion similar to plant cells and somatic cells in vitro and in vivo. A first step in the direction of extrapolating cytogenetic experimental data from mouse to man has been mentioned by the example of calculating the doubling dose for translocation risk in the human population. It was also discussed that, in meiotic prophase, chemical mutagens exert their effects during DNA replication (i.e., in preleptotene), while ionizing radiation acts on all stages of primary spermatocytes, even though the aberration types and yields vary as the cells proceed towards metaphase.

From the given survey, it is evident that male germ cell cytogenetics are too complicated and time-consuming to be applied for screening purposes in chemical mutagenesis. However, once a chemical has been identified as being a mutagen, these methods will provide a tool for determining an effect in actively dividing germ cells.

Literature Cited

Adler, I. -D. 1974. Comparative cytogenetic study after treatment of mouse spermatogonia with mitomycin C. Mutat. Res. 23:369-79.

Adler, I. -D. 1976. Aberration induction by mitomycin C in early primary spermatocytes of mice. Mutat. Res. 35:247-56.

Adler, I. -D. 1977. Stage-sensitivity and dose-response study after $\sqrt{\ }$-irradiation of mouse primary spermatocytes. Int. J. Radiat. Biol. 31:79-85.

Adler, I. -D. 1978. The cytogenetic heritable translocation test. Biol. Zbl. 97:441-51.

Beechey, C. V.; Green, D.; Humphreys, E. R.; Searle, A. G. 1975. Cytogenetic effects of plutonium 239 in male mice. Nature 256:577-78.

Bender, M. A.; Griggs, H. G.; Bedford, J. S. 1974. Mechanisms of chromosomal aberration production. III. Chemicals and ionizing radiation. Mutat. Res. 23:197-212.

Brewen, J. G.; Luippold, H. E. 1971. Radiation-induced human chromosome aberrations: in vitro dose rate studies. Mutat. Res. 12:305-14.

Brewen, J. G.; Preston, R. J. 1973a. Chromosomal interchanges induced by radiation in spermatogonial cells and leukocytes of mouse and Chinese hamster. Nature 244:111-13.

Brewen, J. G.; Preston, R. J. 1973b. Chromosome aberrations as a measure of mutagenesis: comparison in vitro and in vivo and in somatic and germ cells. Environ. Health Perspec. 6:157-66.

Brewen, J. G.; Preston, R. J.; Jones, K. P.; Gosslee, D. G. 1973. Genetic hazards of ionizing radiations: cytogenetic extrapolations from mouse to man. Mutat. Res. 17:245-54.

Brewen, J. G.; Preston, R. J.; Generoso, W. M. 1974. X-ray-induced translocations: comparison between cytologically observed and genetically recovered frequencies. Oak Ridge Nat. Laboratory, Annual Progress Report, Biology Division: 74-75.

Brewen, J. G.; Preston, R. J. 1975. Analysis of X-ray-induced chromosomal translocations in human and marmoset spermatogonial stem cells. Nature 253:468-70.

Brewen, J. G.; Payne, H. S. 1978. Studies on chemically induced dominant lethality. III. Cytogenetic analysis of TEM effects on maturing dictyate mouse oocytes. Mutat. Res. 50:85-92.

Brooks, A. L.; Lengemann, F. W. 1967. Comparison of radiation-induced chromatid aberrations in the testes and bone marrow of the Chinese hamster. Radiation Res. 32:587-95.

Bulsiewicz, H.; Rozewicka, L.; Bajko, J. 1976. Aberrations of meiotic chromosomes induced in mice with insecticides. Folia morphol. (Warsz.) 35:361-68.

Buul, P. van. 1977. Dose-response relationship for radiation-induced translocations in somatic and germ cells of mice. Mutat. Res. 45:61-68.

Cattanach, B. M.. 1975. Comparison of the mutagenic effect of chemicals and ionizing radiation in the spermatogenic cells of the mouse. 5th Rad. Res. Proc. Int. Congr.: 984-92.

Cattanach, B. M.; Heath, C. M.; Tracey, J. M. 1976. Translocation yield from the mouse spermatogonial stem cell following fractionated X-ray treatments: influence of unequal fraction size and of increasing fractionation interval. Mutat. Res. 35:257-68.

Cattanach, B. M.; Murray, I.; Tracey, J. M. 1977. Translocation yield from the immature mouse testis and the nature of spermatogonial stem cell heterogeneity. Mutat. Res. 44:105-17.

Clark, J. M. 1974. Mutagenicity of DDT in mice, *Drosophila melanogaster*, and *Neurospora crassa*. Aust. J. Biol. Sci. 27:427-40.

Cohen, M. M.; Mukherjee, A. B. 1968. Meiotic chromosome damage induced by LSD-25. Nature 219:1072-74.

Cox, B. D.; Lyon, M. F. 1975. The mutagenic effect of triethylenemelamine (TEM) on germ cells of male golden hamsters and guinea pigs. Mutat. Res. 30:293-98.

Egozcue, J.; Irwin, S. 1969. Effect of LSD-25 on mitotic and meiotic chromosomes of mice and monkeys. Humangenetik 8:86-93.

Ehling, U. H.; Neuhäuser, A. 1979. Procarbazine-induced specific locus mutations in male mice. Mutat. Res. 59:245-56.

Evans, E. P.; Breckon, G.; Ford, C. E. 1964. An air-drying method for meiotic preparation from mammalian testis. Cytogenetics 3:289-94.

Evans, E. P.; Ford, C. E.; Searle, A. G.; West, B. J. 1970. Studies on the induction of translocations in mouse spermatogonia. III. Effects of X-irradiation. Mutat. Res. 9:501-6.

Evans, H. J. 1962. Chromosome aberrations induced by ionizing radiation. Int. Rev. Cytol. 13:221-321.

Evans, H. J. 1977. Molecular mechanisms in the induction of chromosome aberrations. *In* Scott, D.; Bridges, B. A.; Sobels, F. H., eds. Progress in Genetic Toxicology. Biomedical Press, Elsevier/North Holland: 57-74.

Ford, C. E.; Searle, A. G.; Evans, E. P.; West, B. J. 1969. Differential transmission of translocations induced in spermatogonia of mice by irradiation. Cytogenetcs 8:447-70.

Gagné, R.; Tanguay, R.; Laberge, C. 1971. Differential staining patterns of heterochromatin in man. Nature 232:29-30.

Generoso, W. M.; Preston, R. J.; Brewen, J. G. 1975. 6-Mercaptopurine, an inducer of cytogenetic and dominant lethal effects in premeiotic and early meiotic germ cells of mice. Mutat. Res. 28:437-47.

Generoso, W. M.; Krishna, M; Sotomayor, R. E.; Cacheiro, N. L. A. 1977. Delayed formation of chromosome aberrations in mouse pachytene spermatocytes treated with triethylenemelamine (TEM). Genetics 85:65-72.

Gerber, G. B; Léonard, A. 1971. Influence of selection, nonuniform cell population, and repair on dose-effect curves of genetic effects. Mutat. Res. 12:175-82.

Goetz, P. 1973. Chromosomal aberrations induced by methyl nitrosourea on the germ cells of male rats. Mutat. Res. 21:34.

Hertwig, P. 1940. Vererbbare Semisterilität bei Mäusen nach Röntgenbestrahlung verursacht durch reziproke Chromosomentranslokationen. Vererbungslehre 79:1-27.

Hilscher, B.; Hilscher, W.; Maurer, W. 1966. Autoradiographische Bestimmung der Generationszeiten und Teilphasen der verschiedenen Spermatogoniengenerationen der Ratte. Naturwiss. 53:415-16.

Hilscher, W. 1967. DNA synthesis, proliferation, and regeneration of the spermatogonia in the rat. Arch. Anat. Microsc. Morphol. Exp. 56S:75-84.

Hoo, S. S.; Bowles, B. 1971. An air-drying method for preparing metaphase chromosomes from the spermatogonial cells of rats and mice. Mutat. Res. 13:85-88.

Ivanov, B.; Léonard A.; Dekundt, G. H. 1973. Blood storage and the rate of chromosome aberrations after in vitro exposure to ionizing radiation. Radiation Res. 55:469-76.

Jacobs, P. A.; Frackiewicz, A.; Law, P. 1972. Incidence and mutation rates of structural rearrangements of the autosomes in man. Ann. Human Genet. 35:301.

Jagiello, G; Polani, P. E. 1969. Mouse germ cells and LSD. Cytogenetics 8:136-47.

Koller, P. C. 1944. Segmental interchange in mice. Genetics 29:247-63.

Lea, D. E.; Catcheside, D. G. 1942. The mechanism of the induction by radiation of chromosome aberrations in Tradescantia. J. Genet. 44:216-45.

Legator, M. S.; Palmer, K. A.; Green, S.; Petersen, K. W. 1969. Cytogenetic studies in rats of cyclohexylmine, a metabolite of cyclamate. Science 165:1139-40.

Léonard, A. 1971. Radiation-induced translocations in spermatogonia of mice. Mutat. Res. 11:71–88.

Léonard, A. 1976. Heritable chromosome aberrations in mammals after exposure to chemicals. Rad. Environm. Biophysics 13:1–8.

Léonard, A.; Deknudt, G. H. 1967. Chromosome rearrangements induced in the mouse by embryonic X-irradiation. I. Pronuclear stage. Mutat. Res. 4:689–97.

Léonard, A.; Deknudt, G. H. 1968. Chromosome rearrangements after low X-ray doses given to spermatogonia of mice. Can. J. Genet. Cytol. 10:119–24.

Léonard, A.; Deknudt, G. H. 1969. Dose-response relationship for translocations induced by X-irradiation in spermatogonia of mice. Radiation Res. 40:276–84.

Léonard, A.; Deknudt, G. H. 1970. Persistence of chromosome rearrangements induced in male mice by X-irradiation of premeiotic germ cells. Mutat. Res. 9:127–33.

Léonard, A.; Deknudt, G. H. 1971. The rate of translocations induced in spermatogonia of mice by two X-irradiation exposures separated by varying time intervals. Radiation Res. 45:72–79.

Luippold, H. E.; Gooch, P. C.; Brewen, J. G. 1978. The production of chromosome aberrations in various mammalian cells by triethylenemelamine. Genetics 88:317–26.

Lyon, M. F.; Cox, B. D. 1975. The induction by X-rays of chromosome aberrations in male guinea pigs, rabbits, and golden hamsters. IV. Dose response for spermatogonia treated with fractionated doses. Mutat. Res. 30:117–27.

Lyon, M. F.; Cox, B. D.; Marston, J. H. 1976. Dose-response data for X-ray-induced translocations in spermatogonia of Rhesus monkeys. Mutat. Res. 35:429–36.

Lyon, M. F.; Morris, T. 1969. Gene and chromosome mutation after large fractionated or unfractionated radiation doses to mouse spermatogonia. Mutat. Res. 8:191–98.

Lyon, M. F.; Morris, T.; Glenister, P.; O'Grady, S. E. 1970a. Induction of translocations in mouse spermatogonia by X-ray doses divided into many small fractions. Mutat. Res. 9:219–23.

Lyon, M. F.; Phillips, R. J. S.; Glenister, P. H. 1970b. Dose-response curve for the yield of translocations in mouse spermatogonia after repeated small radiation doses. Mutat. Res. 10:497–501.

Lyon, M. F.; Phillips, R. F. S.; Glenister, P. H. 1972. Mutagenic effects of repeated small radiation doses to mouse spermatogonia. II. Translocation yield at various dose intervals. Mutat. Res. 15:191–95.

Lyon, M. F.; Phillips, R. J. S.; Glenister, P. H. 1973. The mutagenic effect of repeated small radiation doses to mouse spermatogonia. III. Does repeated irradiation reduce translocation yield from a large radiation dose? Mutat. Res. 17:81–85.

Lyon, M. F.; Smith, B. D. 1971. Species comparisons concerning radiation-induced dominant lethal and chromosome aberrations. Mutat. Res. 11:45–58.

Machemer, L.; Lorke, D. 1975. Method for testing mutagenic effects of chemicals on spermatogonia of the Chinese hamster: results obtained with cyclophosphamide, saccharin and cyclamate. Arzneimittelforsch. 25:1889–96.

Machemer, L.; Lorke, D. 1976. Evaluation of the mutagenic potential of cyclohexamine on spermatogonia of the Chinese hamster. Mutat. Res. 40:243–50.

Malashenko, A. M.; Surkova, N. I. 1974. The mutagenic effect of thio-TEPA in laboratory mice. Communication I. Chromosome aberrations in somatic and germ cells of mice. Soviet Genetics 10:51–58.

Manyak, A.; Schleiermacher, E. 1973. Action of mitomycin C on mouse spermatogonia. Mutat. Res. 19:99–108.

McGaughey, R. W.; Chang, M. C. 1973. Initial chromosomal lesions induced by X-irradiating primary spermatocytes of mice. Can. J. Genet. Cytol. 15:341–49.

Monesi, V. 1962. Autoradiographic study of DNA synthesis and the cell cycle in spermatogonia and spermatocytes of mouse testis using tritiated thymidine. J. Cell Biol. 14:1–18.

Moutchen, J. 1961. Differential sensitivity of mouse spermatogenesis to alkylating agents. Genetics 46:291–99.

Moutchen, J. 1969. Mutagenesis with methyl methanesulfonate in mouse. Mutat. Res. 8:581–88.

Muramatsu, S.; Nakamura, W.; Eto, H. 1971. Radiation-induced translocations in mouse spermatogonia. Japan. J. Genet. 46:281–83.

Muramatsu, S.; Nakamura, W.; Eto, H. 1973. Relative biological effectiveness of X-rays and fast neutrons in inducing translocations in mouse spermatogonia. Mutat. Res. 19:343–47.
Oakberg, E. F. 1956. A description of spermiogenesis in the mouse and its use in analysis of the cycle of the seminiferous epithelium and germ cell renewal. Amer. J. Anat. 99:391–409.
Oakberg, E. F. 1960. Irradiation damage to animals and its effect on their reproductive capacity. J. Dairy Science 43:54–67.
Oakberg, E. F. 1971. A new concept of spermatogonial stem cell renewal in the mouse and its relationship to genetic effects. Mutat. Res. 11:1–7.
Oakberg, E. F. 1974. Response of spermatogonia of the mouse to hycanthone: a comparison with the effect of gamma rays. In Coutinho, E. M.; Fuchs, F. eds. Physiology and genetics of reproduction. Part A. New York: Plenum Publishing.
Oakberg, E. F.; DiMinno, R. L. 1960. X-ray sensitivity of primary spermatocytes of the mouse. Int. J. Radiat. Biol. 2:196–209.
Oftedal, P. 1968. A theoretical study of mutant yield and cell killing after treatment of heterogeneous cell populations. Hereditas 60:177–210.
Pathak, S.; Hsu, T. C. 1977. Monitoring the effects of chemical mutagens with mammalian meiotic cells. Mamm. Chromosome Newsletters 18:42.
Preston, R. J.; Brewen, J. G. 1973. X-ray–induced translocations in spermatogonia. I. Dose response and fractionation response in mice. Mutat. Res. 19:215–23.
Preston, R. J.; Brewen, J. G. 1976. X-ray–induced translocations in spermatogonia. II. Fractionation responses in mice. Mutat. Res. 36:333–44.
Ramaiya, L. K. 1969. The cytogenetic effect of N-nitrosourea, hydroxylamine, and X-rays on the germ cells of male mice. Soviet Genetics 5:188–97.
Rathenberg, R. 1975. Cytogenetic effects of cyclophosphamide on mouse spermatogonia. Humangenetik 29:135–40.
Rathenberg, R.; Schwegler, H.; Miska, W. 1976. Comparative investigations on cytogenetic effects of X-irradiation on the germinal epithelium of male mice and Chinese hamsters. Hum. Genet. 34:171–83.
Röhrborn, G.; Basler, A. 1977. Cytogenetic investigations of mammals. Comparison of the genetic activity of cytostatics in mammals. Arch. Toxicol. 38:35–43.
Savkovic, N. V.; Lyon, M. F. 1970. Dose-response curve for X-ray–induced translocations in mouse spermatogonia. I. Single doses. Mutat. Res. 9:407–9.
Sax, K. 1940. An analysis of X-ray–induced chromosomal aberrations in Tradescantia. Genetics 25:41–68.
Schleiermacher, E. 1966. Über den Einfluß von Treminon^R und Endoxan^R auf die Meiose der männlichen Maus. I. Methoden der Präparation und Analyse meiotischer Teilungen. Humangenetik 3:127–33.
Schleiermacher, E. 1970. Histological and cytogenetic investigation methods on mammalian spermatogenesis. In Vogel, F.; Röhrborn, G. eds. Chemical mutagenesis in mammals and man. New York: Springer-Verlag.
Searle, A. G.; Evans, E. P.; Beechey, C. V. 1971. Evidence against a cytogenetically radioresistant spermatogonial population in male mice. Mutat. Res. 12:219–20.
Searle, A. G.; Evans, E. P.; Ford, C. E.; West, B. J. 1968. Studies on the induction of translocations in mouse spermatogonia. I. The effect of dose-rate. Mutat. Res. 6:427–36.
Searle, A. G.; Evans, E. P.; West, B. J. 1969. Studies on the induction of translocations in mouse spermatogonia. II. Effects of fast neutrons. Mutat. Res. 7:235–40.
Searle, A. G.; Phillips, R. J. S. 1971. The mutagenic effectiveness of fast neutrons in male and female mice. Mutat. Res. 11:97–105.
Searle, A. G.; Beechey, C. V.; Evans, E. P.; Ford, C. E.; Papworth, D. G. 1971. Studies on the induction of translocations in mouse spermatogonia. IV. Effects of acute gamma-irradiation. Mutat. Res. 12:411–16.
Searle, A. G.; Beechey, C. V.; Corp, M. J.; Papworth, D. G. 1972. A dose-rate effect on translocation induction by X-irradiation of mouse spermatogonia. Mutat. Res. 15:89–91.
Searle, A. G.; Beechey, C. V.; Green, D.; Humphreys, E. R. 1976. Cytogenetic effects of protracted exposures to alpha-particles from plutonium 239 and to gamma-rays from cobalt 60 compared in male mice. Mutat. Res. 41:297–310.
Snell, G. D. 1934. The production of translocations and mutations in mice by means of X-rays. Am. Naturalist 68:178.

Snell, G. D.; Picken, D. I. 1935. Abnormal development in the mouse caused by chromosome imbalance. J. Genet. 31:213–35.

Sram, R. J.; Surdova, Z.; Benes, V. 1970. Induction of translocations in mice by TEA. Folia biologica 16:367–68.

Sram, R. J.; Kocisova, J. 1975. Effect of antibiotics on the mutagenic activity induced by chemicals. I. Chromosome aberrations during spermatogenesis in mice. Folia biologica. 21:60–64.

Tates, A. D.; Natarajan, A. T. 1976. A correlative study on the genetic damage induced by chemical mutagens in bone marrow and spermatogonial cells in mice. I. CNU-ethanol. Mutat. Res. 37:267–70.

Tates, A. D.; Natarajan, A. T.; de Vogel, N.; Meijers, M. 1977. A correlative study on the genetic damage induced by chemical mutagens in bone marrow and spermatogonia of mice. II. 1,3-Bis(2 chloroethyl)-3-nitrosourea (BCNU). Mutat. Res. 44:87–95.

Tsuchida, W. S.; Uchida, I. A. 1975. Radiation-induced chromosome aberrations in mouse spermatocytes and oocytes. Cytogenet. Cell Genet. 14:1–8.

Utakoji, T.; Hsu, T. C. 1965. DNA replication patterns in somatic and germ-line cells of the male Chinese hamster. Cytogenetics 4:295–315.

Walker, H. C. 1977. Comparative sensitivities of meiotic prophase stages in male mice to chromosome damage by acute X- and gamma-irradiation. Mutat. Res. 44:427–32.

Wennström, J. 1971. Effects of radiation on the chromosomes in meiotic and mitotic cells. Commentato Biol. 45:1–60.

Welshons, W. J.; Gibson, B. H.; Scandlyn, B. J. 1962. Slide processing for the examination of male mammalian meiotic chromosomes. Stain Technol. 37:1–5.

14

Cytogenetic Analysis of Mammalian Oocytes in Mutagenicity Studies

by J. G. BREWEN† AND R. J. PRESTON*

Abstract

There has been limited use of the mammalian oocyte as a cytogenetic system for the study mutagenesis. This has been due principally to the tedious procedures required to obtain high quality preparations and the large numbers of animals involved in obtaining sufficient numbers of oocytes for quantitative analysis. Recent improvements in technique have largely alleviated the former drawback but the small numbers of oocytes available per female still restrict the use of the method to small laboratory rodents. The most recent technical innovations are discussed along with a summary of the work done to date in the fields of radiation and chemical mutagenesis.

Introduction

The preponderance of mutagenic agents studied to date has been shown to induce gross structural chromosome abnormalities in addition to true intragenic mutations in structural genes. In some instances, e.g., ionizing radiation, these chromosome aberrations probably constitute the great proportion of mutational events observed in higher eukaryotes. Taking these facts into consideration, it seems appropriate that the field of cytogenetics has grown into a substantial subdiscipline of genetics and mutagenesis.

In the course of studying the mutagenic action of an agent, it is usually desirable to be able to determine its effect on germ cells because, in order for a mutation to be transmitted to future generations of sexually reproducing organisms, it must occur in

†Medical Affairs, Allied Chemical Corporation, Morristown, N.J. 07960.
*Biology Division, Oak Ridge National Laboratory, Oak Ridge, Tennessee 37830.

these germ cells. Cytogenetic analysis of mammalian germ cells, however, lagged behind comparable studies in somatic cells due to technical difficulties. As in the case of somatic cell cytogenetics, the advent of dropping a suspension of fixed cells on a wet slide to effect a flat, well-spread preparation provided the impetus for the area of male mammalian germ cell cytogenetics to burgeon. Unfortunately, equivalent techniques for the female germ cells, or oocytes, were not developed until several years later.

Oocyte Maturation

All female mammals are born with their complete set of oocytes, the last oogonial cell division having occurred some time before birth. In most species, including man, the last DNA synthesis, or meiotic S phase, occurs prior to birth, although there are reports of it occurring as late as 14 days postpartum in the rabbit (1) and 4 days postpartum in the Golden hamster (2). Thus, at the time of birth, all female mammals contain their lifetime supply of oocytes, which are resting in a stage of meiotic prophase. In almost every species studied, with the exception of the mouse and rat, this stage has been described as a diffuse diplotene. In the mouse and rat, however, the chromosomes are much more diffuse. In addition, the resting oocytes of the mouse and rat are considerably more radiosensitive than those of other species, and it has been suggested that this may be due to the more diffuse nature of the chromosomes. Thus, the resting oocyte of the mouse and rat differs from that of the other species in two ways.

The maturation process of the mouse oocyte has been studied in detail. It is generally accepted that there are eight distinct morphological stages of oocytes present in the adult ovary. These stages are: (1) diffuse nucleus with only a few follicle cells; (2) diffuse nucleus with a complete layer of flattened follicle cells; (3) single layer of cuboidal follicle cells; (4) two layers of follicle cells; (5) multiple layers of follicle cells; (6) beginning of antrum formation; (7) multiple antra or a single medium-sized antrum; and (8) the mature Graafian follicle (3). The stage 3 follicle has been subdivided into stages 3a and 3b. The work of Oakberg and Tyrrell (3) has clearly shown by labeling with N(acetyl-^{3}H)-D-glucosamine that the zona pellicuda begins to form in stage 3b. Furthermore, it is the stage 3 mouse oocyte that has a nucleus that is morphologically similar to primate oocytes, i.e., contains diffuse diplotene-appearing chromosomes.

Pedersen (4) using [^{3}H]thymidine to study follicle cell growth kinetics, concluded that it takes 19 days for a stage 3 oocyte to mature to ovulation. More recently, however, Oakberg and Tyrrell (3), utilizing the labeling of the zona pellucida of the oocyte with N(acetyl-^{3}H)-D-glucosamine, have concluded that the length of time necessary for a stage 3 oocyte to reach maturity is of the order of 5-6 weeks. This latter technique is undoubtedly more sensitive than the study of follicle cell growth kinetics, and, consequently, it is likely that Oakberg and Tyrrell's estimate is more accurate.

Since only the stage 3b oocytes incorporate the isotope into the zona pellucida, it is possible to estimate the length of the entire stage 3. The Oakberg and Tyrrell data show that, from day 4 after labeling to day 7 after labeling, the frequency of labeled stage 3 oocytes declines from 127/304 to 58/311. Since it is assumed that this rate of loss is constant, it follows that, at the time of labeling (day 1), about 70% (or

210/300) stage 3 oocytes were labeled. Since the labeling frequency of stage 3 oocytes (actually stage 3b) declines to near zero in 14 days, the length of stage 3b is, on the average, 10-14 days; and since this represents 70% of the entire stage 3, the entire stage 3 must be about 14-20 days in duration. This means the immature oocyte requires an average of 6-7 weeks to reach the Graafian follicle, or ovulatory stage, 2-3 weeks of which are spent in the stage that is morphologically similar to the resting human oocyte.

Technical Developments

Early cytogenetic studies on mammalian oocytes were, for the most part, restricted to very few observations on relatively poor preparations. Many factors contributed to this situation among which were (1) only a small number of oocytes can be collected from normally ovulating females; (2) the large amount of cytoplasm and follicle cells in the mature oocyte makes squash preparations difficult to analyze; (3) the majority of oocytes complete the first meiotic division during, or shortly after, ovulation, and so at any one time the desirable metaphase I stage is infrequent. Over the past 10 years, these technical difficulties have been circumvented, and the major problem in doing quantitative cytogenetics in laboratory animals is the number of females required.

The first major improvement in working with oocytes was described by Jagiello (5). She applied the technique of superovulation to enhance the number of maturing Graafian follicles and ovulated oocytes. The procedure was to administer an intraperitoneal dose of pregnant mares' serum, usually 5-10 I.U., followed in 56 hr. by an intrperitoneal dose of human chorionic gonadotropin (HCG) at a level of 5-10 I.U. In the mouse, ovulation usually occurs up to 12 hr. after the HCG treatment. If metaphase I figures are desired, the females are killed 4.5-8.0 hr. after the HCG treatment and the oocytes collected from the largest follicles. If the female is allowed to ovulate, the metaphase II oocytes can be flushed from the oviduct.

Shortly after this, Tarkowski (6) published an article in which he made two improvements in the technique of acquiring high quality preparations of meiotic oocytes. In his description, there is no mention of utilizing superovulation, but it can be used in conjunction with his technique. Tarkowski reported that an injection of colcemid a few hours prior to sacrifice arrested the oocytes in meiosis, although he states that if meiosis is to be studied colcemid is not necessary. The second improvement was the use of a modification of the air-drying technique commonly used for acid-alcohol fixed mammalian cells. The technique allowed for much flatter and more cytoplasm-free preparations, which in turn increased the frequency of analyzable preparations.

The third major innovation in the techniques for preparing meiotic preparations from oocytes was described by Payne and Jones (7). The techniques utilized by Jagiello and Tarkowski required handling only one or a few oocytes at a time and thus was time-consuming, particularly if quantitative data were needed. Payne and Jones coupled superovulation, the use of colchicine to arrest oocytes at metaphase I, and acid-alcohol fixation to a method of working with up to 2000 oocytes at one time. The use of colchicine to arrest the oocytes at metaphase I alleviated the problem of having to isolate them from mature follicles as the oocytes are still in metaphase I after ovulation and can be collected en masse from the oviduct. It also removes the necessity to culture the oocytes to maturity. The ability to work with large numbers of oocytes makes it possible to place hundreds of them on a single slide. These improvements in technique allow the investigator to collect metaphase I oocytes from 50–100 females and make approximately ten slides in only 2–3 hr. The result is usually 150–300 analyzable cells.

Techniques

Single oocytes. Working with single oocytes is advisable if numerical chromosome aberrations are to be studied since the process of fixation is gentle and does not result in rupture of the oocyte and subsequent loss or gain of chromosomes. Young (8–10 weeks of age) female mice are injected with 5 I.U. of pregnant mares' serum late in the afternoon. Forty-eight hr. later, they are injected with 5 I.U. of HCG. If metaphase I cells are desired, an injection of 0.05 ml of 2×10^{-3} M colchicine is given immediately after the HCG; if metaphase II cells are to be analyzed, no colchicine is injected. The morning after the HCG injection, the females are killed and the ovaries and oviducts are dissected out. At this time, the ovulated oocytes are clumped in the ampulla (Fig. 14.1A) of the oviduct and can be removed simply by pricking the ampulla with a small-gauge needle and squeezing the clumped oocytes out into Dulbecco's PBS (8) in a depression slide (Fig. 14.1B). Once the oocytes have been removed from all the oviducts, they are transferred to a solution of 150 I.U. of hyaluronidase per ml of Dulbecco's PBS and allowed to sit until they are free of cumulus cells. The oocytes are then put through several washes of PBS in depression slides to dilute as many cumulus cells as possible (Fig. 14.1C).

Once the oocytes are free of cumulus cells and have been washed, they are placed in a 1% solution of sodium citrate. After 5–10 min., individual or small groups of oocytes are placed on a dry, grease-free slide in a microdrop of the sodium citrate solution. A small (about 0.02 ml) drop of acetic alcohol (3 parts methanol : 1 part glacial acetic acid) is dropped on the oocytes. This procedure is viewed through a dissecting microscope, and, when the oocytes begin to flatten out, several more drops of the fixative are applied one at a time. The slide is allowed to air-dry and then stained with either aceto-orcein or Giemsa stain.

Large numbers of oocytes. The procedures for working with large numbers of oocytes are the same as those for using individual oocytes until after removal from the ampulla, with the exception that 50 or more females are utilized. Once the oocytes are removed from the ampulla, they are transferred en masse to a solution of 150 I.U. of hyaluronidase in 1% sodium citrate. They are left in this solution until the cumulus cells are removed and put through several washes of 1% sodium citrate to dilute away as many of the cumulus cells as possible. The oocytes are then transferred by micropipette to a microcentrifuge tube. This tube is made by cutting the stem off a Pasteur pipette and sealing the end by flaming. It is then inserted into a rubber stopper that fits into a 15 ml centrifuge tube (Fig. 14.1D). The cells are then spun down at moderate speed in a clinical centrifuge and the sodium citrate removed by micropipette. Acetic methanol (3 parts methanol : 1 part glacial acetic acid) is then added and the tube is allowed to sit 30 min. before being gently agitated, spun down again, and the fixative removed. At this time 1 : 1 fixative (1 part methanol : 1 part glacial acetic acid) is added and the cells gently agitated. This procedure is

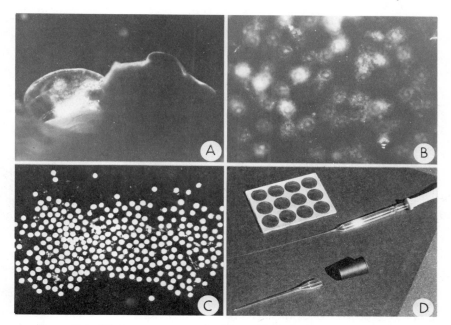

Figure 14.1 (A) A mouse ovary and oviduct with the oocytes clumped in the ampulla; (B) Mouse oocytes surrounded by cumulus cells shortly after removal from the ampulla; (C) Mouse oocytes after removal of the cumulus cells; (D) The micropipette, centrifuge tube apparatus, and depression slide described in text.

repeated twice. During the fixation, care must be taken not to allow the level of fixative to rise above the stem of the centrifuge tube because the oocytes will stick to the wall of the large diameter portion. Once the cells are fixed, they are dropped on wet, grease-free slides from a height of a few centimeters. The drop of fixative containing the cells disperses quickly, effecting spreading. Once the slides are air-dried, they can be stained with any of the conventional chromosome stains.

The major drawback of this procedure is that the large volume of the oocyte coupled with relatively violent spreading effect results in a large area dispersion of the chromosomes in a high frequency of the cells. This fact makes it unadvisable to use this procedure for studies on numerical aberrations, but it is highly useful for studying structural aberrations as hundreds of cells are obtained in a short period of time.

Mutagenicity Studies

Chemical Mutagens

Chemical mutagens fall into two categories: (1) direct-acting chemicals, or radiomimetic chemicals, that induce structural chromosome aberrations in a nonspecific fashion in that they produce them in all stages of the cell cycle, as do ionizing radiations; and (2) indirect-acting chemicals that usually need an intervening round of replicative DNA synthesis in order to induce structural chromosome aberrations. The latter situation is easily recognized in proliferating cells by two facts: (1) the appearance of aberrations as a function of time after treatment is usually delayed and corresponds to the appearance of S phase cells in mitosis; and (2) the aberrations are, initially, solely of the chromatid type. Since the last replicative DNA synthesis in oocytes occurred around the time of birth and the next one occurs only after fertilization, it stands to reason that the only clastogens that will produce structural chromosome aberrations in the oocytes of mature mammals are those that act in a direct fashion. This restriction need not apply to nondisjunction that results from damage to the spindle apparatus of the meiotically dividing oocyte.

Relatively little work has been done on the cytogenetic effects of chemicals on the meiotic chromosomes of female mammals, and the majority of this has been done in three laboratories. The bulk of the data that have been collected are summarized in Table 14.1. The substances that were tested are indicated as being either direct- or indirect-acting clastogens based on studies done on somatic cells either in vitro or in vivo. The substances which showed a strong clastogenic activity are indicated by three plus signs; moderate, two plus signs; weak, one plus sign; and no activity is indicated by a minus sign.

It is seen in Table 14.1 that three substances that are purportedly indirect-acting clastogens did, in fact, produce a moderately high incidence of structural aberrations in meiotic oocytes. It is cautioned, however, that these studies were done on metaphase II oocytes, where the analysis of structural aberrations is highly subjective due to the physical appearance of the chromosomes and the breakdown of sister chromatid attraction. Triethylenemelamine, on the other hand, gave a weakly positive result in oocytes as might be expected; it is also weakly clastogenic in G_1 human leukocytes, although as an alkylating agent it is principally an indirect-acting compound. This is borne out in the one study on triethylenemelamine, as well as methylmethane sulfonate, where fertilized treated ova were analyzed at the first cleavage mitosis and the yields of aberrations were many times higher than in metaphase I oocytes treated at the same stage of maturity (18, 19). Of the direct-acting mutagens tested, all showed an effect on meiotic chromosomes.

Ionizing Radiation

The study of radiation-induced chromosome aberrations is 50 years old, but it has only been in the last 3 years that an appreciable amount of data has been collected on cytogenetic response of mammalian oocytes to ionizing radiation. The great majority of this work has been done by Brewen and colleagues. In a series of three articles, they have studied the variation in sensitivity as oocytes mature, dose-response characteristics of aberration induction, and dose-rate and dose-fractionation phenomena. Furthermore, by careful analysis of their data, they have been able to demonstrate a strikingly close correlation in the patterns of aberration and specific locus mutation induction in the maturing mouse oocyte.

The first published report of a cytogenetic analysis of the meiotic chromosomes of oocytes following irradiation was that of Searle and Beechey (26). They reported a difference in the yields of aberrations as a function of dose and the time after irradiation that the oocytes were ovulated. Shortly after this, Gilliavod and Leonard (27) reported a study on the frequencies of structural aberrations observed 24 hr. after ir-

Table 14.1 Summary of the Cytogenetic Effects of Chemical Substances on the Meiotic Chromosomes of Mammalian Oocytes

	Type of action[a]	Species used	Mode of treatment	Structural aberrations	Numerical aberrations	Reference
Streptonigrin	Direct	Mouse	In vitro	++		9
			In vivo	++		9
SO_2	Direct	Mouse	In vitro	−		10
			In vivo	−	−	10
		Cow	In vitro	+++	+	10
		Ewe	In vitro	+	+	10
Sodium fluoride	Unknown	Mouse	In vitro	+++	+	11
			In vivo	+++		11
		Cow	In vitro	++	+	11
		Ewe	In vitro	+	−	11
Mercury	Unknown	Mouse	In vitro	−	+	12
			In vivo	−	−	12
Actinomycin D	Direct	Mouse	In vitro	+++		13
			In vivo	−		13
Meprobanate	Unknown	Mouse	In vitro	−	−	14
			In vivo	−	−	14
LSD-25	Indirect	Mouse	In vitro	−	−	15
			In vivo	−	−	15
Caffeine	Indirect	Mouse	In vitro	−	−	16
			In vivo	−	−	16
Phleomycin	Direct	Mouse	In vitro	+++		17
			In vivo	++		17
Triethylenemelamine	Indirect	Mouse	In vivo	+		18
Methylmethane sulfonate	Indirect	Mouse	In vivo	−		19
Triaziquone[b]	Indirect	Chinese hamster	In vivo	++	+	20,21
		Mouse	In vivo	++	+	20,21
Cyclophosphamide[b]	Indirect	Mouse	In vivo	++	+	21,22
Methotrexate[b]	Indirect	Mouse	In vivo	++	+	21,22
Cadmium	Unknown	Mouse	In vivo		++	23
Oral contraceptives[c]	Unknown	Several	In vitro		+	24,25

[a] The determination of whether a substance was a direct or indirect clastogen is based on careful somatic cell studies.
[b] The structural aberrations were scored in metaphase II oocytes and the analysis is highly subjective.
[c] Several substances were tested in several species and the reader is referred to the original publications for details.

radiation of females. These two reports strongly indicated that there was a large variation in the sensitivity of oocytes to aberration induction as a function of oocyte maturity, with the most mature, i.e., stage 8 oocytes, being least sensitive.

Subsequently, Brewen et al (28) published a more detailed study in which they demonstrated that the sensitivity of the oocytes to chromosome aberration induction by X-rays varied by as much as a factor of 3–10, depending on the stage irradiated. They also showed that the dose-response characteristics of both deletion and interchange induction approximated a $(dose)^2$ relationship. Based on these data, Brewen et al. (29) did further experiments on oocytes utilizing chronic and fractionated doses of X-rays. The data from these experiments showed that the length of time required for the oocytes to repair the lesions that interact to form chromosome aberrations

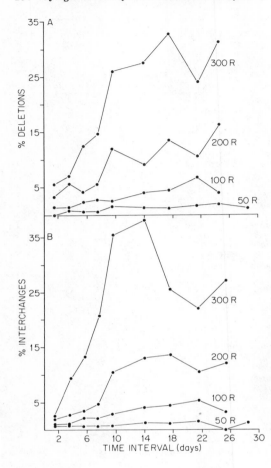

Figure 14.2 (A) The frequency of chromatid aberrations observed in metaphase I oocytes as a function of time after irradiation; (B) The frequency of chromatid aberrations observed in metaphase I oocytes as a function of time after irradiation.

was not influenced by the magnitude of the X-ray dose. They interpreted this to mean that the large reduction in yield of aberrations with chronic exposures was simply due to the reduction of the contribution of the (dose)² component via the mechanism of a continuous rate of repair, and not to a higher efficiency of repair at low doses and dose rates as has been proposed by Russell (30) to explain dose-rate effects on specific locus mutation induction.

Finally, in a very recent study, Brewen and Payne (31) performed a very detailed invesigation at four X-ray doses on the variations in sensitivity of the maturing mouse oocyte. The results of this study are depicted in Fig. 14.2A and B. The interesting aspect of these data is that the same pattern of sensitivity changes is also observed for dominant lethal mutations and specific locus mutations. Since dominant lethal mutations are principally chromosome aberrations, the similar patterns of sensitivity for specific locus mutation induction suggest that they may also be principally small chromosome deletions. In fact, if the specific locus data are analyzed on the basis of oocyte stage, they fit a dose-square model very well and do not show a "saturation phenomena" at high doses (31). This latter observation of saturation led Russell to conclude that the efficacy of repair varied at different doses of X-rays.

The cytogenetic analysis of metaphase I oocytes also provided an explanation of why very few balanced translocations are recovered in the F_1 offspring of irradiated females. In two earlier studies, Russell and Wickham (32) and Searle and Beechey (26) reported that very few, if any, translocation heterozygotes were recovered among the offspring of irradiated mothers. In their original study, Brewen et al. (28) C-banded metaphase I oocytes and analyzed the chromatid interchanges on the basis of whether they were symmetrical or asymmetrical, i.e., would segregate to produce a balanced translocation or a lethal dicentric. It was determined that the frequency of symmetrical interchanges that were present in cells without other lethal aberrations was such that only 0.6% balanced translocations would be expected to be recovered in the F_1 conceived from the oocytes irradiated with the same doses used by Searle and Beechey (26). This value compared very well with the 0.4% actually observed. Thus, the cytogenetic study provided an estimate of the frequency of heritable translocations expected after female irradiation without the necessity of doing the expensive and time-consuming genetic, or heritable translocation, test.

Evaluation

Recent technical developments now make it possible to perform quantitative cytogenetic analysis of the meiotic chromosomes of the mammalian oocyte. Where such analyses once involved tedious labor in preparation of small numbers of cells, the current techniques allow for the mass production of slides containing dozens of high quality preparations in just a short period of time. To date, the majority of the research done on oocytes has utilized the laboratory mouse and, to a lesser extent, the Chinese hamster. It will undoubtedly be some time before other species are used, since fairly large numbers of females are required in order to obtain sufficient numbers of oocytes.

The cytogenetic analysis of meiotic oocytes for assessing potential genetic hazards of direct-acting mutagens in females has certain advantages over the more conventional genetic tests. First, there is no need to breed the females and test their F_1 offspring for semisterility, as is the case in the heritable translocation test. This fact saves considerable time and does not require as large an animal facility. In addition, the cytogenetic analysis allows for an immediate determination of the large deletions that result in dominant lethality, while eliminating the possibility of physiological effects on the female that can result in embryo loss without a concomitant true genetic effect. Second, primary nondisjunction can be scored at metaphase II for the entire genome rather than just one chromosome, as is the case in the genetic test.

In the case of indirect-acting mutagens, the direct analysis of meiotic oocytes most likely will result in false negative results, as there is no replicative DNA synthesis in the oocyte. This difficulty can be circumvented, however, by analyzing first cleavage embryos in parallel experiments. The techniques for slide preparation of embryos are essentially the same as those for oocytes and are detailed elsewhere in this volume.

Overall, the direct cytogenetic analysis of mammalian oocytes holds promise for evaluating mutagenic substances and repair properties in the female. Future research in this area undoubtedly will expand and provide new insights into the underlying mechanisms of chromosome aberration formation.

Literature Cited

1. Peters, H.; Levy, E.; Crone, M. Oogenesis in rabbits. J. Exp. Zool. 158:169-79; 1965.
2. Lemon, J. G.; Morton, W. R. M. Oogenesis in the Golden hamster *(Mesocricetus aunatus)*: a study of the first meiotic prophase. Cytogenetics 7:376-89; 1968.
3. Oakberg, E. F.; Tyrrell, P. D. Labeling the zona pellucida of the mouse oocyte. Biol. Reprod. 12:477-82; 1975.
4. Pedersen, T. Follicle kinetics in the ovary of the cyclic mouse. Acta Endocrinol. 64:304-23; 1970.
5. Jagiello, G. M. A method for meiotic preparations of mammalian ova. Cytogenetics 4:245-50; 1965.
6. Tarkowski, A. K. An air-drying method for chromosome preparations from mouse eggs. Cytogenetics 5:394-400; 1966.
7. Payne, H. S.; Jones, K. P. Technique for mass isolation and culture of mouse ova for cytogenetic analysis of the first cleavage mitosis. Mutat. Res. 33:247-49; 1975.
8. Whittingham, D. G.; Wales, R. G. Storage of 2-celled mouse embryos in vitro. Aust. J. Biol. Sci. 22:1065-68; 1969.
9. Jagiello. G. Streptonigrin: effect on the first meiotic metaphase of the mouse egg. Science 157:453-54; 1967.
10. Jagiello, G. M.; Lin, J. S.; Ducayen, M. B. SO_2 and its metabolite: effects on mammalian egg chromosomes. Environ. Res. 9:84-93; 1975.
11. Jagiello, G.; Lin, J. S. Sodium fluoride as potential mutagen in mammalian eggs. Arch. Environ. Health 29:230-35; 1974.
12. Jagiello, G.; Lin, J. S. An assessment of the effects of mercury on the meiosis of mouse ova. Mutat. Res. 17:93-99; 1973.
13. Jagiello, G. M. Meiosis and inhibition of ovulation in mouse eggs treated with actinomycin D. J. Cell Biol. 42:571-74; 1969.
14. Jagiello, G.; Ducayen, M. B.; Lin, J. S. A meiotic cytogenetic study in mice of a commonly used tranquilizer reported to concentrate in mammalian follicular fluid. Teratology 7:17-22; 1973.
15. Jagiello, G.; Polani, P. E. Mouse germ cells and LSD-25. Cytogenetics 8:136-47; 1969.
16. Jagiello, G.; Ducayen, M.; Lin, J. S. Meiosis suppression by caffeine in female mice. Mol. Gen. Genet. 118:209-14; 1972.
17. Jagiello, G. Action of phleomycin on the meiosis of the mouse ovum. Mutat. Res. 6:289-95; 1968.
18. Brewen, J. G.; Payne, H. S. Studies on chemically induced dominant lethality. III. Cytogenetic analysis of TEM-effects on maturing dictyate mouse oocytes. Mutat. Res. 50:85-92; 1978.
19. Brewen, J. G.; Payne, H. S. Studies on chemically induced dominant lethality in maturing dictyate mouse oocytes. Mutat. Res. 37:77-82; 1976.
20. Hansmann, I.; Neher, J.; Rohrborn, G. Chromosome aberrations in metaphase II oocytes of Chinese hamster *(Cricetulus griseus)*. I. The sensitivity of the preovulatory phase to triaziquone. Mutat. Res. 25:347-59; 1974.
21. Rohrborn, G.; Hansmann, I. Induced chromosome aberrations in unfertilized oocytes of mice. Humangenetik 13:184-98; 1971.
22. Hansmann, I. Chromosome aberrations in metaphase II oocytes: stage sensitivity in the mouse oogenesis to amethopterin and cyclophosphamide. Mutat. Res. 22:175-91; 1974.
23. Watanabe, T.; Shimda, T.; Endo, A. Stage specificity of chromosome mutagenicity induced by cadmium in mouse oocytes. Teratology 16 (1):127; 1977.
24. Jagiello, G.; Lin, J. S. Oral contraceptive compounds and mammalian oocyte meiosis. Am. J. Obstet. Gynecol. 120:390-406; 1974.
25. Rohrborn, G.; Hansmann, I. Oral contraceptives and chromosome segregation in oocytes of mice. Mutat. Res. 26:535-44; 1974.

26. Searle, A. G.; Beechey, C. V. Cytogenetic effects of X-rays and fission neutrons in mouse dictyate oocytes. I. Time and dose relationships. Mutat. Res. 24:171–86; 1974.

27. Gilliavod, N.; Leonard, A. Dose relationship for translocations induced by X-irradiation in mouse oocytes. Mutat. Res. 25:425–26; 1974.

28. Brewen, J. G.; Payne, H. S.; Preston, R. J. X-ray-induced chromosome aberrations in mouse dictyate oocytes. I. Time and dose relationships. Mutat. Res. 35:111–20; 1976.

29. Brewen, J. G.; Payne, H. S.; Adler, I. D. X-ray-induced chromosome aberrations in mouse dictyate oocytes. II. Fractionation and dose-rate effects. Genetics 87:699–708; 1977.

30. Russell, W. L. Radiation and chemical mutagenesis and repair in mice. *In* Beers, R. F., Jr.; Harriot, R. M.; Tilghman, R. C., eds. Proceedings Miles Fifth International Symposium on Molecular Biology: Molecular and Cellular Repair Processes. Baltimore: Johns Hopkins University Press; 1972:239–47.

31. Brewen, J. G.; Payne, H. S. X-ray stage sensitivity of mouse oocytes and its bearing on dose-response curves. Genetics 91:149–61; 1979.

32. Russell, L. B.; Wickham, L. The incidence of disturbed fertility among male mice conceived at various intervals after irradiation of the mother. Genetics 42 (abstr.):392–93; 1957.

15
The Heritable Translocation Test in Mice

by B. M. CATTANACH*

Abstract

Several different types of translocations are known but, of these, only the reciprocal translocation has been used as an end-point in mammalian mutation research. The properties of reciprocal translocations and their induction by ionizing radiation and chemical mutagens have been subject to investigation since the early 1930s. Typically, these translocations have no phenotypic effect but can be detected by semisterility in either sex, by sterility and small testis size in the male, or by the presence of multivalent associations in meiotic cells. The degree of impairment of fertility can be correlated with the cytological findings.

The spontaneous incidence of translocations is extremely low, but they can be induced in all spermatogenic stages of males by ionizing radiation. Recovery from pre-meiotic stages, however, is low. Several chemical mutagens have also been found to induce translocations in male germ cells but sensitivity differences between the different spermatogenic stages appears greater than found with radiation and the distribution of breakpoints may be nonrandom. Translocations are rarely recovered following mutagen treatment of oocytes.

The heritable translocation test is usually applied to mice. The protocol involves the production and testing of F_1 animals following mutagen treatment on one or other parent. For F_1 males, the test procedures may be based upon fertility testing, which may take the form of screening for zygotic deaths in females mated to the test males or for reduced litter size at birth, or upon direct cytological analysis of meiotic cells using the method of Evans (Ford and Evans 1969) or Meredith (1969). Females can also be screened for translocation heterozygosity using fertility test procedures but, although methods for cytological analysis are available, cytological confirmation can be more reliably obtained by examining F_2 sons. Procedures have been developed for testing male and female F_1 animals simultaneously.

Primarly because the test is carried out in a mammal and concerns the type of chromosome damage which forms the source of genetic harm in man, the results are highly relevant for risk assessment. However, disadvantages are that the test is time-consuming, labor-intensive and

*MRC Radiobiology Unit, Harwell, Didcot, Oxon. OX11 ORD.

requires large numbers of animals. It is therefore expensive to operate. Despite this problem, it is of greater practical value than the dominant lethal test once the mutagenicity of a compound has been established. Any advantage over the specific locus method is perhaps limited to tests upon post-meiotic stages.

There is as yet no consensus of opinion on when the test should be used and no definite recommendation can be offered on method of treatment, dose, or size of experiment, but estimates for the latter are reviewed. Estimates on costing are also offered. A summary of the results obtained with chemical agents is presented.

Pioneer mutation work on mouse reciprocal translocations was carried out by Snell (1933, 1935), Snell et al. (1934) and Hertwig (1935, 1938, 1940). They observed that a proportion of the progeny of irradiated males produced abnormally small litters when mated with normal animals. This "semisterility," or "partial sterility" as it was later called by Russell (1954), was transmitted like a dominant gene to about half of the progeny and could be exhibited by either sex. Further investigations revealed that the reduced fertility resulted from zygotic deaths, these usually occurring about the time of implantation. It was inferred from these findings that the semisterile animals were heterozygous for a reciprocal translocation and that the zygotic deaths were due to chromosome imbalance. Confirmation that this was indeed true was subsequently obtained in two ways: (1) Snell (1946) was able to demonstrate linkage between two normally unlinked genes in one of his translocation lines; (2) cytological evidence of translocation was obtained by Koller and Auerbach (1941) and Koller (1944) who observed associations of four chromosomes at synapsis in each of three semisterile lines.

Complete F_1 male sterility was also commonly found in these pioneer experiments, but little information was available as to its cause or nature. Cattanach (1959) was the first to obtain evidence that at least some sterile males carried translocations. Soon after, it was recognized that X-autosome translocations were generally male-sterile, but not until 1966, when Lyon and Meredith studied a series of translocations detected by semisterility in F_1 females, was it established that certain autosome-autosome translocations can cause complete male-sterility.

The properties of reciprocal translocations have since been studied by Cacheiro et al. (1971, 1974), Carter et al. (1955), Cattanach et al. (1968), Generoso et al. (1978a), Searle et al. (1978), and others (see Searle 1974a and Searle et al. 1978). Methods of translocation detection based on the pioneer work have also been refined and developed. Thus, a system of screening for translocations on the basis of live and dead implants has been derived by Carter et al. (1955); a system that relies almost entirely upon reduced litter size and infertility has been developed by Generoso et al. (1978b); and technical advances in methods for making high quality meiotic chromosome preparations (Evans et al. 1964; Meredith 1969) have greatly facilitated the cytological recognition of translocations. In the F_1 translocation test (Léonard 1973), or *heritable translocation test* (Generoso 1978b) as it is variably called, the cytological approach has so far only been used for confirming the presence of translocations detected by other methods, but potentially it could be employed without prescreening. It has found greater application in the spermatocyte test (chapter 13) in which the spermatocytes of treated males are screened for translocations induced in spermatogonial stem cells.

Falconer et al. (1952) were the first to provide evidence that chemical mutagens can induce translocations in mammals. Following treatment of male mice with nitrogen mustard, they detected one semisterile animal among the F_1 progeny and cytological analysis revealed that it carried two translocations. Two sterile males and one gene mutation were also found. More substantial evidence was subsequently obtained by Cattanach (1957, 1959) and Cattanach et al. (1968) using TEM and EMS. These positive results fostered the concept that translocation tests might serve for screening drugs and environmental agents for mutagenicity in mammals. Such an approach has been extensively promoted by Generoso in recent years. Although translocation tests have been employed in a number of mammalian species, future reference will be limited to the mouse.

Properties of Translocation Heterozygotes

In setting up procedures for screening for translocations induced by mutagenic exposure, consideration must be given to the diverse properties and characteristics that typify this class of genetic damage.

Phenotype

Generally, translocations do not cause phenotypic abnormalities, although deleterious effects, such as reduced viability and body weight, are often found in females heterozygous for X-autosome translocations (Eicher 1970). These may be peculiar to rearrangements involving the X because they most probably result from the effective partial monosomy created by the genetic inactivation of autosomal loci in those cells which have the rearranged X chromosome inactive. Coat colour mosaicism may also be observed under given conditions in these heterozygotes and is attributable to the same process. Such position effects have not been commonly found with autosome-autosome translocations. The "Steeloid" phenotype has, however, been found to be associated with translocations which have a breakpoint at the *Steel* locus (Cacheiro and Russell 1974), and Phillips (Phillips 1966; Evans and Phillips 1978) has found the *Agouti suppressor* mutation to be associated with a chromosome inversion. A more common finding is a reduced testis size, breakdown of spermatogenesis, and oligospermia or aspermia in males heterozygous for certain autosomal translocations (Lyon and Meredith 1966; Searle et al. 1978). This effect may be dependent upon the proximal location of at least one translocation breakpoint.

Translocation Semisterility

Translocation semisterility results from the segretation of unbalanced gametes from translocation heterozygotes. At pachytene, each segment of the rearranged chromosomes pairs with the homologous segment of the normal chromosomes and, at anaphase I, the four chromosomes may segregate in a variety of ways. With normal segregation (alternate or adjacent-1), four types of gametes are produced. Two will be chromosomally balanced and two will be chromosomally unbalanced (Fig. 15.1). With reciprocal translocation, the latter carry a duplication of one terminal segment and a deficiency of the other. This has little or no effect upon the gametes themselves and, on syngamy with chromosomally normal gametes, chromosomally unbalanced zygotes are produced. These usually die at about the time of implantation. Only the chromosomally balanced gametes lead to the production of viable zygotes, these being either chromosomally normal or translocation heterozygotes.

A 50% reduction in fertility accords reasonably with observation, but there is considerable variation between different translocations. Thus, the range of average fertility of males was found to be 41–69% in Snell's (1946) radiation experiments, 38–59% in those of Carter et al. (1955), and 43–44% in Generoso et al.'s (1978b) radiation and chemical mutagen experiments. The excess over 50% loss is attributed to segregational events other than those described above. Adjacent-2 segregation, in which homologous centromeres go to the same pole, leads to unbalanced gametes (Fig. 15.1), and nondisjunction events (Fig. 15.1) contribute more. Crossing-over between the rearranged and normal chromosomes in the interstitial segments (between break and centromeres) can increase the total number of gametes produced to 36, of which only 10 are chromosomally balanced (Ford 1969). Percent fertility has been correlated with cytological observation of metaphase/anaphase configurations of the rearranged and normal chromosomes (Koller 1944; Slizynski 1957a). In males, it is also possible that oligospermia may contribute to the reduced fertility of some translocation heterozygotes.

Several of the early studies on translocations indicated that the fertility of female heterozygotes was lower than that of males. Snell (1946) and Carter et al. (1955) suggested that the proportion of aneuploid gametes produced by females could be higher than that produced by males, and recent studies with Robertsonian translocations have indeed shown that nondisjunction events are higher in the female (Winking and Gropp 1976). Russell (1962) has suggested that translocations may affect the fitness of animals and, in the case of females, this could make them poorer mothers.

Russell (1962) has suggested that fertility in excess of 50% may also be accounted for on a segregational basis, but, since alternate and adjacent-1 segregation, with interstitial chiasmata, are equivalent in the mouse (Ford and Clegg 1969), this possibility would seem unlikely. However, viability of one or more of the unbalanced zygotes can create this effect. Several examples of the latter are known. Cattanach (1965) has described the inheritance of a presumptive translocation detected in a F_1 translocation test with TEM only by the appearance of phenotypically abnormal F_2 progeny. Only limited cytological evidence could be obtained, but genetic tests indicated that two mutational events had been induced in the original treated gamete. One caused the phenotypic abnormality and the other suppressed it. Both types were viable and these were interpreted as the unbalanced products of a translocation in which only small segments had been exchanged. A similar case, but with viability of only one unbalanced zygote, has recently been reported by Hollander and Waggie (1977).

Nondisjunction types may also be viable, but, since the frequency of this class of unbalanced zygote is not likely to be very high, any enhancement of fertility this may contribute is unlikely to be great. The first examples of viable tertiary trisomy were obtained with translocations associated with male sterility (Cattanach 1967b; Lyon and Meredith 1966). Further examples derived from mutagen treatment of spermatogonia have since been found (de Boer 1973; Cacheiro et al. 1974; Griffen 1967; Griffen and Bunker 1967). Some phenotypic effects are associated but are generally minor.

Insertions, like reciprocal translocations, should also cause some degree of semisterility. However, although the segregation of the normal and rearranged chromosomes (in the absence of crossing-over within the rearranged segment) will produce the same 50:50 proportion of balanced and unbalanced gametes, the latter will carry duplications *or* deficiencies, unlike the duplication *and* deficient products of reciprocal translocations (Fig. 15.1). As duplications within the genome can often

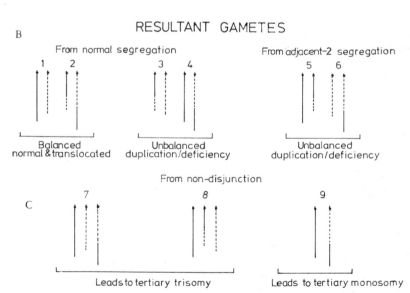

Figure 15.1 Meiotic segregation in translocation heterozygotes. Reciprocal translocation between two nonhomologous chromosomes, meiotic pairing and possible segregation of normal and rearranged chromosomes. From normal disjunction (alternate and adjacent-1) balanced, normal and translocated and two unbalanced, duplication plus deficiency classes of zygotes are produced in near-equal frequencies. When one rearranged chromosome is short, it may fail to synapse and form chiasmata (not shown) with homologous segments, which will lead to a high frequency of nondisjunction. All products of nondisjunction will be unbalanced and will result in tertiary trisomy and monosomy. These may be viable but could suffer from congenital malformations. (Adapted from Searle 1975.)

be tolerated, a 75% fertility might be expected, although crossing-over within the insertion could reduce this. One example is provided by Cattanach's (1961) X-autosome translocation, in which a large piece of chromosome 7 has been inserted into the X. Genetic studies have shown that the duplication female, aided by X-inactivation, is fully viable, and even the duplication male may survive. This translocation was not detected by any loss of fertility. Another insertion has been detected by Searle et al. (1974). This was detected in a heritable translocation test by the fact it was semisterile (according to Carter et al. [1955] criteria). Identification was established by cytogenetic analysis.

Little loss of fertility may be expected with shifts, since unbalanced gametes will only be produced when there is crossing-over within the translocated segment. The situation for Robertsonian translocations is not yet entirely clear. The most extensively studied translocations of this type has been derived from wild *Mus* species. Introduced into laboratory stocks, these translocations have been found to pair regularly with the homologous acrocentric chromosomes, but the disjunction at metaphase I is disorderly (Cattanach and Moseley 1973; Ford 1975; Gropp et al. 1975). As a consequence, monosomic and trisomic gametes are produced and these lead to zygotic deaths and low litter size. A complicating factor is that the nondisjunction frequency differs for each translocation and does not correlate with either chromosome size or centromere position. Whatever the exact cause of the nondisjunction, this property does not appear to be typical of Robertsonian translocations that have risen spontaneously in laboratory mouse stocks. Numerous cases have been discovered within the last two years, and as yet there is little indication that the heterozygotes have impaired fertility or cause any substantial level of nondisjunction. If this proves to be typical of Robertsonian translocations in the laboratory mouse, this type of translocation will not be detectable by semisterility test procedures. Cytologically, however, it is easily recognizable and perhaps should not be neglected in mutation experiments, since it most probably derives from chromosome breakage and reunion and constitutes a significant genetic risk in man.

Translocation Male Sterility

It is now well established that some autosomal translocations cause complete sterility in the male, yet semisterility in the female (Cacheiro et al. 1974; Cattanach 1967; Lyon and Meredith 1966; Searle et al. 1978). The immediate cause is oligospermia or aspermia, but there is considerable variation between translocations in the stage of spermatogenesis primarily affected. Searle et al. (1978) for example, have established a number of male-sterile lines and shown the severity of impairment to range from a reduced sperm-count to the presence only of meiotic cells, with the effect in each line being graded rather than abrupt. A correlated reduction in testis weight was observed and reduced testis weight provides good indication that some kind of chromosomal change may be present.

Some impairment of spermatogenesis is probably also typical of males carrying semisterile translocations. Thus, Generoso et al. have reported that testis weight is more variable among translocation heterozygotes than among normal mice and, in addition, some translocations, such as T6Ca, may cause either male semisterility or male sterility (Cattanach 1967b). This means that translocations cause two distinct effects in the male: spermatogenic impairment and gametic imbalance. Semisterility attributable to gametic imbalance may therefore only be observed when spermatogenic impairment is sufficiently limited. Genetic background may well influence the latter effect (see Lengerova 1970).

The precise mechanisms leading to the impairment are not yet understood, but it is probable that it relates to the location of the breakpoints in the rearranged chromosomes. Lyon and Meredith (1966) have pointed out that, whereas male semisterile translocations most frequently show ring-of-4 associations at metaphase I, male sterile translocations most commonly show chain-of-4 or chain-of-3 plus 1 association, and, in the female, often lead to nondisjunction. Lyon and Meredith (1966) also drew attention to the fact, since confirmed by other investigators (Cacheiro et al. 1974; Cattanach et al. 1968; Searle et al. 1978), that abnormally long and/or abnormally short marker chromosomes could often be detected in such animals. This could suggest a failure of pairing or the presence of univalent chromosomes could be responsible for the sterility (Cattanach et al. 1971; Miklos 1974). On the other hand, a high proportion of sterile translocations have one breakpoint close to a centromere and this has led Cacheiro et al. (1974) and Searle (1974a) to conclude that centric heterochromatin may be involved. Cacheiro et al. (1974) have proposed that the sterility results from a position effect and Searle (1974a) has suggested that it follows from the attachment of homologous centric heterochromatin to synaptonemal complexes of very different lengths, so interfering with attachment to the nuclear membrane. A curious phenomenon noted in several male sterile translocations is a nonrandom contact between the C bands of the X chromosome and the translocation configurations at diakinesis/metaphase I (Forejt 1974; Forejt and Gregorova 1977). Some impairment of X inactivation in the primary spermatocyte brought about by this contact has been proposed as the cause of translocation male sterility.

Other types of reciprocal translocations also cause male sterility. All X-autosome translocations other than Cattanach's (1961) insertional type are female semisterile but male-sterile (Eicher, 1970). Even the latter insertional translocation appears to have some spermatogenic impairment since the males often become sterile soon after maturity. Six of the seven Y-autosome translocations reported in the literature (Cacheiro et al. 1974; Falconer et al. 1952; Léonard and Deknudt 1969a; Searle 1974a) were also discovered in sterile males and, hence, associated with male sterility. In addition, it is now clear that the presence of two or more translocations in an animal may often cause male sterility (Cacheiro et al. 1974; Cattanach 1959; Cattanach et al. 1968), though seldom female sterility. This has been found to be most likely when the translocations involve a common chromosome, as can be demonstrated by crossing two male semisterile lines together (Carter et al. 1955).

The one problem in using male sterility as a criterion for the presence of a translocation is that factors other than translocations may be responsible. Thus, numerical changes, e.g., XXY, XYY, and tertiary trisomies, cause male sterility and such little understood phenomena as X-Y separation and multiple univalents are also associated with this condition (reviewed by Searle et al. 1978).

Cytological Features

Most reciprocal translocations can be detected by cytological analysis. Each segment of the rearranged chromosomes pairs at pachytene with the homologous segment of

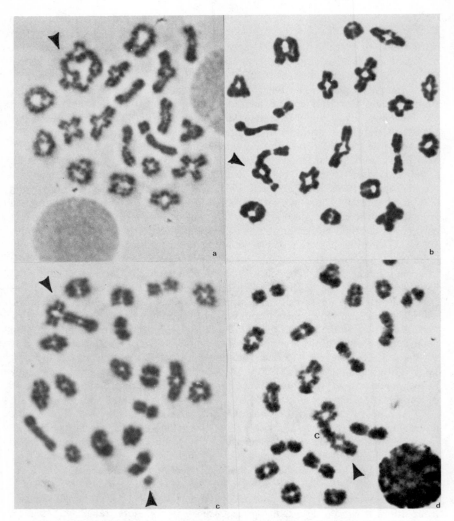

Figure 15.2 Meiotic metaphase I cells showing multivalent associations: (a) 19 elements including large ring-of-4 association, due to a reciprocal translocation; (b) 19 elements including chain-of-4 association, due to a reciprocal translocation; (c) 20 elements including chain-of-3 association plus single univalent, due to reciprocal translocation; (d) 19 elements including chain-of-3 association typically found in heterozygotes for Robertsonian translocations. C = Robertsonian centromere.

the normal chromosomes to give a cross-shaped configuration (Fig. 15.1a), and at diplotene chiasmata may or may not appear in each arm of the cross. As a consequence, the four chromosomes may remain associated at diakinesis and metaphase I and form recognizable configurations. The shape and formation of the translocation configurations are dependent upon two factors: size of the translocated segments and the number and positions of the chiasmata. These are not independent of each other. The smaller the segments, or the nearer the breakpoint is to one or other end of the chromosome, the lower the probability that a chiasma will form between the breakpoint and the chromosome end.

Commonly, the breakpoints lie in the central portions of the chromosomes. Chiasmata therefore can form in each arm of the cross and, at diakinesis/metaphase I, this opens up to form a large ring-of-4 configuration. Only 19, rather than the usual 20, paired elements can then be counted (Fig. 15.2b). When a chiasma does not form in one arm, as commonly occurs when one break is close to the end of one chromosome, a chain-of-4 configuration with a total count of 19 is produced (Fig. 15.2c). If one of the rearranged chromosomes is small, it may fail to form any chiasmata and, as a consequence, a chain-of-3 plus a small univalent is seen. The total count in this case is 20. It is also theoretically possible to find some cells in translocation heterozygotes with 20 bivalents, two of which are asymmetrical, or 19 bivalents plus two univalents, but these alone cannot be taken as evidence that a translocation is present. Most translocation heterozygotes detected by fertility testing show characteristic translocation configurations in the great majority of their metaphase I cells, although frequencies as low as 13% (Léonard and Deknudt 1968) and even 2% (Lang and Adler 1976) have been reported.

Insertional translocations should also be detectable by cytological analysis. Associations of four may be expected to form on the basis of the same criteria outlined for reciprocal translocations. The frequency of metaphase I cells showing translocation and configurations in Cattanach's insertional translocation has been reported to be about 10% (Ohno and Cattanach 1962).

The possibility of detecting shifts during routine cytological screening is remote since, at most, only an abnormal bivalent configuration could be expected. No translocations of this type have yet been found. Robertsonian translocations, on the other hand, should be readily detectable since they derive from whole arm "joinings" which are unlikely to interfere with homologous pairing and chiasma formation. Heterozygotes thus show chain-of-3 configurations (four chromosome arms) and the total number of elements observed is 19 (Fig. 15.2d). Such configurations can, however, be easily confused with chain-of-4 configurations due to reciprocal translocations in individual cells but, since the Robertsonian centromere occupies the middle portion of the chain, any indication of an interstitial chiasma within this region in any cell immediately identifies the translocation as a reciprocal type.

Spontaneous Incidence of Translocations

The spontaneous incidence of reciprocal translocations is sufficiently low that despite quite extensive testing a reliable figure is not yet available. The majority of the data derive from the experiments of Generoso et al. (1978a), which in total provide 4423 analysed F_1. Only one semisterile was detected and confirmed cytologically to carry a translocation, but three completely sterile males were also found though not examined cytologically. At least one and possibly three other translocations have been recorded in 1679 controls for radiation experiments (Léonard and Deknudt 1968; Luning and Searle 1971; Searle et al. 1974), but none in 2568 controls for chemical mutation experiments (referenced in Table 15.6, Section 11) other than those of Generoso et al. (1978a). From these findings, the spontaneous incidence can be calculated to range from 2.3×10^{-4}, when only the two confirmed cases are considered, to 8.1×10^{-4}, when Generoso et al.'s three sterile cases and two unconfirmed semisterile cases (Luning and Searle 1971) are included. These figures are based on a total of 8670 animals, almost all of which have been screened by fertility testing.

Until recently very few Robertsonian translocations were known in the mouse, but within the last two or three years at least ten new cases have been reported (Cattanach and Crocker 1977; Cattanach et al. 1977; Cattanach and Savage 1976; Davisson and Roderick 1978; Lyon et al. 1977; Phillips and Kaufman 1974; Phillips and Savage 1976). In most instances, discovery followed cytological investigation of previously existing Robertsonian translocations or other chromosomal changes. Only one estimate of the spontaneous incidence is obtainable from these data unfortunately, this provided by Phillips and Kaufman (1974) who were investigating the production of XO females by females heterozygous for an X chromosome inversion. Two Robertsonian translocations were detected among 915 offspring (2.2×10^{-3}).

Induction of Translocations

Induction of Translocations in Males by Ionizing Radiation

Ionizing radiation is capable of inducing reciprocal translocations in all stages of spermatogenesis in the male. Germ cell stage sensitivity differences clearly exist, however. Thus, spermatids are approximately four times as sensitive as mature sperm, and postmeiotic stages are more sensitive than premeiotic stages. Cell stage sensitivity correlates with that for dominant lethals (Léonard and Deknudt 1968), a finding consistent with the fact that dominant lethals, like translocations, derive from chromosome breakage events (Brewen et al. 1975).

Recovery of translocations in F_1 offspring from irradiated spermatogonia is particularly low (Searle 1975). This, in part, results from the reduction division of meiosis which allows only one in every four sperm from any spermatogonium carrying a translocation to inherit the balanced translocation. In addition, there is accumulating evidence of selection against cells carrying certain types of balanced translocations. Ford et al. (1969) have found that the F_1 translocation frequency is only half that expected from the frequency observed in the spermatocytes of the treated fathers even after allowance is made for the effect of the reduction division, i.e., only one-eighth of those detected in spermatocytes are recoverable among the F_1. Selection also is consistent with the further observations that translocations which cause male sterility are regularly recovered from postmeiotic germ cells but seldom from spermatogonia. This implies some autonomous action operating at the cell level in cells carrying this type of translocation to bring about meiotic or spermatogonic arrest. Screening F_1 animals is, therefore, an inefficient procedure for detecting translocations induced in spermatogonia. Spermatocyte screening in the treated fathers themselves provides a far more sensitive, rapid, and economical test.

There is relatively little evidence as to whether or not Robertsonian translocations can be induced by ionizing radiation. Ford et al. (1977) have reevaluated their data on spermatogonial translocations detected in spermatocytes and concluded that, although the presence of some Robertsonian translocations could not be excluded, their numbers must have been very few. No other investigators have sought to distinguish chain-of-4 quadrivalents due to reciprocal translocations or chain-of-3 trivalents due to Robertsonian translocations. Ford et al. (1977) also indicated that they found no Robertsonian translocations following spermatozoal irradiation among 531 F_1 animals examined cytologically, although 120 animals with reciprocal translocations were detected. Clearly, if Robertsonian translocations are induced by ionizing radiation, they must occur in negligible numbers relative to reciprocal translocations. This is perhaps not too surprising since there is little limitation on the sites of chromosome breakage for reciprocal exchange, but Robertsonian translocations require both breaks to occur in the centric regions.

Induction of Translocations in Males by Chemical Mutagens

A number of chemical mutagens induce reciprocal translocations in male germ cells. The most extensively studied have been TEM and EMS. Germ cell stage sensitivity differences are evident to the point that they could be described as qualitative rather than quantitative (Cattanach 1975). TEM is the only agent which has so far proven capable of inducing translocations in all stages of spermatogenesis (Cattanach 1959; Cattanach and Pollard 1971; Generoso et al. 1977), but postmeiotic cells are by far the most sensitive. The number of translocations detected following spermatogonial treatment is extremely small despite screening by the spermatocyte test. Among postmeiotic cells, the late spermatid is the most sensitive and, as found with ionizing radiations, the postmeiotic response correlates well with dominant lethal induction (Cattanach 1959; Generoso et al. 1978a). Similar properties may be ascribed to the chemically related compound, TEPA (Epstein et al. 1971; Generoso et al. 1978a; Sram et al. 1970) and probably also to another alkylating agent, cyclophosphamide (Generoso et al. 1978a; Sotomayor and Cumming 1975).

Only postmeiotic stages are affected by EMS and MMS, as judged either by dominant lethal or translocation tests (Cattanach and Pollard 1971; Cattanach et al. 1968; Ehling et al. 1968; Léonard et al. 1971) or even by the specific locus test which screens for point mutations (Cattanach 1971; Ehling 1978). Both compounds are very effective mutagens in these germ cell stages, however, and both produce their greatest effects upon early spermatozoa, i.e., on a somewhat later spermatogenic stage than that most affected by TEM. It is not known how EMS and MMS fail to cause any kind of recoverable genetic effect in premeiotic germ cells.

Almost all chemically induced translocations have been detected by fertility testing procedures and, for reasons already discussed, are therefore unlikely to include Robertsonian translocations. Two "Robertsonian type" translocations have, however, been found in a chemical mutagen experiment (Cacheiro and Swartout 1975). They derived from IMS-treated oocytes, a finding which lends some support to Searle's (1974a) idea that this class of translocation may be recoverable from female germ cells. It may be that they were not true Robertsonian translocations since both were found in sterile males, but it is possible that Robertsonian translocations may be induced by chemical mutagens more readily than by ionizing radiation.

This would follow if chemical mutagens cause more chromosome breakage near the centromeres than ionizing radiation, the distribution of breakpoints for which seems to be more or less random. Breakage events occurring predominantly in the vicinity of the centromere have been observed in studies upon spermatogonia treated with the chemical mutagen MC (Adler 1973), but such nonrandom breakage might also be inferred from the fact that chemical mutagens tend to produce a high proportion of male-sterile translocations (Cacheiro et al. 1974; Cattanach 1959; Cattanach et al. 1968; Generoso et al. 1974; Searle 1974a).

Induction of Translocations in Females

Few translocations are recovered following oocyte irradiation. Searle and Beechey (1974) reported the detection of three reciprocal translocations among 680 F_1 progeny of female mice exposed to 300R of X-rays and, more recently, Krishna and Generoso (1977) have found four among 800 F_1 sons. Earlier studies by Russell and Wickham (1957) and Gilliavod and Léonard (1973) failed to detect any translocations among a total of 505 offspring scored. The overall response must therefore be considered extremely low.

Only one chemical mutagen has been tested for translocation induction in females and no significant response was detected despite extensive testing (Generoso et al. 1978a). The compound was IMS which produces high levels of dominant lethals both in male postmeiotic stages and oocytes (Ehling et al. 1972; Generoso et al. 1971), but Generoso et al. (1978a) could find only one translocation among 1469 F_1 derived from oocytes of different follicular stages. Four sterile F_1 males were also found, but neither of the two animals examined carried a translocation. Cacheiro and Swartout (1975) later reported on the cytology of a total of seven sterile males presumably derived ultimately from this experiment. Two carried the "Robertsonian-type" translocations mentioned in the preceding section. The total data therefore serve to show that some heritable chromosome damage can be recovered from oocytes.

It may be concluded from the radiation and IMS data that the heritable translocation test is not well suited for screening compounds for mutagenicity in oocytes, but the possibility of Robertsonian translocations being induced in this germ cell (Searle, 1974a) clearly warrants attention.

Currently Accepted Protocols for the Heritable Translocation Test

The heritable translocation test customarily requires breeding animals through two generations. The first concerns the production of the F_1 animals for test and the second the testing of the F_1 animals themselves for reduced fertility. For reasons already discussed, the fertility testing procedure can only be expected to detect reciprocal translocations and then only those that effectively cause "semisterility." Unless indicated to the contrary, the term translocation will therefore be used to refer only to the reciprocal type.

Production of F_1 Animals

Either hybrid or random bred mice should be used. The essential requirement is that the treated animals should breed well and produce vigorous healthy F_1 progeny with no inherent disposition to low fertility other than that which may be produced by the treatment.

For reasons discussed previously, treatment is usually confined to males. According to existing knowledge of the test compound, investigation may be limited to only one or two stages of spermatogenesis following a single treatment or extended to include all stages of spermatogenesis either with single or multiple treatments. Table 15.1 shows the germ cell stages that are sampled with matings at different time intervals after treatment is given. Some overlap of stages must of necessity occur, but the duration of spermatogenesis is not known to vary between strains and, since there is no sperm storage as in *Drosophila,* mating frequency does not greatly influence the efficiency of sampling.

A commonly used format is to give each treated male two females for each weekly interval being tested. Mating normally occurs within 3-5 days and litters may be expected 18-21 days later. Alternatively, the females may be checked daily for vaginal plugs and replaced with fresh females immediately after such evidence of mating has been found. The latter procedure is undoubtedly more laborious but allows a finer control over the germ cell stages being sampled and, since a healthy male will mate 0.5 females/day on average, more progeny can be obtained from each male.

If precise information on litter size is to be obtained, e.g., to provide information on dominant lethal induction, the mated females must be allowed to litter in separate cages. Otherwise, greater efficiency of production/cage may be attained by allowing two females which mated at about the same time to litter together.

Since translocations can be detected in F_1 males far more easily than in F_1 females, males only may be retained at weaning (18-21 days). Testing may begin when they reach sexual maturity at about 6-7 weeks.

Procedures for Detecting Translocations in the F_1 Males

No precise format for testing F_1 animals for translocations has yet been established. Procedures vary in different laboratories according to facilities available, but some form of fertility testing is usually employed first. It should therefore be emphasised that screening is limited to reciprocal translocations. Fertility testing can take the form of investigating numbers of live young and numbers of zygotic deaths at midpregnancy in females that have been mated to test males, or simply litter size at birth, or a combination of the two. An essential requirement therefore is a reliable source of females of good average and known fertility. Hybrids or random-breds are to be recommended. Cytological confirmation of translocation heterozygosity in the semisterile and sterile F_1 males so detected should thereafter be sought. Details of the methods are as follows.

Table 15.1 Germ Cell Stages Sampled at Different Mating Intervals after Treatment of the Male[a]

Interval (days)	1-7	8-21	22-42	42+
Germ cell stage	spermatozoa	spermatids	spermatocytes	spermatogonia

[a] Based on data of Oakberg 1957.

Method based on autopsy of tester females. The oldest and most clearly defined test is that described by Carter et al. (1955). Each F_1 male is mated to two or three virgin females. These are killed at mid-pregnancy (12–15 days) and the numbers of live and dead implants scored. The presence of a small proportion of dead implants cannot be taken as evidence that the father is semisterile, since this is a normal feature of the gravid mouse uterus. Equally, the absence of dead implants does not constitute evidence that the father is not semisterile, since this may be a chance distribution effect. Carter et al.'s (1955) method sets up criteria that reduces this problem. Numbers of live and dead implants are evaluated as a proportion of total litter size, and this allows a firm diagnosis (with 99% accuracy) of the fertility status of four out of five males on the basis of first litter data alone. Further testing is then only necessary for the remaining one male in five.

Carter et al.'s semisterility test criteria are set out in Table 15.2. These criteria are routinely employed at Harwell. However, it should be noted that they were determined for Harwell stocks of mice. Before being employed in other laboratories with other mouse stocks, preliminary testing is advised so that any necessary modifications can be made.

The test is simple to operate and offers a high degree of accuracy, but, of necessity, is wasteful of tester animals. Numerous variations of the method that place greater weight on numbers of dead or live implants are in use (see Generoso et al. 1978b; Lang and Adler 1977; Sheu et al. 1978).

Method based on litter size at birth. Many investigators have employed reduction in litter size at birth as the criterion for translocation heterozygosity. The essential requirement before screening can begin is that the reproductive performance of the tester females should be known. As a general guide, two or more litters with an average litter size typical of the average of the stock establishes the absence of a translocation in a test male. Subsequent cytological or Carter et al. (1955) -type testing is then necessary to establish whether or not a translocation is present among suspects.

An example of the methodology that can be used is provided by the sequential analysis procedures developed by Generoso et al. (1978b). He employs hybrid (SEC $\times$ C57BL F_1) females to test his F_1 males. These females can produce 10 litters with

Table 15.2   Semisterility Test Criteria[a]

Dead implantations	Total implantations						
	1–5	6	7	8	9	10	11+
5+	S	S	S	S	S	S	S
4	S	S	S	S	S	S	S
3	I	I	I	I	I	I	I
2	I	I	I	I	I	I	F
1	I	F	F	F	F	F	F
0	I	F	F	F	F	F	F

F = fertile; I = test inconclusive; S = semisterile

[a] From Carter et al. (1955).

an average litter size of 9.6. Each male is placed with a single female and litter size at birth is recorded. Young are discarded immediately. A male is declared fertile and discarded as normal if the size of his first litter is 10 or more. If the first litter is less than 10, a second litter is scored. If the second litter is 10 or more, a male is again declared fertile and discarded; otherwise he is suspected of semisterility and is separated for further testing. Should a third litter be produced, the male is reclassified using the same criteria. Suspect semisterile (and sterile) males are then mated with 2–3 virgin females which are killed at mid-pregnancy and diagnosed as fertile, semisterile, or sterile on the basis of criteria similar to those of Carter et al. (1955). Semisteriles and steriles are taken to carry translocations. Cytological confirmation is considered essential only for sterile males.

Generoso et al. (1978b) has calculated that the cumulative proportions of males that can be declared fertile on the basis of one, two, and three litters are 0.65, 0.88, and 0.91, respectively. Therefore, only 9% of his test males, on average, have to be tested further. Since his tester females are of particularly high reproductive performance, he is able to remate them with several males to produce a maximum of 10 litters. This effects a great economy of females. It is estimated that 5.38 males are, on average, tested by each hybrid female.

Because of the economy in the use of females, screening for semisterility on the basis of litter size offers a distinct advantage for large-scale testing. However, few mouse stocks can be expected to match the reproductive performance of Generoso's SEC × C57BL hybrid. Age and parity effects are likely to introduce unwanted variables into the test system.

Method based on cytological analysis. Cytological analysis is most usually limited to semisterile and sterile males identified by fertility test procedures but, as will be discussed later, it could be used directly on all F_1 males without prescreening.

F_1 males may be used for cytological study as early as 3–4 weeks of age, since by this time spermatogenesis is sufficiently advanced that metaphase I cells are present in numbers adequate for analysis. However, because mature testis size (or weight) provides an indication of the normality or otherwise of spermatogenesis, it is advisable to commence proceedings only after maturity has been reached at about 6–8 weeks.

The testes are removed after sacrifice (or only one may be removed under anaesthesia) and weighed. Low testis weight (one-half to one-third normal) indicates the spermatogenesis is impaired and alone provides good prima facie evidence that a translocation is present. Such small testes require special handling, but for normal-sized testes good quality meiotic chromosome preparations can be made with the air-drying method of Evans et al. (1964). The sequence of steps is as follows (Ford and Evans 1969):

1. Kill mouse and dissect out testes. Place testes in 2.2% weight/vol. trisodium citrate solution in a 2 inch diameter Petri dish.

2. Pierce tunica and expose tubules. Swirl in citrate solution to wash away fat. Time: 15 seconds.

3. Transfer to ca. 3 ml 2.2% citrate solution in a fresh 2 inch Petri dish. Tease tubules with curved forceps, then transfer total contents of dish to a 4 ml centrifuge tube. Pipette gently. Time: 10 minutes.

4. Spin 5 seconds to sediment larger tubule segments. Transfer supernatant to a clean 4 ml centrifuge tube. Discard sediment.

5. Spin for 5 minutes at 500 r.p.m. Discard supernatant. Resuspend pellet in minimum volume of residual supernatant.

6. Add ca. 3 ml of 1.0% trisodium citrate solution, slowly, meanwhile flicking the tube with the forefinger. Divide between two Dreyer tubes (narrow, conical tipped tubes, capacity ca. 2. ml). Total time: 10 minutes.

7. Spin 5 minutes at 500 r.p.m. Remove supernatant. Allow tubes to stand 1 minute, then remove fluid that has drained down from walls by a small pipette.

8. During step 7, make up fixative of 3 parts absolute ethyl alcohol:1 part glacial acetic acid plus a trace of chloroform (say, 45:15:1).

9a. Fixation, variant I. Flick tube to convert pellet to a dense suspension. Allow two drops of fixative to fall directly onto cell suspension. Flick tube vigorously. Add more fixative and continue to flick until tube is ca. three-fourths full. Spin 3 minutes at 500 r.p.m. Change fixative. Spin down and change fixative twice more. Total time: 15 minutes.

9b. Fixation, variant II. Retain pellet intact. Add fixative down side of tube, carefully at first to avoid disrupting pellet, then more rapidly until the tube is full. Remove all fixative immediately. Resuspend pellet and add more fixative until the tube is three-fourths full. Spin 3 minutes at 500 r.p.m. Remove fixative, resuspend pellet, add fresh fixative, and leave to stand 15 minutes then repeat.

10. Preparations. The final suspension of cells in fixative should be dilute and it is best to allow it to stand for 1–3 hours before making preparations. It may then be necessary to resuspend the cells once more before proceeding. Take up cell suspension into a small pipette. (Made from 4 mm diameter soft glass tubing; teats are made from short lengths of polythene tubing, the end being closed by heating followed by pressure from pliers). Allow three evenly spaced drops to fall on a clean slide. Wait until the fluid spreads to its maximum extent and begins to contract. Watch in the light of a bench lamp for interference colours (Newton's rings) to appear, indicating that the remaining film of liquid is very thin. Then blow gently until dry. Repeat until there are a sufficient number of cells on the preparation, as judged by phase contrast examination. The preparation can then be stored, dry, for later examination or stained immediately. Lactic-acetic-orcein is preferred for sharpness; toluidine blue for rapidity and intensity.

The slides are then screened for multivalent associations at diakinesis/metaphase I. Not all cells from translocation heterozygotes will show a multivalent association because, as already discussed, the shape and frequency of such associations is dependent upon chiasmata formation. Ring- or chain-of-4 associations or chain-of-3 plus one univalent in at least a proportion of cells provides unequivocal evidence of translocation heterozygosity. Rings- or chains-of-6 indicate translocations involving three chromosomes, and two quadrivalent associations indicate that two separate translocations are present. Robertsonian translocations will be seen as chains-of-3, but mitotic chromosome analysis may be necessary to distinguish that a Robertsonian, as opposed to a reciprocal exchange, is present.

Slight modification of Evans et al.'s (1964) technique is advised for small testes in which meiotic cells may be few or absent and diagnosis may consequently depend upon analysis of spermatogonial mitosis. Squeezing the unfixed tubules with a rubber roller on a ground glass plate (Brewen and Preston 1978) effects greater separation of the germ cells from the tubules than the standard procedure of mincing with forceps, but mild trypsinization can ensure tha satisfactory numbers of spermatogonial cells are freed from the basement membrane. The procedure recommended by Ford and Evans (1969) is to place the tubules at step 2 into 2 ml of 0.25% trypsin solution in phosphate-buffered saline and to pipette gently for 5-6 minutes, then to add 5 ml of tissue culture medium 199 and to spin down at 500 r.p.m. for 5 minutes. The pellet is then resuspended in 1.0% citrate solution as in step 6 and the normal schedule followed. Injection of the mice prior to sacrifice with colcemid (0.01 ml of 0.04% weight/volume solution per gm body weight for 1½ hrs.) can also increase the numbers of spermatogonial metaphases.

An alternative technique that can be recommended for situations in which mitotic as well as meiotic cells from small testes may need to be studied is that developed by Meredith (1969). This procedure is less time-consuming than that of Evans et al. (1964) and can give a better recovery of cell tissue mass, but the quality of the preparation is seldom so high. The procedures involved are as follows:

1. Kill animals and remove testes. Place in 1% trisodium citrate to wash.
2. Remove the tunica; transfer the tubules to fresh 1% citrate.
3. Tease the tubules apart gently to allow penetration of the hypotonic solution.
4. Transfer the tubules to fresh 1% citrate and leave for 8-12 minutes.
5. Remove the tubules from the citrate and remove excess fluid by touching them against filter paper and/or glass plate.
6. Swirl the tubules in 3:1 absolute alcohol/glacial acetic acid fixative contained in a shallow dish. Leave for a few minutes and then transfer to more fixative in a screw-top bottle until slides are required. The material can be stored in the refrigerator for some time without adverse effect.
7. Transfer a few tubules to a Dreyer tube about half full with 60% acetic acid. Leave for a few minutes till the tubules go transparent, tapping the tube to aid the process.
8. When a suspension of cells can be seen, place a slide on a hotplate (at about 40-60°C) and, using a micropipette, draw up suspension from lower part of tube and place small drops of suspension all over the slide. Remove the drops almost immediately by tapping the edge of the slide against a filter paper.
9. Repeat process until slide is covered with cells. Try to avoid creating rings of cells.
10. Stain and examine.

In the absence of meiotic cells, spermatogonial metaphases can be scored for chromosome number and screened for the presence of abnormally large or small rearranged chromosomes. Only the latter provide definitive evidence that a translocation is present. Chromosome counts of 41 rather than the normal 40 indicate that some trisomic condition, e.g., XYY, is responsible for the sterility. A total absence of germ cells most probably can be attributed to an XXY chromosome constitution. Testis size may be expected to be lower than that of translocation steriles although the ranges may overlap (15-25 mg as compared to 20-50 mg).

The cytogenetic preparation techniques and microscopic analysis of slides have generally been regarded as too laborious for use on all F_1 animals. Slide preparation does indeed require about 1½ hrs of work, though preparations for up to four animals can be made concurrently. Slide analysis can be achieved relatively quickly, however. Studies by Léonard (1971) and Lang and Adler (1977) have shown that the translocation multivalent frequency in individual males can range from 2% up to 100%, with the average in the two experiments being 80% and 51% respectively. Using these values, Adler (1978) has calculated the probabilities of detecting a cell carrying a translocation multivalent in a translocation heterozygote for varying numbers of cells scored. The results are shown in Table 15.3. It can be seen that, by analysing only five cells from each F_1 male, the frequency of missing a translocation carrier would be negligible if the multivalent frequency were 80%, but it would be only 3% if the multivalent frequency was 50%. She concluded that 25 cells must be scored in order to reduce to 10% the chance of missing an animal with a multivalent frequency of 10%. The error in classifying a normal animal as a translocation carrier by such cytological analysis is essentially zero since the frequency of translocation multivalents in spermatocytes of normal males is less than 0.01% (Jaszczak 1975; Léonard and Deknudt 1969b 1970; Searle et al. 1969; Murumatsu 1974; Murumatsu et al. 1973).

Because of the speed, low cost, and high efficiency with which most types of translocations, including Robertsonian translocations, can be detected by cytological analysis alone, this procedure should be considered for screening purposes.

Procedures for Detecting Translocations in Females

F_1 females can be screened for translocation semisterility on the basis of litter size, just as males, but again criteria for semisterility have to be established from the reproductive performance of normal females of the same genetic background. Note should also be taken of the lower fertility of female translocation heterozygotes as compared with the male.

The protocol is the same as for males. Females are mated on reaching maturity with normal males and allowed to produce three or possibly four litters. Criteria established by Carter et al. (1955) for Harwell stocks defined suspect semisteriles as those which produced less than 12 progeny in three litters. More detailed criteria, shown in Table 15.4, were later developed by Lyon and Meredith (1966). Searle and Beechey (1974) adopted Carter et al.'s (1955) criterion in their irradiated female experiment but also widened the screening to include females with less than 9 progeny in the first litter, less than 12 in two litters and less than 15 in three litters. Few

Table 15.3 Probabilities of Detecting a Multivalent for Varying Numbers of Cells Scored[a]

Multivalent frequency	Number of cells scored	Probability of detection
80%	5	0.999
50%	5	0.969
	10	0.999
10%	25	0.928
	50	0.995

[a] From Adler (1978).

Table 15.4 Semisterility Test Criteria for Female Mice[a]

Litter order	Size of individual litter	Sum of all litters
1	6	6
2	7	11
3	7	16
4	7	21

[a] From Lyon and Meredith, Autosomal translocations causing male sterility and viable aneuploidy in the mouse, Cytogenetics 5:335-54; 1966. Reprinted with permission of S. Karger AG, Basel and the authors.

translocations were recovered in this experiment, but all were first identified by the original Carter et al. (1955) criterion. Generoso et al. (1978b) has used a litter size of 9 as the minimum for classifying a female as normal in his stocks. This rationale was based on the observation that normal females of this genetic background mated to proven translocation heterozygotes produce 9 or more progeny in only 3% of their litters. By comparison, 59% of litters to normal males were of this size.

Since many factors other than translocation heterozygosity can reduce litter size of individual females, it is essential that all suspect semisteriles be submitted to further testing. Cytological analysis of diakinesis/metaphase I oocytes is possible using the techniques of Edwards (1962) and Tarkowski (1966), but their reliability for females that have already produced three or four litters might be questioned. Progeny testing, i.e., screening F_2 sons for translocations, is therefore favoured. The protocol is identical to that for testing F_1 males.

Simultaneous Screening of Male and Female F_1 Progeny

Since screening of females for semisterility is possible, it has recently been realized that it must be far more efficient to screen male and females simultaneously by mating them *inter se*. This approach is currently being investigated by Adler and Neuhauser (1978) and Generoso et al. (1978b). The same criteria, based on litter size at birth, are used as in the F_1 male tests but scaled according to the reproductive performance of chromosomally normal F_1 females.

The system has the problem that once semisterility is identified in a mating, the parents have to be separated and retested to confirm the presence of a translocation in one (or both) of them. On the other hand there are at least two advantages: (1) twice as many progeny are scored as when only males are tested; (2) production of tester females is not required.

Importance of the Heritable Translocation Test

It is now well established that chromosome anomalies form a substantial part of the human genetic load (Searle 1975). Translocations form an important source of this genetic harm because some of the aneuploid products which result from meiotic segregation and nondisjunction in heterozygotes may survive to late foetal stages before dying or may give rise to severe congenital malformations. Evaluation of the cytogenetic effects of drugs and environmental agents is therefore of great importance.

The heritable translocation test screens for chromosome breakage events, as do several other tests, but it has the following features which make it particularly valuable in mutagenicity testing.

1. It screens for the type of chromosome damage that forms the source of genetic harm in man.

2. The genetic damage scored is transmissible from generation to generation once induced.

3. The test screens for damage in: (a) a mammal, which allows easier extrapolation to man: (b) germ cells, which have a relatively anoxic environment compared with most somatic cells. Genetic damage may therefore be induced in these cells by some compounds that are not mutagenic in other cell systems, and vice versa. Again, this makes for valid extrapolation to man.

4. It provides a finite endpoint which can be employed directly for risk assessment. Because of the low spontaneous occurrence of reciprocal translocations: (a) it allows definitive statistical evaluation; and (b) it can be applied to screen for the effects of low or chronic mutagen exposures.

Point 1 above refers to reciprocal translocation detected through fertility testing. With cytogenetic screening on all F_1 males, Robertsonian translocations, which form a further source of genetic harm in man, would also be detected.

These points summarize the main advantages of the test but they have to be set against a number of disadvantages:

1. The test is time-consuming and laborious whatever screening system is used.

2. It requires large numbers of animals and, particularly when fertility testing is undertaken, extensive animal facilities are needed.

Translocation testing is therefore unquestionably expensive.

Comparison with Other In Vivo Mammalian Tests

Dominant Lethal Test

The dominant lethal test, in common with the heritable translocation test, screens for chromosome breakage events in mammalian germ cells. It has a number of features which give it an advantage over the translocation tests, however:

1. It screens directly for zygotic death in the F_1. No further testing is necessary. Since all kinds of chromosome breakage that cause zygotic deaths are scored, the frequency of the mutational events scored is high and experiment size can therefore be quite small. The dominant lethal test is therefore far quicker and cheaper to operate. It requires far less in the way of animal facilities, therefore overhead costs are also substantially less.

2. It can detect genetic damage in a greater range of germ cell types than can the heritable translocation test. Thus, while both tests can detect chromosome breakage

in postmeiotic cells of the male, dominant lethals can be recovered in high frequencies from spermatocytes and oocytes, whereas significant increases in translocations cannot easily be detected. This has been best demonstrated with the chemical mutagens 6-MP and MC in spermatocytes and with IMS in oocytes. Neither test has great application for screening for genetic damage in spermatogonial stem cells.

3. Since the dominant lethal test can very quickly provide information on germ cell stage sensitivity differences, it can identify the cell stage that can be investigated most usefully with the heritable translocation test.

In summary, therefore, the dominant lethal test is the quicker, cheaper and generally more effective test and therefore of greatest use in "prescreening" for mutagenic potential.

The translocation test, on the other hand, has at least two advantages over the dominant lethal assay:

1. As already discussed, it provides a finite endpoint, not available from dominant lethal data, that can be used directly for risk assessment.
2. When applied to postmeiotic cells, it is the more sensitive test. The difference is due to the low and relatively stable frequency of spontaneously occurring translocations, which allows small increases in frequency following mutagen treatment to be detected. By contrast, the dominant lethal test is handicapped by the high natural wastage of zygotes, which is influenced by factors such as strain, age, parity, and litter size of female, as well as being subject to external environmental influences. Dominant lethal mutation can therefore only be detected when the induced frequency exceeds the spontaneous level; below this level, dominant lethals only replace the spontaneous zygotic losses.

Generoso et al. (1978a) has demonstrated the greater sensitivity of the heritable translocation test with two compounds, TEM and EMS. He found that significant increases in translocations could be detected at doses one-third to one-quarter of those necessary to give clear-cut increases in dominant lethal mutations. Similar results have been obtained both by Generoso et al. (1978b) and Sheu et al. (1978) using an extended treatment protocol with TEM. However, it should be stressed that the greater sensitivity of the heritable translocation test in part reflects only the greater scale of this type of experiment. Increasing the size of a dominant lethal test might be expected to increase its sensitivity up to a point defined by the high spontaneous dominant lethal rate.

In summary, therefore, the heritable translocation test is of greater practical value once the mutagenicity of a compound is established but also provides a more sensitive screening system for postmeiotic cells that can be used with chemicals that have given ambiguous results in a dominant lethal assay.

Specific Locus Mutation Test

The specific locus test screens for point mutations and small deletions detectable in the F_1, though dominant mutations can also, incidentally, be scored. In common with the heritable translocation test, therefore, it screens for transmissible genetic damage induced in germ cells and provides a satisfactory end-point for use in risk assessment.

The specific locus test has at least two advantages over the heritable translocation test:

1. It is a far simpler test to operate. The mutational events can be recognized by direct observation of the F_1 animals at or before weaning age and no further analysis is essential.
2. It can be applied to any germ cell. Since chromosome breakage events need not be involved, cells carrying specific locus mutations are less liable to meiotic selection. Specific locus mutations have been recovered from oocytes treated with TEM and MC; from postmeiotic stages in the male treated with TEM, EMS, and MMS; and from spermatocytes and spermatogonia treated with MS, natulan and, with less effect, TEM (Cattanach 1971; Ehling 1978). Generally, the response correlated well with dominant lethal induction.

The principal disadvantage of the specific locus test is that it scans for mutations at only six or seven loci out of many thousands, so that relatively large numbers of progeny have to be scored. This difficulty is partly offset by the fact that the progeny only have to be observed briefly and then discarded. Hence, when screening for genetic damage in spermatogonia, treated males are allowed to breed with tester females for life. No further manipulations of parents or progeny are necessary. When screening other spermatogenic stages or oocytes, however, many animals have to be exposed to the test compounds in order to accumulate the large numbers of progeny necessary from a selected germ cell stage.

The specific locus test has been most extensively used for screening for genetic damage induced in spermatogonial stem cells by ionizing radiations (Searle 1974b). It has found equivalent use for investigating proven mutagens which cause genetic damage primarily in this germ cell (e.g., MC and natulan, reviewed by Ehling 1978). So far, however, it has been little used for compounds which cause chromosome damage in postmeiotic germ cells, although the results of experiments with TEM, EMS, and MMS (Cattanach 1971; Ehling 1978) suggest that high rates of specific locus mutations may also be recoverable from the most sensitive cell stage (Cattanach 1967a).

Application of the Heritable Translocation Test

Usage

From the foregoing sections, it should be clear that the heritable translocation test has its greatest application only in postmeiotic cells of the male. Beyond this, there is as yet no consensus of opinion as to when the test should be employed. However, it may be used effectively in two distinct situations: (1) When previous testing has indicated that a compound is mutagenic and there are adequate reasons why it should be further investigated. The heritable translocation test can be used to provide evidence on its potential for causing genetic harm in man, i.e., for risk assessment. (2) When preceding dominant lethal tests have given ambiguous results and perhaps other cytogenetic or *Drosophila* tests have indicated mutagenicity then the heritable translocation test can be recommended because of its greater sensitivity.

Methods of Treatment/Dose

If previous testing has demonstrated cell stage sensitivity differences, it is clearly an advantage to apply the heritable translocation test to the most sensitive germ cell stage. Treatment may then consist of a single acute dose. This protocol may be most applicable for drug testing.

However, when germ cell stage sensitivity differences are not known, when previous tests have given ambiguous results, or, if it is not practical to collect enough offspring which include those from the most sensitive stage, then a different treatment regime may be recommended. Repeated or continuous treatments may be given over a period long enough to ensure that all germ cell stages, or perhaps only all postmeiotic germ cell stages, are treated (6-8 weeks and 3 weeks, respectively) and progeny derived from matings within the succeeding 1-2 weeks are then tested. The test chemicals may be mixed with the food or drinking water or injected intraperitoneally daily, as relevant. The dose should be the maximum that can be tolerated throughout the period of treatment. At minimum, additive effects may be expected from each postmeiotic germ cell stage treated (Generoso et al. 1978b; Sheu et al. 1978). This protocol may be appropriate either for drug testing or for investigating environmental agents.

Size of Experiment

No definite recommendations can be made concerning the number of F_1 progeny that should be scored in any experiment for this will depend upon a number of factors. These include: (1) the efficiency of screening, particularly if reliance is placed entirely upon fertility testing; (2) the acceptance of male sterility as a criterion for translocation heterozygosity; (3) the spontaneous rate, whether use is made of an historical control rate or whether concurrent controls are run; (4) the level of effect to be detected; (5) the significance level required; and (6) the probability of detecting any chosen rate of induction.

Estimations of the numbers of F_1 progeny that need to be scored have been made by Generoso et al. 1978 a and b) and Wiemann and Lang (1978). Generoso's calculations are based on the following assumptions: 1) the spontaneous rate is known, and for this he used his pooled control data from his many extensive experiments; 2) false classification of semisteriles by his fertility testing criteria is low (0.02 or less); and 3) sterile as well as semisterile males are translocation heterozygotes. His conclusions are shown in Table 15.5. These indicate the minimum number of translocations which must be observed in a sample of males in order to declare that the observed rate, π, is significantly greater than the spontaneous rate. The probability of falsely claiming significance is at most α. Sample sizes were calculated so that the actual level of significance was close to 0.05 and so that the probability of detecting a given translocation rate (π) was at least 0.95. It can be seen that 400 F_1 progeny is the maximum number that must be scored in order to detect a translocation rate of 1.5% with a probability of at least 95%. Somewhat higher numbers were calculated by Wiemann and Lang (1978) but without the use of an historical control and excluding

Table 15.5 Minimum Number of Male Progeny Needed for Testing to Detect (with a Probability of at Least 0.95) a Translocation Rate of π^c

Number	0^a	α^b	π
300	2	0.031	0.020
400	2	0.052	0.015
700	3	0.026	0.010
1800	5	0.025	0.005
2600	6	0.033	0.004

[a] Minimum number of translocations needed to declare that the observed rate is significantly larger than Generoso et al.'s spontaneous rate of 0.000904.
[b] Actual level of significance for a nominal level of 0.05.
[c] From Generoso et al. (1978a).

male steriles in their scoring. Lower numbers on the order of 100–200 (but with several different dose groups) have been used in a number of experiments (Cattanach 1962; Sheu et al. 1978; Jorgenson et al. 1978a, b and c; Léonard et al. 1975).

The need for and scale of controls may vary in different experiments. In the absence of sound historical control values, a negative control is clearly needed for large scale experiments with a low expected response. A positive control can be coded into such experiments but would not need to be large scale. However, for experiments with a high expected response, both positive and negative controls are less necessary.

Costing

Costing the heritable translocation test is an extremely difficult task since no precise format can be laid down and many variables are likely to occur. These will include the dominant lethal rate which will specify the size of the F_1 production component of the test, the magnitude of the response, the size of the test itself, whether both positive and negative controls are run concurrently, and the variation in the methods of F_1 testing which may range from thorough fertility testing with little or no cytological analysis to the opposite extreme. To illustrate some of these points and indicate how the test might be costed, it is to some advantage to use facts and figures from an actual experiment. Jorgenson et al.'s (1975) extensive study with nitrilotriacetic acid (NTA) has been selected for this purpose. This employed both positive (TEM) and negative controls, and the published report indicates the numbers of F_1 of the fertility test procedure employed. The following exercise attempts to cost this experiment as it might be done at Harwell. Procedures and hence cost may therefore be considerably different from those actually encountered.

Costing can be presented usefully in terms of manpower and cage space occupied per unit time. In a small laboratory, greatest economy can be achieved by splitting the experiment into small units and setting these off at regular intervals. This avoids wastage of animals and utilizes manpower more effectively. Productivity/cage is decidedly higher if trios of two females to one male are used rather than pairs. Under these conditions production/cage can be roughly estimated by $N = n(t-1)$, where N is the total number of weaned offspring/female, n is the mean litter size of the stock and is the number of months (up to nine) a cage is occupied. At Harwell, the C3H × 101 F_1 hybrid mouse might be employed with an expected production/female of about seven offspring/month. Production time from mating to weaning will be taken as two months and a further month will be allotted for maturing animals to a mating age.

Jorgenson et al.'s (1975) experiment utilized 400 test-series F_1 males, 400 untreated controls, and 100 positive (TEM) controls, for a total of 900 males. To produce these, 100 males were treated with NTA for 7 weeks. An untreated 100 produced the control group and 40 TEM treated males produced the positive control. Each male was placed with two females and the F_1 male progeny were collected and subjected to test. Evaluation of the fertility status of the F_1 males was based on openings of two females/male. Suspect translocation heterozygotes (27 NTA, 35 control, and 23 TEM) were tested again with the result that only 8 NTA, 8 control, and 20 TEM derived F_1 males remained suspects. These 36 animals were then subjected to cytological analysis.

Minimum costing could therefore be as follows:

	cage-month
1. To produce original 240 males for treatment and 480 females to weaning age: 80 cages for 2 months	160
2. To mature these and subject males to treatment, grouped as 5/cage	400
3. To produce 900 plus F_1 progeny from 240 trios to weaning age	480
4. To mature F_1	200
5. To produce 1800 tester females to weaning age	560
6. To mature tester females	380
7. To test 900 F_1 males, say 1 month	900
8. To retest 85 F_1 males	85
Total	3165

Estimating the manpower required for such an experiment is difficult, since this will vary from laboratory to laboratory according to animal facilities, deployment, and quality of labor, etc. and is also dependent upon the amount of cytogenetic analysis. At Harwell, one animal technician is responsible for feeding and changing 400 cages, clean cages being provided by maintenance staff. With assistance from a scientific assistant, he is also responsible for setting up matings, litter recording, etc. The scientific assistant might also undertake all cytogenetic analyses and, in total, such duties might take up about one-quarter of his time. Scientific responsibilities would include planning, supervision, and perhaps also cytogenetic analyses.

On this basis, the manpower required for the heritable translocation tests is little different from the specific locus tests but an ability to carry out cytogenetic work is also needed. Costing of cage space for 3165 cage-months (minimum) is about half that of Searle's (1977) estimate for a large-scale specific locus test for spermatogonial mutations (5500 cage-months). Reductions in the figure for translocation testing could be possible if fertility status were established by litter size rather than on dissections at mid-pregnancy. Each female might be able to test several males as Generoso has indicated, but this entails considerable juggling of mice between cages and detailed record keeping.

An alternative approach would be to analyse cytogenetically all F_1 males. This removes costing steps 5-8 but entails extensive laboratory work. However, if the experiment is spread out over a period of time, this need not require numerous suitably qualified staff. One skilled technician or scientific assistant should be able to make cytological preparations from at least eight animals and complete analysis of up to 25 cells from each in the course of a normal working day.

Summary of Results with Chemical Agents

Table 15.6 summarizes the results of mutagenicity experiments with chemicals employing the heritable translocation test. The data have been broken down to show the germ cell stages tested, but because the range of doses, routes of administration, duration of treatment, etc., vary so much between experiments, details of positive results are not given. The data obtained for weakly positive, ambiguous, and negative results are shown to allow some degree of evaluation.

It can be seen that almost all the clearly positive results derive from experiments with powerful alkylating agents applied to postmeiotic germ cell stages of the male. The several exceptions require comment. Positive results were obtained following TEM-treatment of spermatocytes (Generoso et al. 1977). There is little to question about this result for the 274 F_1 tested were conceived 25-28 days posttreatment and 10 proved to be semisterile and 3 sterile. Generoso et al. (1978a) also found 2 semisteriles among 1633 F_1 progeny derived from treated spermatogonia. This response is not significantly greater than his historical control (4/4423) but, together with Cattanach and Pollard's (1971) result with the spermatocyte test, may indicate a weakly positive result. Greater doubt must be given to the extremely high response obtained by Sram et al. (1970) with TEPA-treatment of spermatogonia. In view of the fact that their findings conflict so entirely with other investigations and the response was even higher than they obtained in the same experiment from postmeiotic stages, it seems probable that some technical error, such as the existence of a preexisting translocation in the stock, may be responsible. Sotomayor and Cumming's (1975) observation of two translocations in 185 F_1 from cyclophosphamide-treated spermatocytes may be taken as a weakly positive result and supports the TEM data in showing that translocations can be recovered from spermatocytes following treatment with chemical mutagens. A positive result is also claimed for MC-treatment of spermatocytes (Adler and Neuhäuser 1978), but clearly more data are required.

The test on the food additives and contaminants, pesticides, and environmental pollutants generally appear to have given negative results. Possible exceptions are Knudsen and Meyer's (1977) study with feeding of nitrite-treated salt pork and Jorgenson et al.'s (1978c) study with Captan. However, since only one translocation was detected in each experiment further investigation is warranted.

Conclusions and Discussion

The heritable translocation test potentially is a highly sensitive method for detecting genetic damage, even though it requires large numbers of animals and extensive animal facilities. It screens for transmissible damage that can be induced in mammalian germ cells and provides a finite end-point for risk assessment. However, its application is undoubtedly limited to postmeiotic germ cells. Translocations induced in spermatogonia can be more efficiently detected by the spermatocyte test.

The test is considerably more costly and time-consuming than the dominant lethal method and may therefore only be recommended when dominant lethal or other mutagenicity tests indicate its use is warranted. It appears to compare favourably with the specific locus test, as far as cost is concerned, but is more cumbersome to operate because of the need for F_1 testing. However, this direct comparison is not entirely justified because the two tests are normally applied to different spermatogenic stages. Were the specific locus test applied to postmeiotic cells, like the heritable translocation test, animal and labor costs would be higher because of the need to produce large numbers of progeny from only a limited range of germ cell stages. On the other hand, large scale experiments might be less necessary since the sensitivity of postmeiotic stages to specific locus mutation-induction could be extremely high.

A basic problem of the heritable translocation test is that the mutational event scored has a diverse range of properties and characteristics and, at this level, several questions regarding the screening techniques have yet to be resolved. Should, for instance, screening techniques be adapted to detect as wide a range of translocations as possible or should it be accepted that they should detect only certain translocations with certain properties? Thus, how should male sterile F_1 animals be scored? It is known that a high proportion of male steriles which derive from mutagen-treated postmeiotic germ cells carry translocations: some may carry one translocation; some may carry two. But how should male steriles that arise in low response or control groups, or those that derive from treated premeiotic stages be scored? Male sterility can arise from nondisjunctional events, genic changes, or a variety of environmental causes, such as injury, and the lower the translocation rate the greater is the risk that these rare events may cause scoring errors.

Cytological investigation using Meredith's (1969) technique or a suitably modified version of Evans et al.'s (1964) method is clearly warranted for male steriles, especially for those that derive from low response or control groups. Evidence of spermatogenic impairment should also be sought. Reduced testis weight immediately indicates a disturbance of spermatogenesis; meiotic analysis will demonstrate whether or not translocations are present; and, if meiotic cells are absent, analysis of spermatogonial mitoses will distinguish if trisomy is involved or, possibly, provide obvious karyotypic evidence of a translocation. An absence of germ cells provides presumptive evidence of XXY or related chromosome constitutions. At the present state of knowledge, it would seem justified to do full karyotype analysis of all male-steriles deriving from low response or control groups when no overt reason for the sterility can be found. The opposing view, that male-steriles be ignored (Wiemann and Lang 1978) would seem totally unwarranted, especially if the distinction from translocation male semisterility should be influenced by genetic background differences.

The same types of questions require resolving for the semisterile animals. As has been discussed, the reduction in fertility caused by different translocations and different types of translocation varies. Therefore, should fertility testing procedures be so rigourous that only those translocations which cause evident semisterility be scored, and is cytogenetic analysis then necessary? Or, should fertility testing be accepted only as a prescreening procedure to eliminate obvious normals and greater

Table 15.6 Summary of Results with Chemical Agents

Compound	Stage treated	Response	Reference
TEM	Postmeiotic	+	Cattanach (1957, 1959)
		+	Generoso et al. (1978a)
		+	Sheu et al. (1978)
		+	Staub and Matter (1977)
	Spermatocytes	+	Generoso et al. (1977)
	Spermatogonia	2/1633	Generoso et al. (1978a)
	All stages	+	Pecevski et al. (1978)
Trenimon	Postmeiotic	+	Datta et al. (1970)
	Postmeiotic	+	Epstein et al. (1971)
		+?	Sram et al. (1970)
TEPA	Spermatocytes	0/59	Epstein et al. (1971)
	Spermatogonia	0/40	Epstein et al. (1971)
		0/1031	Generoso et al. (1978a)
		18/36?	Sram et al. (1970)
Thio-TEPA	Postmeiotic	+	Malashenko and Surkova (1974)
			Malashenko et al. (1978)
			Semenov and Malashenko (1977)
Nitrogen Mustard	Postmeiotic	+	Falconer et al. (1952)
EMS	Postmeiotic	+	Cattanach et al. (1968)
		+	Generoso et al. (1974)
MMS	Postmeiotic	+	Jackson et al. (1964)
		+	Lang and Adler (1977)
		+	Léonard (1973)
IMS	Oocytes	1/1464	Generoso et al. (1978a)
Cyclophosphamide	Postmeiotic	+	Datta et al. (1970)
		+	Generoso et al. (1978b)
		+	Sotomayor and Cumming (1975)
	Spermatocytes	2/185	Sotomayor and Cumming (1975)
	Spermatogonia	1/1148	Generoso et al. (1978a)
		0/202	Sotomayor and Cumming (1975)

emphasis be given to cytogenetic analysis? Some cytogenetic analysis seems essential; it eliminates all risk of scoring false positivies which could create serious error in low response or control groups and identifies F_1 animals carrying more than one translocation, a phenomenon that may be expected in high response groups.

There remains the further option of screening all F_1 animals cytogenetically without recourse to any fertility testing. Would such procedure be feasible for large scale testing? It probably is. In a cytogenetics laboratory geared to this sort of work, two skilled technicians should easily be able to score 100 animals/week and the microscope work would not be any more laborious or tedious than that of other routine cytological studies. This approach would also offer a number of advantages

Table 15.6 Continued

Compound	Stage treated	Response	Reference
MC	Spermatocytes	1?/362	Adler and Neuhauser (1978)
6-Mercaptopurine	Spermatocytes	1/615	Generoso et al. (1978a)
Caffeine	All stages	0/201	Cattanach (1962)
Nitrilotriacetic acid	All stages	0/400	Jorgenson et al. (1975)
bis(2-chloroethyl) and bis(2-chloroisopropyl)ethers	All stages	0/300	Jorgenson et al. (1978b)
Manganese sulphate Sodium acid pyrophosphate Sodium erthorbate	All stages	0/100 0/100 0/100	Jorgenson et al. (1978a)
N-methyl-N'-nitro-N-guanidine	All stages	0/200	Jorgenson et al. (1978c)
Nitrite-treated salt pork	All stages	1/50	Knudsen and Meyer (1977)
Dimethylnitrosamine	All stages	0/56	Savkovic et al. (1978)
Afloxtin B1		0/185	Léonard et al. (1975)
Streptomycin	Not specified	Zero indicated	Savkovic et al. (1974)
Sodium bisulphite	All stages	0/858	Generoso et al. (1978c)
Captan	All stages	1/200	Jorgenson et al. (1978d)
Mebendazole	Postmeiotic	0/30	Léonard et al. (1974)
Cadmium	Not specified	Zero indicated	Léonard (1973)

over current protocol: it would have the potential for detecting a greater range of reciprocal translocations; insertions might well be recoverable; and Robertsonian translocations could certainly also be found. Without the need for extensive animal facilities for fertility testing, laboratories with more limited animal housing and which routinely carry out other cytogenicity assays could adopt the method. There could be scope for automation of slide processing and perhaps also of slide screening. In view of the genetic harm to man caused by chromosome rearrangements other than the reciprocal translocations which can be identified in the mouse by fertility tests, the possibility of widening the scope of the heritable translocation test by the use of full cytogenetic investigation of all F_1 animals merits attention.

Acknowledgments

I would like to thank Dr. M. F. Lyon F.R.S. and Dr. A. G. Searle for critically reading the manuscript and Dr. J. Wassom (Biol. Div. Oak Ridge) for producing an E.M.I.C. print out of references on the heritable translocation test in rodents.

Literature Cited

Adler, I. D. 1973. Cytogenetic effects of mitomycin C on mouse spermatogonia. Mutat. Res. 21: 20–21.
Adler, I. D. 1978. The cytogenetic heritable translocation test. Bio. Zbl. 97: 441–51.
Adler, I. D.; Neihäuser, A. 1978. Heritable translocations induced in mouse primary spermatocytes. Mutat. Res. 53: 143–44.
de Boer, P. 1973. Fertile tertiary trisomy in the mouse (*Mus musculus*). Cytogenet. Cell Genet. 12: 435–42.
Brewen, J. G.; Payne, H. S.; Jones, K. P.; Preston, R. J. 1975. Studies on chemically induced dominant lethality. I. The cytogenetic basis of MMS-induced dominant lethality in postmeiotic male germ cells. Mutat. Res. 33: 239–50.
Brewen, J. G.; Preston, R. J. 1978. Chromosome aberration analysis in mammalian germ cells. *In*, Hollander, A.; de Serres, F.J. eds. Chemicals mutagens: principles and methods for their detection. Vol. 5. 127–60. New York: Plenum Press.
Bridges, C. B. 1923. The translocation of a section of chromosome II upon chromosome III in *Drosphila*. Anat. Rec. 24: 426–27.
Cacheiro, N. L. A. 1971. Cytological studies of sterility in sons of mice treated with mutagens. Genetics 68: 8–9.
Cacheiro, N. L. A.; Russell, L. B. 1974. Possibility that *Sl* is in chromosome 10. Mouse News Letter 50: 52.
Cacheiro, N. L. A.; Russell, L. E.; Swartout, M. S. 1974. Translocations. The predominant cause of total sterility in sons of mice treated with mutagens. Genetics 76: 73–91.
Cacheiro, N. L. A.; Swartout, M. S. 1975. Cytological studies of sterility in sons of mice treated at spermatogonial or oocyte stages with mutagenic chemicals. Genetics 80: S18.
Carter, T. C.; Lyon, M. F.; Phillips, R. J. S. 1955. Gene-tagged chromosome translocations in eleven stocks of mice. J. of Genet. 53:154–66.
Carter, T. C.; Lyon, M. F.; Phillips, R. J. S. 1956. Further genetic studies of eleven translocations in the mouse. J. Genetics 54: 462–73.
Cattanach, B. M. 1957. The induction of translocations in male mice by triethylenemelamine. Nat. Lond. 180: 1364–66.
Cattanach, B. M. 1959. The sensitivity of the mouse testis to the mutagenic action of triethylenemelamine. Z. Verbungsl. 90: 1–6.
Cattanach, B. M. 1961. A chemically induced variegated-type position effect in the mouse. Z. Verbungsl. 92: 165–82.
Cattanach, B. M. 1962. Genetical effects of caffeine in mice. Z. Verbungsl. 93: 215–19.
Cattanach, B. M. 1965. Snaker: a dominant abnormality caused by chromosomal imbalance. Z. Verbungsl. 96: 275–84.
Cattanach, B. M. 1967a. Induction of paternal sex-chromosome losses and deletions and of autosomal gene mutations by the treatment of mouse postmeiotic germ cells with triethylenemelamine. Mutat. Res. 4: 73–82.
Cattanach, B. M. 1967b. A test of distributive pairing between two specific nonhomologous chromosomes in the mouse. Cytogenetics 6: 67–77.
Cattanach, B. M. 1971. Specific locus mutations in mice. *In* Hollaender, A., ed. Chemical mutagens: principles and methods for their detection. Vol. 2. New York: Plenum Press; 535–39.
Cattanach, B. M. 1975. Comparison of the mutagenic effect of chemicals and ionizing radiation in the spermatogenic cells of the mouse. *In* Nyfaard, O. F.; Adler, H. I.; Sinclair,

W. K., eds. Radiation Research: biomedical, chemical, and physical perspectives. London: Academic Press Inc.; 984-92.

Cattanach, B. M.; Crocker, A. J. M. 1977. New Robertsonian translocations. Mouse News Letter 57: 16.

Cattanach, B. M.; Moseley, H. 1973. Nondisjunction and reduced fertility caused by the tobacco mouse metacentric chromosomes. Cytogenet. Cell Genet. 12: 264-87.

Cattanach, B. M.; Murray, I.; Bigger, T. R. L. 1977. New Robertsonian translocations. Mouse News Letter 56: 37.

Cattanach, B. M.; Pollard, C. E. 1971. A search for chromosomal aberrations induced in mouse spermatogonia by chemical mutagens. Mutat. Res. 13: 371-75.

Cattanach, B. M.; Pollard, C. E.; Hawkes, S. G. 1971. Sex-reversed mice: XX and XO males. Cytogenetics 10: 318-37.

Cattanach, B. M.; Pollard, C. E.; Isaacson, J. H. 1968. Ethylmethanesulfonate-induced chromosome breakage in the mouse. Mutat. Res. 6: 297-307.

Cattanach, B. M.; Savage, J. R. K. 1976. A new Robertsonian translocation. Mouse News Letter 54: 38.

Cumming, R. B.; Walton, M. 1971. Genetic effects of cyclophosphamide in the germ cells of male mice. Genetics 68: S14.

Davisson, M. T.; Roderick, T. H. 1978. Robertsonian chromosomes. Mouse News Letter 58: 48.

Edwards, R. G. 1962. Meiosis in ovarian oocytes of adult mammals. Nature 196: 446-50.

Ehling, U. H. 1978. Specific-locus mutations in mice. In Hollaender, A.; de Serres, F., eds. Mutagens: principles and methods for their detection. Vol. 5 New York: Plenum Press; 233-56.

Ehling, U. H.; Cumming, R. B.; Malling, H. V. 1968. Induction of dominant lethal mutations by alkylating agents in male mice. Mutat. Res. 5: 417-28.

Ehling, U. H.; Doherty, D. G.; Malling, H. V. 1972. Differential spermatogenic response of mice to the induction of dominant-lethal mutations by n-propyl methanesulphonate and iso-propyl methanesulphonate. Mutat. Res. 15: 175-84.

Eicher, E. M. 1970. X-autosome translocations in the mouse: total inactivation versus partial inactivation of the X chromosome. Adv. in Genetics 15: 175-259.

Epstein, S. S.; Bass, W.; Arnold, E.; Bishop, Y.; Joshi, S.; Adler, I. D. 1971. Sterility and semisterility in male progeny of male mice treated with the chemical mutagen TEPA. Toxicol. Appl. Pharmacol. 19: 134-46.

Evans, E. P.; Breckon, G.; Ford, C. E. 1964. An air-drying method for meiotic preparations from mammalian testes. Cytogenetics 3: 289-94.

Evans, E. P.; Phillips, R. J. S. 1978. A phenotypically marked inversion (In(2)2H). Mouse News Letter 58: 44-45.

Falconer, D. S.; Slizynski, B. M.; Auerbach, C. 1952. Genetical effects of nitrogen mustard in the house mouse. J. Genetics 51: 81-88.

Ford, C. E. 1964. Autosomal abnormalities. In Second international conference on congenital malformations, New York City, July 14-19, 1963: 22-31. International Medical Congress, New York.

Ford, C. E. 1969. Meiosis in mammals. In Benirschke, K., ed. Comparative mammalian cytogenetics. New York: Springer-Verlag; 91-106.

Ford, C. E. 1975. The time in development at which gross genome unbalance is expressed. In Balls and Wild, eds. The early development of mammals. Cambridge: Cambridge University Press.

Ford, C. E.; Clegg, H. M. 1969. Reciprocal translocations. Br. Med. Bull. 25: 110-14.

Ford, C. E.; Evans, E. P. 1969. Meiotic preparations from mammalian testes. In Benirschke, K., ed. Comparative mammalian cytogenetics. New York: Springer-Verlag; 461-64.

Ford, C. E.; Evans, E. P.; Searle, A. G. 1977. Failure of irradiation to induce Robertsonian translocations in germ cells of male mice. ZLM conference on mutations: their origin and nature and potential relevance to genetic risk in man. Freiburg, Breisgau.

Ford, C. E.; Searle, A. G.; Evans, E. P.; West, B. J. 1969. Differential transmission of translocations induced in spermatogonia of mice by irradiation. Cytogenetics 8: 447-70.

Forejt, J. 1974. Nonrandom association between a specific autosome and the X chromosome in meiosis of the male mouse: possible consequence of homologous centromeres separation. Cytogenet. Cell Genet. 13: 369-83.

Forejt, J.; Gregorova, S. 1977. Meiotic studies of translocations causing male sterility in the mouse. Cytogenet. Cell Genet. 19: 159-79.

Generoso, W. M.; Cain, K. T.; Huff, S. W. 1978a. Inducibility by chemical mutagens of heritable translocations in male and female germ cells. In Flamm, W. Gary; Mehlman, M. A., eds. Advances in Modern Toxicology. Vol. 5 Washington, D.C.: Hemipheres Publ. Co.; 101-29.

Generoso, W. M.; Cain, K. T.; Huff, S. W.; Gosslee, D. G. 1978b. Heritable translocation test. In Hollaender, A., ed. Chemical mutagens: principles and methods for their detection. Vol. 5. New York: Plenum Press; 55-77.

Generoso, W. M.; Huff, S. W.; Cain, K. T. 1975. Comparative inducibility by 6-mercaptopurine of dominant lethal mutations and heritable translocations in early meiotic male germ cells and differentiating spermatogonia of mice. Mutat. Res. 31: 341-42.

Generoso, W. M.; Huff, S. W.; and Cain, K. T. 1978. Tests on induction of chromosome aberrations in mouse germ cells with sodium bisulfite. Mutat. Res. 56: 363-65.

Generoso, W. M.; Huff, S. W.; Stout, S. K. 1971. Chemically induced dominant lethal mutations and cell-killing in mouse oocytes in the advanced stages of follicular development. Mutat. Res. 11: 411-20.

Generoso, W. M.; Krishna, N.; Sotomayor, R. E.; Cachiero, N. L. A. 1977. Delayed formation of chromosome aberrations in mouse pachytene spermatocytes treated with triethylenemelamine (TEM). Genetics 85: 65-72.

Generoso, W. M.; Russell, W. L.; Huff, S. W.; Stout, S. K.; Gosslee, D. G. 1974. Effects of dose on induction of dominant lethal mutations and heritable translocations with ethyl methanesulfonate in male mice. Genetics 77: 741-52.

Gilliavod, N.; Léonard, A. 1973. Sensitivity of the mouse oocyte to the induction of translocations by ionizing radiations. Can. J. Genet. Cytol. 15: 363-66.

Griffen, A. B. 1967. A case of tertiary trisomy in the mouse, and its implications for the cytological classification of trisomics in other mammals. Can. J. Genet. Cytol. 9: 503-10.

Griffen, A. B.; Bunker, M. C. 1967. Four further cases of autosomal primary trisomy in the mouse. Proc. Nat. Acad. Sci. USA 58: 1446-52.

Gropp, A.; Kolbus, U.; Giers, D. 1975. Systematic approach to the study of trisomy in the mouse. Cytogenet. Cell Genet. 14: 42-62.

Hertwig, P. 1935. Sterilitätserscheinungen bei röntgenbestrahlten Mäusen. Z. Indukt. Abstamm u. Vererbsl. 70: 517-23.

Hertwig, P. 1938. Unterschiede in der Entwicklungsfähigkeit von F_1 Mäusen nach Röntgen-Bestrahlung von spermatogonien, fertigen und unfertigen spermatozoen. Biol. Zbl. 58: 273-301.

Hertwig, P. 1940. Vererbbare semisterilitat bei Mäusen nach Röntgen-Bestrahlung verursacht durch reziproke chromosomentranslokationen.

Hollander, W. F.; Waggie, K. S. 1977. Gnome and other effects of a small translocation in the mouse. J. Hered. 68: 41-47.

Jackson, H.; Partington, M.; Walpole, A. L. 1964. Production of heritable partial sterility in the mouse by methyl methanesulfphonate. Brit. J. Pharmacol. 23: 521-28.

Jaszczak, K. 1975. Influence of parental age on the genetic quality of the progeny of mice and level of spontaneous and induced chromosome translocations in the male germ cells. Genet. Pol. 16: 109-15.

John, B.; Freeman, M. 1975. Causes and consequences of Robertsonian exchange. Chromosoma 52: 123-36.

Jorgenson, T. A.; Newell, G. W.; Scharpf, L. G.; Gribling, P.; O'Brien, M.; Chu, D. 1975. Study of the mutagenic potential of nitrilotriacetic acid (NACANTA) in mice by the translocation test. Mutat. Res. 31: 337-38.

Jorgenson, T. A.; Rusherook, C. J.; Newell, G. W.; Green, S. 1978a. Study of the mutagenic potential of three GRAS chemicals in mice by the heritable translocation test. Mutat. Res. 53: 124-25.

Jorgenson, T. A.; Rusherook, C. J.; Newell, G. W.; Tardiff, R. G. 1978b. Study of the mutagenic potential of bis(2-chloroethyl) and bis(2-chloroisopropyl) ethers in mice by the heritable translocation test. Mutat. Res. 53: 124.

Jorgenson, T. A.; Rusherook, C. J.; Newell, G. W.; Green, S. 1978c. A study of the mutagenic potential of N-methyl-N'-nitro-N-nitrosoguanidine in mice by the translocation test. Mutat. Res. 53: 205–6.
Jorgenson, T. A.; Rusherook, C. J.; Newell, G. W.; Waters, M. 1978d. Study of the mutagenic potential of Captan by the heritable translocation test in mice. Mutat. Res. 53: 206.
Knudsen, I.; Meyer, O. A. 1977. Mutagenicity studies on rats and mice given canned, heated, nitrite-treated pork. Mutat. Res. 56: 177–84.
Koller, P. C. 1944. Segmental interchange in mice. Genetics 29: 247–63.
Koller, P. C.; Auerbach, C. 1941. Chromosome breakage and sterility in the mouse. Nature 148: 501–2.
Krishna, M.; Generoso, W. M. 1977. X-ray induction of heritable translocations in mouse dictyate oocytes. Genetics 86: 36–37.
Lang, R.; Adler, I. D. 1977. Heritable translocation test and dominant-lethal assay in male mice with methyl methanesulfonate. Mutat. Res. 48: 75–88.
Lengerova, A. 1970. Incorporation of the T6 chromosome marker into C57BL/10 strain. Mouse News Letter 42: 35–36.
Léonard, A. 1971. Radiation-induced translocations in spermatogonia of mice. Mutat. Res. 11: 71–81.
Léonard, A. 1973. Observations on meiotic chromosomes of the male mouse as a test of the potential mutagenicity of chemicals in mammals. In Hollaender, A., ed. Chemical mutagens: Principles and methods for their detection. Vol. 3. New York: Plenum Press; 21–56.
Léonard, A. 1976. Heritable chromosome aberrations in mammals after exposure to chemicals. Rad. and Environm. Biophys. 13: 1–8.
Léonard, A.; Deknudt, G. 1968. The sensitivity of various germ cell stages of the male mouse to radiation induced translocations. Can. J. Genet. Cytol. 10: 495–507.
Léonard, A.; Deknudt, G. 1969a. Étude cytologique d'une translocation chromosomes Y-autosome chez la souris. Experientia 25: 876–77.
Léonard, A.; Deknudt, G. 1969b. Dose-response relationship for translocations induced by X-irradiation in spermatogonia of mice. Rad. Res. 40: 276–84.
Léonard, A.; Deknudt, G. 1970. Persistence of chromosome rearrangements induced in male mice by X-irradiation of premeiotic germ cells. Mutat. Res. 9: 127–33.
Léonard, A.; Deknudt, G.; Linden, G. 1971. Failure to detect meiotic chromosome rearrangement in male mice given chemical mutagens. Mutat. Res. 13: 89–92.
Léonard, A.; Deknudt, G.; Linden, G. 1975. Mutagenicity tests with aflatoxins in the mouse. Mutat. Res. 28: 137–39.
Léonard, A.; Vandesteene, R.; Marseoom, R. 1974. Mutagenicity tests with meendazole in the mouse. Mutat. Res. 26: 427–30.
Luning, K. G.; Searle, A. G. 1971. Estimates of the genetic risks from ionizing radiation. Mutat. Res. 12: 291–304.
Lyon, M. F.; Mason, I. M.; Bigger, T. R. L. 1977. New Robertsonian translocations. Mouse News Letter 56: 37.
Lyon, M. F.; Meredith, R. 1966. Autosomal translocations causing male sterility and viable aneuploidy in the mouse. Cytogenetics 5: 335–54.
Malashenko, A. M. 1968. Chemical mutagenesis in laboratory mammals. Sov. Genet. 4: 538–43.
Malashenko, A. M.; Semenov, K. K.; Selenzneva, G. P.; Surkova, N. I. 1978. Studies of mutagenic effect of chemical compounds in laboratory mice. Genetika 14: 52–61.
Malashenko, A. M.; Surkova, N. I. 1974. Mutagenic effect of thio-TEPA in laboratory mice. I. Chromosome aberrations in somatic and germ cells of male mice. Sov. Genet. 10: 51–58.
Meredith, R. 1969. A simple method for preparing meiotic chromosomes from mammalian testes. Chromosoma 26: 254–58.
Miklos, G. L. G. 1974. Sex-chromosome pairing and male fertility. Cytogenet. Cell Genet. 13: 558–77.
Muramatsu, S. 1974. Frequency of spontaneous translocations in mouse spermatogonia. Mutat. Res. 24: 21–32.

Muramatsu, S.; Nakamura, W.; Eto, H. 1973. Relative biological effectiveness of X-rays and fast neutrons in inducing translocations in mouse spermatogonia. Mutat. Res. 19: 343–47.

Oakberg, E. F. 1957. Duration of spermatogenesis in the mouse. Nature 180: 1137–39.

Ohno, S.; Cattanach, B. M. 1962. Cytological study of an X-autosome translocation in *Mus musculus*. Cytogenetics 1: 129–40.

Pecevski, J.; Maric, N.; Savkovic, N.; Radivojevic, D.; Green, S. 1978. An analysis of meiotic chromosomes of inbred male mice and their F_1 sons after long-term treatment of sires with triethylenemelamine. Mutat. Res. 54: 55–60.

Phillips, R. J. S. 1966. A cis-trans position effect at the A locus of the house mouse. Genetics 54: 485–95.

Phillips, R. J. S.; Kaufman, M. H. 1974. Bare-patches, a new sex-linked gene in the mouse, associated with a high production of XO females. Genet. Res. 24: 27–41.

Phillips, R. J. S.; Savage, J. R. K. 1976. A new Robertsonian translocation. Mouse News Letter 55: 14.

Russell, L. B. 1962. Chromosome aberrations in experimental mammals. Prog. Med. Genet. 2: 230–94.

Russell, L. B.; Wickham, L. 1957. The incidence of disturbed fertility among male mice conceived at various intervals after irradiation of the mother. Genetics 42: 392.

Russell, W. L. 1954. Genetic effects of radiation of mammals. In Radiation Biology I. New York: McGraw-Hill; 825–59.

Savkovic, N.; Maric, N.; Pecevski, J.; Radivojevic, D.; Green, S. 1978. Effect of dimethylnitrosamine on the induction of chromosomal aberrations in male mice and their F_1 offspring. Toxicol. Lett. 1: 179–82.

Savkovic, N.; Pecevski, J.; Djelineo, A. 1975. Cytogenetic analysis of meiotic chromosomes of irradiated mice and their progeny after treatment with streptomycina and dihydrodeoxystreptomycin. Mutat. Res. 29: 206.

Searle, A. G. 1974a. Nature and consequences of induced chromosome damage in mammals. Genetics 78: 173–86.

Searle, A. G. 1974b. Mutation induction in mice. Adv. in Rad. Biol. 4: 131–207.

Searle, A. G. 1975. Radiation-induced chromosome damage and assessment of genetic risk. In Emery, A. E. H., ed. Modern trends in human genetics. Vol. 2. London & Boston: Butterworths; 83–110.

Searle, A. G. 1977. The specific locus test in the mouse. In Kilbey, B. J.; Legator, M.; Nichols, W.; Ramel, C., eds. Handbook of mutagenicity test procedures. Oxford: Elsevier Scientific Publishing Co.; 311–24.

Searle, A. G.; Beechey, C. V. 1974. Cytogenetic effects of X-rays and fission neutrons in female mice. Mutat. Res. 24: 171–86.

Searle, A. G.; Beechey, C. V.; Evans, E. P. 1978. Meiotic effects in chromosomally derived male sterility of mice. Ann. Biol. Anim. Bioch. Biophys. 18: 391–98.

Searle, A. G.; Evans, E. P.; West, B. J. 1969. Studies on the induction of translocations in mouse spermatogonia. II. Effects of fast neutron-irradiation. Mutat. Res. 7: 235–40.

Searle, A. G.; Ford, C. E.; Evans, E. P.; Beechey, C. V.; Burtenshaw, M. D.; Clegg, H. M. 1974. The induction of translocations in mouse spermatozoa. I. Kinetics of dose response with acute X-irradiation. Mutat. Res. 22: 157–74.

Semenov, K. K.; Malashenko, A. M. 1977. Search of translocation heterozygous female mice among the progeny of males treated with a chemical mutagen (thio-TEPA). Tsitol. Genet. 11: 454–56.

Sheu, C. W.; Moreland, F. M.; Oswald, E. J.; Green, S.; Flamm, W. G. 1978. Heritable translocation test on random-bred mice after prolonged triethylenemelamine treatment. Mutat. Res. 50: 241–50.

Slizynski, B. M. 1957a. Cytological analysis of translocations in the mouse. J. Genetics 55: 122–30.

Slizynski, B. M. 1957b. Chromosomal mechanism in translocation infertility. Proc. Roy. Soc. Phys. Soc. Edinburgh 26: 49–60.

Snell, G. D. 1933. X-ray sterility in the house mouse. J. Exp. Zool. 65: 421–41.

Snell, G. D. 1935. The induction of X-rays of hereditary changes in mice. Genetics 20: 545–67.

Snell, G. D. 1946. An Analysis of Translocations in the Mouse. Genetics 31: 157–80.

Snell, G. D.; Bodemann, E.; Hollander, W. 1934. A translocation in the house mouse and its effect on development. J. Exp. Zool. 67: 93–104.

Sotomayor, R. E.; Cumming, R. B. 1975. Induction of translocations by cyclophosphamide in different germ cell stages of male mice: cytological characterization and transmission. Mutat. Res. 27: 375–88.

Sram, R. J.; Zudova, Z.; Benes, V. 1970. Induction of translocations in mice by TEPA. Folia Biol. (Prague) 16: 367–68.

Staub, J. E.; Matter, B. E. 1976, 1977. Heritable reciprocal translocations and sperm abnormalities in the F_1 offspring of male mice treated with triethylenemelamine (TEM). Archiv. für Genetik 49/50: 29–41.

Tarkowski, A. K. 1966. An air-drying method for chromosome preparations from mouse eggs Cytogenetics 5: 394–400.

Tettenborn, U.; Gropp, A. 1970. Meiotic nondisjunction in mice and mouse hybrids. Cytogenetics 9: 272–83.

White, M. J. D. 1954. Animal cytology and evolution. 2nd ed. London: Cambridge University Press.

Wiemann, H.; Lang, R. 1978. Strategies for detecting heritable translocations in male mice by fertility testing. Mutat. Res. 53: 317–26.

Winking, H.; Gropp, a. 1976. Meiotic nondisjunction of metacentric heterozygotes in oocytes versus spermatocytes. In Crosignxni, P. G.; Mishell, D. R., eds. Ovulation in the human. London: Academic Press; 47–56.

16
Application of Flow Cytometry to Cytogenetic Testing of Environmental Mutagens

by LARRY L. DEAVEN*

Abstract

The available and potential applications of flow cytometry to cytogenetic analysis are discussed and evaluated in this chapter. A brief history of the development of flow instrumentation and a description of the varieties of instruments currently available are included; major emphasis is given to those instrumental variations that contribute to cytogenetic applications. Methods for the preparation and staining of samples for flow analysis are presented in sufficient detail for the reader to acquire an elementary working knowledge of the available techniques. Results from studies using flow instruments to detect cytogenetic changes in mammalian cells or isolated metaphase chromosomes are reviewed and compared with results from conventional approaches. Additional developments are needed in instrumental design or quality and in sample processing in order for flow analysis to be applied to clinical and industrial screening programs. These necessary improvements and the current attempts to expedite them are discussed in relation to the future prospects for flow cytogenetic analysis.

Introduction

Interest in genetic toxicology has intensified during the past decade for a number of reasons. Estimates of the genetic contribution to our total health burden now exceed 25% (Lederberg 1971). An increasing number of suspicious chemicals pervade the general environment and bring to question the health and safety of certain workers. Our current energy dilemma forces a consideration of massive reliance on fuel sources that are known to contain clastogens, mutagens, and carcinogens. These

*Cellular and Molecular Biology Group, Los Alamos Scientific Laboratory, University of California, Los Alamos, New Mexico 87545.

reasons as well as others, combined with an ever-increasing environmental awareness in the general public, create a need for genetic testing and screening of unprecedented scope.

A significant component of current and future genetic testing programs will be cytogenetic analysis. The putative relationship between chromosome changes and mutagenesis or carcinogenesis and the ease with which samples for cytogenetic study can be obtained from human subjects suggest that this test may be the focal point of many industrial screening programs. This potential has been strengthened in recent years by innovations in cytogenetic technique that permit the detection of subtle chromosome changes which previously went unnoticed by the cytogeneticist. However, these advantages must be contrasted with certain deficiencies in this approach. One objectionable feature is the cost of cytogenetic analysis. It is questionable whether industry and government could afford massive chromosome screening programs even if trained personnel were available to do the work. Other disadvantages are accuracy and precision as applied to chromosome scoring. Although many attempts have been made to standardize breakage analysis, none has been completely successful, and laboratory-to-laboratory variabilities continue to cloud the results. Likewise, most cytogeneticists would like to have data on 10,000 cells but usually settle for scores from 100–200 cells because of time constraints. When these factors are combined with known differences in breakage rate as a result of medium or serum choice and time of cell harvest, the validity of cytogenetic testing seems questionable.

While increased attention to the need for standardized protocols will solve some of these problems, the need for increased data acquisition cannot be solved by conventional cytogenetic techniques. This has been recognized by others and has resulted in the construction of computer-oriented image-analysis machines originally designed to karyotype reliably the human chromosome complement. The discovery of chromosome banding has made many of these systems obsolete, and interest in this area of machine technology has declined before making a significant impact on cytogenetics.

Independent efforts by biophysicists and engineers have resulted in the development of a series of instruments designed to analyze and sort cells according to differences in biochemical and physical properties. Because the cells are made to flow in a liquid medium and fluorescence intensity is a commonly used detection parameter, these instruments are called flow microfluorometers (FMF) or, more recently, flow cytometers (FCM). Major advantages of these instruments are that a large number of cells (10^5) can be measured in a few minutes and that the resulting statistical precision reveals new information on subtle cell-to-cell differences. If these differences are due to chromosome changes, cytogenetic information may be obtained indirectly, while direct measurements may be made on chromosomes isolated from metaphase cells. Certain modifications of flow instrument design ultimately may allow breakage analysis on large numbers of cells in a few minutes. This chapter is a discussion of the available and potential applications of flow cytometry to clastogen testing and screening.

Technology

Flow instrumentation has been described in extensive detail in several previous articles (Holm and Cram 1973; Herzenberg et al. 1976; Arndt-Jovin and Jovin 1978;

Melamed et al. 1979); therefore, this discussion will be limited to a brief and general historical background of instrumental development, with major emphasis on those variations in design that contribute to cytogenetic applications. The principles of flow analysis originated with particulate analysis of aerosols in the 1940s. In this application, light scattered by airborne dust particles was detected photoelectrically and collected into scatter distributions for analysis. These methods were applied to biological studies with the construcion of erythrocyte counters which incorporated the use of liquid-sheath flow to direct the suspended cells to the center of the flow stream (Crosland-Taylor 1953). A few years later, a new technique for cell enumeration (i.e., electrical resistance counting) was described which gave rise to the first commercially produced instruments for cell number and size determinations (Coulter 1956). The first cell sorter was constructed in 1965. This was a device designed to sort cells on the basis of cell volume but, in principle, could be used to separate cells on the basis of any electronically measurable parameter (Fulwyler 1965).

A major improvement in flow technology came in 1969 with the addition of an argon-ion laser as the source of cellular illumination (Van Dilla et al. 1969). This permitted sufficient power for fluorescence measurements to be made on single cells. Adaptation of available fluorescent stains resulted in unprecedented DNA measurements of single cells in terms of numbers and precision. Parallel but independent research efforts resulted in instruments that measure cells in a liquid jet rather than in a chamber (Hulett et al. 1969), use an incoherent light source rather than a laser (Dittrich and Göhde 1969), and utilize slit-scan devices capable of differentiating nuclear from cytoplasmic fluorescence as well as other morphological features (Wheeless and Patten 1971).

A basic flow instrument for analyzing suspended particles consists of a flow chamber to define accurately the conditions where the sample stream and illumination source intersect, a detector (usually a photomultiplier tube), and a device to analyze and store the electronic data. Figure 16.1 is a schematic illustration of a Los Alamos flow cytometer. Fluorescently stained cells or chromosomes flow through a focused laser beam where they emit a pulse of fluorescent light that is detected and quantitated, and the results are stored as frequency distributions. Figure 16.2 shows oscilloscope images of pulses made by cells flowing through the instrument. Each pulse represents a single cell. Some pulses are bimodal and represent cell doublets. These bimodal pulses can be detected and eliminated for cell analysis but can be useful for chromosome measurements, as will be discussed later. Particle detection in the sorting machines is similar, but these instruments also include droplet generators, charging electrodes, and deflection plates. The liquid stream is broken into droplets, some of which contain particles of interest. When a chromosome or cell of a desired type flows through the detector, an electronic time delay is triggered that charges the group of droplets containing the cell or chromosome of interest. These charged droplets are then deflected by an electric field between two deflection plates. The charged particles are collected in one container, while those that did not trigger the charging mechanism pass undeflected into a separate container.

Modifications of these systems include instruments that make two or more measurements simultaneously (i.e., fluorescence, light scatter, or cell volume), instruments that illuminate samples with two laser beams in a sequential manner, and instruments with flow chamber design modifications. The first of these, multiparameter analysis (Steinkamp 1977), has contributed significantly to studies of

328 Flow Cytometry for Cytogenetic Testing

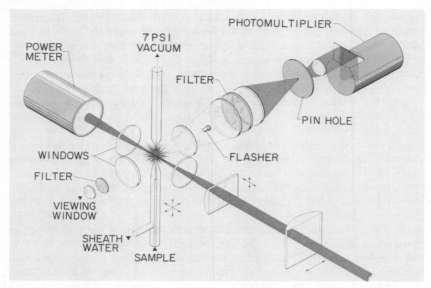

Figure 16.1 Schematic diagram of a Los Alamos flow cytometer. Cells or chromosomes flow through a focused laser beam and emit a pulse of fluorescent light that is detected by a photomultiplier tube.

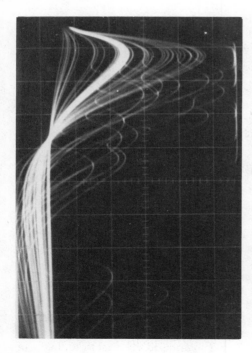

Figure 16.2 Oscilloscope images of pulses generated by the passage of fluorescently stained cells through the laser beam of a flow cytometer.

whole cells but has not been extensively applied to chromosome studies. This is due, at least in part, to limitations in staining methodology and chromosome isolation technique. The machine capability to measure DNA, RNA, or protein simultaneously in single chromosomes is available, but chromosome preparations with these components quantitatively stained have not been achieved.

Dual-laser systems are relatively recent developments (Dean and Pinkel 1978; Steinkamp et al. 1979), but early results have suggested that they may contribute significantly to flow cytometric studies (Gray et al. 1979). In these systems, samples can be illuminated simultaneously or sequentially with lasers tuned to different wavelengths. Stains with different excitation maxima and with affinities for the GC- or AT-rich regions of DNA may be analyzed for their relative binding to chromosomes.

One of the limiting factors of flow-system design especially applicable to chromosome analysis is light-collection efficiency. Some systems are perfectly adequate to measure the intensity of fluorescent light pulses from whole cells, but the signals from isolated chromosomes do not have sufficient intensity to give optimum resolution. Several approaches have been attempted to resolve this problem. An obvious one is simply to use higher-powered lasers with existing design to increase the fluorescence output. While this method is successful in cases where illumination is a limiting factor, it has the disadvantage of additional cost to instruments that are already expensive.

Another approach has been to alter flow chamber design to increase the light-collection efficiency. In most flow instruments, the suspension of particles to be analyzed flows perpendicularly to the path of the light beam. However, in one system (ICP Flow Cytometer, Ortho Instruments, Westwood, MA), the chamber is designed to cause the particle suspension to flow along the optical axis (Dittrich and Göhde 1970). This system uses the same objective lens for incident light excitation and for fluorescence emission collection. This design reduces variations in particle illumination and allows the use of an objective with a large solid angle for fluorescence collection. Fluorescent signals of sufficient intensity for chromosome measurements can be obtained using a conventional super-pressure mercury lamp.

The flow cytometer illustrated in Figure 16.1 has a light-collection efficiency of 3–4%. Two specialized flow chambers have been designed and constructed at Los Alamos which considerably increase this efficiency. One of these, an ellipsoidal chamber, is shown in Figure 16.3. This chamber has a collection efficiency of about 60% due to the light-collecting characteristics of an ellipsoid. The laser beam intersects the cell stream at the primary focus, and fluorescence is collected and measured at the secondary focus (Skogen-Hagenson et al. 1977). A more recent development is a parabolic flow chamber that collects even more photons than the ellipsoid (78% efficiency), has increased stability because of a much shorter flow path between orifices, and has parallel light output. Both of these chambers increase the resolution of chromosomal DNA measurements, as will be discussed later.

One other instrumental modification should be mentioned that is being developed and improved currently and that is most promising for single-chromosome analysis. Most flow cytometers measure the total fluorescence of an object as it passes the laser beam or other light source. The distribution of fluorescence intensity within an object is a source of additional information that can be measured in these new systems. These instruments measure fluorescence along the length of a chromosome

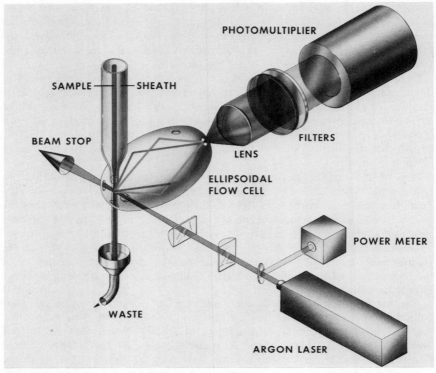

Figure 16.3 Schematic diagram of a high-efficiency flow cytometer. The cells intersect the laser beam at the primary focus, and fluorescence is measured at the secondary focus.

in a time-sequential manner. This has been accomplished by two methods. Either the laser beam must be highly focused to provide imaging in the object plane (Gray et al. 1979), or a good quality image of the object must be magnified and projected onto a variable-width slit positioned in the image plane (Cram et al 1979). Such instruments produce pulse-shape profiles that define the fluorescence asymmetry of an object. Figure 16.4 is a diagram of a pulse-shape profile of a chromosome as produced by the latter of these instruments. Measurable parameters include pulse width (P.W.), pulse area (P.A.), pulse height (P.H.), or a combination of these dimensions.

Preparation and Staining Protocols

The quality of data from flow instruments is dependent on a number of parameters. These include sample preparation and staining, machine quality and adjustment, and electronic signal processing. In spite of progress over the past five years, at this writing, machine development and electronic data processing are more advanced than sample preparation and staining. Early procedures for cell and chromosome preparation and staining were directed toward materials fixed on slides, and most of

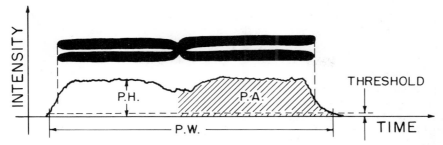

Figure 16.4 Diagram of a metacentric chromosome and its expected pulse-shape profile as measured by a slit-scan flow cytometer. Measurable parameters include pulse height (P.H.), pulse area (P.A.), and pulse width (P.W.).

these techniques are not compatible with cell suspensions. Nevertheless, a number of methods have been developed and properly validated, and those most applicable to cytogenetic analysis will be discussed here. The reader is referred to Parts 3 and 4 of *Flow Cytometry and Sorting* (Melamed et al. 1979) for a more comprehensive discussion of these topics.

Preparation and Staining of Single Cells

Single-cell dispersions can be prepared from cultured monolayers with relative ease. A standard protocol has been developed at Los Alamos that is applicable without modification to a large number of cell lines (Kraemer et al. 1971). This method provides large numbers of single cells with a minimum amount of cellular debris and cell clumps.

Solutions. Stock solution I (g/liter): Glucose, 1.1; NaCl, 8.0; KCl, 0.4; $Na_2HPO_4 \cdot 12 H_2O$, 0.39; and KH_2PO_4, 0.5. Stock solution II (g/liter): $MgSO_4 \cdot 7 H_2O$, 1.54; $CaCl_2 \cdot 2 H_2O$, 0.16. Phenol red stock: 1 g/liter phenol red in saline. Puck's saline G: 900 ml of solution I plus 100 ml of solution II plus 1 ml of phenol red stock. Puck's saline GM: 1 liter solution I plus 1 ml phenol red stock. Trypsin stock: 10 mg/ml in 0.001 N HCl (Worthington, crystallized three times). Dispersing solution: saline GM, 0.5 mM EDTA, 0.1 mg/ml trypsin. Neutralization solution: saline G containing 0.01 mg/ml DNase (Worthington), 0.2 mg/ml soybean trypsin inhibitor (Worthington) and 1 mg/ml bovine serum albumin. Fixing solution: 95% ethanol.

Monolayer treatment. Pour growth medium from monolayers and rinse cells gently three times with saline GM. Add 10 ml of the dispersing solution and incubate at 37°C for 10 min. Draw the detached suspension through a serological pipette several times and add 10 ml of the neutralization solution. The suspension is again drawn through a serological pipette 3–4 times and then gently sedimented with low-speed centrifugation. The pellet of cells is resuspended in 5 ml of normal saline and fixed by adding 15 ml of 95% ethanol, 5 ml at a time, with pipetting after each addition.

Staining. The first successful stains used for cellular DNA studies with flow systems were auramine-O and acriflavine (Kraemer et al. 1972). Feulgen procedures were developed to apply these stains to cells in liquid suspension and resulted in the first high-quality DNA distributions from flow instruments. These methods were lengthy and required relatively large numbers of cells, many of which were lost in a series of required hydrolysis and centrifugation steps. Improved staining came with the development of procedures for using the DNA intercalators ethidium bromide and propidium iodide (Crissman et al. 1975). These techniques are considerably shorter than the Feulgen method but require RNase treatment and occasionally result in cell clumping. Further improvements include the use of the antibiotic mithramycin (Pfizer). Mithramycin selectively binds to DNA, does not interact significantly with RNA (Ward et al. 1965), and can be used as a DNA stain simply by adding it to a properly fixed cell suspension in the presence of Mg^{++} (Crissman and Tobey 1974).

Cells fixed in the manner described above are suitable for mithramycin staining. The cells are sedimented, the fixative is removed by aspiration, and the cells are resuspended in mithramycin solution (100 mg/ml mithramycin, 15 mM $MgCl_2$, 25% ethanol) (Crissman and Tobey 1974). This staining method has been validated for cell-cycle-analysis studies. However, when applied to cells from a broad range of species with varying nucleotide composition, the results should be interpreted with caution because of known binding specificities of mithramycin to GC-rich DNA (Ward et al. 1965). For studies of this type, the longer and more difficult staining procedures using acriflavine or the DNA intercalators, ethidium bromide or propidium iodide, should be used along with mithramycin (Deaven et al. 1977).

A modification of the propidium iodide staining technique that is rapid and gives excellent results has been described by Krishan (1975). Cultured cells or peripheral blood lymphocytes are resuspended in a hypotonic propidium iodide solution (0.05 mg/ml in 0.1% sodium citrate) and examined by FCM after 5-10 min. This method avoids cell loss by the centrifugations necessary for RNase digestion and can provide flow analaysis of the nucleated cells from 1-2 drops of blood. If desired, lymphocytes can be isolated before staining by the Ficoll-hypaque gradient method (Boyum 1968). Cultured cells with large amounts of cytoplasm do not stain as precisely as lymphocytes with the hypotonic propidium iodide method, and mithramycin should be used to avoid peak broadening and elevated coefficients of variation.

Tissue disaggregation. Tissue disaggregation is an important and active area of cell preparation for flow analysis. Current techniques include enzymatic digestion, physical methods, and use of chelating agents. A general method applicable to many types of tissues is not available, and a detailed descriptive account of the many types of approaches to this task is beyond the scope of this chapter (see Part 3 of *Flow Cytometry and Sorting*, Melamed et al. 1979). At the present time, the most useful tissues for clastogen testing by flow systems are bone marrow and testicular tissues. Good results from bone marrow samples have been obtained simply by resuspending the marrow samples in hypotonic propidium iodide solution (Krishan 1975). Testicular tissues may be disaggregated by placing a testis in Hanks' solution, cutting it into small pieces with scissors, and drawing the pieces through a Pasteur pipette. The resulting suspension is placed in a centrifuge tube and allowed to sediment for 30 min. At this time, the supernatant is carefully removed to a separate tube and gently centrifuged, and the resulting pellet of cells is treated with dispersion and neutralization solutions as described above. If the cells are fixed in ethanol and stained with mithramycin, good resolution of the ploidy levels of spermatogonial cells will result (L. L. Deaven and T. C. Hsu, unpublished data).

Isolation and Staining of Single Chromosomes

Isolated chromosomes for flow analysis ideally should be morphologically intact, free from clumps of chromosomes and debris, stable for a reasonable length of time, and contain unaltered macromolecular components. The isolation method should be rapid and repeatable and should provide quantitative recovery of all chromosome types from the cells being studied. While none of the available isolation techniques fulfills all of these conditions, several can provide chromosomes suitable for flow studies of clastogen effects. The most important criteria for these studies are good morphology, absence of clumping, quantitative DNA recovery, and speed and simplicity of the isolation technique.

The method of choice for most flow karyotype studies to date has been the isolation technique devised by Wray and Stubblefield (1970). This method functions by swelling mitotic cells in an isolation buffer and then breaking the cell membrane by physical forces. The chromosome isolation buffer consists of 1.0 M hexylene glycol, 0.5 mM $CaCl_2$, and 0.1 mM PIPES buffer, pH 6.5 (Calbiochem). Mitotic cells blocked with Colcemid can be collected from monolayer cultures by replacing the medium with trypsin solution and shaking the flask gently (Stubblefield et al. 1967), by mechanical dislocation using a shaker device (Petersen et al. 1968), or by a rapid rotation of the culture bottles in a culture rotator (Talandic Research Corporation, Pasadena, Calif.). The mitotic cells are resuspended in culture medium and cooled to 4°C, centrifuged, and washed with isolation buffer at 4°C. The cells are then gently resuspended in 50 volumes of isolation buffer (4°C) and incubated in a 37°C water bath. At this point, the cells should be monitored by placing drops of the cellular suspension on slides and overlaying with coverslips for microscopic observation. The cells swell and form blebs on the membrane. When these blebs are apparent, the chromosomes should be condensed and of distinct morphology. At the time of bleb formation and chromsome condensation, the cells can be sheared by gentle syringing from a 22-gauge needle. Large volumes of cells can be homogenized with a Dounce glass homogenizer from Kontes (Wray 1973). The optimal time of 37°C incubation for CHO cells is $\cong$ 7 min. Shearing before this time results in many elongated or stretched chromosomes; if shearing is performed too late, many cells will not lyse and chromosome clumping is increased. The optimal time for shearing may vary from one cell type or line to another.

Improvements to this method since the original descriptions include the use of nitrogen cavitation for cell rupture (Wray, personal communication), a pretreatment of the mitotic cells in hypotonic KCl (75 mM) (Gray et al. 1975a), and using 25 mM Tris-HCl as a buffer in place of PIPES (Carrano et al. 1978a). The use of nitrogen cavitation gives more uniform cell rupture and chromosomes with less distortion than the syringe treatment. Preincubating with hypotonic KCl makes disruption easier and gives the technique better applicability to a variety of cell lines than the original method. The change to Tris-HCl buffer increases the buffering capacity and stability of the isolated chromosomes and lowers the fluorescence background.

A very simple and rapid technique for chromosome isolation has been used by Stubblefield et al. (1975). Mitotic cells are treated in growth medium diluted with three volumes of water to induce cell swelling. After 5 min., the cells are sedimented by low-speed centrifugation and resuspended in 1 ml of the hypotonic growth medium. Ten volumes of 50% acetic acid are added to the cell suspension, and the cells are passed through a 22-gauge needle on a syringe until the cells are broken. The resulting chromosome suspension can be sedimented by centrifugation and resuspended in hexylene glycol chromosome isolation buffer for staining and FCM analysis. Chromosome morphology is good following these treatments, but the need to pellet the chromosomes introduces the possibility of partial and selective chromosome loss.

Another technique designed specifically for flow studies combines chromosome isolation and staining into one procedure (F. Otto, personal communication). This method requires two solutions: a hypotonic staining solution (20 mM Tris-HCl buffer at pH 7.5, 40 mM $MgCl_2$, 25 μM ethidium bromide, 25 μM mithramycin), and an isotonic staining solution (hypotonic staining solution plus 240 mM NaCl). Mitotic cells are collected from 1-3 T-75 culture flasks and centrifuged to a pellet. The cell pellet is resuspended in 1 ml of the hypotonic staining solution and kept at room temperature for 10 min. At this time, Triton X-100 is added to the suspension to a final concentration of 0.25% and mixed by tapping the tube with a finger. After 10 min. at room temperature, chromosomes are isolated by drawing the solution through a 27-gauge needle 5-10 times. The isolation process should be monitored by microscopic examination. The final step is the addition of 4 ml of the isotonic staining solution. The chromosomes are ready to be analyzed immediately, and good resolution has been obtained by Otto and coworkers using an ICP Flow Cytometer with a mercury lamp.

This method is fast and reproducible and gives excellent results from a variety of cell lines. Chinese hamster chromosomes isolated by this method and analyzed at 488 nm on FMF II at Los Alamos gave a resolution similar to those isolated by the hexylene glycol buffer technique. Opossum chromosomes gave better resolution when isolated by this method than could be obtained by either the hexylene glycol or acetic acid method (Keith et al. 1979). Disadvantages are that chromosomes isolated by this method are not as stable as those isolated by other methods (maximum lifetime 24-48 hr.) and that many of the isolated chromosomes have attached bits of cytoplasm or debris.

The choice of stain for isolated chromosomes is dependent on several alterable components of the flow instruments. These include illumination power and wavelength availability, phototube sensitivity, and light-collection capacity. The ICP Flow Cytometer is most versatile with respect to these requirements. The mercury lamp light source provides a broad range of excitation wavelengths, and the chamber design provides sufficient sensitivity of fluorescence detection. Laser-illuminated systems have additional constraints because of the limited number of excitation lines. Many lasers lack useful illumination capabilities below 350 nm and have only a limited number of lines with sufficient power above 350 nm. Therefore, stains must be selected that are excited at or near the available laser lines. Greater versatility is available through the use of high-power UV lasers in combination with continuously tunable dye lasers. However, these options require additional developmental work and entail considerable expense. In all systems, phototubes must be sufficiently sensitive in the emission spectrum to measure precisely the emitted fluorescence. Light-collection capability can be improved with special chamber designs, as previously described.

In addition to these requirements, fluorescent staining of chromosomes must be DNA-specific and must bind to the DNA in a stoichiometric manner with minimal chemical or physical pretreatments. The stain should exhibit enhanced fluorescence when bound to DNA to minimize background nonspecific emissions and, because some chromosomes contain very small amounts of DNA, the stain must have high quantum yield. Finally, in keeping with the overall requirements for clastogen screening, the staining technique must be simple and fast.

Most of these exacting requirements are met by the DNA intercalators ethidium bromide and propidium iodide. Chromosomes isolated by the hexylene glycol or acetic acid method give good flow spectra when mixed with ethidium bromide at 10^{-4} M final concentration. The stain is dissolved in the hexylene glycol buffer, and the chromosome suspensions are brought up to the proper stain concentration. Propidium iodide gives similar results. These stains have excitation maxima that are close to the strong argon-ion laser lines (488 nm) available on many commercial instruments and thus have wide applicability. Ethidium bromide is not sufficiently excited by mercury lamp emission, but this problem can be remedied using a combination of stains as in the Otto procedure. In this case, the ethidium bromide is excited by energy transfer from mithramycin, which increases the separation of emission and excitation wavelengths and permits better resolution (Zante et al. 1976).

Mithramycin staining does not give adequate resolution on the Los Alamos FMF II equipped with a 4-watt Spectra-Physics Model 164-03 argon-ion laser. However, high resolution has been obtained with a closely related stain, chromomycin A_3, using a laser with three times more power (Spectra-Physics Model 171-05) in the excitation range (457 nm) of these stains (Jensen et al. 1977). This probably reflects the relationship between resolution of the chromosome peaks and fluorescence signal intensity. Using ethidium bromide, Jensen and coworkers have demonstrated unequivocally this relationship. They measured the coefficient of variation of a single chromosome peak over a range of decreasing signal intensities. The coefficient of variation increased from 3.8–5.4% as the signal intensity was decreased 50-fold.

New dual-laser systems permit flow studies of chromosomes stained with more than one fluorochrome and with different excitation wavelengths. Using this technology, human chromosomes in Tris-buffered hexylene glycol isolation medium were stained with Hoechst 33258 (2 µg/ml) and with chromomycin A_3 (80 µg/ml) and excited with the appropriate lines for each stain by separate lasers. The result was the best resolution for human chromosomes yet achieved (20 groups) (Gray et al. 1979). Chinese hamster chromosomes treated in this manner showed less resolution than if they were stained only with Hoechst 33258. These observations may reflect differences in base composition between Chinese hamster and human DNAs. Previous studies have indicated that Hoechst 33258 produces Q-band type patterns (Raposa and Natarajan 1974) and that chromomycin A_3 produces R-band patterns (Van de Sande et al. 1977).

Results

Whole-Cell Studies

Actual clastogen testing by flow systems is essentially nonexistent at the time of this writing. However, a number of experiments have been done which demonstrate real or potential application to this end, and these will be reviewed in this section.

The original study, which combined cytogenetics with flow analysis compared chromosome number and DNA content per cell in normal and heteroploid cell populations (Kraemer et al. 1971). This investigation and subsequent studies demonstrated that, although heteroploid cell populations have an elevated variability in cell-to-cell chromosome number per cell, the distribution of DNA content is no greater than in the euploid cells from which they were derived (Kraemer et al. 1972; Deaven et al. 1975). Although these studies were addressed to elucidating the basic mechanisms involved in heteroploidy, they also resulted in developing several techniques that have direct bearing on clastogen testing. Because their results were at variance with previous conclusions concerning DNA content in heteroploid populations, Kraemer and coworkers applied rigorous validation tests to the flow instruments and to cell processing procedures, and the resulting data were the first with sufficient precision for detecting clastogen effects.

A typical DNA distribution derived from a population of exponentially growing mammalian cells is shown in figure 16.5. Examination of this illustration reveals a number of parameters that can be used to detect changes in chromosome constitution. The G_1 peak of euploid cells falls at the 2C value for DNA content. Loss of one or more chromosomes results in a measurable loss in DNA content and a corresponding shift in the G_1 mode. Using this approach, aneuploid CHO Chinese hamster cells were shown to contain 4% less DNA than euploid Chinese hamster cells (Deaven and Peterson 1973)—a measurement in good agreement with previous estimates based on armlength measurements (Kao and Puck 1969). Further refinement of technique, including the introduction of hypotonic propidium iodide staining of lymphocytes, permitted the use of this type of analysis to be applied to the

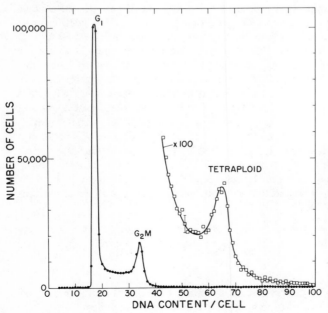

Figure 16.5 Flow-analysis distribution of suspension-cultured CHO cells. Channels 43–100 are amplified 100-fold to show the presence of a peak of tetraploid cells at the 8C position. The absence of a 6C peak indicates monodispersity of the stained cell preparation.

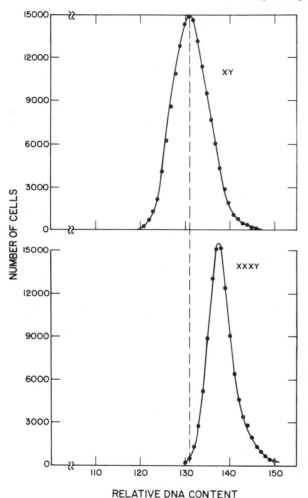

Figure 16.6 Flow measurements of lymphocytes from a normal male and from a patient with a 48 (xxxy) chromosome constitution. The peak modes differ by about 5%. Lymphocytes were isolated by the Ficoll-hypaque technique and stained with hypotonic propidium iodide for FCM analysis.

detection of human aneuploidy (Callis and Hoehn 1976; Cram and Lehman 1977; Hoehn et al. 1977). FCM measurements of lymphocytes from a normal male (xy) and from a patient with a 48 (xxxy) chromosome constitution are shown in Figure 16.6. The average increase in DNA content, as measured by flow analysis, was 5%. This is in excellent agreement with previous measurements by the CYDAC scanning cytophotometer (5.2%) (Mendelsohn et al. 1973) or by armlength measurements (5.6%) (Penrose 1964). Current estimates are that DNA differences of less than 1% of the human genome can be reliably detected by flow analysis, permitting accurate diagnosis of aneuploidies involving the smallest human chromosomes (e.g., trisomy 21, xyy). Figure 16.7 shows a comparison of modal values for the DNA content of male and female lymphocytes. DNA content differences between individuals of the same sex are not completely understood but may be related to differences in con-

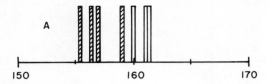

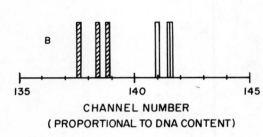

Figure 16.7 Modal DNA values of peripheral lymphocytes from seven normal males (crosshatched bars) and six normal females (clear bars) stained with propidium iodide. The two sets of data were made on different days at different instrumental gain settings, as shown by channel number differences for (A) and (B).

stitutive heterochromatin content. If these experimental results can be obtained consistently, FCM analysis could decrease the need for conventional cytogenetics in mass screening for human aneuploidies.

The tetraploid peak in Figure 16.5 is another cytogenetic parameter that can be measured with great sensitivity by flow analysis. The peak in Figure 16.5 is amplified 100 times and represents 1% of the cells in the population. Changes in ploidy levels also affect the G_1 and $G_2 + M$ peaks of the diploid population by decreasing and increasing them respectively, but the greatest sensitivity is at the 8C region where frequencies are low in most cultured diploid cells. Kraemer et al. (1972) studied the changes in flow profiles following the addition of Colcemid to cultured CHO cells. After one day most cells appeared in the 4C peak, at two days octaploid cells appeared, and by day 4 a few 16-ploid cells appeared; the modal DNA values were very close to the expected progression of 2C, 4C, 8C, and 16C. FCM analysis also has been used by Tobey and Crissman (1975) to detect changes in ploidy levels following treatment of CHO cells with 3-nitrosourea analogs. An FCM profile of CHO cells treated with 10 μg/ml of 1,3-bis(2-chloroethyl)-1-nitrosourea and then returned to normal medium for four days is shown in Figure 16.8. The redistribution of cells from a normal cell-cycle traverse is obvious, as well as the increase in tetraploid (G_2) and octaploid (G_1) cells. The cells were stained with the acriflavine-Feulgen technique (open circles) and with mithramycin (dark circles). The differences between the profiles reflect the presence of fragile cells that are broken during the cellular manipulations necessary for acriflavine staining.

Further consideration of the G_1 peak in Figure 16.5 reveals an additional measurement that can be applied to cytogenetic analyses. The peak modal value represents the average DNA content per G_1 cell and can be used to detect stable aneuploidies, as shown in Figure 16.6. The width of the peak represents the distribution of DNA between cells and can be used to detect aneuploidy induced by nondisjunction. In an ideal situation, a measurement of euploid cells would show no cell-to-cell dispersion

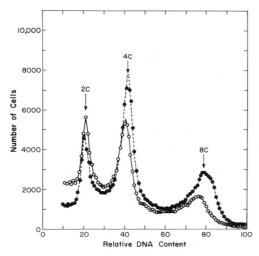

Figure 16.8 Flow profiles of CHO cells treated with 10 μg/ml of 1,3-bis(2-chloroethyl)-1-nitrosourea for 2 hr., then grown in normal medium for 95 hr. Cells were stained with mithramycin (solid circles) or by the acriflavine-Feulgen technique (open circles). The manipulations necessary for acriflavine staining result in the breakage of fragile cells, and the loss of these cells accounts for the difference between the two profiles. The large increase in the 4C and 8C peaks, compared to exponentially growing CHO cells (Fig. 16.5), is indicative of mitotic arrest and tetraploidization as a result of drug treatment.

of DNA content, and the peak would appear as a straight line; the coefficient of variation would be close to zero. In actual practice, staining inconsistencies and dispersions of instrumental origin result in small contributions to peak width, resulting in typical coefficients of variation of from 1–3%. This base-line value is well within the limits necessary to detect the loss or gain of DNA by nondisjunction.

Studies on the induction of nondisjunction by Colcemid have demonstrated that cells treated with low doses of the drug and then allowed to grow in drug-free medium completed an abnormal cell division with a considerable number of nondisjunctive errors (Cox and Puck 1969). The rates of nondisjunction also were found to be in proportion to the amount of time the cells were exposed to the drug (Kato and Yosida 1971). This experiment was repeated for flow studies and resulted in the flow profiles shown in Figure 16.9. Monolayers of CHO cells were given Colcemid (0.06 μg/ml) and shaken to remove mitotic cells at 2., 4., and 6 hr. After each shake, medium with Colcemid was added back to the cells. The mitotic cells were placed in drug-free medium and allowed to complete division. After 8 hr., the cells were prepared for FCM analysis. The symmetrical broadening at the base of the G_1 peaks reflects the increasing rates of nondisjunction with time of exposure to Colcemid (Deaven and Petersen 1974).

The data in Figure 16.9 suggest that flow analysis could be used very easily and efficiently to screen compounds for nondisjunction activity. This supposition unfortunately is not completely true. The unequal distribution of DNA following division of cells that contain DNA fragments may broaden the G_1 peaks in the same way as nondisjunction. High doses of a clastogen, where up to 80% of the exposed cells contain broken chromosomes, result in flaring of the entire G_1 peak (Deaven et al. 1978; Tobey et al. 1978). At low doses, where only a few percent of the cells sustain chromosome breakage, the G_1 peaks are flared at the base very similar to those in Figure 16.9. Flow detection of nondisjunction can be a useful protocol if this limitation is kept in consideration. This is especially true for agents suspected of inducing low levels of nondisjunction and that do not have clastogenic activity. The large

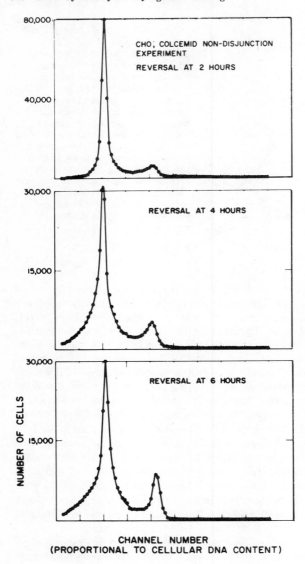

Figure 16.9 Flow profiles of Chinese hamster cells following treatment with Colcemid. Monolayer cultures were given 0.06 µg/ml of Colcemid and shaken at 2, 4, and 6 hr. to remove mitotic cells. These cells were then grown in drug-free medium for 8 hr. and prepared for FCM analysis by staining with the acriflavine-Feulgen technique. The broadening at the base of the G_1 peaks reflects the loss or gain of chromosomes as a result of Colcemid-induced nondisjunction.

number of cells that can be analyzed in flow systems make this test far less time-consuming and expensive than conventional cytogenetic detection. When applied to germinal tissues, the flow approach could be a powerful tool for in vivo nondisjunction detection.

Studies on Isolated Chromosomes

The first experiments on flow studies of isolated mammalian chromosomes were described by Gray et al. (1974, 1975a). Chromosomes were isolated from Chinese hamster cells (M3-1) by the Wray and Stubblefield technique and stained with ethidium bromide for FCM analysis. The chromosomes were resolved into nine peaks and were sorted onto microscope slides for visual analysis. Measurements of DNA values for individual chromosomes were made from metaphase spreads on slides by a scanning cytophotometer and compared with the peak means from the flow profiles. Peak areas were compared with the relative frequencies of each chromosome in whole cells. All of these comparisons were in good agreement, indicating that flow analysis of isolated chromosomes is a powerful new tool for cytogenetic studies. An ample demonstration of this fact was contained in the 1975a paper of Gray et al. Coefficients of variation for chromosome peaks varied from 2.2-3.3%, and a subtle alteration of chromosome number 1 amounting to a DNA content difference of about 6×10^{-14} g was resolved in the flow profiles.

This initial work was verified using a different flow system and a Chinese hamster line with unaltered chromosomes (Don-6) (Stubblefield et al. 1975). In this work, a different verification system was used but with essentially the same results. Chromosomes were isolated with the acetic acid-methanol fixation technique, stained with ethidium bromide, and analyzed on the Los Alamos FMF II. Some of the isolated chromosomes were layered on a 20-40% linear sucrose gradient and separated according to size by centrifugation at 2500 rpm in a Sorvall Model SZ-14 zonal rotor. Fractions of chromosomes were collected, stained with ethidium bromide, and analyzed by FCM. Aliquots of the fractions were prepared for visual analysis so that peak locations on the FCM profiles could be related to specific chromosome content. Although the resolution of mixed chromosomes in this study was not as good as that described by Gray et al. (1975a), certain peaks in the chromosomes partially separated by gradient centrifugation could be resolved which were not distinct in mixed chromosomes in either study. This suggested that refinements in technique could result in differentiation of all Chinese hamster chromosomes, resulting in 11 or 12 separate peaks for euploid cells.

These initial studies were extended to flow analyses of other species with simple karyotypes (Carrano et al. 1976) and led to steady progress in the resolution of the relatively complex human karyotype (Gray et al. 1975b, 1979; Carrano et al. 1979). The initial separation of human chromosomes was done using ethidium bromide as the stain and resulted in eight distinct groups. This resolution was increased to 15 peaks and shoulders using Hoechst 33258 stain and to 20 groups using both Hoechst 33258 and chromomycin A_3 staining and analysis in a dual-laser system. The last two analyses involving stains with affinities for specific base pairs are especially interesting and may represent major advances toward the goal of routine karyotype analysis by flow systems.

A significant problem associated with chromosome separation by DNA content is that many karyotypes contain nearly identical chromosomes in this respect (e.g., human "c" group, *Mus musculus* complement). This problem was alleviated to some extent by the results of Carrano et al. (1979). They found that, while Hoechst-stained chromosomes produce a fluorescence intensity close to that expected on the basis of DNA content, there are several disparities that aid in chromosome resolution. Gray et al. (1979) extended this approach by staining Chinese hamster and human chromosomes with Hoechst 33258 and chromomycin A_3. When Chinese hamster chromosomes were stained with Hoechst 33258, they were resolved into 16 groups. When stained with both stains and resolved by dual-laser excitation, only 10 groups could be separated. Human chromosomes stained with Hoechst 33258 were resolved

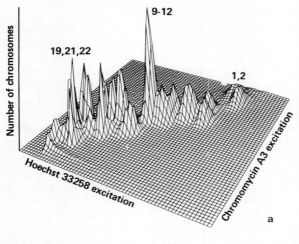

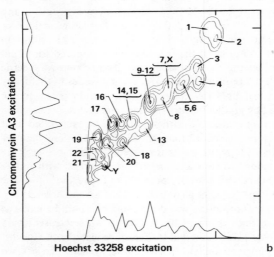

Figure 16.10 (a) Isometric display of the two-parameter frequency distribution of fluorescence from Hoechst 33258 and chromomycin A_3 stained human chromosomes. The chromosomes were analyzed in a dual-laser flow system with one laser tuned to 458 nm and the other in the UV range. (b) A contour display of the two-parameter data shown in (a). The assignments of chromosomes to the peaks in the two-parameter distribution were inferred from peak locations in one-parameter studies.

into 12 groups; when stained with both dyes, they formed 20 groups. Figure 16.10 illustrates this latter resolution. Continued progress in playing the specificities of one stain against another probably will result in increased flow karyotype resolution and in greater insight into the specific chemical composition of individual chromosomes.

Another approach to increasing chromosome resolution is the development of specialized flow chambers, as described in the technology section of this chapter. This has some advantage in that it obviates the need for more powerful and expensive laser systems. Chromosomes isolated from Chinese hamster (CHO) cells by the Wray and Stubblefield technique and stained with ethidium bromide resolve into eight distinct peaks when analyzed on the Los Alamos FMF II (Figure 16.11). If these same chromosomes are analyzed on a system with an ellipsoidal chamber, an

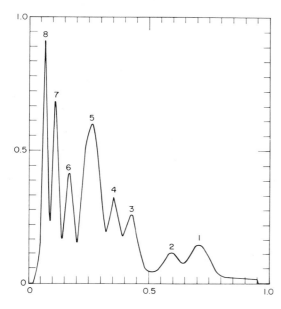

Figure 16.11 FCM spectrum of isolated aneuploid CHO chromosomes stained with ethidium bromide and analyzed on a Los Alamos Flow Cytometer. The peak numbers do not correspond to chromosome numbers in all cases. Relative fluorescence intensity (abscissa) and relative number of chromosomes (ordinate).

additional peak is resolved (Fig. 16.12); the same preparation resolves into 12 peaks and shoulders when a parabolic chamber is used (Fig. 16.13). Perhaps the best example of the effects of light-collection efficiency comes from the studies of Otto and Oldiges (1977). Using a mercury lamp-equipped ICP Flow Cytometer and chromosomes isolated and stained by the Otto procedure, several Chinese hamster cell lines and clones have been resolved into profiles consisting of 9 to 11 peaks. It should be explained that the number of peaks in chromosome profiles is not a quantitative measure of resolution when comparing the results from a number of Chinese hamster lines with different karyotypes. The expected number of peaks can vary from 11 in euploid female cells to 18 or more in lines with extensive rearrngements. This parameter is used because it gives some comparative measurements which, in some cases, are the only data available.

Studies using isolated chromosomes and directed at detecting aberrant karyotypes fall into two categories: (1) the detection of stable aberrations identical in every cell; and (2) the detection of unstable, usually transient, aberrations (breaks) that are distributed at random in a cell population. Some results have been obtained from studies in each category; however, this area of flow analysis needs increased research activity.

Gray et al. (1975b) analyzed a variant clone of the Chinese hamster M3-1 line. The clone contained a reciprocal translocation between chromosome numbers 1 and 4 as determined by Giemsa banding analysis. The rearranged chromatin was easily detected by flow analysis, with a loss or gain in numbers of chromosomes in the involved chromosome peaks. Similarly, a temperature-sensitive mutant of the CHO line with a chromosome deletion in the short arm and an additional chromosome number 10 was easily identified by flow profile changes in the appropriate peaks (Deaven et al. 1976). A test for mosaicism was designed by Van Dilla et al. (1976) by

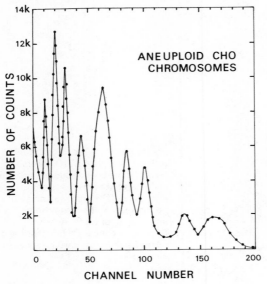

Figure 16.12 FCM spectrum of aneuploid CHO chromosomes stained with ethidium bromide and analyzed with a flow instrument equipped with an ellipsoidal chamber. The light-collection characteristics of an ellipsoid result in the increased resolution of chromosomes as compared to Figure 16.11.

mixing cells from two clones of M3-1 cells. One clone contained the reciprocal translocation described above; the other clone did not have that rearrangement. By decreasing the concentration of cells with the translocation, they were able to show the low limit of detection to be 5%.

The only reported attempt at detecting unstable aberrations is a notable but unsuccessful effort by Carrano et al. (1978b). They attempted to exploit the fact that random chromosome breakage and rejoining would change the continuum underlying an FCM profile (i.e., the valleys between chromosome peaks would increase as breakage increased). Chinese hamster cells (M3-1) were X-irradiated and collected by mitotic block 15-20 hr. later. Chromosomes were isolated, stained with ethidium bromide, and analyzed by FCM, and sorts from each peak and valley were collected on slides for visual analysis. X-ray doses from 25-800 rad increased the profile backgrounds as a function of aberration frequency, as expected. However, tests for goodness-of-fit indicated that experiment-to-experiment variations are too great for a quantitative analysis of the breakage rates. The authors concluded that this indirect method for random breakage detection is not sufficiently sensitive for analyzing cells exposed to low levels of clastogens.

A more promising method for the detection of random breakage lies in the continued development of slit-scan instruments. The flow instruments in use today were designed and developed primarily for the analysis of whole cells; slit-scan devices may represent the beginning of a new generation of machines designed specifically for the analysis of isolated chromosomes. Figure 16.14 shows pulses generated by Chinese hamster number 1 chromosomes in one of these systems (Cram et al. 1979). The decrease in height at the center of the pulses corresponds with the centromere location. Similar results have been reported by Gray et al. (1979) using chromosomes of the Indian muntjac. Multicentric or acentric chromosomes could be recognized and differentiated from normal chromosomes with this level of resolution. These in-

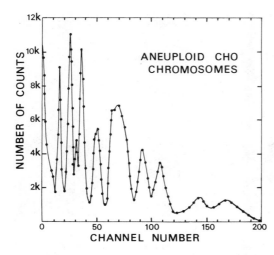

Figure 16.13 FCM spectrum of aneuploid CHO chromosomes stained with ethidium bromide and analyzed with a flow instrument equipped with a parabolic chamber.

struments also should be capable of identifying inversions that do not involve loss or gain of DNA per chromosome. These rearrangements are not detected by conventional flow systems but could be recognized by slit-scan instruments by changes in the centromeric index.

Future Prospects

Predicting the future role of flow cytogenetics in environmental mutagen testing and screening programs is a difficult task. The results reviewed in this chapter are remarkable achievements, especially if consideration is given to the rate of progress. The first high-quality DNA per cell data were collected less than 10 years ago and the first DNA per chromosome readings 5 years ago. Yet in spite of this progress, the relationship between flow-systems analysis and cytogenetic screening is not firm. As a research tool, the future of flow cytogenetics is more secure, and progress undoubtedly will continue in areas such as chromosome sorting, cell-cycle–related chromosome phenomena, and in-depth studies of cell and chromosome perturbations by external insults. However, in terms of clinical and industrial screening applications, adequate results have been more difficult to achieve and unwarranted promises or overstatements must be avoided. The cytogenetics community already has witnessed the development of machines for metaphase finding and chromosome analysis which appear to function properly in research laboratories but not in clinical settings. On the other hand, the basic characteristics of flow instruments of precise measurement and speed are ideally suited to current cytogenetic problems. In this section, I will discuss some of the needs and desires that could change the outlook from cautious optimism to certain application.

Instrumentation

It would be desirable to have greater participation from more laboratories in the search for improved protocols for flow cytogenetics. Major impediments to this goal

346 Flow Cytometry for Cytogenetic Testing

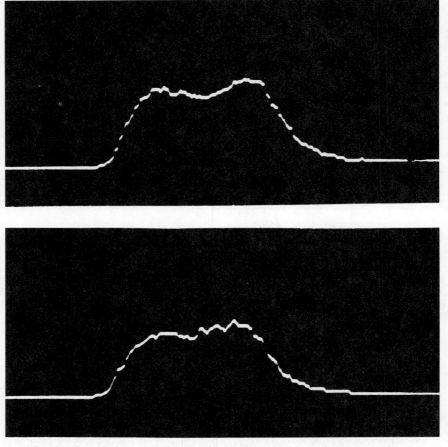

Figure 16.14 Pulses generated by the passage of Chinese hamster number 1 chromosomes through a slit-scan flow cytometer. The decrease in height near the center of the pulse corresponds to the centromere location.

are the high cost and technical support requirements necessary to purchase or construct and maintain flow instruments. This cost and manpower constraint may be alleviated by a new type of flow instrument that can be assembled by a nonspecialist with readily available components (Steen and Lindmo 1979). This system consists of a commercially available nozzle to produce a pressurized jet of water (cell stream), a fluorescence microscope, a photomultiplier tube, and a multichannel pulse-height analyzer. The cell stream is formed on a coverslip on the microscope stage and positioned by the object guide of the microscope. The photomultiplier tube is mounted on the microscope phototube and connected to the multichannel pulse-height analyzer. The system is reported to be very stable and has measured the DNA content of rat thymocytes with a coeffecient of vaiation of less than 1%. This compares favorably with the best systems available today. The imaging capabilities of this system and the hydrodynamic focusing of objects in the flow stream suggest that it may be possible to adapt it for slit-scan analysis. Systems such as this should encourage and enable other investigators to participate in flow cytogenetics research.

Continual development of the slit-scan flow systems is highly desirable. The capability of these instruments to detect directly certain types of aberrations (rings, dicentrics, deletion fragments) would eliminate any need for further consideration of the indirect methods of conventional machines. Although the capability of these instruments to detect chromosome profiles has been demonstrated, little is known about the accuracy and precision with which this detection is accomplished. Although it is known that some profiles are measurable, good estimates of the percentage of chromosomes that are properly oriented and large enough for profile detection are not available. All studies to date have been limited to large chromosomes of the Chinese hamster or Indian muntjac. It is not clear whether these instruments will be capable of detecting and measuring changes in all chromosomes of the human karyotype. Much work remains in terms of instrumental validation, but preliminary results should stimulate interest in this direction.

More attention should be given to the utilization of existing instruments for cytogenetic studies. The ICP Flow Cytometer does not require as much technical support as the larger, more complex, laser systems and has some cost advantage. The high resolution achieved with whole-cell analyses and with limited studies of isolated chromosomes suggests that it may play a role in cytogenetic screening; this potential should be further explored and exploited. The same suggestions apply to the parabolic flow chamber system. Finally, biophysicists and engineers should examine the possibilities of combining these existing systems with other devices such as slit-scan detection to improve chromosome resolution.

Specimen Preparation and Staining

Intact cells. Preparative techniques and staining procedures for measuring DNA in whole cells are adequate at the present time to permit cytogenetic studies with flow instruments, as outlined earlier in this chapter. These methods undoubtedly will improve in the future because of the widespread interest in flow analysis. Stains that are specific for mutagen binding sites or for areas of DNA in need of repair are highly desirable and would add new dimensions to the general area of detecting mutagen effects by flow systems. Studies on the detection of nondisjunction should be further refined and developed into test protocols. In spite of the limitations of this approach, as described earlier, it could be developed into an effective test for agents with nondisjunctive activity. Studies of aneuploidy detection by lymphocyte analysis suggest that this procedure should be applied to samples in human cytogenetics clinics. A substantial portion of the normal workload of these clinics is to confirm or exclude the diagnosis of aneuploidy, and this could be accomplished much faster with the use of flow systems. In special cases, where epidemiological data are sought to confirm the suspicion of increased rates of nondisjunction in a geographical area or subpopulation, this method could be of considerable value to neonatal screening programs.

Isolated chromosomes. Chromosome isolation procedures that are faster and more reliable than those in present use will be necessary for successful flow-oriented screening programs. The Otto technique is promising in that it is rapid and combines isolation and staining in one procedure. It also appears to have wide applicability to a variety of cell lines. However, the problems with chromosome stability and debris require further investigation. Isolation techniques must be developed that provide large numbers of isolated chromosomes from peripheral lymphocytes. None of the available methods appear to be suitable for lymphocyte samples, which are an important source of material for monitoring clastogen-exposed individuals.

Perhaps the largest potential for future developments in flow cytogenetics lies in cytochemical technique. Studies of stains with specific nucleotide affinities should be continued and expanded. This could lead to stains for specific nucleotide sequences or to the development of staining procedures for telomere or centromere regions. The staining of chromosomal proteins for flow analysis is an unexplored area that might aid in chromosome resolution, especially if coupled with DNA staining. Stains for mutagen binding sites or for sites of DNA repair are other possibilities. Exploitation of cytochemistry could make it possible to perform rapid breakage analyses with existing flow instrumentation.

Acknowledgments

This work was performed under the auspices of the United States Department of Energy and the United States Environmental Protection Agency under interagency agreement EPA-IAG-D5-E681.

I am indebted to the following people for the use of figures: Dr. M. J. Skogen-Hagenson (Fig. 16.3); Drs. L. S. Cram, D. J. Ardt-Jovin, B. G. Grimwade, and T. M. Jovin (Figs. 16.4 and 16.14); Dr. L. S. Cram (Figs. 16.6 and 16.7); Dr. R. A. Tobey (Fig. 16.8); and Drs. J. W. Gray, R. G. Langlois, A. V. Carrano, and M. A. Van Dilla (Fig. 16.10). Drs. L. S. Cram, J. H. Jett, A. G. Saponara, and D. F. Petersen provided helpful criticisms on the content and organization of the manuscript. I also want to thank G. J. Whitmore and E. M. Sullivan for secretarial and editorial help in manuscript preparation.

Literature Cited

Arndt-Jovin, D. J.; Jovin, T. M. 1978. Automated cell sorting with flow systems. Ann. Rev. Biophys. Bioeng. 7:527–58.

Boyum, A. 1968. Isolation of mononuclear cells and granulocytes from human blood. Isolation of mononuclear cells by one centrifugation, and of granulocytes by combining centrifugation at 1 g. Scand. J. Clin. Lab. Invest. 21:51–76.

Callis, J.; Hoehn, H. 1976. Flow-fluorometric diagnosis of euploid and aneuploid human lymphocytes. Am. J. Hum. Genet. 28:577–84.

Carrano, A. V.; Gray, J. W.; Moore, D. H., II; Minkler, J. L.; Mayall, B. H.; Van Dilla, M. A.; Mendelsohn, M. L. 1976. Purification of the chromosomes of the Indian muntjac by flow sorting. J. Histochem. Cytochem. 24:348–54.

Carrano, A. V.; Van Dilla, M. A.; Gray, J. W. 1978a. Flow cytogenetics: a new approach to chromosome analysis. In Melamed, M. R.; Mullaney, P. F.; Mendelsohn, M. L., eds. Flow cytometry and sorting. New York: John Wiley & Sons, Inc.; 421–51.

Carrano, A. V.; Gray, J. W.; Van Dilla, M. A. 1978b. Flow cytogenetics: Progress towards chromosomal aberration detection. In Evans, H. L.; Lloyd, D. C., eds. Mutagen-induced chromosome damage in man. Edinburgh, Scotland: Edinburgh University Press; 326–38.

Carrano, A. V.; Gray, J. W.; Langlois, R. G.; Buckhart-Schultz, K. J.; Van Dilla, M. A. 1979. Measurement and purification of human chromosomes by flow cytometry and sorting. Proc. Nat. Acad. Sci. USA 76:1382–84.

Coulter, W. H. 1956. High-speed automatic blood cell counter and cell size analyzer. Proc. Nat. Electron. Conf. 12:1034-42.
Cox, D. M.; Puck, T. T. 1969. Chromosomal nondisjunctions: the action of Colcemid on Chinese hamster cells in vitro. Cytogenetics 8:158-69.
Cram, L. S.; Lehman, J. M. 1977. Flow microfluorometric DNA content measurements of tissue culture cells and peripheral lymphocytes. Hum. Genet. 37:201-6.
Cram, L. S.; Arndt-Jovin, D. J.; Grimwade, B. G.; Jovin, T. M. 1979. Capabilities of a flow sorter that combines image slit scanning and high resolution DNA analysis. Biophys. J. (submitted).
Crissman, H. A.; Tobey, R. A. 1974. Cell cycle analysis in 20 minutes. Science 184:1297-98.
Crissman, H. A.; Mullaney, P. F.; Steinkamp, J. A. 1975. Methods and applications of flow systems. In Prescott, D. M., ed. Methods in cell biology. Vol. 9. New York: Academic Press Inc.; 179-246.
Crosland-Taylor, P. J. 1953. A device for counting small particles suspended in a fluid through a tube. Nature 171:37-38.
Dean, P. N.; Pinkel, D. 1978. High resolution and dual laser flow cytometry. J. Histochem. Cytochem. 26:622-27.
Deaven, L. L.; Petersen, D. F. 1973. The chromosomes of CHO, an aneuploid Chinese hamster cell line: G-band, C-band, and autoradiographic analysis. Chromosoma 41:129-44.
Deaven, L. L.; Peterson, D. F. 1974. Measurements of mammalian cellular DNA and its location in chromosomes. In Prescott, D. M., ed. Methods in cell biology. Vol. 8. New York: Academic Press Inc.; 179-204.
Deaven, L. L.; Sanders, P. C.; Grilly, J. L.; Kraemer, P. M.; Petersen, D. F. 1975. Chromosome G-banding and DNA constancy in aneuploid cell populations. In Richmond, C. R.; Petersen, D. F.; Mullaney, P. F.; Anderson, E. C., eds. Mammalian cells: probes and problems ERDA Symposium Series CONF-731007. National Technical Information Service, Springfield, VA: 212-27.
Deaven, L. L.; Stubblefield, E.; Jett, J. H. 1976. Karyotype analysis of Chinese hamster chromosomes by flow microfluorometry. In Mendelsohn, M. L., ed. Automation of cytogenetics, asilomar workshop ERDA Symposium Series CONF-751158. National Technical Information Service, Springfield, VA: 165-69.
Deaven, L. L.; Vidal-Rioja, L.; Jett, J. H.; Hsu, T. C. 1977. Chromosomes of *Peromyscus* (Rodentia, Cricetidae). VI. The genomic size. Cytogenet. Cell Genet. 19:241-49.
Deaven, L. L.; Oka, M. S.; Tobey, R. A. 1978. Cell-cycle-specific chromosome damage following treatment of cultured Chinese hamster cells with 4'-[(9-acridinyl)-amino]methanesulphon-m-anisidide-HCl. J. Nat. Cancer Inst. 60:1155-61.
Dittrich, W.; Göhde, W. 1969. Impulsflorometrie bei Einzelzellen in Suspensionen. Z. Naturforsch. 24b:360-61.
Dittrich, W.; Göhde, W. 1970. Phase progression in two dose responses of Ehrlich ascites tumour cells. Atomkernenergie 15:174-76.
Fulwyler, M. J. 1965. An electronic particle separator with potential biological application. Science 150:371-72.
Gray, J. W.; Carrano, A. V.; Steinmetz, L. L.; Van Dilla, M. A.; Mendelsohn, M. L. 1974. Chromosome measurement and sorting using flow systems. J. Cell Biol. 63:120a.
Gray, J. W.; Carrano, A. V.; Steinmetz, L. L.; Van Dilla, M. A.; Moore, D. H., II; Mayall, B. H.; Mendelsohn, M. L. 1975a. Chromosome measurement and sorting by flow systems. Proc. Nat. Acad. Sci. USA 72:1231-34.
Gray, J. W.; Carrano, A. V.; Moore, D. H., II; Steinmetz, L. L.; Minkler, J.; Mayall, B. H.; Mendelsohn, M. L.; Van Dilla, M. A. 1975b. High-speed quantitative karyotyping by flow microfluorometry. Clin. Chem. 21:1258-62.
Gray, J. W.; Langlois, R. G.; Carrano, A. V.; Van Dilla, M. A. 1979. High resolution chromosome analysis: one and two parameter flow cytometry. Chromosoma. In press.
Herzenberg, L. A.; Sweet, R. G.; Herzenberg, L. A. 1976. Fluorescence-activated cell sorting. Sci. Am. 234:108-17.
Hoehn, H.; Johnston, P.; Callis, J. 1977. Flow-cytogenetics: sources of DNA content variation among euploid individuals. Cytogenet. Cell Genet. 19:94-107.
Holm, D. M.; Cram, L. S. 1973. An improved flow microfluorometer for rapid measurement of cell fluorescence. Exp. Cell Res. 80:105-10.

Hulett, H. R.; Bonner, W. A.; Barrett, L. A.; Herzenberg, L. A. 1969. Cell sorting: automated separation of mammalian cells as a function of intracellular fluorescence. Science 166:747-49.
Jensen, R. H.; Langlois, R. G.; Mayall, B. H. 1977. Strategies for choosing a deoxyribonucleic acid stain for flow cytometry of metaphase chromosomes. J. Histochem. Cytochem. 25:954-64.
Kao, F.; Puck, T. T. 1969. Genetics of somatic mammalian cells. IX. Quantitation of mutagenesis by physical and chemical agents. J. Cell. Physiol. 74:245-57.
Kato, H.; Yosida, T. H. 1971. Isolation of aneusomic clones from Chinese hamster cell line following induction of nondisjunction. Cytogenetics 10:392-403.
Keith, D. H.; Deaven, L. L.; Goodpasture, C. E.; Teplitz, R. L.; Riggs, A. D. 1979. Isolation and purification of X chromosomes derived from an opossum kidney cell line. Exp. Cell Res. (submitted).
Kraemer, P. M.; Petersen, D. F.; Van Dilla, M. A. 1971. DNA constancy and heteroploidy and the stem line theory of tumors. Science 174:714-17.
Kraemer, P. M.; Deaven, L. L.; Crissman, H. A.; Van Dilla, M. A. 1972. DNA constancy despite variability in chromosome number. In DuPraw, E. J., ed. Advances in cell and molecular biology. Vol. 2. New York: Academic Press, Inc.; 47-108.
Krishan, A. 1975. Rapid flow cytofluorometric analysis of mammalian cell cycle by propidium iodide staining. J. Cell Biol. 66:188-93.
Lederberg, J. 1971. The mutagenesis of pesticides: Concepts and evolution. Cambridge, Mass. MIT Press, X-XI.
Melamed, M. R.; Mullaney, P. F.; Mendelsohn, M. L., eds. 1979. Flow cytometry and sorting. New York: John Wiley & Sons, Inc.
Mendelsohn, M. M.; Mayall, B. H.; Bogart, E.; Moore, D. H.; Perry, B. H. 1973. DNA content and DNA-based centromeric index of the 24 human chromosomes. Science 179:1126-29.
Otto, F.; Oldiges, H. 1977. Preconditions and performance of chromosomal DNA measurements for rapid karyotype analysis in mammalian cells. In Proc. 3rd International Symposium on Pulse Cytophotometry. Vienna, Austria. March 30-April 1, 1977; 60-67.
Penrose, L. A. 1964. A note on the mean measurements of human chromosomes. Ann. Hum. Genet. 28:195-96.
Petersen, D. F.; Anderson, E. C.; Tobey, R. A. 1968. Mitotic cells as a source of synchronized cultures. In Prescott, D. M. ed. Methods in cell physiology. Vol. 3. New York: Academic Press Inc.; 347-70.
Raposa, T.; Natarajan, A. T. 1974. Fluorescence banding pattern of human and mouse chromosomes with a benzimidazol derivative (Hoechst 33258). Humangenetik 21:221-26.
Skogen-Hagenson, M. J.; Salzman, G. C.; Mullaney, P. F.; Brockman, W. H. 1977. A high-efficiency flow cytometer. J. Histochem. Cytochem. 25:784-89.
Steen, H. D.; Lindmo, T. 1979. Flow cytometry: a high-resolution instrument for everyone. Science 204:403-4.
Steinkamp, J. A. 1977. Multiparameter analysis and sorting of mammalian cells. In Catsimpoolas, N., ed. Methods of cell separation. Vol. 1. New York: Plenum; 251-306.
Steinkamp, J. A.; Orlicky, D. A.; Crissman, H. A. 1979. Dual-laser flow cytometry of single mammalian cells. J. Histochem. Cytochem. 27:273-76.
Stubblefield, E.; Klevecz, R.; Deaven, L. L. 1967. Synchronized mammalian cell cultures. I. Cell replication cycle and macromolecular synthesis following brief Colcemid arrest of mitosis. J. Cell. Physiol. 69:345-53.
Stubblefield, E.; Cram, L. S.; Deaven, L. L. 1975. Flow microfluorometric analysis of isolated Chinese hamster chromosomes. Exp. Cell Res. 94:464-68.
Tobey, R. A.; Crissman, H. A. 1975. Comparative effects of three nitrosourea derivatives on mammalian cell cycle progression. Cancer Res. 35:460-70.
Tobey, R. A.; Deaven, L. L.; Oka, M. S. 1978. Kinetic response of cultured Chinese hamster cells to treatment with 4'-[(9-acridinyl)-amino]methanesulphon-m-anisidide-HCl. J. Nat. Cancer Inst. 60:1147-53.
Van de Sande, J. H.; Lin, C. C.; Jorgenson, K. F. 1977. Reverse banding on chromosomes produced by a guanosine-cytosine specific DNA binding antibiotic: olivomycin. Science 195:400-402.

Van Dilla, M. A.; Trujillo, T. T.; Mullaney, P. F.; Coulter, J. R. 1969. Cell microfluorometry: a method for rapid fluorescence measurement. Science 163:1213-14.
Van Dilla, M. A.; Carrano, A. V.; Gray, J. W. 1976. Flow karyotyping: current status and potential development. *In* Mendelsohn, M. L., ed. Automation of cytogenetics, asilomar workshop. ERDA Symposium Series CONF-751158. Springfield, VA: National Technical Information Service; 145-64.
Ward, D. C.; Reich, E.; Goldberg, I. H. 1965. Base specificity in the interaction of polynucleotides with antibiotic drugs. Science 149:1259-63.
Wheeless, L. L.; Patten, S. F. 1971. Definition of available cellular parameters with a slit-scan optical system. Acta Cytol. 15:111-12.
Wray, W. 1973. Isolation of metaphase chromosomes, mitotic apparatus, and nuclei. *In* Prescott, D. M., ed. Methods in cell biology. Vol. 6. New York: Academic Press, Inc.; 283-306.
Wray, W.; Stubblefield, E. 1970. A new method for the rapid isolation of chromosomes, mitotic apparatus, or nuclei from mammalian fibroblasts at near neutral pH. Exp. Cell Res. 5:469-78.
Zante, J.; Schumann, J.; Barlogie, B.; Göhde, W.; Buchner, T. 1976. New preparation and staining procedures for specific and rapid analysis of DNA distributions. *In* Göhde, W.; Schumann, J.; Buchner, T., eds. Pulse-Cytophotometry. Vol. 2. Ghent, Belgium: European Press; 97-106.

17
Premature Chromosome Condensation for the Detection of Mutagenic Activity

by WALTER N. HITTELMAN*

Abstract

This chapter is concerned with the use of the phenomenon of premature chromosome condensation (PCC) for the detection of mutagenic activity at the chromosome level. By visualizing the chromosomes of the interphase cell after fusion with mitotic cells, one can monitor the chromatin for structural changes resulting from a variety of types of mutagenic insult. The events leading up to the discovery of the phenomenon of PCC is discussed first. This is followed by detailed descriptions for the propagation of Sendai virus and the cell fusion technique. Since the PCC technique allows the visualization of interphase chromosomes, this phenomenon is useful in cytogenetic studies where the interphase karyotype can be determined and compared with the cells that reach mitosis. This allows for the cytogenetic analysis of mixed cell and/or nondividing cell populations. Also, since the morphology of the PCC is correlated with stage in the cell cycle, the PCC technique is useful for the cytokinetic analysis of cell populations. This approach has proven especially productive in the study of human leukemia. The PCC technique is also quite useful for assessment of chromosome damage in cell populations after insult with a variety of clastogens, including radiation and drugs. Chromosome aberrations can be monitored from cells treated in vitro or in vivo, and the technique is especially helpful for measuring damage and repair in nondividing or slowly dividing populations where few cells reach mitosis, especially after clastogenic insult. The PCC technique therefore has considerable potential for use as a sensitive method for the detection of mutagenic activity.

Introduction

With the realization that damage to cellular genetic material is related to a multitude of human disorders, including cancer and birth defects, increasing attention has been focused on the detection of potential mutagens and carcinogens in the environment.

*Department of Developmental Therapeutics, University of Texas System Cancer Center, M. D. Anderson Hospital and Tumor Institute, Houston, Texas

One active area of research is the development of both short-term assays for potentially active agents and sensitive methods to quantitate genomic damage in persons already exposed to purported mutagens (e.g., see Ames 1979). As discussed in other chapters in this volume, cytogenetic techniques have already been proven quite sensitive in the detection of environmental mutagens. The purpose of this chapter is to review the phenomenon of premature chromosome condensation (PCC) and to describe the additional benefits of this cytogenetic technique in the detection of chromosome damage and malignant transformation in living cells, both in vitro and in vivo, after mutagen exposure.

Most cytogenetic assays used to detect genomic damage require that the target cell reach mitosis where the chromatin is neatly packaged into individual chromosomes and can be scored for chromosome damage. Conventional cytogenetic techniques, however, are limited by the fact that considerable chromosome damage can be repaired prior to mitosis and the most damaged cells are delayed in their progression to mitosis. Thus, aberration levels determined in mitotic preparations underestimate the amount of genomic damage suffered by the cell in interphase. A second limitation to the conventional mitotic technique is that only dividing cell populations can be studied.

The premature chromosome condensation technique (PCC) involves the fusion of mitotic inducer cells with interphase cells resulting in the condensation of the diffuse interphase chromatin into chromosomal units. The morphology of the PCC reflects the cell's stage in the cell cycle; therefore, this technique is useful for cell cycle analysis of cell populations. The PCC also reflect the karyotype of the cell, thus chromosome damage can be detected directly in the interphase cell. With regard to mutagen testing in particular, the PCC technique offers two distinct advantages over conventional mitotic chromosome techniques. First, a more accurate estimate of the amount of chromosome damage can be determined directly in the target cell; and second, aberration frequencies can be determined in nondividing cell populations. In addition, recent reports have suggested that noncycling normal and transformed cells can be distinguished on the basis of the morphology of their PCC. The PCC technique might therefore be useful in the early detection of transformation in resting populations of cells.

The first part of this chapter is concerned with the discovery of the phenomenon of premature chromosome condensation and a discussion of the nature of the PCC reaction. The second section describes the PCC methodology, including propagation of virus, cell preparation, and the fusion process. The third section describes the usefulness of the PCC technique as a cytogenetic and cytokinetic tool, including a discussion of how this technique might be sensitive in the detection of malignant transformation. The fourth section is concerned with the use of the PCC technique as a sensitive means to detect chromosome damage and rearrangements after mutagen treatment.

The PCC technique is still in its infancy with regard to its use in the detection of mutagens, since only a handful of laboratories in the world are actively pursuing this approach. As a result, there is scant information with which to compare this technique with others concerned with mutagen testing. The purpose of this chapter, therefore is first to describe the PCC technique in detail so that more laboratories can become involved, and second, to illustrate the potential usefulness of this technique in mutagen testing.

The Phenomenon of Premature Chromosome Condensation

The purpose of this section is to familiarize the reader with the reaction that follows fusion between a mitotic and an interphase cell. Two excellent reviews (Rao and Johnson 1974a; Sperling and Rao 1974a) have described the early studies in detail, so this section will only provide highlights of this area of research. More recent investigational approaches will also be discussed.

In the early days, premature chromosome condensation, like cancer, was a phenomenon without a well-understood etiology. Now one can induce both PCC and cancer at will; however, the molecular mechanisms involved in the reactions are still not fully described. The PCC phenomenon was commonly detected in hybrid cells where one would observe a set of intact chromosomes alongside chromosomes of a different condensation state. For example, during meiosis in pollen mother cells of Triticum/Secale hybrids, Bleier (1930) occasionally found binucleate cells with one set of condensed chromosomes and one set of elongated, single-stranded chromosomes. Similarly, after infection of human cells with measles virus, Nichols et al. (1965) observed a mixture of normally condensed chromosomes alongside pulverized-appearing chromosomes. This latter phenomenon was observed after infection by a variety of viruses, all of which are known to promote cell fusion (Roizman 1962; Poste 1972).

Odd-looking chromosomes were also detected in cells which were not exposed to virus. Stubblefield (1964) observed the presence of subsets of extended chromosomes in multinucleated Chinese hamster cells accumulated in mitosis with Colcemid. In addition, Kato and Sandberg (1968a) observed morphologically abnormal chromosomes arising from micronuclei which were asynchronous with the major nucleus in a human cell line derived from a patient with Burkitt's lymphoma.

In an effort to understand these findings, several investigators set out to probe systematically the phenomenon by means of virus-induced fusion of cells. Initial studies indicated the following: (1) an asynchrony of nuclear elements is required for the reaction to occur (Kato and Sandberg 1967, 1968a); (2) the phenomenon can be observed as early as 30 minutes after treatment of the cells with virus (Kato and Sandberg 1968b); and (3) the reaction can be induced when one of the nuclei was in S or G_2 and one of the nuclei has reached mitosis (Kato and Sandberg 1968c). Based on such studies, these investigators suggested that after virus-induced fusion of mitotic and interphase cells, elements of the mitotic cell caused the break-up of the chromatin and nuclear membranes of G_1, S, or G_2 cells (Takagi et al. 1969; Sandberg et al. 1970) and thus the phenomenon was called "chromosome pulverization."

At the similar point in time, Rao and Johnson were using the cell fusion technique to study regulation of the cell's progression through the reproductive cycle. In particular, Rao and Johnson (1970) were measuring the mitotic accumulation pattern among binucleate HeLa cells which had been formed by fusion between S and G_2 cells. In a proportion of the fused cells, they observed a phenomenon similar to that described above (i.e., a normal set of mitotic chromosomes associated with a pulverized set of chromosomes). To investigate this phenomenon further, they fused mitotic cells with synchronized populations of G_1, S, or G_2 cells and found a rapid condensation of the interphase chromatin into chromosomes, and thus they called the phenomenon "premature chromosome condensation" (Johnson and Rao 1970).

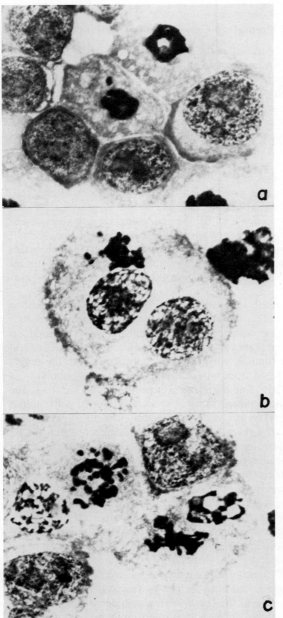

Figure 17.1 The phenomenon of premature chromosome condensation by cell fusion using Sendai virus. (a) Mixture of mitotic and interphase cells prior to fusion. (b) Intermediate stage of fusion and PCC formation. The chromatin of the interphase component is in the early stages of condensation and the nuclear membrance has begun to break down. (c) Late stage of PCC formation. The interphase chromatin is condensed in the cells fused with mitotic cells, whereas the chromatin is diffused in nonfused cells. Note the continued association of nucleolar remnants with the PCC.

The prematurely condensed chromosomes (PCC) obtained from G_1 and G_2 nuclei did not yield a pulverized appearance, so Johnson and Rao suggested that instead of pulverization, this phenomenon represented an induced condensation of interphase chromatin by factors from the mitotic component.

The induction of PCC resembles a process similar to that observed as a cell enters into mitosis, and thus the phenomenon is also called "prophasing" (Matsui et al. 1972). Early after fusion with a mitotic cell, the interphase nucleus shows slightly enhanced heterochromatic regions followed soon after by conspicuous condensation of chromatin and visualization of chromatin networks (Fig. 17.1). This is followed by disruption of the nuclear membrane (Aula 1970) and continued chromosome condensation. While the nucleolus remains intact, it takes on a more granular appearance. The whole process takes place within 20 minutes after fusion. As the nuclear membrane of the interphase cell begins to break down, substantial amounts of RNA and protein are transferred out of the interphase nucleus into the cytoplasm of the fused cells (Matsui et al. 1972). Further, as the interphase chromatin condenses, RNA synthesis (Stenman 1971) and DNA synthesis (Johnson and Rao 1970) are reduced in the interphase components.

The molecular factors involved in this process of premature chromosome condensation are still not understood. The following lines of evidence suggest that the mitotic component provides the factors for condensation. First, induction of PCC always requires the presence of a mitotic cell in the fusion reaction. Second, mitotic cells used in the fusion reaction which were accumulated in the presence of inhibitors of protein and RNA synthesis inhibitors (e.g., puromycin, cycloheximide, and actinomycin D) are less capable of inducing PCC (Matsui et al. 1971). This suggests that the factors are synthesized late in G_2 phase. Third, the higher the ratio of mitotic cells to interphase cells, the higher the PCC induction frequency (Johnson and Rao 1970). Fourth, if the mitotic inducer cells are accumulated in the presence of radioactive amino acids prior to fusion, labeled protein can be observed to pass from the mitotic component and become associated with the condensing PCC. This phenomenon is most specific with cells prelabeled with tryptophan in G_2 phase prior to accumulation in mitosis (Rao and Johnson 1974b) implying that nonhistone proteins play an important role in the condensation process.

Numerous attempts have been made in recent years to isolate the mitotic factors responsible for the condensation reaction. Successful induction of PCC by the use of mitotic extracts or mitoplasts (Sunkara et al. 1977) have been limited. Recently a breakthrough has occurred in this regard. It had been previously shown that progesterone-stimulated mature ovarian oocytes contained factors which, when injected into immature oocytes, could induce breakdown of the nuclear envelope (germinal vesicle) and chromosome condensation (Masui et al. 1971; Drury and Schorderet-Slatkine 1975; Wasserman and Masui 1976). Since this process resembles the prophasing reaction observed in the induction of PCC, Sunkara, Wright and Rao (1979) succeeded in inducing the maturation process by injecting mitotic cytoplasmic factor into *Xenopus* oocytes. Using synchronized cell populations for obtaining extract, they found that mitotic and late G_2 HeLa cells contained active factors, whereas extracts from G_1 and S cells had little activity. These investigators are now in the process of characterizing these maturation-inducing factors from mitotic HeLa cells which might be similar to the factors which induce PCC.

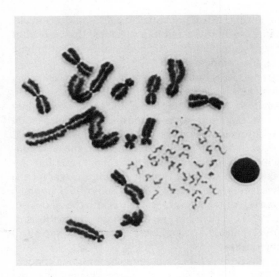

Figure 17.2 G_1 PCC obtained by fusion of fish gill cells with mitotic CHO cells. Note the difference in chromosome size between CHO and fish cells. (Photograph courtesy of David Wildrick.)

The ability of mitotic cells to induce PCC in interphase cells appears to cross species barriers. Human and Chinese hamster mitotic cells have been shown to be capable of inducing PCC in interphase cells of rat kangaroo, *Xenopus*, mosquito, horse, chick, bovine (Johnson et al. 1970), and fish (Wildrick and Hittelman 1978) (Figure 17.2). PCC can be induced in cells growing in tissue culture or in cells obtained directly from the living organism whether growing or differentiated. While the PCC reaction is rather universal, the efficiency of fusion can be influenced by a number of factors, including polyamines, salts, estradiol 17β, cyclic AMP, heparin, pH, the time of accumulation of inducer cells in mitosis (Rao and Johnson 1971; Obara et al. 1973), and the surface characteristics of the cells (e.g., availability of virus receptors).

One interesting observation has been that fusion between a mitotic cell and an interphase cell does not always result in PCC formation. In fact, after fusion, a battle ensues over which cellular component will dominate the reaction. If the mitotic component dominates the reaction, PCC will be induced in the interphase component (prophasing). If the interphase component dominates the reaction, the interphase nucleus will remain intact, whereas double membranes will form around the metaphase chromosomes, the mitotic chromosomes will decondense, and nucleoli will reform. Since this latter reaction resembles the mitotic-G_1 transition, this process has been called "telophasing" (Obara et al. 1973; Obara et al. 1974a). At higher pH levels (e.g., pH 8.0), the telophasing reaction dominates, while at lower pH levels (e.g., pH 6.6), prophasing dominates.

The balance between PCC induction and the telophase reaction also appears to depend on the comparative sizes of the fusing cells. As mentioned above, the direction of the reaction depends on the mitotic : interphase ratio (Johnson and Rao 1970; Obara et al. 1974b). Similarly, if the interphase cell is larger than the mitotic cell, telophasing occurs with a higher frequency. In one study to determine whether mouse trophoblast cells (which contain manyfold the amount of DNA of a diploid

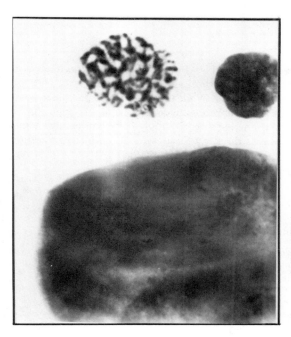

Figure 17.3 "Telophasing" of a mitotic HeLa cell when fused with mouse trophoblast cell. Note the formation of a nuclear membrane around the HeLa chromosomes.

cell) exhibit polyteny or polyploidy, we fused mouse trophoblasts with mitotic HeLa cells in order to induce PCC and visualize the chromosomes. In all cases, wherever fusion occurred between the trophoblast and the HeLa mitotic cell, the HeLa chromosomes were promptly surrounded by membranes and began to decondense (Fig. 17.3) (Hittelman and Rao, unpublished observations).

As mentioned earlier in this section, the PCC reaction can occur without cell fusion. PCC can develop whenever a multinucleate situation occurs and one of the nuclear components asynchronously reaches mitosis. For example, after clastogenic treatment, many cells are found to contain micronuclei. In these cases, if the main nucleus achieves mitosis prior to the maturation of micronuclei, PCC will be formed from the micronuclei. This phenomenon has been observed both in culture after clastogenic treatment (Kurten and Obe 1975; Obe and Beek 1975) and in vivo after chemotherapy and radiation (Kurten and Obe 1975a; Kurten and Obe, 1975b; Matter and Jaeger 1975; Witkowski and Anger 1976). While not the topic of this chapter, this process has also proven useful as a mutagen assay.

Methodology

The PCC technique requires the fusion of mitotic inducer cells with interphase cells. I will briefly review the following techniques: A) propagation of Sendai virus; B) accumulation of mitotic cells, and C) cell fusion procedure. The methods described here should not be taken as dogma; they are just those that have worked consistently for our laboratory.

Virus Propagation

Cell fusion can be accomplished with a variety of fusogens including virus (e.g., Sendai virus, Newcastle disease virus), lysolecithin, and polyethylene glycol. We use Sendai virus routinely in our own laboratory. While this virus can be purchased commercially it is easy to propagate under appropriate conditions and enough can be grown in three successive weeks to provide a year's supply for even the most active laboratory. A good batch of virus is the key to successful fusion. The procedure described here is a modification of that of Harris et al. (1966).

Required materials.

1. 10 day-old chick embryos.
2. Active Sendai virus.
3. Phosphate-buffered saline (sterile).
4. Sterile alcohol swabs.
5. Heavy tape (not Scotch).
6. Sharp-pointed scissors.
7. 1.0 ml syringe with several 20-gauge needles.
8. Beaker on which to set eggs.
9. Sterile 50 ml conical centrifuge tubes.
10. Sterile 50 ml steel centrifuge tubes.
11. Sterile, short (5¾") pasteur pipettes stoppered with cotton.
12. Ice bucket.
13. Egg incubator with water pan for maintaining humidity. The temperature should be set at 36°C.
14. Light box for candling eggs.

Procedure.

Day 1. Candle embryos using light box. Select living embryos and discard the dead ones. Mark the air sac and eyespot of the embryo on the eggshell. When ready to inject virus into egg, swab area of shell with sterile alcohol pad. Punch a small hole in the shell above the air sac with sharp pointed scissors. Make the hole just large enough for a 20-gauge needle to fit through. Inject two hemagglutinating units of active virus in 0.1 ml through the air sac membrane into the allantoic fluid avoiding the embryo eyespot. Place a square of tape over the injection hole and place the embryo into the incubator.

Day 2. Candle the embryos for viability and discard the dead ones.

Day 3. Candle the embryos for viability. Collect dead eggs and place in the refrigerator for harvesting on day 5. Reinject live embryos with four hemagglutination units of active virus in 0.1 ml and reincubate.

Day 4. Place all remaining eggs in refrigerator.

Day 5. Harvest virus from the eggs in the following manner:

    a. Peel away eggshell of refrigerated eggs above air sac and gently peel off membrane covering allantoic fluid. Collect allantoic fluid with pasteur pipets and place fluid in ice cooled, 50 ml conical glass centrifuge tubes. Avoid collecting membranes or yolk. Collect only clear or milky white fluid.

b. Centrifuge fluid at 2,000 rpm for 15 min. at 4°C. Place the supernantant into cooled steel centrifuge tubes and discard the pellet.

c. Centifuge the supernatant fluid at 15,000 rpm (27,000 g) for 30 min. at 4°C. Aspirate off the resultant supernatant, and resuspend the virus pellet in PBS at pH 7.2.

d. Check an aliquot of the virus suspension for fusion index and hemagglutination titer. Determine fusion index at various virus dilutions to find the concentration at which predominantly binucleate and trinucleate cells are formed.

e. Freeze the remaining virus suspension in 2 ml quantities at -70°C or in liquid nitrogen. The virus will remain active indefinitely at these temperatures.

Inactivation and dilution of virus. When needed, the active frozen virus should be thawed quickly. Occasionally some particulate material will appear after thawing; this material can be removed be centrifugation at 2,000 rpm, discarding the pellet. The virus can be inactivated by exposure to ultraviolet radiation from a germicidal tube to a dose of 540 joules/m². This dose is chosen so that the reproductive capacity of the virus is destroyed while retaining most of the fusion activity (Okada 1969). An alternative means of virus inactivation is the use of 0.01% β-propiolactone treatment (Wainberg et al. 1971).

The inactivated concentrated virus is then diluted in Hank's balanced salt solution (Hanks' BSS) without glucose at pH 6.8 with 20 mM PIPES. The low pH is favorable to the induction of PCC as discussed elsewhere in this chapter. The most useful final dilution is that which gives mostly bi- and trinucleate cells after fusion. If the virus is too strong, considerable cell lysis will result and adequate chromosome preparations will be difficult to attain.

Preparation of Cells for Fusion

Mitotic Cells. The PCC reaction appears to cross species barriers, so that any convenient cell line can be used for accumulation of mitotic inducer cells. In general, one should choose cells where the mitotic cells are larger than the interphase cells. This increases the chances for PCC induction. We routinely use two cell lines for mitotic inducer cells, i.e., CHO and HeLa, and the procedure for accumulating mitotic cells is as follows for CHO cells.:

One day prior to fusion, CHO cells are subcultured and plated onto petri dishes such that the dishes will approach confluence the next day. We routinely employ very confluent cultures for trypsinization, since most of these cells are in late G_1 and will yield a wave of mitoses 16–20 hours after trypsinization. This increases the mitotic yield. On the day of fusion, loose cells are removed from the dishes by vigorous pipetting, and fresh medium containing 0.5 μg/ml Colcemid is added back to the dishes. The dishes are incubated for approximately 3½ hours, at which time the accumulating mitotic cells are rounded and can be selectively detached from the dishes by gentle pipetting. This technique consistently yields greater than 95% mitotically pure populations.

For HeLa cells, a different procedure is used: Subconfluent cultures of actively growing HeLa cells are treated for 24 hours with 3 mM thymidine. At the end of this treatment, the cultures are washed twice with fresh media to remove excess thymidine and loose cells and are reincubated with fresh media. After a couple of hours, the media is changed again to remove loose cells, fresh media is added and the dishes are placed in a pressure tank at 37°C with 85 psi nitrous oxide gas and enough CO_2 to maintain pH. After 16 hours under high pressure, the mitotic cells will have rounded up and can be gently but selectively detached. This process yields large numbers of mitotically pure populations (Rao 1968). Colcemid treatment can replace the nitrous oxide procedure.

Interphase cells. A single cell suspension is desirable for fusion in suspension. For cells growing on petri dishes, we have found that removal of cells with cold 0.05% trypsin/0.02% EDTA buffered with 20 mM PIPES to pH 6.8 is adequate. Generally, the shortest treatment time yields the best results. Thus, the trypsin-EDTA solution can be removed prior to cell detachment and the cells can be removed from the dish with Hanks' BSS. Bone marrow cells are easily dispersed by mechanical methods so there is no need for enzymatic treatment. Solid tissue material is more difficult to dissociate; however, we have found that collagenase yields better fusion results than does trypsin. Apparently extensive trypsin treatment can remove the virus-binding sites on the cell membrane. The dispersed cells are then washed twice in Hanks' BSS.

Cell Fusion and Induction of PCC

Approximately equal numbers of mitotic inducer cells and interphase cells are mixed together and washed twice by centrifugation in Hanks' BSS. After the final centrifugation, the supernatant is aspirated and the cell pellet resuspended in 0.5 ml of UV-inactivated Sendai virus diluted to the appropriate concentration in Hanks' BSS without glucose. The cell fusion mixture is placed in a 4°C water bath for 15 min., then 1 drop of 20 mM $MgCl_2$ and 1 drop of 5 µg/ml Colcemid are added to the cell suspension, and the cells are placed in a 37°C water bath for 45 min. For best PCC results, the pH of the cell suspension should be maintained at 6.8–7.0 during the whole process.

At the end of the warm incubation, the fusion mixture is resuspended into 12 ml of 0.075 M KCl for 10 min. to achieve hypotonic swelling of the cells. The cells are then fixed twice in methanol : glacial acetic acid (3 : 1). We have found that a final fixative wash in methanol : glacial acetic acid (1 : 1) yields improved spreads. The fixed cells can then be dropped on clean wet slides and stained with Giemsa or by other preferred stains.

Polyethylene Glycol as an Alternative Fusogen

This laboratory has routinely used Sendai virus as a fusogen because it has yielded fairly consistent results. More recently, several laboratories have employed polyethylene glycol (PEG) to induce cell fusion for somatic cell genetics studies. Lau et al. (1976) have shown that PCC can be induced in either suspension or monolayer cultures using PEG or PEG with dimethyl sulfoxide (DMSO). The advantages of PEG techniques are twofold. First, PEG is easily available commercially; this would eliminate the need for propagation of virus. Second cells with few or no available membrane virus receptors can be fused since PEG fusion does not require site-specific recognition. For example, many laboratories have tried to induce PCC in spermatids using Sendai virus without success. Recently, however, Drwinga, Hsu, and Pathak (personal communication) apparently have succeeded in this regard with PEG. Figure 17.4 shows PCC resulting from the fusion of CHO mitotic cells and mouse testis cells. The PCC in these photographs show a haploid number of G_1 chromosomes. It is interesting to note that the centromeric regions of the mouse PCC appear despiralized. This despiralization at mouse centromere regions has also been observed in somatic cell fusions (Hittelman, unpublished observations).

The PEG technique also has disadvantages, however. First, Colcemid accumulated mitotic cells tend to be excluded from cell fusion, possibly due to the lack of microvilli on the cell surface which are required for PEG fusion (Hansen and Stabler 1977). This can be partially overcome with the use of lectins. Second, the PEG reaction is somewhat hard to control on a routine basis. Cell lysis and giant multinucleate cell formation are common problems. However, careful control of concentration and molecular weight of the PEG, the proper incubation solution, controlled timing of the reaction, and quick and effective removal of the PEG can help overcome this problem (Davidson and Park 1976; Davidson et al. 1976; Vaughan et al. 1976; Gefter et al. 1979; Schneiderman et al. 1979; Anders et al. 1978). A third disadvantage is that maximum PEG-induced cell fusion is attained after a number of hours (Lau et al. 1977). For a variety of experiments, it would be more useful to obtain PCC immediately since prolonged fusion time might confuse some results. For these reasons, at this time, this laboratory prefers the use of Sendai virus for PCC induction in all cases except those where fusion cannot be obtained by viral techniques.

The PCC Technique as a Cytokinetic and a Cytogenetic Tool

The morphology of the prematurely condensed chromosomes reflects the phase of the interphase component in the cell cycle at the time of fusion (Johnson and Rao 1970). PCC induced in G_1 cells exhibit a single chromatid per chromosome while G_2 PCC exhibit two chromatids per chromosome (Fig. 17.5). A more pulverized appearance is obtained in S phase PCC, yet both single and double chromatid fragments can be observed representing prereplicative and postreplicative chromatin, respectively.

Since the morphologies of the PCC reflect the various phases of the cell cycle, the PCC technique is useful for determining the cell cycle distribution of cell populations. While two reports have suggested that cells in various phases of the cell cycle have different efficiencies of PCC induction (Stadler and Adelberg 1972; Schnedl et al. 1975), Rao, Wilson, and Puck (1977) have found that G_1, S, and G_2 cells are equally susceptible to virus-induced fusion with mitotic cells, and thus the PCC method for cell cycle analysis is both practical and accurate.

It had been proposed by Mazia that cells go through a chromosome condensation cycle. The chromosomes at mitosis represent the most condensed state, and early S phase represent the least condensed state of the chromatin (Mazia 1963). This notion is supported by studies using such diverse chromatin condensation probes as actinomycin D binding (Pederson and Robbins 1972), DNAse susceptibility (Pederson 1972), acridine orange fluorescence measurements (Ashihara et al. 1978; Darzynkiewicz et al. 1977, 1979), ethidium bromide fluorescence intensity (Nicolini et al. 1977b), texture analysis (Nicolini et al. 1977a), and heparin sensitivity (Hildebrand

Figure 17.4 PCC resulting from fusions between mouse testis cells and CHO mitotic cells using polyethylene glycol. Since there are only 20 chromosomes visible in the G_1 PCC, these fusions probably involved mouse spermatids. Note the decondensed (lightly stained) centromere regions of the mouse PCC. (Photograph courtesy of Helen Drwinga.)

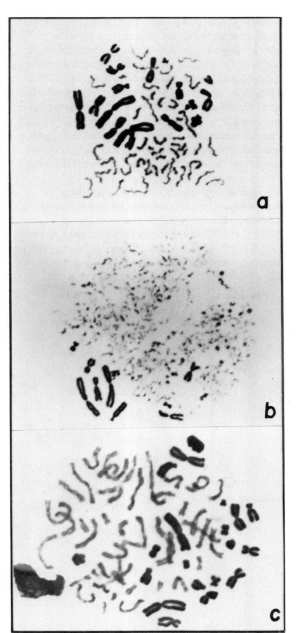

Figure 17.5 PCC of human bone marrow cells from various stages of the cell cycle after fusion with mitotic CHO cells. (a) G_1 PCC exhibiting a single chromatid per chromosome. (b) S PCC exhibiting a pulverized appearance with both single (prereplicative) and double (postreplicative) elements. (c) G_2 PCC exhibiting two chromatids per chromosome.

366 Premature Chromosome Condensation

and Tobey 1975). The PCC technique has also proven useful in determining the position of cells within each cell cycle phase based on the degree of condensation of the PCC. In studies using cell populations synchronized at mitosis and then released into G_1 phase, it was found that early G_1 cells give rise to very condensed G_1 PCC, while late G_1 PCC are very extended (Hittelman and Rao 1978a) (Fig. 17.6). Similarly, unstimulated human peripheral blood lymphocytes exhibit relatively condensed G_1 PCC, while stimulated lymphocytes give rise to increasingly elongated G_1 PCC as they pass through G_1 toward S phase (Hittelman and Rao 1976). Interestingly, using this technique one can show that quiescent normal cells accumulate in early G_1 phase while quiescent transformed cell populations accumulate in late G_1 phase (Hittelman and Rao 1978a).

S PCC present a pulverized appearance at first glance, yet the morphology of the pulverized chromatin yields information about the cells. As the cell passes through S phase, the fraction of G_2 elements (fragments with two chromtids per chromosome) increases, thus the morphology of the S PCC reflects the position of the cell in S phase (Lau and Arrighi 1978). Interestingly, heterochromatic regions which replicate very late in S phase do not appear pulverized until the time of their DNA synthesis

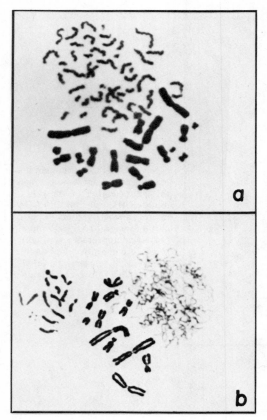

Figure 17.6 G_1 PCC of human bone marrow cells. (a) Early G_1 PCC. (b) Late G_1 PCC.

Figure 17.7 S PCC of a CHO cell in which DNA synthesis was inhibited. Note the apparent continuity of the chromosomes with single-chromatid regions contiguous with double-chromatid regions.

(Sperling and Rao 1974b). Even though the chromosomes appear fragmented during S phase, chromosome continuity is suggested by autoradiographic analysis of radioactively labeled S PCC (Rhone 1975). The continuity of the S phase chromosome is even further apparent in the PCC when DNA synthesis has been inhibited prior to induction of PCC (Sunkara et al. 1979; Hittelman, unpublished observations) (Fig. 17.7).

The chromosome condensation cycle continues as cells pass through G_2 toward mitosis. With the PCC technique as a probe, early G_2 PCC are more extended than late G_2 PCC (Sperling and Rao 1974b). While the G_2 PCC exhibit two chromatids per chromosome like mitotic chromosomes, the G_2 PCC stain lighter and are much more elongated than mitotic chromosomes, and the two chromatids are always closely aligned.

The use of the PCC technique as a cytokinetic tool is not yet widespread. However, early studies indicate its promise in this regard. As described above, normal quiescent cells in culture yield condensed G_1 PCC, while transformed quiescent cells exhibit extended G_1 PCC. This phenomenon also seems to hold true in vivo, whether in solid tumors (Grdina et al. 1977) or in the bone marrow of patients with leukemia (Hittelman and Rao 1978b, Hittelman et al. 1979a). In fact, the fraction of G_1 bone marrow cells in late G_1 is high upon leukemic presentation, intermediate during chemotherapy maintained remission, and jumps back to high levels a few months prior to clinical evidence of relapse. Based on these observations, one could hypothesize that the PCC technique might be useful in the detection of malignant transformation after treatment with environmental mutagens. This remains to be tested.

The number of PCC induced equals the number of chromosomes in the cell. Thus, the PCC technique allows the cytogenetic analysis of cell populations even when the cells are not in mitosis. For example, Yaneshevsky and Carrano (1975) showed that cells with tetraploid DNA content in senescent WI38 populations are G_1 cells with the tetraploid number of chromosomes rather than diploid cells accumulating in G_2 phase. Similarly, in a leukemia patient who showed a population of cells in her bone

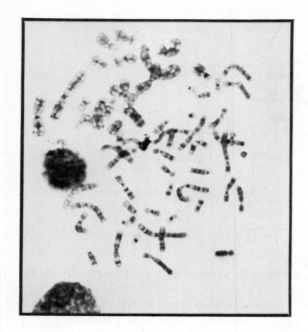

Figure 17.8 G-banded G_2 PCC of human bone marrow cell.

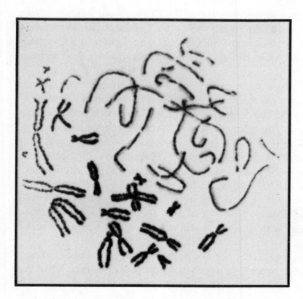

Figure 17.9 G_2 PCC of CHO cell exhibiting differential sister chromatid staining after two rounds of BUdR incorporations.

marrow with a subdiploid DNA content, PCC analysis revealed a proportion of the G_1 cells with 44 chromosomes which did not appear to be cycling into mitosis (Barlogie et al. 1977).

The PCC can also be banded by conventional banding techniques (Unakul et al. 1973; Aula 1973). While both G_1 and G_2 PCC can be banded, G_2 PCC appears to be more suitable for G-banding (Fig. 17.8). In my own experience, only early G_1 condensed PCC are suitable for banding. Apparently, chromosome decondensation in the cell progressing toward S phase results in the disappearance of the structural basis for the induction of G-banding. At the PCC level, the chromosome becomes uncoiled and this eliminates the longitudinal heterogeneity in condensation along the chromosome. In a similar manner, early G_2 PCC are more difficult to band than late G_2 PCC.

Another cytogenetic manipulation that is applicable to the PCC technique is that of sister-chromatid differential staining after cellular incorporation of bromodeoxyuridine. As described by Latt et al. in this volume, this technique is useful for the detection of sister chromatid exchanges and has proven to be a sensitive indicator for damage to the chromatin after treatment with a variety of mutagens. Using incubation conditions similar to those described in the literature, sister chromatid differential staining patterns can be induced in the PCC as well as mitotic chromosomes (Lau et al. 1976) (Fig. 17.9). Sister chromatid exchanges (SCEs) could be visualized in the S and G_2 PCC, and SCEs from prior generations are observed in the G_1 PCC as a switch from light to dark staining properties along the length of the G_1 chromosomes.

The PCC Technique for the Measurement of Chromosome Aberrations

The fact that chromosome aberration determinations are useful in the detection of mutagenic activity is supported by the existence of this volume. Traditionally, the detection of chromosome aberrations was possible only in dividing cell populations since chromosomes are normally visible only at mitosis. As a result, chromosome aberration studies required proliferating cell cultures or dividing cell populations (e.g., bone marrow) and stimulable populations obtained from the organism (e.g., mitogen-stimulated lymphocytes or liver cells stimulated by partial hepatectomy). The requirement represents a drawback when nondividing cells are the target of a mutagen in vivo.

Soon after the PCC technique was shown to be feasible for cytogenetic studies, it was felt that it might be useful for the detection of chromosome aberrations in interphase cells. In our own initial studies, we treated Chinese hamster ovary cells with a variety of clastogens and immediately fused these treated cells with synchronized mitotic cells to determine if chromosome aberrations could be detected immediately in the G_2 PCC. Chromatid gaps, breaks, and exchanges could be observed in the G_2 PCC immediately after treatment with ionizing radiation (Hittelman and Rao 1974a). On the other hand, chromatid breaks and exchanges were not observed immediately after treatment with ultraviolet radiation and alkylating agents (i.e., nitrogen mustard and trenimon) (Hittelman and Rao 1974b). These results were to be expected since ionizing radiations are thought to produce aberrations directly in treated cells. However, after UV-irradiation or alkyating agent treatment, cells must

370 Premature Chromosome Condensation

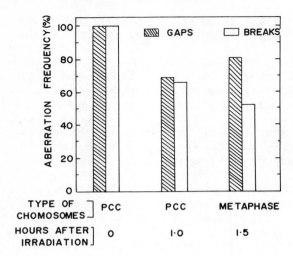

Figure 17.10 The frequency of gaps and breaks in PCC as compared to those in metaphase chromosomes after γ-irradiation. The number of aberrations (gaps and breaks) scored in PCC at one hr. after irradiation and in metaphase chromosomes are expressed as a percent of that observed in PCC immediately after irradiation. (Reproduced from Hittelman and Rao 1974a, with permission of Elsevier.)

pass through S phase in order for DNA lesions to be translated into chromosome aberrations (see Evans 1962; Bender 1974; Brinkley and Hittelman 1975).

The sensitivity of the PCC technique was tested by comparing the aberration levels observed in G_2 cells immediately after ionizing radiation treatment with that measured when these cells reach mitosis. Higher aberration frequencies were observed in the G_2 PCC than in mitotic preparations (Fig. 17.10). At the doses used many of these differences in aberration frequencies could be accounted for by chromosome repair prior to entry into mitosis. In fact, the rate of chromosome repair can be measured directly in the G_2 cells using the PCC technique. We found chromatid exchanges could not be repaired, whereas about half the chromatid breaks and gaps could be repaired within an hour (Fig. 17.11).

While breaks and exchanges were not apparent in the G_2 PCC immediately after treatment with UV radiation or alkylating agents, chromatid gaps were observed. Interestingly, after UV treatment the gap frequency in the G_2 PCC was higher than in the mitotic preparations, whereas the reverse was true for alkylating agents. These results suggested that two types of chromosome gaps exist; both are detected in mitotic chromosomes, but only one can be observed in PCC. The second type of gap might reflect the difference in condensation states of PCC and mitotic chromosomes.

Waldren and Johnson (1974) simultaneously reported similar findings using a different approach. These investigators treated synchronized HeLa cells in G_1 phase with X- and UV-irradiation and determined the aberration frequencies in G_1 PCC. X-rays induced immediate breakage of the G_1 chromosomes and the break frequency was linearly related to the dose. In the case of G_1 cells, the break frequency is determined by counting the total number of chromosome fragments per cell after treatment and subtracting the control chromosome count. In these experiments, depending on the dose, up to 41% of the induced breaks could be repaired within a two hour post-treatment incubation period. Chromosome repair after X-irradiation could not be associated with significant unscheduled DNA synthesis. This is typically the case with ionizing radiation.

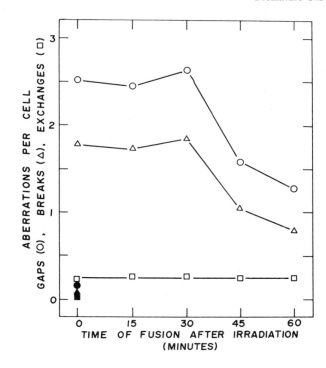

Figure 17.11 Measurement of the rate of chromosome aberration repair after γ-irradiation. The frequency of chromatid aberrations in G_2 PCC are expressed as a function of the time elapsed between γ-irradiation and cell fusion.

The G_1 PCC did not show appreciable fragmentation after UV-irradiation. This finding is consistent with the notion that the UV-irradiated cells must pass through S phase for aberrations to appear. On the other hand, upon posttreatment incubation, the G_1 PCC were observed to uncoil and become elongated more quickly than the unirradiated controls as the cells progressed toward S phase. Unscheduled DNA synthesis was associated with the uncoiling of the chromosomes. Both the rate of unscheduled DNA synthesis and the rate of elongation of chromatid fibers were dose dependent.

In subsequent work, it was found that hydroxyurea enhanced the ability of UV-irradiation to promote the decondensation of chromosomes during G_1, G_2, and mitosis (Schor, Johnson, and Waldren 1975). Repair of UV damage is thought to involve unwinding or decondensation of chromatin, and it was felt that hydroxyurea allowed initiation of repair (nicking at DNA dimer sites) yet blocked completion of repair (Collins, Schor, and Johnson 1977; Burg, Collins, and Johnson 1977). Using the PCC technique to identify the location of unscheduled DNA synthesis within chromosomes, Johnson and Sperling (1978) recently observed a nonrandom chromosomal distribution in diploid and heteroploid human cells. In fact, autoradiographically defined unscheduled DNA synthesis appeared to occur preferentially over telomeric and centromeric regions.

The PCC technique has been used to visualize chromosome damage in interphase cells after a variety of other agents. Bleomycin has been shown to produce chromatid damage immediately in G_2 cells and about half of the chromatid breaks are repaired

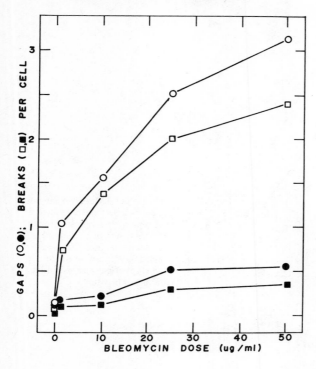

Figure 17.12 Bleomycin dose response for chromatid gaps and breaks in CHO cells treated in G_2 and aberrations measured either in G_2 PCC (open symbols) or mitotic preparations (closed symbols). Note the increased sensitivity of determining aberration levels in G_2 PCC.

within an hour (Hittelman and Rao 1974c). On the other hand, exchanges once formed are not repaired. As in the case of ionizing radiation, the aberration frequency observed in the G_2 PCC was higher than that observed when the G_2 cells reached mitosis. Some of the difference in aberration frequency could be accounted for by repair of chromosome damage prior to mitosis. In these bleomycin experiments, a block in cell progression to mitosis was observed; in fact, some of the cells treated in G_2 did not reach mitosis for many hours. These results suggested that the more damaged cells remain in G_2 phase for an extended period of time while the less damaged cells can attain mitosis.

It is useful to expand on this point because it illustrates one of the advantages of the PCC technique over conventional mitotic techniques. In one series of experiments, chromosome damage was determined as a function of bleomycin dose in Chinese hamster ovary cells. Aberration frequencies were determined either immediately after treatment in G_2 PCC or in mitotic figures of the cells treated while in G_2. As shown in figures 17.12 and 17.13, many more gaps, breaks, and exchanges were observed in the G_2 PCC than in the mitotic figures, and the ratio of the G_2 aberration frequency to the mitotic aberration frequency increased with dose. These results suggest that the aberration levels observed in mitosis grossly underestimate the amount of damage incurred in the cell. While some of the difference reflects repair, the chromatid exchange curve yields additional information. Since chromatid exchanges are not found to be repaired, the difference in the two curves reflects the inability of the more damaged cells to reach mitosis. The same observation was made

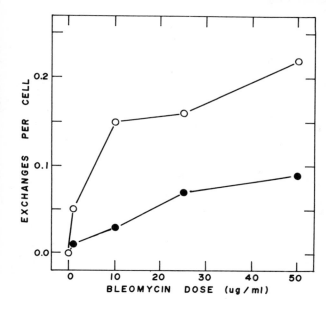

Figure 17.13 Bleomycin dose response for chromatid exchanges induced in G_2 CHO cells and measured in G_2 PCC (open symbols) or mitotic preparations (closed symbols).

by Waldren and Johnson (1974) where the aberration frequencies observed in G_1 PCC after X-irradiation were approximately 10 times higher than that expected when these cells reach mitosis.

The relationship between G_2 progression delay and chromosome damage is not unique to bleomycin and X-rays; actually, this phenomenon is routinely observed after clastogen treatment. In fact, Rao and Rao (1976) found a strong correlation for the degree of G_2 accumulation and the fraction of extensively damaged G_2 PCC in the population after treatment of HeLa cells with a variety of agents, including VM 26, nitrosoureas, cis-acid, and neocarzinostatin. The reason for the G_2 block is not well understood. However, the failure of cells to reach mitosis appears to be either a direct or indirect result of chromosome damage. The G_2 cells can be induced into PCC to the same degree as untreated G_2 cells, so the G_2 block is not due merely to inability of the chromatin to condense (Mitchell and Bedford 1978). However, the protein factors and chromatin maturation required for condensation of chromosomes into mitotic figures might involve more than that required for induction of PCC. In any case, if untreated G_2 cells are fused with G_2-blocked cells, the untreated cells can pull the treated cells into mitosis (Al-Bader et al. 1978). This fact combined with the observation that certain proteins appear to be missing from blocked G_2 cells suggest that the damaged G_2 cells are unable to make the mitotic protein factors, and the untreated G_2 cells can supply these factors upon cell fusion.

On the other hand, very damaged cells can reach mitosis with normal kinetics if the cells are allowed first to repair much of the reparable damage. For example, after low-dose irradiation of Chinese hamsters for extended periods of time, their PHA-stimulated lymphocytes will show extremely high chromosome exchange frequencies yet show normal progression to mitosis (Brooks 1979). This finding and the observation that severely damaged cells can attain mitosis after an extended period of time

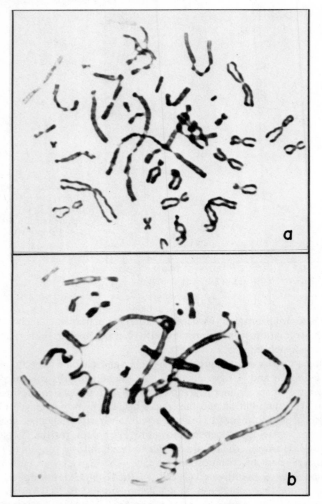

Figure 17.14 Chromatid damage in CHO cells after adriamycin treatment. (a) G_2 PCC. (b) Mitotic preparation.

suggest the alternative hypothesis that G_2 delay is related to the decondensation process involved in repair of DNA and chromosome damage. Once the repair is complete, the chromosome can then be condensed and the cell can proceed to mitosis.

The chromosome repair process after bleomycin treatment has been recently studied in somewhat more detail using the PCC technique. It has been found that chromosome repair occurs despite the presence of the metabolic inhibitors, hycanthone (RNA synthesis) and hydroxyurea (DNA synthesis). However, cycloheximide and streptovitacin A (protein synthesis inhibitors) effectively block chromosome repair (Sognier, Hittelman, and Rao 1979). On the other hand, none of the metabolic inhibitors significantly block the repair of bleomycin-induced DNA damage as measured by the alkaline elution technique (Sognier and Hittelman 1979). Interestingly, the rate of repair of DNA damage was found to be much quicker ($t\frac{1}{2} \sim 15$ min.) than the rate of repair of chromosome damage ($t\frac{1}{2} \sim 60$ min.).

The picture is somewhat different after adriamycin treatment. As with bleomycin, chromosome damage could be detected in G_2 PCC immediately after adriamycin treatment, and this was associated with a prolongation of the S and G_2 periods but not of the G_1 period (Hittelman and Rao 1975). On the other hand, at higher doses, the frequency of exchange-type aberrations appeared to be higher in mitotic figures than they were in G_2 as measured by the PCC technique (Fig. 17.14). These results were consistent with the notion that tightly bound intercalating agents continue to induce chromosome damage (Hsu et al. 1975). This effect is most apparent in the exchange frequencies, since exchanges are not reparable while breaks and gaps can be repaired.

One advantage of the PCC technique, therefore, is its increased sensitivity over the mitotic technique in determining the initial amount of chromosome damage. A second major advantage of the PCC technique is that it allows the measurement of chromosome damage in nondividing cells. Normal resting cell populations appear to accumulate in early G_1 and exhibit condensed G_1 PCC (Hittelman and Rao 1978a). Thus, chromosome aberration frequencies can be determined by counting the number of chromosome pieces per cell (Waldren and Johnson 1974). Further, chromosome exchange rates can be determined if the centromeres of the G_1 PCC are enhanced by C-banding.

We recently tested this notion by treating quiescent normal human PA2 cells with neocarzinostatin and measuring the aberration frequencies in the G_1 PCC. Neocarzinostatin immediately induced chromosome breakage in the quiescent cells (Fig. 17.15), and the break frequency was dose-dependent. As shown in Figure 17.16, the NCS-induced breaks could be repaired with a half-time on the order of five hours. By 43½ hours after NCS treatment, nearly all the chromosome breaks were repaired. The frequency of exchange was not determined in this experiment.

The unique ability of the PCC technique to assay chromosome damage in quiescent cells makes it potentially very useful for monitoring the in vivo effect of mutagens. Current in vivo assays require the use of dividing cell populations (e.g., bone marrow) for the visualization of chromosome damage or determination of micronuclei production, or require the use of cell populations such as lymphocytes which can be stimulated to divide by mitogens. The PCC technique, however, allows any biopsiable cell population to be assayed for chromosome damage, dividing or not. This approach should provide a fertile ground for future mutagen testing for it combines all the advantages (pharmacological) and disadvantages (cost) of in vivo assays, with the additional ability to assay a greater variety of tissues.

In recent years, we have found the PCC technique to be quite useful for monitoring the bone marrow cells of patients with leukemia for response of this disease to chemotherapy (Hittelman and Rao 1978b; Hittelman et al. 1979). The effect of chemotherapy can be observed as chromosome damage in the PCC of the bone marrow cells. Continued response to chemotherapy is correlated with continued evidence of chromosome damage, and resistance of disease to chemotherapy is correlated with a lack of chromosome damage (Hittelman et al. 1980). The effect of chemotherapy can be quite dramatic on the bone marrow cells. Two patient examples will help illustrate this point. The first patient, suffering from untreated acute myelogenous leukemia, was found to have a ring chromosome in his bone marrow

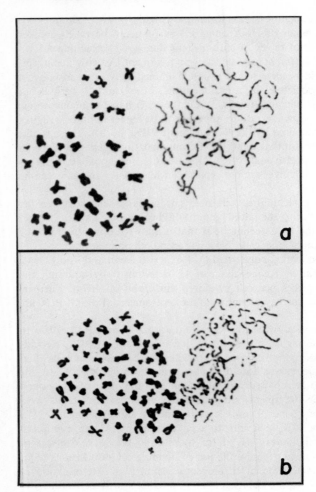

Figure 17.15 Condensed G_1 PCC of quiescent normal human PA2 cells. (a) G_1 PCC from untreated cell. (b) G_1 PCC from cell treated with neocarzinostatin exhibiting chromosome damage.

cells which could be visualized both in G_1 and G_2 PCC (Fig. 17.17a,b). The patient was started on combination chemotherapy treatment consisting of adriamycin, cytosine arabinoside, vincristine, and prednisone. On day 3 of therapy, PCC analysis of a bone marrow specimen showed extreme chromosome damage both in G_1 and G_2 cells (Fig. 17.17c,d). The high frequency of exchanges in these cells probably reflected the action of adriamycin. Figure 17.18 is an example of the type of damage observed in the bone marrow cells of a second leukemia patient after treatment with AMSA. Notice the amount of chromatid exchange in the G_2 PCC. The effects of chemotherapy can be long-lived as well. Cells can sustain a lot of chromosome damage and yet survive as long as the cells are not required to progress through cell division. In fact, we have visualized significant chromosome damage in the bone marrow cells of patients who have been off therapy for many months. It is no

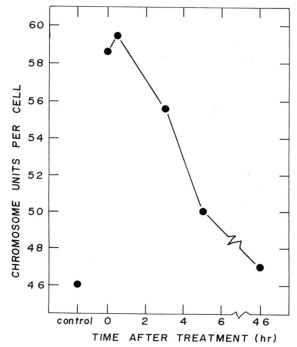

Figure 17.16 Chromosome aberration repair kinetics after treatment of quiescent normal human PA2 cells with 25 μg/ml neocarzinostatin.

wonder then that second malignancies are not uncommon in patients who have received chemotherapy and radiotherapy for their disease.

The PCC technique can therefore be quite useful for monitoring the long-term clastogenic effect of environmental mutagens on the nondividing cells of the body, providing these cells can be obtained from the body. Two hypothetical situations will illustrate this point. Workers are often exposed to toxic materials in their jobs, yet it is often difficult to determine the effect and degree of exposure of the cells at risk. For example, it is currently impossible to determine the amount of lung cell damage in workers who have accidentally inhaled plutonium since (1) lung cells are nondividing for the most part, and (2) peripheral blood lymphocytes which can be obtained and assayed for damage might not have been exposed to the plutonium which is lodged in the lung. The PCC technique offers the unique ability to detect chromosome damage in the lung cells because (1) lung cells can be obtained by bronchial washing; (2) alpha particles emitted from plutonium induce chromosome damage directly in cells; and (3) chromosome damage can be determined in the G_1 PCC.

Alternatively, the PCC technique might be useful for determining the organ specificity of various clastogens after mutagenic treatment in vivo. Using this experimental approach, the test animal could be treated with the desired mutagen, and then cells from various tissues could be assayed for chromosome damage by the PCC

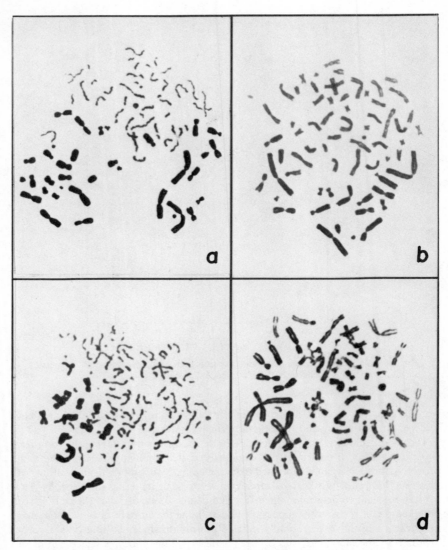

Figure 17.17 Chromosome damage observed in PCC of bone marrow cells obtained from a leukemia patient undergoing remission induction therapy. (a) G_1 PCC and (b) G_2 PCC prior to initiation of therapy. Note ring chromosome. (c) G_1 PCC and (d) G_2 PCC obtained one day after initiation of therapy with adriamycin, cytosine arabinoside, vincristine, and prednisone. Note extensive chromosome damage in the cells after therapy.

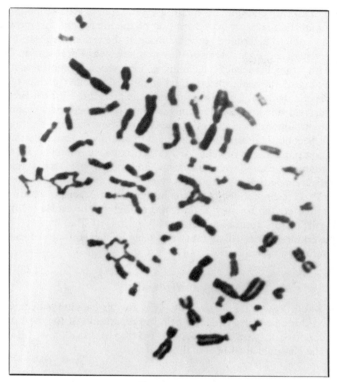

Figure 17.18 Chromosome damage observed in G_2 PCC of bone marrow cell of a leukemia patient treated with AMSA. Note the high frequency of chromatid exchanges.

technique. The advantage of this approach is that the mutagenic treatment is performed in vivo and the organ target specificity can be correlated with the amount of chromosome damage in the cells of each organ.

One disadvantage with the use of the PCC technique to monitor chromosome damage in nondividing cells is that, at present, it is only useful for determining the effect of direct clastogens, i.e., agents which produce chromosome damage directly in the cell. Agents which require passage of the damaged cell through S phase for the formation of aberrations (e.g., alkylating agents, ultraviolet light) would be missed in the G_1 PCC of noncycling cells. Future improvements, however, might allow the artificial translation of DNA damage into chromosome damage (Natarajan and Obe 1978) which could then be visualized by the PCC technique. This ability would greatly increase the sensitivity of the PCC technique for determining the mutagenic effect of a variety of agents on cells in vivo.

Summary

The phenomenon of premature chromosome condensation (PCC) has been shown to be a potentially useful tool in the detection of mutagenic activity in the environment.

The PCC technique involves the fusion of target interphase cells with mitotic inducer cells resulting in the premature condensation of the interphase chromosomes. This reaction allows one to determine both the phase of the cell in the reproductive cycle as well as to visualize the karyotype of the interphase cell. Combination of the PCC technique with other cytogenetic techniques (i.e., chromosome banding and differential staining of the sister chromatids) enhance the ability of this technique to detect cytogenetic damage. Studies reported in the literature so far suggest the PCC technique has two advantages over conventional cytogenetic techniques. First, it allows a more accurate measurement of the chromosome damage residing in the interphase cell. Second, it allows the determination of damage levels in nondividing as well as dividing populations of cells. Any target tissue can therefore be assayed for cytogenetic damage caused by clastogenic and mutagenic agents regardless of whether exposure occurred in vitro or in vivo.

While the PCC technique requires little in the way of expensive hardware, its use has been limited so far to only a few laboratories in the world. Hopefully improvements in fusion technology and elaboration of the factors involved in chromosome condensation will simplify this technique even more so that it will be used as a routine cytogenetic tool.

Acknowledgments

The author would like to thank Dr. Potu Rao for his encouragement and helpful guidance, and Barbara Wilson for her help in preparation of the manuscript. This work was supported in part by Grants CA 14528, CA 11520, CA 27931 and GM 23252 from the National Institutes of Health.

Literature Cited

1. Al-Bader, A. A.; Orengo, A.; Rao, P. N. G_2-phase-specific proteins of HeLa cells. Proc. Nat. Acad. Sci. USA 75:6064–68; 1978.

2. Ames, B. N. Identifying environmental chemicals causing mutations and cancer. Science 204:587–93; 1979.

3. Anders, G. J. P. A.; Wierda, J.; Nienhaus, A. J.; Idenburg, U. Time and cell systems as variables in fusion experiments with polyethylene glycol. Hum. Genet. 42:319–22; 1978.

4. Ashihara, T.; Traganos, F.; Baserga, R.; Darzynkiewicz, Z. A comparison of cell-cycle–related changes in postmitotic and quiescent AF8 cells as measured by cytofluorometry after acridine orange staining. Cancer Res. 38:2514–18; 1978.

5. Aula, P. Electron-microscopic observations on Sendai virus-induced chromosome pulverization in HeLa cells. Hereditas 65:163–70; 1970.

6. Aula, P. Virus-induced premature chromosome condensation (PCC) in single cells and G bands of PCC-chromatin. Hereditas 74:81–88; 1973.

7. Barlogie, B.; Hittelman, W.; Spitzer, G.; Trujillo, J.; Hart, J.; Smallwood, L.; Drewinko, B. Correlation of DNA distribution abnormalities with cytogenetic findings in human adult leukemia and lymphoma. Cancer Res. 37:4400–4407; 1977.

8. Bender, M. A.; Griggs, H. G.; Bedford, J. S. Mechanisms of chromosomal aberration production. III. Chemicals and ionizing radiation. Mutat. Res. 23:197–212; 1974.

9. Bleier, H. Verhalten der verschiedenen kernkomponenten bei der Reduktionsteilung von Bastarden. Cellule 40:87–111; 1930.

10. Brinkley, B. R.; Hittelman, W. N. Ultrastructure of mammalian chromosome aberrations. International Review of Cytology 42:49–101; 1975.

11. Brooks, A. L.: Low dose and dose-rate effects on cytogenetics. *In* Meyn, R.; Withers, R., eds. Radiation biology in cancer research. New York: Raven Press; 1979: 263–76.

12. Burg, K.; Collins, A. R. S.; Johnson, R. T. Effects of ultraviolet light on synchronized Chinese hamster ovary cells: potentiation by hydroxyurea. J. Cell Science 28:29-48; 1977.
13. Collins, A. R. S.; Schor, S. L.; Johnson, R. T. The inhibition of repair in UV-irradiated human cells. Mutat. Res. 42:413-32; 1977.
14. Darzynkiewicz, Z.; Traganos, F.; Andreeff, M.; Sharpless, T.; Melamed, M. R. Different sensitivity of chromatin to acid denaturation in quiescent and cycling cells as revealed by flow cytometry. J. Histochem. and Cytochem. 27:478-85; 1979.
15. Darzynkiewicz, Z.; Traganos, F.; Sharpless, T.; Melamed, M. R. Cell cycle-related changes in nuclear chromatin of stimulated lymphocytes as measured by flow cytometry. Cancer Res. 37:4635-40; 1977.
16. Davidson, R. L.; O'Malley, K. A.; Wheeler, T. B. Polyethylene glycol-induced mammalian cell hybridization: effect of polyethylene glycol molecular weight and concentration. Somatic Cell Genet. 2:271-80; 1976.
17. Davidson, R. L.; Park, G. S. Improved techniques for the induction of mammalian cell hybridization by polyethylene glycol. Somatic Cell Genet. 2:165-76; 1976.
18. Drury, K. C.; Schorderet-Slatkine, S. Effects of cycloheximide on the "autocatalytic" nature of the maturation promoting factor (MPF) in oocytes of *Xenopus laevis*. Cell 4:269-74; 1975.
19. Evans, H. J. Chromosome aberrations induced by ionizing radiations. *In* Bourne, G. H.; Danielli, S. F. eds. International review of cytology. Vol 13. New York: Academic Press; 1962: 221-321.
20. Gefter, M. L.; Margulies, D. H.; Scharff, M. D. A simple method for polyethylene glycol-promoted hybridization of mouse myeloma cells. Somatic Cell Genet. 3:231-36; 1976.
21. Grdina, D.; Hittelman, W. N.; White, R. A.; Meistrich, M. L. Relevance of density, size, and DNA content of tumour cells to the lung colony assay. Brit. J. Cancer 36:659-69; 1977.
22. Hansen, D.; Stadler, J. Increased polyethylene glycol-mediated fusion competence in mitotic cells of a mouse lymphoid cell line. Somatic Cell Genet. 3:471-82; 1977.
23. Harris, H.; Watkins, J. F.; Ford, C. E.; Schoefl, G. I. Artificial heterokaryons of animal cells from different species. J. Cell Science 1:1-34; 1966.
24. Hildebrand, C. E.; Tobey, R. A. Cell-cycle-specific changes in chromatin organization. Biochem. Biophys. Res. Commun. 63:134-39; 1975.
25. Hittelman, W. N.; Broussard, L. C.; McCredie, K. Premature chromosome condensation studies in human leukemia. I. Pretreatment characteristics. Blood 54:1001-14; 1979.
25a. Hittelman, W. N.; Broussard, L. C.; McCredie, K.; Murphy, S. G. Premature chromosome condensation studies in human leukemia. II. Proliferative potential changes after induction therapy for AML patients. Blood 55:457-65; 1980.
26. Hittelman, W. N.; Rao, P. N. Visualization of X-ray-induced chromosome damage in interphase cells. Mutat. Res. 23:251-58; 1974a.
27. Hittelman, W. N.; Rao, P. N. Premature chromosome condensation. II. The nature of chromosome gaps produced by alkylating agents and ultraviolet light. Mutat. Res. 23:259-66; 1974b.
28. Hittelman, W. N.; Rao, P. N. Bleomycin-induced damage in prematurely condensed chromosomes and its relationship to cell cycle progression in CHO cells. Cancer Res. 34:3433-39; 1974c.
29. Hittelman, W. N.; Rao, P. N. The nature of adriamycin-induced cytotoxicity in Chinese hamster cells as revealed by premature chromosome condensation. Cancer Res. 35:3027-35; 1975.
30. Hittelman, W. N.; Rao, P. N. Premature chromsome condensation: conformational changes of chromatin associated with phytohemagglutinin-stimulation of peripheral lymphocytes. Exp. Cell Res. 100:219-22; 1976.
31. Hittelman, W. N.; Rao, P. N. Mapping G_1 phase by the structural morphology of the prematurely condensed chromosomes. J. Cell Physiol. 95:333-41; 1978a.
32. Hittelman, W. N.; Rao, P. N. Predicting response or progression of human leukemia by premature chromosome condensation of bone marrow cells. Cancer Res. 38:416-23; 1978b.
33. Hsu, T. C.; Pathak, S.; Kusyk, C. J. Continuous induction of chromatid lesions by DNA-intercalating compounds. Mutat. Res. 33:417-20; 1975.
34. Johnson, R. T.; Rao, P. N. Mammalian cell fusion: induction of premature chromosome condensation in interphase nuclei. Nature 226:717-22; 1970.

35. Johnson, R. T.; Rao, P. N.; Hughes, S. D. Mammalian cell fusion. III. A HeLa cell inducer of premature chromosome condensation active in cells from a variety of animal species. J. Cell Physiol. 77:151-58; 1970.
36. Johnson, R. T.; Sperling, K. Pattern of ultraviolet-light-induced repair in metaphase and interphase chromosomes. Int. J. Radiat. Biol. 34:575-82; 1978.
37. Kato, H.; Sandberg, A. A. Chromosome pulverization in human binucleate cells following Colcemid treatment. J. Cell Biology 34:35-45; 1967.
38. Kato, H.; Sandberg, A. A. Chromosome pulverization in human cells with micronuclei. J. Nat. Cancer Inst. 40:165-79; 1968a.
39. Kato, H.; Sandberg, A. A. Chromosome pulverization in Chinese hamster cells induced by Sendai virus. J. Nat. Cancer Inst. 41:1117-23; 1968b.
40. Kato, H.; Sandberg, A. A. Cellular phase of chromosome pulverization induced by Sendai virus. J. Nat. Cancer Inst. 41:1125-31; 1968c.
41. Kurten, S.; Obe, G. Premature chromosome condensation in the bone marrow of Chinese hamsters after application of bleomycin in vivo. Mutat. Res. 27:285-94; 1975a.
42. Kurten, S.; Obe, G. Premature chromosome condensation in the bone marrow of Chinese hamsters after whole body irradiation with Co^{60} γ-rays in vivo. Humangenetik 28:97-102; 1975b.
43. Lau, Y. R.; Arrighi, F. E. Interphase chromosome replication patterns. J. Cell Biology 79:130a 1978.
44. Lau, Y. F.; Brown, R. L.; Arrighi, F. E. Induction of premature chromosome condensation in CHO cells fused with polyethylene glycol. Exp. Cell Res. 110:57-61; 1977.
45. Lau, Y. F.; Hittelman, W. N.; Arrighi, F. E. Sister chromatid differential staining pattern in prematurely condensed chromosomes. Experientia 32:917-18; 1976.
46. Masui, Y.; Market, C. L. Cytoplasmic control of nuclear behavior during meiotic maturation of frog oocytes. J. Exp. Zool. 177:129-46; 1971.
47. Matsui, S.; Weinfeld, H.; Sandberg, A. A. Dependence of chromosome pulverization in virus-fused cells on events in the G_2 period. J. Nat. Cancer Inst. 47:401-11; 1971.
48. Matsui, S.; Yoshida, H.; Weinfeld, H.; Sandberg, A. A. Induction of prophase in interphase nuclei by fusion with metaphase cells. J. Cell Biology 54:120-132; 1972.
49. Matter, B. E.; Jaeger, I. Premature chromosome condensation, structural chromosome aberrations, and micronuclei in early mouse embryos after treatment of paternal postmeiotic germ cells with triethylenemelamine: possible mechanisms for chemically induced dominant-lethal mutations. Mutat. Res. 33:251-60; 1975.
50. Mazia, D. Synthetic activities leading to mitosis. J. Cell Comp. Physiol. (Suppl. 1) 62:123-40; 1963.
51. Mitchell, J. B.; Bedford, J. S. Chromosome condensation and radiation-induced G_2 arrest studied by the induction of premature chromosome condensation following cell fusion. Int. J. Radiat. Biol. 34:349-57; 1978.
52. Natarajan, A. T.; Obe, G. Molecular mechanisms involved in the production of chromosomal aberrations. I. Utilization of *Neurospora* endonuclease for the study of aberration production in G_2 stage of the cell cycle. Mutat. Res. 52:137-49; 1978.
53. Nichols, W. W.; Levan, A.; Aula, P.; Norrby, E. Chromosome damage associated with the measles virus in vitro. Hereditas 54:101-18; 1965.
54. Nicolini, C.; Giaretti, W.; DeSaive, C.; Kendall, F. The G_0-G_1 transition of WI38 cells. II. Geometric and densitometric texture analyses. Exp. Cell Res. 106:119-25; 1977a.
55. Nicolini, C.; Kendall, F.; Baserga, R.; DeSaive, C.; Clarksen, B.; Fried, J. The G_0-G_1 transition in WI38 cells. I. Laser flow microfluorometric studies. Exp. Cell Res. 106:111-18; 1977b.
56. Obara, Y.; Chai, L. S.; Weinfeld, H.; Sandberg, A. A. Synchronization of events in fused interphase-metaphase binucleate cells: progression of the telophase-like nucleus. J. Nat. Cancer Inst. 53:247-59; 1974a.
57. Obara, Y.; Chai, L.; Weinfeld, H.; Sandberg, A. A. Prophasing of interphase nuclei and induction of nuclear envelopes around metaphase chromosomes in HeLa and Chinese hamster homo- and heterokaryons. J. Cell Biology 62:104-13; 1974b.
58. Obara, Y.; Yoshida, H.; Chair, L. S.; Weinfeld, H.; Sandberg, A. A. Contrast between the environmental pH dependencies of prophasing and nuclear membrane formation in interphase-metaphase cells. J. Cell Biology 58:608-17; 1973.

59. Obe, G.; Beek, B. The human leukocyte test system. VII. Further investigations concerning micronucleus-derived premature chromosome condensation. Humangenetik 30:143-54; 1975.

60. Obe, G.; Beek, B.; Vaidya, V. G. The human leukocyte test system. III. Premature chromosome condensation from chemically and X-ray-induced micronuclei. Mutat. Res. 27:89-101; 1975.

61. Okada, Y. Factors in fusion of cells by HVJ. Current Topics Microbiol. Immunol. 48:102-28; 1969.

62. Pederson, T. Chromatin structure and the cell cycle. Proc. Nat. Acad. Sci. USA 69:2224-28; 1972.

63. Pederson, T.; Robbins, E. Chromatin structure and the cell division cycle. J. Cell Biology 55:322-27; 1972.

64. Poste, G. Mechanisms of virus-induced cell fusion. International Review of Cytology 33:157-252; 1972.

65a. Rao, P. N. Mitotic synchrony in mammalian cells treated with nitrous oxide at high pressure. Science 160:774-76; 1968.

65b. Rao, A. P.; Rao, P. N. The cause of G_2-arrest in Chinese hamster ovary cells treated with anticancer drugs. J. Nat. Cancer Inst. 57:1139-43; 1976.

66. Rao, P. N.; Johnson, R. T. Mammalian cell fusion. I. Studies on the regulation of DNA synthesis and mitosis. Nature (London) 225:159-64; 1970.

67. Rao, P. N.; Johnson, R. T. Mammalian cell fusion. IV. Regulation of chromosome formation for interphase nuclei by various chemical compounds. J. Cell Physiol. 78:217-24; 1971.

68. Rao, P. N.; Johnson, R. T. Induction of chromosome condensation in interphase cells. In DuPraw, E. J. ed. Advances in cell and molecular biology. Vol. 3. New York: Academic Press; 1974a: 135-89.

69. Rao, P. N.; Johnson, R. T. Regulation of cell cycle in hybrid cells. In Control of proliferation in animal cells. Vol. 1. Cold Spring Harbor Symposium on Quantitative Biology; 1974b: 785-800.

70. Rao, P. N.; Wilson, B.; Puck, T. Premature chromosome condensation and cell cycle analysis. J. Cell Physiol. 91:131-42; 1977.

71. Rhone, D. Evidence suggesting chromosome continuity during the S phase of Indian muntjac cells. Hereditas 80:145-49; 1975.

72. Roizman, B. Polykaryocytosis. Cold Spring Harbor Symposium on Quantitative Biology. 27:327-40; 1962.

73. Sandberg, A. A.; Aya, T.; Ikeuchi, T.; Weinfeld, H. Definition and morphologic features of chromosome pulverization: a hypothesis to explain the phenomenon. J. Nat. Cancer Inst. 45:615-23; 1970.

74. Schnedl, W.; Czaker, R.; Ammerer, G.; Schwarzacher, H. G. Induction of premature chromosome condensation (PCC) depending on the cell cycle phase. Cytobiologie 12:140-44; 1975.

75. Schneiderman, S.; Farber, J. L.; Baserga, R. A simple method for decreasing the toxicity of polyethylene glycol in mammalian cell hybridization. Somatic Cell Genet. 5:263-69; 1979.

76. Schor, S. L.; Johnson, R. T.; Waldren, C. A. Changes in the organization of chromosomes during the cell cycle: response to ultraviolet light. J. Cell Science 17:539-65; 1975.

77. Sognier, M. A.; Hittelman, W. N. The repair of bleomycin-induced DNA damage and its relationship to chromosome aberration repair. Mutat. Res. 62:517-27; 1979.

78. Sognier, M. A.; Hittelman, W. N.; Rao, P. N. Effect of DNA repair inhibitors on the induction and repair of bleomycin-induced chromosome damage. Mutat. Res. 60:61-72; 1979.

79. Sperling, K.; Rao, P. N. The phenomenon of premature chromosome condensation: its relevance to basic and applied research. Humangenetik 23:235-58; 1974a.

80. Sperling, K.; Rao, P. N. Mammalian cell fusion. V. Replication behavior of heterochromatin as observed by premature chromosome condensation. Chromosoma (Berlin) 45:121-31; 1974b.

81. Stadler, J. K.; Adelberg, E. A. Cell cycle changes and the ability of cells to undergo virus-induced fusion. Proc. Nat. Acad. Sci. USA 69:1929-33; 1972.

82. Stenman, S. Depression of RNA synthesis in the prematurely condensed chromatin of pulverized HeLa cells. Exp. Cell Res. 69:372-76; 1971.

83. Stubblefield, E. DNA synthesis and chromosomal morphology of Chinese hamster cells cultured in media containing N-deacetyl-N-methylcolchicine (Colcemid). *In* Harris, R. J. C., ed. Cytogenetics of cells in culture. Vol. 3. New York: Academic Press; 1964: 223-48.

84. Sunkara, P. S.; Al-Bader, A. A.; Rao, P. N. Mitoplasts: mitotic cells minus the chromosomes. Exp. Cell Res. 107:444-48; 1977.

85. Sunkara, P. S.; Pargac, M. B.; Nishioka, K.; Rao, P. N. Differential effects of inhibition of polyamine biosynthesis on cell cycle traverse and structure of the prematurely condensed chromosomes of normal and transformed cells. J. Cell Physiol. 98:451-58; 1979.

86. Sunkara, P. S.; Wright, D. A.; Rao, P. N. Mitotic factors from mammalian cells induce germinal vesicle breakdown and chromosome condensation in amphibian oocytes. Proc. Nat. Acad. Sci. USA 76:2799-2802; 1979.

87. Takagi, N.; Aya, T.; Kato, H.; Sandberg, A. A. Relation of virus-induced cell fusion and chromosome pulverization to mitotic events. J. Nat. Cancer Inst. 43:335-47; 1969.

88. Unakul, W.; Johnson, R. T.; Rao, P. N.; Hsu, T. C. Giemsa banding in prematurely condensed chromosomes obtained by cell fusion. Nature (London) New Biology 242:106-7; 1973.

89. Vaugh, V. L.; Hansen, D.; Stadler, J. Parameters of polyethylene glycol-induced cell fusion and hybridization in lymphoid cell lines. Somatic Cell Genet. 2:537-44; 1976.

90. Wainberg, M. A.; Hjorth, R. N.; Howe, C. Effect of β-propiolactone on Sendai virus. Appl. Microbiol. 22:618-21; 1971.

91. Waldren, C. A.; Johnson, R. T. Analysis of interphase chromosome damage by means of premature chromosome condensation after X- and ultraviolet-irradiation. Proc. Nat. Acad. Sci. USA 71:1137-41; 1974.

92. Wasserman, W. J.; Masui, Y. A cytoplasmic factor promoting oocyte maturation: its extraction and preliminary characterization. Science 191:1266-68; 1976.

93. Wildrick, D. M.; Hittelman, W. N. The use of premature chromosome condensation as a tool for obtaining fast chromosome counts in fishes. Abstracts of the Am. Soc. Ichthyologists and Herpetologists Annual Meeting, 1978.

94. Witkowski, R.; Anger, H. Premature chromosome condensation in irradiated man. Hum. Genet. 34:65-68; 1976.

95. Yaneshevsky, R.; Carrano, A. V. Prematurely condensed chromosomes of dividing and nondividing cells in aging human cell cultures. Exp. Cell Res. 90:169-74; 1975.

18

The Epidemiological Approach: Chromosome Aberrations in Persons Exposed to Chemical Mutagens

by ERICH R. E. GEBHART*

Abstract

The analysis of chromosome aberrations in peripheral lymphocyte cultures from persons exposed to chemical mutagens is a valuable approach to detect possible in vivo effects of mutagens in man. For getting reliable results, a series of technical particularities of this system has to be taken into consideration. It is, however, far from being ideal as a routine procedure, as it does not allow systematic experimental testing and is restricted to studies on selected and specifically exposed groups of persons. The data obtained as yet by this method on alkylating agents, antimetabolites, antimicrobials, psychoactive drugs, other therapeutics, and environmental mutagens, on the other hand, document well its usefulness and reliability. Positive results give us evidence for the genetic and cancer risks of chemical mutagens in man, which can help to establish safer levels of exposure for individuals having frequent contact with mutagens.

Introduction

Testing mutagenic actions of environmental chemicals has become an important part of modern environmental science and prophylactic medicine. As shown by this book, a wide range of test methods and organisms for the evaluation of cytogenetic damage induced by chemical mutagens is now available. However, all the results from analyses of the mutagenicity of drugs and environmental chemicals should eventually be used for mankind's welfare. Therefore, it is most desirable to include data from studies made directly on humans.

One approach which has been under consideration is the analysis of chromosome

*Institut für Humangenetik und Anthropologie, der Universität, Erlangen-Nürnberg

damage in chemically loaded persons. Although such studies are confronted with several problems, a good deal of effort has been exerted in this area since the first papers of Conen and Lansky (1961) and Arrighi et al. (1962). The number of publications has increased steadily over the past years (Fig. 18.1).

Since structural chromosomal changes observed in chemically exposed individuals are of the same types found in persons with chromosome instability syndromes or malignancies (Mitelman and Levan 1978), chromosome aberrations detected in chemically exposed persons can be advantageously employed as an indicator to measure the toxic effects of the agents and their metabolites on the genetic apparatus of cells in vivo. Furthermore, cytogenetic effects on persons with chronic exposure to chemicals can also be estimated by the so-called longitudinal study, that is, consecutive analyses of the same persons over a given period of time.

The best initial approach to detect possible in vivo effects of mutagens was the study on chromosome damage of patients treated with chemotherapeutic agents using lymphocyte cultures as the material. However, industrial environments have become health problems because they may seriously increase the mutation load of the human population. Workers with frequent exposures to noxious chemicals may have accumulated genetic damage. Kilian and Picciano (1976) pointed out that cytogenetic monitoring of individuals working around potentially clastogenic (chromosome damaging) substances is a logical hygiene philosophy. Even though the current techniques of cytogenetic monitoring cannot solve all of the pressing problems of mutation research (e.g., they cannot detect point mutations), chromosome analysis appears to be the procedure of choice for objective observation and measurement of possible insult to human genetic material.

One must realize that the present methodology has a number of shortcomings:

1. The experimental approach cannot be used for man.
2. Cytogenetic abnormalities found in exposed persons represent the most obvious expression of genetic damage while more subtle genetic changes, conceivably more numerous, are not detected.
3. Peripheral lymphocytes in culture, the commonest target cells for clastogen analyses of man, are not in the cell cycle at the time the chemicals exert their in-

Figure 18.1 Number of publications on chromosome damage in chemically loaded individuals, which were available for the author.

fluence. In other words, only the chemicals which can cause primary genetic lesions in the G_0 phase can manifest their detrimental effects. Furthermore, the effects of such agents must be relatively strong. Agents that cause genetic damage in the S phase, or weaker clastogens, may or may not exhibit cytogenetic damage in lymphocyte cultures. Therefore, caution should be exercised in interpreting negative findings. Nevertheless, the spectrum of substances to be shown in this article as clastogenic using the current method demonstrates that at least the great majority of known strong mutagens can be detected in the lymphocyte system.

As will be shown later, mutagenicity testing in man may be extended to tissues other than cultured lymphocytes. For example, bone marrow aspirates may be available for cytogenetic analyses. In exceptional cases, the activity of a mutagenic agent on gonads can be studied. Meiotic cells from testicular biopsies can be utilized for the estimation of cytogenetic damage in spematogenesis, but such material is not available for routine procedure. As far as the author is aware, only one report (Hulten et al. 1968) is available.

Methodology

In the earlier years of cytogenetic studies on chemically loaded persons, efforts were focussed primarily on patients treated with chemotherapeutic agents. In cases of monotherapy, the activity of a single substance (or its metabolic derivatives in vivo) on chromosomes could be estimated with a certain degree of precision. Furthermore, additional factors which may affect the drug's activity were available in clinical records. During recent years, cytogenetic investigations have been extended to individuals exposed to chemicals in industrial environments.

Since the methods for chromosome studies in man have been discussed extensively (Evans and O'Riordan 1975; Kilian and Picciano 1976), our considerations will be restricted to the following basic problems:

The size of the test group. It appeared at first that this was not a serious problem since a sufficient number of patients were available. However, some studies used very small samples, e.g., less than 10, while others used more than 30 test individuals (Table 18.1). Thus, some contradictory results may be attributable to statistical problems. Moreover, in some studies (nearly 30%), the results were based on observations of less than 100 metaphases per individual. The sample size was certainly not sufficiently large for critical evaluations. It is recommended that prior to such investigations, a statistician should be consulted to devise a proper protocol. Whorton et al. (1979) offer a number of helpful instructions on required sample size and other considerations.

Control individuals. Of 250 published papers, more than 20% gave no indication that a control group had been analyzed (Table 18.1). In nearly 50% of the 250 data from normal, unexposed persons (in some cases matched for age and sex) were used for comparison. In about 15% of the total cases, control individuals were selected with geographic and occupational considerations or the data were compared from the same individuals before and after exposure. The last mentioned system seems to be the most reliable and should be followed whenever possible, but unfortunately it

Table 18.1 Methodological Parameters Drawn from Available Publications

Parameter	Total number of analyzed publications	Number of publications in category	Percentage
Number of test individuals	270	Less than 10	32.5
		10 to 30	40.7
		More than 30	24.4
		No information	7.5
Control group	269	No control	20.8
		Average population[a]	44.2
		Same area[b]	12.6
		Same individual[c]	14.9
		No information	7.5
Target cells	272	Lymphocytes	89.3
		Bone marrow	9.6
		Other cells	1.1
Culture time of lymphocytes	243	40 to 52 hr.	23.9
		65 to 72 hr.	58.9
		Other time	1.2
		No information	16.0
Number of evaluated metaphases	174	Less than 50	11.5
		50 to 99	15.5
		100	58.6
		More than 100	14.4

[a] Average population means normal (untreated) people from the average population (sometimes matched for age and sex).
[b] Same area means normal (untreated) individuals from the same environment or untreated patients from the same disease group.
[c] Same individual means control examination was performed before treatment or load and was compared with data from the same individual during or after treatment.

can only be applied to clinical cases. For occupational health studies, especially on individuals with mutagenic exposure for many years, matching control individuals must be selected with utmost caution.

As mentioned, in about 90% of all studies on chemically loaded persons, cultured lymphocytes were used as the target cells. In less than 10% of cases, bone marrow cells were used as additional (or exclusive) material, whereas fibroblasts or gonadal cells were sporadically employed. This discussion will therefore concentrate on lymphocyte cultures.

Many investigators favor the 48 hr. harvest time because most of the transformed lymphocytes should have entered their first mitosis. The data on 48 hr. samples should give a highly representative picture of chromosome damage in vivo. At 72 hr. of culture, second or even third cell cycles may have been completed; therefore, selection and differential multiplication rates between damaged and healthy lymphocytes may obscure the true picture. On the other hand, a considerably higher number of metaphases can be obtained from 72 hr. samples than from 48 hr. samples. It should be added that recent data from studies using BUDR-labeling did

Table 18.2 Amount of First, Second, and Third Metaphases Differentiated by BUDR-labelling in PHA-stimulated Human Lymphocyte Cultures after 3 Days of Culture

Number of individuals	Number of analyzed metaphases	Percentage of metaphases			References
		1st	2nd	3rd	
10	1,000	33.1	44.6	22.3	Crossen and Morgan 1977a
30	3,000	47.6	45.2	7.2	Crossen and Morgan 1977b
	400	49.3	39.7	11.0	Beek and Obe 1979
56	11,000	36.2	38.1	25.7	Gebhart, unpublished

not seem to alter significantly the aberration rates (Table 18.2). The proportion of the first, second, and third cell cycles in 72 hr. samples appears to be dependent on the culture medium used (Obe et al. 1975). Even in 48 hr. cultures, some authors reported a considerable number of metaphases in the second cell cycle: from 5% to as much as 40% in some cases (Crossen and Morgan 1977; Beck and Obe 1979).

A number of investigators (Bauchinger and Schmid 1969; Nielsen et al. 1969; Forni et al. 1971; Kilian and Picciano 1976; Purchase et al. 1978) claimed that the chromosome aberration rates between the 48 hr. and the 72 hr. samples showed no significant difference. Other investigators observed some minor differences (Table 18.3). In some cases a higher aberration rate was found in the 48 hr. samples, but in others the reverse (higher aberration rate in 72 hr. samples) was found. It appears, therefore, the choice of sampling time is not a serious consideration and the 72 hr. harvest is certainly more advantageous (Gebhart, 1980b).

The classification of chromatid and chromosome aberrations have been much discussed. Figure 18.2 presents the types of aberrations (Gebhart 1970). This subject will not be dealt with here since readers can consult the articles of Evans and O'Riordan (1975) and Savage (1975).

During recent years, it has been emphasized that sister chromatid exchanges (SCE) can be used as an indicator of the mutagenic activity of chemicals. In most cases, a good correlation can be obtained between induced chromosome breakage and SCE rate, and the concentration of chemicals required for SCE can be considerably reduced (Evans 1977; Galloway 1977). Therefore, employment of SCE as a test method for mutagenicity has been popular. Although a number of substances

Table 18.3 Examples of Average Aberration Rates in Peripheral Lymphocytes of Chemically Loaded Persons after 48 hr. and 72 hr. in Culture

Number of individuals	Number of analyzed metaphases	Percentage of aberrant metaphases		Breakage rate		References
		48 hr.	72 hr.	48 hr.	72 hr.	
5	850	20.8	15.9	.33	.32	Bauchinger and Schmid 1969
16	2,034	.23	2.56	.002	.02	Falek et al. 1972
3	600	43.0	48.6	.35	.42	Gebhart et al. 1974
10	1,850	9.6	12.6	.05	.06	Gebhart 1975
57	11,400	19.4	12.3	.19	.14	Gebhart, unpublished

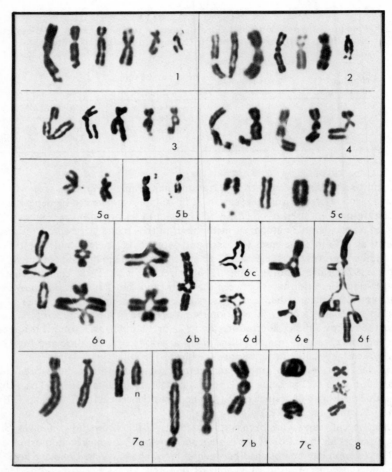

Figure 18.2 Types of chromatid and chromosome aberrations induced by chemical agents. (From Gebhart 1970.) (1) Chromatid gaps; (2) Isochromatid gaps; (3) chromatid breaks; (4) isochromatid breaks; (5a) minute (interstitial deletion compared to a small terminal break); (5b) "double minutes"; (5c) acentric fragments; (6 a–f) different types of chromatid interchanges; (7a) marker chromosomes (atypical chromosomes) compared with a normal D chromosome (n); (7b) dicentric chromosomes; (7c) ring chromosomes; (8) "pulverisation" (PCC).

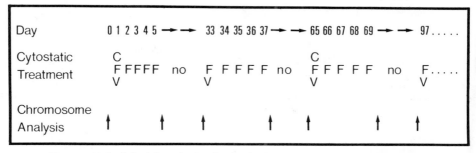

Figure 18.3 Schedule of the interval therapy with methyl-CCNU (C), 5-fluorouracil (F), and vincristin (V). (Wopfner, personal communication.)

(bleomycin, ethidium bromide, cytosine arabinoside, chloramphenicol, cycloheximide, nitrosated derivatives of carbamate triazine pesticides) induce chromosome breakage without inducing an excess amount of SCE, the sister chromatid exchange method should be included in all cytogenetic investigations on chemically loaded individuals in order not to neglect a valuable tool (Fig. 18.3).

Results

Chemotherapeutic Agents

Since alkylating chemicals had been shown to be mutagenic in nearly all test systems, it was not surprising to find that even therapeutic doses of these cytostatic agents were sufficient to induce chromosome damage in patients (Table 18.4). In contradiction to most reports, however, Schinzel and Schmid (1976) concluded from their study on cytostatically treated patients that increased incidence of chromatid breaks and exchanges was not a typical finding in lymphocyte cultures of persons exposed in vivo to chromosome-breaking agents. It should be noted that in their study, the cytogenetic examination was conducted one to several weeks after the cessation of therapy.

Nucleic acid antimetabolites were shown to be not only mutagenic in most test systems but also, with a few exceptions, clastogenic when administered for cytostatic or immunosuppressive therapy (Table 18.4). It should be added that the results with antimetabolites are not as conclusive as those with alkylating agents. In experimental animals and cell cultures, the clastogenic activity of antimetabolites was found to require higher concentrations than alkylants. Therefore, therapeutic doses in some cases might not have reached the level necessary for the induction of chromosome damage. Furthermore, most of these substances are known to act preferentially on cells in the S phase, which is absent in peripheral lymphocytes. Thus, bone marrow in vivo may be a better target than cultured lymphocytes (Krogh-Jensen 1967, 1970).

Only a limited amount of data is available for cytostatic antibiotics (Table 18.4). In both experimental animals and human lymphocyte cultures, these antiobiotics were found to be clastogenic. However, most antibacterial and antiparasitic antibiotics have failed to show distinct clastogenic effects on human cells in vivo.

A group of therapeutics, the psychotropic agents, has gained importance in many industrialized countries. Again as shown in Table 18.4, only a few of these

Table 18.4 Data from the Literature and Personal Unpublished Data from Chromosome Studies in Persons under Chemotherapy (or Abuse)

Substance	Number of test individuals	Cell system	Results	References
Cytostatics and immunosuppressives:				
Cyclophosphamide	> 180	L	+ (−) SCE+	Bauchinger and Schmid 1969; Dobos et al. 1974; Tolchin et al. 1974; Goetz et al. 1975; Schinzel and Schmid 1976; Fischer et al. 1979
Busulphan	38	L, BM	+	Gebhart et al. 1974
Chlorambucil	> 30	L, BM	+ −	Stevenson and Patel 1973; Reeves et al. 1975
Thiotepa	63	BM	+	Blank 1979
Nitrosomethylurea	32	L	−	Selezneva and Korman 1973
CCNU	51	L	+ SCE+	Lambert et al. 1978; Gebhart et al., 1980
5-Fluorouracil	> 60	L	+ SCE+	Gebhart et al., 1980
6-Mercaptopurine	25	L, F	+	Fedortzeva et al. 1973
Cytosinearabinoside	> 20	BM, L	+ SCE−	Bell et al. 1966; Raposa 1978
Azathioprin	> 100	L, BM	− (+)	Krogh-Jensen 1967, 1970; Eberle et al. 1968; Friedrich and Zeuthen 1970; Ganner et al. 1973; Rossi et al. 1973; Apelt et al. 1981
Amethopterin	> 70	L, BM, Go	− (+)	Krogh-Jensen 1967; Locher and Fränz 1967; Schinzel and Schmid 1976; Lawler and Walden 1979
Hydroxyurea	11	L	+ −	Rossi et al. 1973
Actinomycin D	12	L	−	Cohen et al. 1971
Adriamycin	11	L	(+)	Schinzel and Schmid 1976
Daunomycin	20	L	+ −	Whang-Peng et al. 1969; Schinzel and Schmid 1976
Bleomycin	> 10	L	+	Bornstein et al. 1971; Dresp et al. 1978; Schinzel and Schmid 1976
Vincristine	25	L	(+) −	Gebhart et al. 1969; Schinzel and Schmid 1976
Antibiotics and antiparasitics:				
Penicillines	11	L	−	Gebhart, unpublished
Lincomycin	15	L	−	Gebhart, unpublished
Bactrim (Eusaprim)	34	L	−	Gebhart 1973, 1975; Stevenson et al. 1973
Metronidazol	34	L	−	Mitelman et al. 1976, 1980; Hartley-Asp 1979
Hycanthone	?	L	−	Frota-Pessoa et al. 1975
Oxamniquine	24	L	−	Monsalva et al. 1976
Isoniazide	70	L	−	Bauchinger et al. 1978

Abbreviations and symbols
L = lymphocytes
BM = bone marrow
F = fibroblasts
Go = gonadal cells
Ab = abortus material

+ = positive result
− = negative result
(+) = weakly positive
(−) = weakly negative
SCE+ = SCE test positive

Table 18.4 Continued

Substance	Number of test individuals	Cell system	Results	References
Psychotropic substances:				
Perphenazine	> 50	L	− (+)	Nielsen et al. 1969; Gabilondo et al. 1971; Gilmour et al. 1971; Cohen et al. 1972; Matsuyama and Jarvik 1975
Chlorpromazine	35	L	−	Nielsen et al. 1969; Roman 1969; Gabilondo et al. 1971; Cohen et al. 1972; Matsuyama and Jarvik 1975
Diazepam	50	L	−	Stenchever et al. 1970; White et al. 1974; Matsuyama and Jarvik 1975
Chlordiazepoxide	10	L	−	Cohen et al. 1969; Matsuyama and Jarvik 1975
Clozapine	20	BM	(+)	Knuutila et al. 1977
Lithium salts	> 100	L	−	Genest and Villeneuve 1971; Jarvik et al. 1971; Bille et al. 1975; De la Torre and Krompotic 1976; Banduhn et al. 1980; Matsuyama and Jarvik 1975
Antiepileptic comb.	> 60	L	(+)	Grosse et al. 1972
Diphenylhydantoine	62	L, BM	−	Marquez-Monter et al. 1970; Herha and Obe 1976; Knuutila et al. 1977; Alving et al. 1977
Primidon	39	L	− (+) SCE+	Galle 1978
Carbamazepine	10	L	(+)	Herha and Obe 1976
Dipropylacetate	10	L	−	Kotlarek and Faust 1978
Methadone	> 100	L	− (+)	Amarose and Norusis 1976; Matsuyama et al. 1978
Amphetamine	12	L	−	Fu et al. 1978
Mescalin	50	L	−	Dorrance et al. 1975
Cannabis	150	L	−	Dorrance et al. 1970; Matsuyama et al. 1973, 1977; Herha and Obe 1974; Martin et al. 1974; Nichols et al. 1974; Stenchever 1974; Matsuyama and Jarvik 1975
LSD	> 300	L	− (+)	Cohen et al. 1967; Matsuyama and Jarvik 1975; Robinson et al. 1974
Heroin	> 35	L	+ −	Amarose and Norusis 1976
Opiates	16	L	(+)	Falek et al. 1972; Matsuyama and Jarvik 1975
Drug combinations	> 40	L, BM	+ −	Amarose and Schuster 1971; Gilmour et al. 1971
Alcohol abuse	> 25	L	(+)	De Torok 1972; Obe and Herha 1975
Other drugs:				
Phenylbutazone	110	L, BM	− (+)	Stevenson et al. 1971; Walker et al. 1975; Crippa et al. 1976

continued

Table 18.4 Continued

Substance	Number of test individuals	Cell system	Results	References
Other drugs: (cont'd.)				
Allopurinol	19	L	−	Stevenson et al. 1976
Colchicine	> 40	L	−	Cohen et al. 1977
Sulfonylureas	23	L	(+)	Watson et al. 1976
Cyclamate	35	L	(+) −	Bauchinger et al. 197 ; Dick et al. 1974
Arsenic	14	L	+ − SCE+	Burgdorf et al. 1977; Nordenson et al. 1979
Psoralen + UVA	> 20	L	− SCE −	Brogger et al. 1978; Mourelatos et al. 1977; Wolff-Schreiner et al. 1977
Cyproheptadine	40	L	−	Gebhart 1971; Murken 1971
Oral contraceptives	> 1000	L, Ab	−	McQuarrie et al. 1970; De Gutierrez and Lisker 1973; Fitzgerald et al. 1973; Bishun et al. 1975; Littlefield et al. 1975; Müller, Ritter 1978
HMG and HCG	48	Ab	(+)	Boue and Boue 1973

substances have been shown to be weakly clastogenic. In a series of studies involving combinations of antiepileptic drugs, however, Grosse et al. (1972) discovered chromosome aberrations even in newborn infants who had contact with these drugs in utero.

Although such halucinogens as lysergic acid diethylamide (LSD) were first alleged to be clastogenic, subsequent studies with pure LSD yielded negative results. It is probable that drug addicts in most cases use several drugs in combination or in sequence and these drugs contain far more impurities than substances administered in a controlled study. While a higher level of chromosome changes was found in heroin and opiate addicts than that in the controls, marijuana and mescalin have thus far yielded negative data.

The results of chromosome studies in patients loaded with other chemotherapeutics are also included in Table 18.4. It is noteworthy that in one investigation (Watson et al. 1976) antidiabetic sulfonyl ureas were shown to be clastogenic.

Industrial Chemicals

During recent years, a great deal of interest has been focussed on environmental mutagens (Table 18.5). The substance tested most extensively has been vinyl chloride monomer. Studies on the peripheral lymphocytes of persons occupationally in contact with this agent yielded conflicting results. In one-half of the cases, an increase in chromosome aberration rate was discovered, while in the other half, there was no significant increase. Different environmental conditions (i.e., concentration, additional factors) as well as differences in techniques for culture and cytogenetic analysis, might play a role here.

From a series of investigations, it was found that epichlorohydrine, styrene oxide,

Table 18.5 Data from the Literature on Chromosome Studies in Persons Exposed to Industrial Chemicals

Substance	Number of test individuals	Cell system	Results	References
Vinylchloride	> 500	L	(+) −SCE+	Ducatman et al. 1975; Czeizel et al. 1977; Picciano et al. 1977; Szentesi et al. 1977; Hansteen et al. 1978; Purchase et al. 1978; Kucerova et al. 1979 Anderson et al. 1980
Epichlorohydrine	128	L	(+)	Kucerova et al. 1977; Picciano 1979
Chloroprene	> 56	L	(+)	Zhurkov et al. 1977
Styrene oxide	26	L	+ −	Meretoja et al. 1977; Thiess and Fleig 1978
Chloromethyl ether	12	L	+	Zudova and Landa 1977
Trichloroethylene	28	L	(+) −	Konietzko et al. 1978
Acrylonitril	18	L	−	Thiess and Fleig 1978
Dialkylcarbamoylchloride	10	L	−	Fleig and Thiess 1978
o-Phthalodinitrile	20	L	−	Fleig and Thiess 1979
Spray adhesives	40	L	−	Cervenka and Thorn 1974; Pawlowitzki et al. 1974; Garson and Blackstock 1976; Lubs et al. 1976
Benzene	> 190	L, BM	+	Tough and Court Brown 1965; Tough et al. 1970; Forni et al. 1971; Hartwich and Schwanitz 1972; Khan and Khan 1973; Prost et al. 1976
Toluene	56	L	−	Forni et al. 1971; Donner et al. 1981
Lead	> 250	L	−(+)	Bauchinger et al. 1972, 1976, 1977; Schmid et al. 1972; Forni and Secchi 1972; Deknudt et al. 1973; O'Riordan and Evans 1974; Schwanitz et al. 1975; Bijlsma and De France 1976; Garza Chapa et al. 1977; Nordenson et al. 1978
Lead + Cadmium	47	L	(+)	Deknudt and Leonard 1975
Cadmium	40	L	−	O'Riordan et al. 1978
Chromium		L	+	Bigaliev et al. 1977
Mercury	71	L	+ −	Skerfving et al. 1974; Verschaeve et al. 1976, 1978
Arsenic	33	L	(+)	Nordenson et al. 1978
Organophosphates	> 180	L	(+)	Van Bao et al, 1974; Kiraly et al. 1977
Benomyl	20	L	−	Ruzicska et al. 1975
DDT	33	L	−	Nazareth-Rabello et al. 1975
Zineb	15	L	+	Pilinskaya 1974
Ziram	9	L	+	Pilinskaya 1970
Pesticide combinations	184	L	(+) SCE (+)	Yoder et al. 1973; Pilinskaya and Zhurkov 1977; Georgieva 1977; Crossen et al. 1978

Abbreviations and symbols: see Table 18.4.

and chloromethyl ethers induced chromosome damage in persons having occupational contact with these substances. The observed aberration rates, however, were much lower than those after cytostatic therapy. Also, persons with a heavy load of benzene or after benzene intoxication exhibited increased rates of chromosome aberrations.

Most studies on persons occupationally in contact with lead yielded negative results. Exceptions may be attributed to the influence of other metals. For example, in workers exposed to lead and zinc, a higher chromosome aberration rate was found than in those exposed to lead and cadmium; but, according to Deknudt and Léonard (1975), the chromosome damage recorded in the latter group was of a more serious type than that in the former. Recently, Nordenson et al. (1978) reported elevated rates of chromosome aberrations and spontaneous abortions in persons working in or at least living near a smelter in Northern Sweden that emits arsenic, lead, and other substances. Funes-Cravito et al. (1977) reported significantly increased frequencies of chromosome breakage and SCE rates in workers (as well as in their children) in chemical laboratories and in a rotoprinting factory. In fishermen who had consumed mercury-contaminated fish, an increase in chromosome damage was observed and the rate was shown to be correlated with the mercury content in the blood (Skerfving 1974).

Only limited information is available concerning persons exposed to pesticides. In Russia, chromosome damage was found in persons occupationally in contact with Ziram and Zineb, but in Hungary the increase was only transient with organic phosphate pesticides. In operators occasionally exposed to high concentrations of pesticides, Yoder et al. (1973) found higher aberration rates during the heavy spraying period than in midwinter. Georgieva (1977) reported a slight but significant increase in aberration rates in pesticide workers, but Crossen et al. (1978) observed no overall difference in SCE rate between a control group and 57 pesticide sprayers (except in 5 of the latter group whose lymphocytes exhibited a significantly elevated SCE rate).

Antimitotic Action

Thus far, we have devoted our discussion to structural chromosomal aberrations (chromosome mutations) as valid indicators of genetic damage induced by chemicals in man. Numerical deviations of chromosomes (genomic mutations) are of as profound (if not more) practical importance. Unfortunately, only a limited amount of

Table 18.6 Substances Inducing Deviations of Chromosome Number in Man In Vivo

Substance	Number of test individuals	Cell system	Aneuploidy	Polyploidy
Antiepileptic drugs	76	L	+ −	+ −
Oral contraceptives	> 500	Ab	−	− +
HCG/HMG	> 470	Ab	(+)	?
Colchicine	41	L	− +	− +
Vinca-alkaloids	7	L	−	(+)

Abbreviations and symbols: see Table 18.4.

data is available. Table 18.6 presents a few examples in which substances were shown to induce aneuploidy or polyploidy in vivo.

Specific Problems

Although dose-effect relations are helpful and important data to verify and to support the conclusions on mutagen or clastogen effects, most in vivo cytogenetic studies in man have not stressed this point. Various technical difficulties and limitations have hampered the progress in this direction. However, cytogenetic studies on patients receiving chemotherapy should be able to throw some light on this problem. We have tried to find out, in patients treated with busulphan, a well-known alklyating cytostatic, whether a dose-effect relationship existed for chromosome damage (Gebhart et al. 1974). In these patients, a clear correlation was established between the frequency of chromosome aberrations and the total dose administered over a long period of time.

As mentioned, Schinzel and Schmid (1976) pointed out that their study on patients treated with cytostatics revealed no induction of chromosome aberrations even after the application of known strong mutagens. These authors concluded that cytogenetic analysis of this type should be abandoned because it is an insentive and unreliable method. Their conclusion, among other reasons, prompted us to analyze the possibilities and limitations of cytogenetic analyses in chemically loaded individuals. Patients with colorectal carcinoma undergoing a postoperative cytostatic interval therapy (Fig. 18.4) constituted the main source of our material (Gebhart et al. 1980, a, b).

A total of 229 blood samples were taken for chromosome breakage analyses from 43 patients (19 male, 24 female) prior to and in the course of therapy and 92 of which were also taken for SCE analyses. All slides were coded. A maximum of 100 metaphases were evaluated for breakage and a minimum of 20 metaphases were evaluated for SCE.

A distinct increase in breakage frequency was noted when samples before and after chemotherapy were compared (Fig. 18.5). Pooling all data before therapy (control) and after therapy, the results again indicate a significant increase in chromosome damage (Tables 18.7 and 18.8). Even four weeks after the last therapy, the aberration rates were not distinctly reduced (Table 18.9).

One of the purposes of this study was to compare breakage and SCE rates (Gebhart et al. 1980b). A steady increase of both types of aberrations was observed as the number of therapy steps increased. The number of SCEs exceeded that of breaks by a factor of 100 (Fig. 18.6). Although this observed difference seemed clear-cut when pooled data were used, a wide variability existed when individual cases were considered (Fig. 18.3).

It should be pointed out that the BUDR-labeling technique for SCE enables cytogeneticists to distinguish first (no differentiated chromatids), second (clearly differentiated chromatids) and third (25% darkly stained chromatids) metaphases in culture. This phenomenon may be utilized to distinguish the aberration rates between cell generations. In our preliminary data, we found a positive correlation between the breakage rate and the number of first metaphases in 72 hr. cultures, but a negative correlation between the breakage rate and the number of third metaphases (Fig. 18.7). Thus, our data support the notion that, with substances which do not obviously delay the cell cycle, 48 hr. should be the preferred harvest time, whereas with

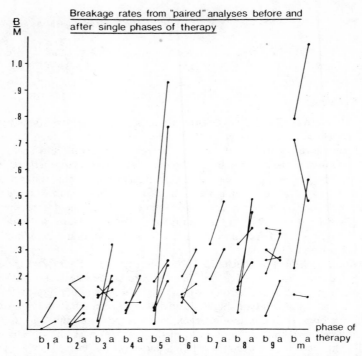

Figure 18.4 Individual data gained before and after single therapy phases (connected points are from the same patient): b = before; a = after; 1, 2, etc. = 1st, 2nd, etc. phase of therapy.

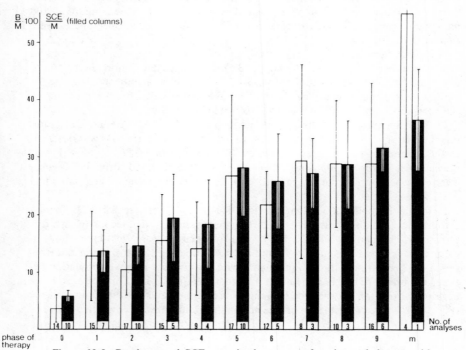

Figure 18.5 Breakage and SCE rates in the course of an interval therapy with methyl-CCNU, fluorouracil, and vincristin.

Table 18.7 Data from Individual Patients before and after an Interval Therapy with Methyl-CCNU, Fluorouracil, and Vincristin

Patient	Stage of therapy	Analyzed metaphases	Aberrant metaphases %	Metaphases containing breaks %	Breaks per metaphase
Hi	0	125	2.4	1.6	.02
	after 9[a]	100	48	27	.36
He	0	200	5.5	4	.06
	after 9	100	32	20	.26
Pr	0	200	3.5	1.5	.02
	after 9	100	30	20	.37
Dö	0	100	3	2	.02
	after 8	150	25	15	.25
La	0	200	6	2.5	.04
	after 7	200	36.5	30.5	.66
Gr	0	100	9	5	.05
	after 6	200	25.5	21.5	.28
Kr	0	200	3.5	1	.01
	after 6	100	21	12	.17
Ge	0	100	4	3	.03
	after 5	100	23	18	.26
Gl	0	200	4	3.5	.04
	after 5	100	20	16	.20
Re	0	100	9	2	.04
	after 5	200	27	23.5	.25
Ni	0	200	9.5	4	.04
	after 2	100	9	5	.09
Total	0	1,725	5.4	2.7	.03
	after therapy	1,450	27.1	18.9	.28

[a] See Figure 18.4.

Table 18.8 Data from a Group of Patients with Cytostatic Interval Therapy Immediately after a 5-Day Therapy Phase Compared with That of a Group before Therapy

Group	Number of individual analyses	Number of analyzed metaphases	Aberrant metaphases %	Metaphases containing breaks %	Breaks per metaphase	SCE per metaphase
Control	28	2,800	5.3	2.6	.03	5.8
After therapy	72	6,898	23.3	17.3	.27	25.6

Table 18.9 Comparison of Data from the Same Patients Immediately after and 4 Weeks after a 5-Day Therapy Phase (from 72 hr. Cultures)

Weeks after therapy phase	Number of individual analyses	Number of analyzed metaphases	Aberrant metaphases %	Metaphases containing breaks %	Breaks per metaphase
0	21	2,094	19.3	14.6	.22
4	21	2,010	17.2	12.2	.18

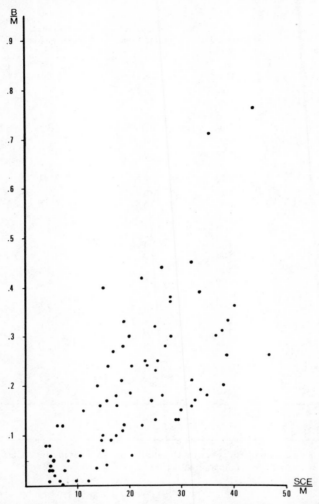

Figure 18.6 Correlation of breakage and SCE rate in individual analyses from the cytostatic interval therapy study.

substances which may prolong or delay cell proliferation, 72 hr. sample will be more suitable.

Many investigators have pointed out the phenomenon of rather high interindividual variations in studies on chemically loaded persons. We confirm this conclusion in our analyses (Fig. 18.8). The phenomenon (or problem) of heterogeneity only reflects a reality which would have been neglected if data on all mutagenicity studies were collected from experiments employing test materials selected for their low variability.

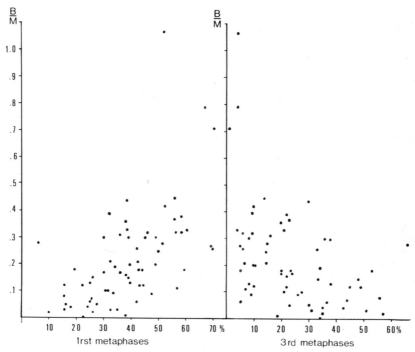

Figure 18.7 Correlation between the breakage rate and amount of first and third metaphases in culture.

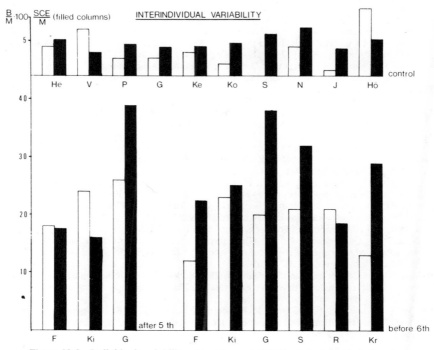

Figure 18.8 Individual variability of breakage and SCE rate in patients before (upper line), after 5th, and before 6th phase of a cytostatic interval therapy according to the schedule given in Figure 18.3.

Conclusion

The discussion presented in this chapter concerning methodology and available data on humans exposed to chemical mutagens has shown that the cytogenetic test system is certainly far from ideal as a routine procedure. On the other hand, the method is still useful, informative, and appropriate for estimating the mutagenicity of environmental chemicals in man. Positive results give us evidence for the genetic and cancer risk of chemical mutagens, which can help to establish safer levels of exposure for individuals having frequent contact with the mutagens.

Acknowledgements

My thanks are due to Dr. T. C. Hsu for encouraging me to prepare this chapter. The extremely reliable and fruitful cooperation of Dr. F. Wopfner, (Surgical Clinic of the University Erlangen-Nürnberg) for the interval therapy is greatly appreciated and the skillful and valuable technical assistance of Miss C. Werner, Miss I. Ottenberger, and Miss R. Bach is gratefully acknowledged.

Literature Cited

Alving, J.; Jensen, M. K.; Meyer, H. Chromosome studies of bone marrow cells and peripheral blood lymphocytes from diphenylhydantoine-treated patients. Mutat. Res. 48:361-66; 1977.
Amarose, A. P.; Schuster, C. R. Chromosomal analyses of bone marrow and peripheral blood in subjects with a history of illicit drug use. Arch. Gen. Psychiatr. 25:181-86; 1971.
Amarose, A. P.; Norusis, M. J. Cytogenetics of methadone-managed and heroin-addicted pregnant women and their newborn infants. Am. J. Obstet. Gynecol. 124:635-40; 1976.
Anderson, D.; Richardson, C. R.; Weight, T. M.; Purchase, I. F. H.; and Adams, W. G. F.: Chromosomal analyses in vinyl chloride exposed workers. Mutation Res. 79:151-62; 1980.
Apelt, F.; Kolin-Gerresheim, J.; Bauchinger, M.: Azathioprine, a clastogen in human somatic cells? Mutation Res. 88:61-72; 1981.
Arrighi, F. E.; Hsu, T. C.; Bergsagel, D. E. Chromosome damage in murine and human cells following cytoxan therapy. Texas Rep. Biol. Med. 20:545-49; 1962.
Banduhn, N.; Obe, G.; Müller-Oerlinghausen, B.: Is lithium mutagenic in man? Pharmakopsych.Neuro-Psychopharmakol. 13:218-27; 1980.
Bauchinger, M.; Schmid, E. Cytogenetische Veränderungen in weißen Blutzellen nach Cyclophosphamidtherapie. Z. Krebsforsch. 72:77-87; 1969.
Bauchinger, M.; Schmid, E.; Pieper, M.; Zöllner, N. Zytogenetische Wirkung von Cyclamat in menschlichen peripheren Lymphozyten in vivo. Dtsch. Med. Wschr. 95:2220-23; 1970.
Bauchinger, M.; Schmid, E.; Schmidt, D. Chromosomenanalyse bei Verkehrspolizisten mit erhöhter Bleilast. Mutat. Res. 16:407-12; 1972.
Bauchinger, M.; Schmid, E.; Einbrodt, H. J.; Dresp, J. Chromosome aberrations in lymphocytes after occupational exposure to lead and cadmium. Mutat. Res. 40:57-62; 1976.
Bauchinger, M.; Dresp, J.; Schmid, E.; Englert, N.; Krause, C.: Chromosome analyses of children after ecological lead exposure. Mutat. Res. 56:75-80; 1977.
Bauchinger, M.; Gebhart, E.; Fonatsch, C.; Schmid, E.; Müller, W.; Obe, G.; Beek, B.; Göbel, D.; Radenbach, K. L. Chromosome analyses in man in the course of chemoprophylaxis against tuberculosis and of antituberculosis chemotherapy with isoniazid. Hum. Genet. 42:31-43; 1978.
Beek, B.; Obe, G. Sister chromatid exchanges in human leukocyte chromosomes: spontaneous and induced frequencies in early- and late-proliferating cells in vitro. Hum. Genet. 49:51-61; 1979.
Bell, W. R.; Whang, J. J.; Carbone, P. P.; Brecher, G.; Block, J. B. Cytogenetic and morphologic abnormalities in human bone marrow cells during cytosine arabinoside therapy. Blood 27:771-81; 1966.

Bigaliev, A. B.; Turebaev, M. N.; Bigalieva, R. K.; Elemesova, M. S. Cytogenetic examination of workers engaged in chrome production. Genetika 13:545-47; 1977.
Bijlsma, J. B.; De France, H. F. Cytogenetic investigations in volunteers ingesting inorganic lead. Int. Arch. Occup. Environ. Hlth. 38:145-48; 1976.
Bille, P. E.; Jensen, M. K.; Jensen, J. P. K.; Poulsen, J. C. Studies on the haematologic and cytogenetic effect of lithium. Acta Med. Scand. 198:281-86; 1975.
Bishun, N.; Mills, J.; Parke, D. V.; Williams, D. C. A cytogenetic study in women who had used oral contraceptives and in their progeny. Mutat. Res. 33:299-310; 1975.
Blank, M. A.: Cytogenetic analysis of the damaging effect of thiophosphamide on bone marrow of patients with malignant tumors. Vopr. Onkol. 25:72-73; 1979.
Bornstein, R. S.; Hungerford, D. A.; Haller, G.; Engstrom, P. F.; Yarbro, J. W. Cytogenetic effects of Bleomycin therapy in man. Cancer Res. 31:2004-7; 1971.
Boue, J. G.; Boue, A. Increased frequency of chromosomal anomalies in abortions after induced ovulation. Lancet 1:679-80; 1973.
Brøgger, A.; Waksvik, H.; Thune, P. Psoralen/UVA treatment and chromosomes. II. Analyses of psoriasis patients. Arch. Dermatol. Res. 261:287-94; 1978.
Burgdorf, W.; Kurvink, K.; Cervenka, J. Elevated sister chromatid exchange rate in lymphocytes of subjects treated with arsenic. Hum. Genet. 36:69-72; 1977.
Cervenka, J.; Thorn, H. L. Chromosomes and spray adhesives. New Engl. J. Med. 190:543-45; 1974.
Cohen, M. M.; Hirschhorn, K.; Frosch, W. A. In vivo and in vitro chromosomal damage induced by LSD-25. New Engl. J. Med. 277:1043-49; 1967.
Cohen, M. M.; Hirschhorn, K. Frosch, W. A. Cytogenetic effect of tranquilizing drugs in vivo and in vitro. J. Am. Med. Assn. 207:2425-26; 1969.
Cohen, M. M.; Gerbie, A. B.; Nadler, H. L. Chromosomal investigations in pregnancies following chemotherapy for choriocarcinoma. Lancet 2:219; 1971.
Cohen, M. M.; Lieber, E.; Schwartz, H. N. In vivo cytogenetic effect of perphenazine and chlorpromazine: a negative study. Brit. Med. J. 3:21-23; 1972.
Cohen, M. M.; Levy, M.; Ehakim, M. A cytogenetic evaluation of long-term colchicine therapy in the treatment of familial Mediterranean Fever. Am. J. Med. Sci. 274:147-52; 1977.
Conen, P. E.; Lansky, G. S. Chromosome damage during nitrogen mustard therapy. Brit. Med. J. 2:1055-57; 1961.
Crippa, L.; Klein, D.; Linder, A. Etude cytogénétique et statistique d'un groupe de patients traités à la butazolidine (phenylbutazone). J. Génét. hum. 24:1-13; 1976.
Crossen, P. E.; Morgan, W. F. Analysis of human lymphocyte cell cycle time in culture measured by sister chromatid differential staining. Exp. Cell Res. 104:453-57; 1977.
Crossen, P. E.; Morgan, W. F. Proliferation of PHA- and PWM-stimulated lymphocytes measured by sister chromatid differential staining. Cell Immunol. 32:432-38; 1977.
Crossen, P. E.; Morgan, W. F. Cytogenetic studies of pesticide and herbicide sprayers. N.Z. Med. J. 88:192-95; 1978.
Czeizel, A.; Szentesi, I.; Hornyak, E.; Ungvari, G.; Bognar, Z.; Timar, M. Genetic study on PVC workers. Mutat. Res. 46:215-16; 1977.
De Gutierrez, A. C.; Lisker, R. Longitudinal study of the effects of oral contraceptives on human chromosomes. Ann. Génét. 16:259-62; 1973.
Deknudt G.; Léonard, A.; Ivanov, B. Chromosome aberrations observed in male workers occupationally exposed to lead. Environ. Physiol. Biochem. 3:132-38; 1973.
Deknudt G.; Léonard, A. Cytogenetic investigations on leucocytes of workers from a cadmium plant. Environ. Physiol. Biochem. 5:319-27; 1975.
De la Torre, R.; Krompotic, E. The in vivo and in vitro effects of lithium on human chromosomes and cell replication. Teratol. 13:131-38; 1976.
De Torok, D. Chromosomal irregularities in alcoholics. Ann. N.Y. Acad. Sci. 197:90-100; 1972.
Dick, C. E.; Schniepp, M. L.; Sonders, R. C.; Wiegand, R. G. Cyclamate and cyclohexylamine: lack of effect on the chromosomes of man and rats in vivo. Mutat. Res. 26:199-203; 1974.
Dobos, M.; Schuler, D.; Fekete, G. Cyclophosphamide-induced chromosomal aberrations in nontumorous patients. Humangenetik 22:221-27; 1974.

Donner, M.; Husgafvel-Pursiainen, K.; Mäki-Paakkanen, J.; Sorsa, M.; and Vainio, H.: Genetic effects of in vivo exposure to toluene. Mutation Res. 85:293-94; 1981.

Dorrance, D.; Janiger, O.; Teplitz, R. L. In vivo effects of illicit hallucinogens on human lymphocyte chromosomes. J. Am. Med. Assn. 212:1488-91; 1970.

Dorrance, D.; Janiger, O.; Teplitz, R. L. Effect of peyote on human chromosomes. Cytogenetic study of the Huichol Indians of Northern Mexico. J. Am. Med. Assn. 234:299-302; 1975.

Dresp, J.; Schmid, E.; Bauchinger, M. The cytogenetic effect of bleomycin on human peripheral lymphocytes in vitro and in vivo. Mutat. Res. 56:341-53; 1978.

Ducatman, A.; Hirschhorn, K.; Selikoff, I. J. Vinyl chloride exposure and human chromosome aberrations. Mutat. Res. 31:163-68; 1975.

Eberle, P.; Hunstein, W.; Perings, E. Chromosomes in patients treated with Imuran. Humangenetik 6:69-73; 1968.

Evans, H. J.; O'Riordan, M. L. Human peripheral blood lymphocytes for the analysis of chromosome aberrations in mutagen tests. Mutat. Res. 31:135-48; 1975.

Evans, H. J. What are sister chromatid exchanges? In De la Chapelle, A.; Sorsa, M., eds. Chromosomes Today. Vol. 6. Amsterdam: Elsevier; 1977.

Falek, A.; Jordan, R. B.; King, B. J.; Arnold, P. J.; Skelton, W. D. Human chromosomes and opiates. Arch. Gen. Psychiatr. 27:511-15; 1972.

Fedortzeva, R. F.; Dygin, V. P.; Mamaeva, S. E.; Goroshchenko, Y. L. Cytogenetical analysis of the effect of 6-mercaptopurine on human chromosomes. I. Effects on blood cells of acute leukemia patients. Tsitologiya 15:1172-73; 1973.

Fischer, P.; Nacheva, E.; Pohl-Rüling, J.; Krepler, P. Cytogenetic effects of chemotherapy and cranial irradiation on the peripheral blood lymphocytes of children with leukaemia. In Evans, H. J.; Lloyd, D. C., eds. Mutagen-induced chromosome damage in man. Edinburgh University Press, 1978.

Fitzgerald, P. H.; Pickering, H. F.; Ferguson, D. N.; Hamer, J. W. Long-term use of oral contraceptives: a study of chromosomes and lymphocyte transformation. Aust. N.Z.J. Med. 3:572-75; 1973.

Fleig, I.; Thiess, A. M. Chromosome investigations of persons exposed to dimethylcarbamoyl chloride and diethylcarbamoyl chloride. J. Occup. Med. 20:745-46; 1978.

Fleig, I.; Thiess, A. M.: Chromosome study of workers exposed to orthophthalodinitrile. Zbl. Arbeitsmed.Arbeitsschutz Proph. 29:127-29; 1979.

Forni, A.; Pacifico, E.; Limonta, A. Chromosome studies in workers exposed to benzene or toluene or both. Arch. Environ. Hlth. 22:373-78; 1971.

Forni, A.; Secchi, G. Chromosome changes in preclinical and clinical lead poisoning and correlation with biochemical findings. Proc. Inter. Symp. Environ. Hlth. Asp. of Lead. Amsterdam, 1972.

Friedrich, U.; Zeuthen, E. Chromosomenabnormitäten und Behandlung mit Imuran (Azathioprin) nach Nierentransplantationen. Humangenetik 8:289-94; 1971.

Frota-Pessoa, O.; Ferraira, N. R.; Pedroso, M. B.; Moro, A. M.; Otto, P. A.; Chamone, D. A. F.; Da Silva, L. C. A study of chromosomes of lymphocytes from patients treated with hycanthone. J. Toxicol. Environ. Hlth. 1:305-7; 1975.

Fu, T. K.; Jarvik, L. F.; Matsuyama, S. S. Amphetamine and human chromosomes. Mutat. Res. 53:127-28; 1978.

Funes-Cravioto, F.; Kolmodin-Hedman, B.; Lindsten, J.; Nordenskjöld, M.; Zapata-Gayon, C.; Lambert, B.; Norberg, E.; Olin, R. Chromosome aberrations and sister chromatid exchange in workers in chemical laboratories and a rotoprinting factory and in children of women of laboratory workers. Lancet 2:322-25; 1977.

Gabilondo, F.; Cobo, A.; Lisker, R. Anormalidades cromosomicas en pacientes tratados con chlorpromazina y perfenazina. Rev. Invest. Clin. 23:177-80; 1971.

Galle G. Cytogenetische und biochemische Untersuchungen bei Kindern mit Primidon-Monotherapie. M.D. Thesis. Erlangen, 1978.

Galloway, S. M.: What are sister chromatid exchanges? In Nichols, W. W.; Murphy, D. G. eds. DNA repair processes. Miami, Florida, 1977.

Ganner, E.; Osment, J.; Dittrich, P.; Huber, H. Chromosomes in patients treated with azathioprine. Humangenetik 18:231-36; 1973.

Garson, O. M.; Blackstone, A. M. Chromosome studies in users of spray adhesives. Med. J. Aust. 2:837-38; 1976.

Garza Chapa, R.; Leal Garza, C. H.; Molina Ballesteros, G. Analisis cromosomico en personas profesionalmente expuestas a contaminacion con plomo. Arch. Invest. Med. 8:11-20; 1977.
Gebhart, E. The treatment of human chromosomes in vitro: results. *In* Vogel, F.; Röhrborn, G. eds. Chemical mutagenesis in mammals and man. Berlin, Heidelberg, New York: Springer, 1970.
Gebhart, E. Chromosomenuntersuchungen bei Nuran^R-Therapie. Z. Kinderheilk. 111:109-17; 1971.
Gebhart, E. Chromosomenuntersuchungen bei Bactrim^R-Therapie. Med. Klin. 68:878-81; 1973.
Gebhart, E.; Schwanitz, G.; Hartwich, G. Chromosomenaberrationen bei Busulfan-Behandlung. Dtsch. Med. Wschr. 99:52-56; 1974.
Gebhart, E. Chromosomenuntersuchungen bei Bactrim^R-behandelten Kindern. Z. Kinderheilk. 119:47-52; 1975.
Gebhart, E.; Lösing, J.; Wopfner, F. Chromosome studies on lymphocytes of patients under cytostatic therapy. I. Hum. Genet. 55:53-63; 1980.
Gebhart, E.; Windolph, B.; Wopfner, F. Chromosome studies on lymphocytes of patients under cytostatic therapy. II. Hum. Genet. 56:157-67; 1980.
Genest, P.; Villeneuve, A. Lithium, chromosomes, and mitotic index. Lancet 1:1132; 1971.
Georgieva, V. L. Cytogenetic investigations in agricultural workers in occupational contact with pesticides. *In* Szabo, G.; Papp, Z., eds. Medical genetics. Proc. Symp. Debrecen-Hajduszoboszlo. Excerpta Medica, Amsterdam, 1977.
Gilmour, D. G.; Bloom, A. D.; Lele, K. P.; Robbins, E. S.; Maximilian, C. Chromosomal aberrations in users of psychoactive drugs. Arch. Gen. Psychiatr. 24:268-72; 1971.
Goetz, P.; Sram, R. J.; Dohnalova, J. Relationship between experimental results in mammals and man. I. Cytogenetic analysis of bone marrow injury induced by a single dose of cyclophosphamide. Mutat. Res. 31:247-54; 1975.
Grosse, K. P.; Schwanitz, G.; Rott, H. D.; Wissmüller, H. F. Chromosomen-untersuchungen bei Behandlung mit Antikonvulsiva. Humangenetik 16:209-16; 1972.
Hansteen, I. L.; Hillestad, L.; Thiis-Evensen, E.; Heldaas, S. S. Effects of vinyl chloride in man. A cytogenetic follow-up study. Mutat. Res. 51:271-78; 1978.
Hartley-Asp, B. Absence of chromsomal damage in the lymphocytes of patients treated with metronidazole for Trichomoniasis vulgaris. Toxocology Letters 4:15-19; 1979.
Hartwich, G.; Schwanitz, G. Chromosomenuntersuchungen nach chronischer Benzol-Exposition. Dtsch. Med. Wochenschr. 97:45-49; 1972.
Herha, J.; Obe, G. Chromosomal damage in chronic users of Cannabis: in vivo investigation with two-day leukocyte cultures. Pharmakopsychiatr. Neuro-Psychopharmakol. 7:328-37; 1974.
Herha, J.; Obe, G. Chromosomal damage in epileptics on monotherapy with carbamazepine and diphenylhydantoin. Hum. Genet. 34:255-63; 1976.
Hulten, M.; Lindsten, J.; Lidberg, L.; Ekelund, H. Studies on mitotic and meiotic chromosomes in subjects exposed to LSD. Ann. Génét. 11:201-10; 1968.
Jarvik, L. F.; Bishun, N. P.; Bleiweiss, H.; Kato, T.; Moralishvili, E. Chromosome examinations in patients on lithium carbonate. Arch. Gen. Psychiatr. 24:166-68; 1971.
Khan, H.; Khan, M. H. Cytogenetische Untersuchungen bei chronischer Benzolexposition. Arch. Toxicol. 31:39-49; 1973.
Kilian, D. J.; Picciano, D. J. Cytogenetic surveillance of industrial population. *In* A. Hollaender, ed. Chemical mutagens. Vol. 4. New York: Plenum Press, 1976; 321-339.
Kiraly, J.; Czeizel, A.; Szentesi, I. Genetic study on workers producing organophosphate insecticides. Mutat. Res. 46:224; 1977.
Knuutila, S.; Siimes, M.; Simell, O.; Tammisto, P.; Weber, T. Long-term use of phenytoin: effects on bone-marrow chromosomes in man. Mutat. Res. 43:309-12; 1977.
Knuutila, S.; Helminen, E.; Knuutila, L.; Leist, S.; Siimes, M.; Tammisto, P.; Westermarck, T. Role of clozapine in the occurrence of chromosomal abnormalities in human bone-marrow cells in vivo and in cultured lymphocytes in vitro. Hum. Genet. 38:77-89; 1977.
Konietzko, H.; Haberlandt, W.; Heilbronner, H.; Reill, G.; Weichardt, H. Cytogenetische Untersuchungen an Trichloräthylen-Arbeitern. Arch. Toxicol. 40:201-6; 1978.
Kotlarek, F.; Faust, J. Chromosomal investigations in children with pyknolepsy on dipropylacetate monotherapy. Hum. Genet. 43:329-31; 1978.

Krogh-Jensen, M. Chromosome studies in patients treated with azathioprine and amethopterin. Acta Med. Scand. 182:445-55; 1967.
Krogh-Jensen, M. Effect of azathioprine on the chromosome complement of human bone marrow cells. Int. J. Cancer 5:147-51; 1970.
Kucerova, M.; Zhurkov, V. S.; Polivkova, Z.; Ivanova, J. E. Mutagenic effect of epichlorohydrin. II. Analysis of chromosomal aberrations in lymphocytes of persons occupationally exposed to epichlorohydrin. Mutat. Res. 48:355-60; 1977.
Kucerova, M.; Polivkova, Z.; Batora, J. Comparative evaluation of the frequency of chromosomal aberrations and the SCE numbers in peripheral lymphocytes of workers occupationally exposed to vinyl chloride monomer. Mutat. Res. 67:97-100; 1979.
Lambert, B.; Ringborg, U.; Harper, E.; Lindblad, A. Sister chromatid exchanges in lymphocyte cultures of patients receiving chemotherapy for malignant disorders. Cancer Treat. Rep. 62:1413-19; 1978.
Littlefield, L. G.; Lever, W. E.; Miller, F. L.; Goh, K. Chromosome breakage studies in lymphocytes from normal women, pregnant women, and women taking oral contraceptives. Am. J. Obstet. Gynecol. 121:976-79; 1975.
Lubs, H. A.; Verma, R. S.; Summitt, R. L.; Hecht, F. Re-evaluation of the effect of spray adhesives on human chromosomes. Clin. Genet. 9:302-6; 1976.
Marquez-Monter, H.; Ruiz-Fargoso, E.; Velasco, M. Anticonvulsant drugs and chromosomes. Lancet 2:426-27; 1970.
Martin, P. A.; Thorburn, M. J.; Bryant, S. A. In vivo and in vitro studies of the cytogenetic effects of Cannabis sativa in rats and men. Teratol. 9:81-86; 1974.
Matsuyama, S.; Yen, F. S.; Jarvik, L. F.; Fu, T. K. Marijuana and human chromosomes. Genetics 74:175 (Abstr.); 1973.
Matsuyama, S. S.; Jarvik, L. F. Cytogenetic effects of psychoactive drugs. In Mendlewicz, J. ed. Genetics and psychopharmacology. Basel: Karger, 1975.
Matsuyama, S. S.; Charuvastra, V. C.; Jarvik, L. F.; Fu, T. K.; Sanders, K.; Yen, F. S. Chromosomes in patients receiving methadone and methadyl acetate. Arch. Gen Psychiatr. 35:989-91; 1978.
Matsuyama, S. S.; Yen, F. S.; Jarvik, L. F.; Sparkes, R. S.; Fu, T. K.; Fisher, H.; Reccius, N.; Frank, I. M. Marijuana exposure in vivo and human lymphocyte chromosomes. Mutat. Res. 48:255-66; 1977.
McQuarrie, H. G.; Scott, C. D.; Ellsworth, H. S.; Harris, J. W.; Stone, R. A. Cytogenetic studies on women using oral contraceptives and their progeny. Am. J. Obstet. Gynecol. 108:659-65; 1970.
Meretoja, T.; Vainio, H.; Sorsa, M.; Härkönen, H. Occupational styrene exposure and chromosomal aberrations. Mutat. Res. 56:193-97; 1977.
Mitelman, F.; Hartley-Asp, B.; Ursing, B. Chromosome aberrations and metronidazole. Lancet 2:802; 1976.
Mitelman, F.; Levan, G. Clustering of aberrations to specific chromosomes in human neoplasms. III. Hereditas 89:207-32; 1978.
Mitelman, F.; Strombeck, B.; Ursing, B.: No cytogenetic effect of metronidazole. Lancet 1:1249-50; 1980.
Monsalve, M. V.; Frota-Pessoa, O.; Garcia Campos, A. M.; Sette, H. A study of chromosomes of Schistosomiasis patients under oxamniquine (UK 4271) treatment. J. Toxicol. Environ. Hlth. 1:1023-26; 1976.
Mourelatos, D.; Faed, M. J. W.; Gould, P. W.; Johnson, B. E.; Frain-Bell, W. Sister chromatid exchanges in lymphocytes of psoriatics after treatment with 8-methoxypsoralen and long wave ultraviolet radiation. Brit. J. Dermatol. 97:649-54; 1977.
Müller, R.; Ritter, C. Zytogenetische Untersuchungen von Frauen während und nach der Einnahme hormonaler Kontrazeptiva. Zentralbl. Gynäkol. 100:347-54; 1978.
Murken, J. D. Chromosomenschäden durch Appetitanreger? Dtsch. Med. Wochenschr. 96:1696-97; 1971.
Nichols, W. W.; Miller, R. C.; Heneen, W.; Bradt, C.; Hollister, L.; Kanter, S. Cytogenetic studies on human subjects receiving marihuana and 9-tetrahydrocannabinol. Mutat. Res. 26:413-17; 1974.
Nielsen, J.; Friedrich, U.; Tsuboi, T. Chromosome abnormalities in patients treated with chlorpromazine, perphenazine, and lysergide. Brit. Med. J. 3:634-36; 1969.

Nordenson, I.; Beckman, G.; Beckman, L.; Nordström, S. Occupational and environmental risks in and around a smelter in northern Sweden. I. Chromosomal aberrations in workers exposed to arsenic. Hereditas 88:47-50; 1978.

Nordenson, I.; Beckman, G.; Beckman, L.; Nordström, S. Occupational and environmental risks in and around a smelter in northern Sweden. IV. Chromosomal aberrations in workers exposed to lead. Hereditas 88:263-67; 1978.

Nordenson, I.; Salmonsson, S.; Brun, E.; Beckman, G. Chromosome aberrations in psoriatic patients treated with arsenic. Hum. Genet. 48:1-6; 1979.

Obe, G.; Beek, B.; Dudin, G. The human leukocyte test system. V. DNA synthesis and mitoses in PHA-stimulated 3-day cultures. Humangenetik 28:295-302; 1975.

Obe, G.; Herha, J. Chromosomal damage in chronic alcohol users. Humangenetik 29:191-200; 1975.

O'Riordan, M. L.; Evans, H. J. Absence of significant chromosome damage in males occupationally exposed to lead. Nature 247:50-53; 1974.

O'Riordan, M. L.; Hughes, E. G.; Evans, H. J. Chromosome studies on blood lymphocytes of men occupationally exposed to cadmium. Mutat. Res. 58:305-11; 1978.

Pawlowitzki, I. H.; Bartsch-Sandhoff, M.; Okimoto, S. Chromosomenuntersuchungen nach Benutzung von Sprühklebern. Dtsch Med. Wochenschr. 99:1927-28; 1974.

Picciano, D. J.; Flake, R. E.; Gay, P. C.; Kilian, D. J. Vinyl chloride cytogenetics. J. Occup. Med. 19:527-30; 1977.

Picciano, D. Cytogenetic investigation of occupational exposure to epichlorohydrine. Mutat. Res. 66:169-73; 1979.

Pilinskaya, M. A. Chromosome aberrations in the persons contacted with Ziram. Genetika 6:157-63; 1970.

Pilinskaya, M. A. Results of cytogenetic examination of persons occupationally in contact with the fungicide zineb. Genetika 10/5:140-46; 1974.

Pilinskaya, M. A.; Zhurkov, V. S. The frequency of chromosome aberrations in persons who live in the districts with different expenditure of pesticides. Genetika 13/1:158-61; 1977.

Prost G.; Barthelemy, C.; Casset, J. C. Étude cytogénétique chez des ouvriers exposés aux hydrocarbures benzeniques. Arch. Mal. Prof. 37:544-48; 1976.

Purchase, I. F. H.; Richardson, C. R.; Anderson, D.; Paddle, G. M.; Adams, W. G. F. Chromosomal analyses in vinyl chloride-exposed workers. Mutat. Res. 57:325-34; 1978.

Rabello, M. N.; Becak, W.; DeAlmeida, W. F.; Pigati, P.; Ungaro, M. T.; Murata, T.; Pereira, C. A. B. Cytogenetic study on individuals occupationally exposed to DDT. Mutat. Res. 28:449-54; 1975.

Raposa, T. Sister chromatid exchange studies for monitoring DNA damage and repair capacity after cytostatics in vitro and in lymphocytes of leukaemic patients under cytostatic therapy. Mutat. Res. 57:241-51; 1978.

Reeves, B. R.; Pickup, V. L.; Lawler, S. D.; Dinning, W. J.; Perkins, E. S. A chromosome study of patients with uveitis treated with chlorambucil. Brit. Med. J. 4:22-23; 1975.

Robinson, J. T.; Chitham, R. G.; Greenwood, R. M.; Taylor, J. W. Chromosome aberrations and LSD. A controlled study in 50 psychiatric patients. Brit. J. Psychiatr. 125:238-44; 1974.

Roman, I. C. Rat and human chromosome studies after promazine medication. Brit. Med. J. 4:172; 1969.

Rossi, A.; Sebastio, L.; Ventruto, V. Studio cromosomico in 14 soggetti psoriasici trattati con idrossiurea o azathioprine. Minerva Med. 64:1728-32; 1973.

Ruzicska, P.; Peter, S.; Czeizel, A.: Studies on the chromosomal mutagenic effect of Benomyl in rats and humans. Mutation Res. 29:201; 1975.

Savage, J. R. K. Classification and relationships of induced chromosomal structural changes. J. Med. Genet. 12:103-22; 1975.

Schinzel, A.; Schmid, W. Lymphocyte chromosome studies in humans exposed to chemical mutagens. The validity of the method in 67 patients under cytostatic therapy. Mutat. Res. 40:139-66; 1976.

Schmid, E.; Bauchinger, M.; Pietruck, S.; Hall, G. Die cytogenetische Wirkung von Blei in menschlichen peripheren Lymphocyten in vitro und in vivo. Mutat. Res. 16:401-6; 1972.

Schwanitz, G.; Gebhart, E.; Rott, H. D.; Schaller, K. H.; Essing, H. G.; Lauer, O.; Prestele, H. Chromosomenuntersuchungen bei Personen mit beruflicher Bleiexposition. Dtsch. Med. Wochenschr. 100:1007-11; 1975.

Selezneva, T. G.; Korman, N. P. Analysis of chromosomes of somatic cells in patients treated with antitumour drugs. Genetika 9/12:112-18; 1973.
Skerfving, S.; Hansson, K.; Mangs, C.; Lindsten, J.; Ryman, N. Methyl-mercury-induced chromosome damage in man. Environ. Res. 7:83-98; 1974.
Stenchever, M. A.; Frankel, R. S.; Jarvis, J. A. Effect of diazepam on chromosomes of human leukocytes in vivo. Am. J. Obstet. Gynec. 107:456-60; 1970.
Stenchever, M. A.; Kunysz, T. J.; Allen, M. A. Chromosome breakage in users of marihuana. Am. J. Obstet. Gynecol. 118:106-13; 1974.
Stevenson, A. C.; Bedford, J.; Hill, A. G. S.; Hill, H. F. H. Chromosomal studies in patients taking phenylbutazone. Ann Rheum. Dis. 30:487-500; 1971.
Stevenson, A. C.; Clarke, G.; Patel, C. R.; Hughes, D. T. D. Chromosomal studies in vivo and in vitro of trimethoprim and sulphamethoxazole. Mutat. Res. 17:255-60; 1973.
Stevenson, A. C.; Patel, C. Effects of chlorambucil on human chromosomes. Mutat. Res. 18:333-51; 1973.
Stevenson, A. C.; Silcock, S. R.; Scott, J. T. Absence of chromosome damage in human lymphocytes exposed to allopurinol and oxipurinol. In vivo and in vitro studies. Ann. Rheum. Dis. 35:143-47; 1976.
Szentesi, I.; Hornyak, E.; Czeizel, A. Chromosome analysis in PVC workers. *In* Szabo, G.; Papp, Z. eds. Medical Genetics. Proc. Symp. Debrecen-Haiduszoboszlo. Excerpta Medica, Amsterdam, 1977.
Thiess, A. M.; Fleig, I. Chromosome investigations on workers exposed to styrene/polystyrene. J. Occup. Med. 20:747-49; 1978.
Thiess, A. M.; Fleig, I. Analysis of chromosomes of workers exposed to acrylonitrile. Arch. Toxicol. 41:149-52; 1978.
Tolchin, S. F.; Winkelstein, A.; Rodnan, G. P.; Pan, S. F.; Nankin, H. R. Chromosome abnormalities from cyclophophamide therapy in rheumatoid arthritis and progressive systemic sclerosis (scleroderma). Arthritis Rheum. 17:375-82; 1974.
Tough, I. M.; Court Brown, W. M. Chromosome aberrations and exposure to ient benzene. Lancet 1:684-85; 1965.
Tough, I. M.; Smith, P. G.; Court Brown, W. M.; Harnden, D. G. Chromosome studies on workers exposed to atmospheric benzene. The possible influence of age. Europ. J. Cancer 6:49-55; 1970.
Van Bao, T.; Szabo, I.; Ruzicska, P. Czeizel, A. Chromosome aberrations in patients suffering acute organic phosphate insecticide intoxication. Humangenetik 24:33-57; 1974.
Verschaeve, L.; Kirsch-Volders, M.; Susanne, C.; Groetenbriel, C.; Lecomte, A.; Roossels, D. Genetic damage induced by occupationally low mercury exposure. Environ. Res. 12:1976.
Verschaeve, L.; Kirsch-Volders, M.; Hens, L.; Susanne, C. Chromosome distribution studies in phenyl mercury acetate exposed subjects in age-related controls. Mutat. Res. 57:335-47; 1978.
Walker, S.; Price Evans, D. A.; Benn, P. A.; Littler, T. R.; Halliday, L. D. C. Phenylbutazone and chromosomal damage. Ann. Rheum. Dis. 34:409-15; 1975.
Watson, W. A. F.; Petrie, J. C.; Galloway, D. B.; Bullock, I.; Gilbert, J. C. In vivo cytogenetic activity of suplhonylurea drugs in man. Mutat. Res. 38:71-80; 1976.
Whang-Peng, J.; Leventhal, B. G.; Adamson, J. W.; Perry, S. The effect of daunomycin on human cells in vitro. Cancer 23:113-121; 1969.
White, B. J.; Driscoll, E. J.; Tijo, J. H.; Smilack, Z. H. Chromosomal absorption rates and intravenously given diazepam. A negative study. J. Am. Med. Ass. 230:414-17; 1974.
Whorton, E. B.; Bee, D. E.; Kilian, D. J. Variations in the proportion of abnormal cells and required sample sizes for human cytogenetic studies. Mutat. Res. 64:79-86; 1979.
Wolff-Schreiner, E. C.; Carter, D. M.; Schwarzacher, H. G.; Wolff, K. Sister chromatid exchanges in photochemotherapy. J. Invest. Dermatol 69:387-91; 1977.
Yoder, J.; Watson, M.; Benson, W. W. Lymphocyte chromosome analysis of agricultural workers during extensive occupational exposure to pesticides. Mutat. Res. 21:335-40; 1973.
Zhurkov, V. S.; Fichidzhyan, B. S.; Batikyan, G. G.; Arutyunyan, R. M.; Zil'fyan, V. N. Cytogenetic examination of persons in contact with chloroprene under industrial conditions. Tsitol. Genet. 11:13-15; 1977.
Zudova, Z.; Landa, K. Genetic risk of occupational exposures to haloethers. Mutat. Res. 46:242-43; 1977.

19
Short-term Cytogenetic Tests in Modern Society

by SHELDON WOLFF*

It hasn't been too many years since we thought that man had 48 chromosomes. The major breakthrough that disabused us of that notion was T. C. Hsu's discovery that exposure of cells to hypotonic solutions before fixation caused the chromosomes to separate from one another. This advance, along with T. T. Puck's discovery of how to grow mammalian cells in culture as if they were bacteria and P. C. Nowell's discovery that exposure of human lymphocytes to phytohemagglutinin causes the cells to undergo mitosis, has led to an explosion of research in mammalian cytogenetics.

Before these technical advances, most investigators interested in chromosomes worked with plants or insects, in which the cytology was far more favorable than it was in mammalian cells. Because certain species of plants, in particular, have their DNA organized into a small number of rather large chromosomes that are very favorable for analysis, many of the basic principles of cytogenetics were discovered in experiments with plant cells. With the advent of the new techniques facilitating mammalian cell studies, many plant cytogeneticists turned their attention to animal cytogenetics, as did some biophysicists and radiobiologists who were interested in how ionizing radiations brought about their effects. It should be noted that before this, radiation biophysics already had prospered mightily from chromosomal studies, which were particularly felicitous for the development of target theory and the kinetics of chromosome aberration induction. The main reason for this was that we no longer had to rely on inferences from surviving cells that were not hit by radiation, but could now study the hit targets themselves, which were large chromosomes visible under the microscope. As a result, from the very beginning cytogenetics provided a tool by which it was possible to detect detrimental effects of insults to the genes and chromosomes.

*Laboratory of Radiobiology and Department of Anatomy, University of California, San Francisco, San Francisco, CA 94143.

Further interest in mammalian studies, with a concomitant increase in the number of people involved in this type of research, came from the discovery that many human malformations and diseases have a chromosomal or cytogenetic basis. Thus, within a very short time, we have observed a transition whereby investigators interested in chromosomes now move directly into mammalian cytogenetics without first studying plant or insect cytology. Generally speaking, the mammalian cell has now become the preferred system for the study of basic as well as many applied aspects of cytology and cytogenetics.

In the technologically advanced civilization in which we live, man, as well as the rest of the biosphere, is exposed to a large number of chemicals, many of which are mutagens and carcinogens. The relation between the two has contributed to the development of a new field called genetic toxicology. Although this field has a large basic science component, it also has great practical importance, for many of the chemicals have overt chromosomal effects and the study of mutagenic carcinogens provides a very fast and economical means for the identification of potential carcinogens.

There are two very different needs for the development of short-term tests for genotoxic compounds. The first is to provide a system that can be used to screen the very large numbers of chemicals to which we are exposed in order to see if they are potentially dangerous. The second is to estimate the degree of hazard these chemicals present to man. Philosophically different approaches are required for these needs. For the first, we require a system that is extremely sensitive to the effects of the chemicals, so that all potentially dangerous compounds can be identified; for the second, we require a system that has the sensitivity and repair capacity of man so that we can estimate what the human effects might be. It should be noted that we also need to determine if the cytogenetic phenomena have purely genetic effects that will affect future generations or have somatic effects that could lead, for instance, to cancer in the exposed individuals.

The material in this book gives vivid testimony to the diversity of cytogenetic tests currently being used for short-term screening of chemicals and even of exposed people. The need for such a large number of tests becomes evident when one considers the many different questions that must be addressed. Among these are questions regarding: (a) whether or not the chemicals need metabolic activation to be converted from a promutagenic or procarcinogenic form to a proximate or ultimate mutagenic form; (b) whether the chemicals are effective at any part of the cell cycle or induce lesions that only lead to damage when the chromosomes replicate in S; (c) whether there are in vivo–in vitro differences in action; (d) whether there are cell-specific or tissue-specific effects; and (e) whether there are effects that can pass through the meiotic sieve, i.e., are not selected against before they produce viable gametes and zygotes.

By a proper selection of tests and their correct use, such as the ascertainment that cells treated in all parts of the cycle are observed at their first metaphase after treatment, many of these questions can be answered in several of the tests carried out in somatic cells. Other questions dealing with transmissible effects can be obtained only with tests in germ cells.

To date, the use of short-term tests to determine if a chemical is potentially dangerous has been largely empirical. Compounds have been administered to cells with or without a metabolic activation system and the numbers of mutations, chromosome aberrations, sister chromatid exchanges, etc., have been observed. In some cases, in vivo tests have been performed in which the chemicals were administered to an adult animal or an embryo. Although it is certainly possible to continue in this way, many more insights about the nature of mutagenesis will come from basic research in which mutation is related to the chemical structure of the various classes of compounds. Only in this way will we discover the chemical rules regarding whether or not a compound is mutagenic and, parenthetically, carcinogenic. Only with such rules will we be able to predict whether or not a compound will be genotoxic.

All in all, the tests we have still provide important ways to screen for potentially dangerous chemicals and also give background information necessary for making risk estimates for man. In the area of risk estimate, however, quantitative differences between man and the test organism in physiology, tissue specificity, cell kinetics, and genetic repair become very important. Nevertheless, we need to rely on the data from the short-term tests (imperfect as they might be) because we must make reasonable estimates of the risks before we discard a potentially beneficial chemical.

Some of the cytogenetic tests, such as the human lymphocyte tests, can be used to estimate the degree of damage to humans exposed to chemicals. Because the data obtained from exposed people can vary widely and be compromised by the way in which they are gathered, standard proper protocols are required for any test system used to determine the hazards involved in human exposure, and responsible sophisticated investigators are needed to carry out the tests and to interpret them. Epidemiologically sound methods must be used to ascertain that any observed effect is not caused by other phenomena such as radiation, chemotherapy, or viruses. Because of the increased reliance on such short-term tests by regulatory agencies, there is also a need for sophisticated scientific cytogenetic acumen on the part of regulators to interpret the results before acting precipitously. In particular, the events at the Love Canal in New York have proven that regulators and some of those working in cytogenetics should become aware that the presence of a few cells with cytogenetic abnormalities does not mean a given individual will develop neoplastic disease, have spontaneous abortions, or give birth to malformed children. The presence of such aberrations can be an indication that the population has been exposed, as in the case of radiation exposure of the survivors of the atom bomb explosions in Japan. But, although there has been an increase in neoplastic disease in this exposed population, the individuals with cancer cannot be related to those with increased numbers of chromosomally aberrant cells.

We have entered an era in which genetic toxicology will become of ever-increasing importance, perhaps largely because of the relation between mutagenesis and carcinogenesis. Less than 25 years ago, H. J. Muller, the discoverer of the phenomenon of induced mutation, made a plea at the Fifth International Conference on Radiobiology that we not express concerns about the genetic effects of chemicals because it would divert attention from our arguments regarding the hazards of radiation. At the time, one could argue that although the position was not scientifically defensible, it was socially responsible. Today we are long past that point, for now the hazards of chemicals seem far to outweigh those of radiation. The degree of exposure has made us aware that we must protect the genetic material not only in germ plasm but also in somatic cells. The methods described in this book, which have been developed in a short span of time and will certainly be supplemented with newer methods, give us the tools to test potentially dangerous substances quickly and easily and to monitor ourselves for hazardous exposures. It is hoped that the data obtained from such short-term cytogenetic tests will enable us to go even farther and determine the degree of risk engendered by exposure to genotoxic compounds.

Risk estimation, like the question of how to predict whether a chemical will be mutagenic and/or carcinogenic, is an area that will require fundamental studies on the effects of chemicals in mutagenesis and carcinogenesis. In particular, theoretical aspects will have to be considered in the development of biologically meaningful models by which we will be able to interpolate between the lowest effective dose levels measured experimentally and the levels found with zero exposure. It is in just such a dose range, where we cannot determine the shape of the dose-response curve empirically, that most human exposures will occur. Since very large numbers of people can be exposed, even low individual risk could lead to large numbers of affected persons in the population. To estimate how many requires intimate knowledge of the shapes of dose-curves. This will be but a first step, for after being armed with this information, we will have to make far-ranging decisions about the acceptability of the risk. Here we will need input from many segments of society, but because decisions about the relative effects of various technological alternatives must be made, the input of knowledgeable scientists will be indispensable.

The short-term tests we have today constitute state-of-the-art methods by which we can test potentially dangerous chemicals and also begin to monitor exposed people. The main needs of the future, however, will require that we buttress these with additional tests that might be even more useful for determining genotoxicity, and that we apply the principles of basic research to gain an understanding of the fundamental mechanisms involved. Only then will we be able to make accurate estimates of the degree of hazard and be in a position to decide subtle questions of risk versus benefit.

Name Index

Name Index

Acton, A. B., 165, 172
Adelberg, E. A., 363
ADLER, I.-D., 250, 257, 259, 262, 265, 266, 267, 269, 271, 297, 300, 302, 306, 307, 314
Al-Bader, A. A., 373
Allen, J. W., 126, 172
Altenberg, E., 1
Altenburg, L. C., 123
Ames, B. N., 12, 162, 204, 207, 240, 243, 354
Anders, G. J. P. A., 363
Anderson, D., 204
Andersson, A., 20
Andrews, A. W., 204
Anger, H., 359
Arndt-Jovin, D. J., 326
Arrighi, F. E., 123, 366, 386
Ashburner, M., 110
Ashihara, T., 363
Atkin, N. B., 216
AU, W., 204, 207, 212, 215
Auerbach, C., 290
Aula, P., 369

Bachmann, K., 91
Baetcke, K. P., 91
Baim, A. S., 117
Baker, R. H., 111
Bakulima, E. D., 163
Barker, C. J., 176
Barlogie, B., 369
Barnes, J. M., 98
Basler, A., 257
Bauchinger, M., 389
Bauknecht, T., 66
Becchetti, A., 25
Bedford, J. S., 373
Beechey, C. V., 264, 282, 285, 300, 306
Beek, B., 359, 389
Beermann, W., 125
Belling, J., 119
Belyaeva, V. N., 163
Bender, M. A., 62, 66, 266, 370
Benedict, W. F., 208
Bianchi, N. O., 117, 118
Bissell, D. M., 243
Bleier, H., 355
BLOOM, S. E., 39, 61, 139, 166, 169, 171, 172, 173, 177
Blumenthal, A. B., 117
Bodell, W. J., 19
Bonney, R. J., 243

Bovery, T., 107, 162
Bowles, B., 250
Boyce, R. P., 12
Boyd, G. A., 121
Boyes, J. W., 112, 116
Boyum, A., 332
BREWEN, J. G., 232, 234, 261–65, 267, 271, 282, 283, 284, 298, 305
Bridges, C. B., 108
Brinkley, B. R., 214, 370
Brookes, P., 243
Brooks, A. L., 259, 373
Brooks, M. A., 117
Brown, S. W., 116
Bryant, E. M., 65
BUCKTON, K. E., 232
Bulsiewicz, H., 268
Bunker, M. C., 293
Burdette, W. L., 162
Burk, P. G., 242
Butler, M. A., 208

Cacheiro, N. L. A., 290, 291, 293, 294, 295, 299, 300
Cairns, J., 22, 25
Callis, J., 337
Carlson, J. G., 118, 119, 120
Carothers, E. B., 108
Carrano, A. V., 60, 63, 333, 341, 344, 367
Carrier, W. L., 12, 15, 238
Carter, T. C., 290, 292, 294, 295, 302, 303, 306, 307
Caspari, E. W., 112
Caspersson, T., 123, 124, 125
Cassidy, J. D., 112
Catcheside, D. G., 81, 261
CATTANACH, B. M., 263, 264, 266, 290, 291, 292, 294, 295, 297–300, 310, 312, 314
Cerutti, P. A., 14, 16, 18
Chaganti, R. S. K., 162
Chang, M. C., 269
Chen, T. R., 167
Clark, J. M., 268
Clarkson, J. M., 24
Clayson, D. B., 203
Clayton, F. E., 110
Cleaver, J. E., 12, 22, 238
Clegg, H. M., 292
Cochran, D. G., 108, 110, 111
Cohen, M. M., 205
Cohen, S., 110
Cole, A., 14, 19

415

Name Index

Cole, R. S., 22
Collins, A. R. S., 371
Coluzzi, M., 111
Comings, D. E., 6
Conen, P. E., 386
Conger, A. D., 94, 121
Connell, J. R., 216
Coombs, M. M., 204
Coquerelle, T., 14
Corry, P. M., 14, 19
Coulter, W. H., 327
Countryman, P. I., 205
Cox, B. D., 264, 266
Cox, D. M., 339
Cox, R., 19
Cozzi, R., 118
Cram, L. S., 326, 330, 337, 344
Crissman, H. A., 332, 338
Crocker, A. J. M., 298
Crosland-Taylor, P. J., 327
Cross, D. P., 118
Crossen, P. E., 389, 396
Crozier, R. H., 119, 123
Cumming, R. B., 299, 314

Damjanov, I., 19
Darlington, C. D., 94, 116, 119
Darzynkiewicz, Z., 363
Das, B. C., 124
DasGupta, U. B., 25
Davidson, R. L., 363
Davisson, M. T., 298
Day, M. F., 117
Dean, P. N., 329
DEAVEN, L. L., 332, 333, 336, 339, 343
Debec, A., 118
de Boer, P., 293
Deknudt, G. H., 261, 295, 297, 298, 306, 396, 403
De Marco, A., 118
Demerec, M., 162
Denton, T. E., 166
de Serres, F. J., 204, 208
Deutsch, W. A., 20
DiMinno, R. L., 268, 269
Dittrich, W., 327, 329
Djordjevic, B., 12, 22
Dolfini, S. F., 118, 123, 126, 127
Dosik, H., 108
Drury, K. C., 357
Dutrillaux, B., 171

Eagle, H., 239
Echalier, G., 117, 118
Edenberg, H., 20, 24
Edwards, R. G., 307
Egolina, N. A., 171
Ehling, U. H., 267, 299, 300, 310
Ehrenberg, L., 16

Eichen, E. M., 291, 295
England, J. M., 242
Epstein, S. S., 130, 161, 299
Eto, H., 264
Evans, E. P., 123, 250, 252, 261, 289, 290, 291, 303, 304, 305, 315
EVANS, H. J., 83, 92, 98, 108, 165, 172, 268, 269, 271, 370, 387, 389
Ewig, R. A. G., 24

Fairchild, L. M., 94, 121
Falconer, D. S., 291, 295
Fedoroff, S., 117
Fontana, P. G., 124
Ford, C. E., 264, 289, 292, 298, 299, 303, 305
Ford, E. H. R., 120
Forejt, J., 295
Forni, A., 389
Foster, G. G., 110
Fox, A. S., 117
Fox, D. P., 124
Francke, U., 123, 125
Freed, J. J., 239
Freedlender, E. F., 171
French, W. L., 119
Frew, J. G. H., 116
Friedberg, E. C., 20
Frost, H. B., 116
Fujiwara, Y., 22
Fulwyler, M. J., 327
Funes-Cravito, F., 396
Furukawa, M., 234
Fye, R. L., 127

Gagné, R., 257
Galloway, S. M., 389
Garner, J. V., 243
Garner, R. C., 243
Gatti, M., 123, 124, 127
Gaudin, D., 244
GAULDEN, J. G., 118, 119, 120, 121, 127, 128
GEBHART, E., 389, 397
Gefter, M. L., 363
Generoso, W. M., 271, 290, 292, 294, 297–300, 302, 303, 307, 309, 311, 314
Georgieva, V. L., 396
Gerber, G. B., 263
Gerchman, L. L., 22
German, J., 65
Giles, N. H., 81
Gilliavod, N., 282, 300
Godoy, H. M., 243
Goetz, P., 271
Goh, K., 205
Göhde, W., 327, 329
Gold, J. R., 166
Goldschmidt, R., 116

Goncalves, L. S., 112
Goth-Goldstein, R., 25
Goto, K., 171
Gottlieb, F. J., 112
Grace, T. D. C., 117
Grauwiler, J., 228
Gray, J. W., 329, 330, 333, 335, 340, 341, 344
Grdina, D., 367
Gregorova, S., 295
Greilhuber, J., 124, 125
Grell, E. H., 110
Griffen, A. B., 293
Gropp, A., 292, 294
Grosse, K. P., 394
Grunberger, D., 18

Habazin, V., 16, 24
Halfer, C., 117, 118, 123
Han, A., 16, 24
Hanawalt, P., 20, 22
Hansen, D., 363
Hanson, C. V., 118
Hanson, R. S., 243
Hariharan, P. V., 16
Harm, H., 20
Harris, H., 360
Hart, R. W., 162
Hayflick, L., 205
Heddle, J. A., 62
Heindorff, K., 104
Hertwig, P., 260, 290
Herzenberg, L. A., 326
Higginson, J., 161
Hildebrand, C. E., 363, 366
Hilscher, B., 256
Hilscher, W., 256
HITTELMAN, W. N., 358, 359, 363, 366, 367, 369, 370, 372, 374, 375
Hoehn, H., 337
Hollander, W. F., 292
Holliday, R. A., 67
Holm, D. M., 326
Holmquist, G., 123, 124
Hoo, S., 250
Horikawa, M., 117
Howard-Flanders, P., 12, 24
HSU, T. C., 39, 123, 139, 148, 204, 205, 212, 215, 268, 333, 375, 409
Huang, C. C., 205, 234
Hulett, H. R., 327
Hulten, M., 387
Humphrey, R., 214
Hunt, L. M., 113

Imai, H. T., 124, 125
Ishidate, M., Jr., 204, 207, 217
Ishii, Y., 62, 66

Jacobs, A. J., 242
Jacobs-Lorena, M., 118
Jaeger, I., 359
Jagiello, G. M., 256, 279
Janssens, F. A., 108
Jaszczak, K., 306
Jensen, R. H., 335
John, B., 120, 123
Johnson, C. D., 112
Johnson, R. T., 355, 357, 358, 363, 370, 371, 373, 375
Jones, G. H., 126
Jones, K. P., 279
Jorgenson, T. A., 312, 313, 314
Jovin, T. M., 326

Kaina, B., 104
Kakpakov, V. T., 117
Kang, Y. S., 166
Kao, F., 336
Karran, P., 20, 22
Kato, H., 37, 62, 66, 93, 139, 171, 339, 358
Kaufman, M. H., 298
Kaufmann, B. P., 110
Keith, D. H., 334
Kieser, D., 244
KIHLMAN, B. A., 25, 85, 88, 89, 91–96, 98, 99, 101
Kilian, D. J., 386, 387, 389
Kim, M. A., 34
King, M., 123, 124
King R. C., 108
Klasterska, I., 124
KLIGERMAN, A. D., 166, 169, 171, 177
Knudsen, I., 314
Koch, C. J., 19
Kocisova, J., 266
Kohn, K. W., 24
Koller, P. C., 260, 290, 292
Korenberg, J. R., 171
Kraemer, P. M., 331, 336, 338
Kriek, E., 18
Krishan, A., 332
Krishna, M., 300
Krogh-Jensen, M., 391
Kronborg, D., 89, 93, 95
Kubota, M., 124
Kurita, Y., 216
Kurten, S., 359
Kurtti, T. J., 117, 118
Kusyk, C. J., 205

Labrecque, G. C., 127
La Cour, L. F., 94, 119
Lagler, K. F., 161
Laishes, B. A., 240, 243
Lambert, B., 205
Lanar, D., 118
Landureau, J. C., 118

Lang, R., 297, 302, 306, 311, 315
Lansky, G. S., 386
LATT, S. A., 64, 123, 126, 139, 171, 172, 369
Lau, Y. F., 363, 366, 369
Lawley, P. D., 243
Lea, D. E., 81, 261
Lederberg, J., 325
Lefevre, G., Jr., 123
Legator, M. S., 130, 256
Lehman, J. M., 337
Lehmann, A. R., 14, 19, 24
Lemeunier, F., 123
Lengemann, F. W., 259
Lengerova, A., 295
Lennartz, M., 14
Léonard, A., 261, 263, 265, 282, 290, 295, 297, 298, 300, 306, 312, 396
Lesley, M. M., 116
Lester, D. S., 124
Lett, J. T., 19
Levan, A., 1, 82
Levan, G., 216, 386
Lewis, E. B., 116, 119
Lewis, K. R., 120
LIANG, J. C., 119
Lieberman, M. W., 242
Lin, R. C., 243
Lindahl, T., 20
Lindmo, T., 346
Lindsley, D. L., 110
Littlefield, L. G., 205
Livneh, Z., 20
Lo, L. W., 244
Lokki, J., 109
Longnecker, D. S., 173
Longwell, A. C., 164, 165
Lorke, D., 256
LOVEDAY, K. S., 67
Ludlum, D. B., 22
Luippold, H. E., 257, 265, 266, 271
Luning, K. G., 298
Lyon, M. F., 261, 263, 264, 266, 290, 291, 292, 294, 295, 298, 306

Machemer, L., 256
Madle, S., 208
Maeki, K., 112
Maier, P., 228
Malashenko, A. M., 265
Malling, H. V., 243
Manyak, A., 257
Marks, E. P., 117
Marshall, A., 243
Martin, C. N., 238, 242
Martin, P., 118
Masui, Y., 357
Matsui, S., 357
Matsuoka, A., 207, 216

Matsushima, T., 165
Matter, B. E., 228, 359
Maudlin, I., 116, 118, 124
Mazia, D., 363
McCann, J., 12, 162, 204
McClintock, B., 119
McClung, C. E., 108
McGaughey, R. W., 269
McGill, M., 214
McGrath, R. A., 14, 19, 238
McLean, A. E. M., 243
McLeish, J., 98
Melamed, M. R., 327, 331, 332
Mendelsohn, M. M., 337
Meneghini, R., 24
Merchant, D. J., 239
Meredith, R., 289, 290, 291, 292, 294, 295, 305, 306, 315
Meyer, O. A., 314
Michaelis, A., 82, 83, 101, 104
Michalopoulos, G., 243
Migalovskaya, V. N., 163, 164
Milani, R., 112
Miller, E. C., 161, 162
Miller-Faures, A., 67
Mitchell, J. B., 373
Mitelman, F., 216, 386
Mohr, O. L., 108
Monesi, V., 256
Montgomery, T. H., 107
Moore, P. D., 67
Moorhead, P. S., 205
Morgan, T. H., 108
Morgan, W. F., 389
Morris, T., 261, 263
Mosbacher, 112
Moseley, H., 294
Moutchen, J., 268, 270
Mukherjee, A. B., 117
Mullen, H. A., 241
Muller, H. J., 1, 108, 411
Muramatsu, S., 261, 264, 306

Nakamura, W., 264
Natarajan, A. T., 25, 204, 207, 257, 266, 335, 379
Naylor, A. F., 112, 116
Neal, G. E., 243
Neifakh, A. A., 163
Neuhäuser, A., 267, 307, 314
Nettesheim, P., 232
Neuhaus, P., 124
Nevstad, N. P., 58
Nichols, W. W., 355
Nicolini, C., 363
Nicoloff, H., 104
Nielsen, J., 389
Nordenson, I., 396
Nowell, P. C., 216, 409

Name Index 419

Oakberg, E. F., 249–50, 256, 268, 269, 278
Obara, Y., 358
Obe, G., 208, 359, 379, 389
Ockey, C. H., 216
O'Conner, P. J., 243
Odashima, S., 204, 207
Ohanessian, A., 117, 118
Ohno, S., 297
Oishi, K., 165
Okada, S., 19
Okada, Y., 361
Oldiges, H., 343
Oliver, N., 123, 125
Oppermann, K., 163
O'Riordan, M. L., 387, 389
Ormerod, M. G., 14, 19
Otto, F., 334, 343, 348

PAINTER, R. B., 12, 19, 22, 24, 238
Painter, T. S., 123
Palitti, F., 25
Pankova, N., 163
Parida, B. B., 127
Park, G. S., 363
Patrick, M. H., 20
Patten, S. F., 327
Patterson, M. C., 65
Payne, H. S., 271, 279, 284
Pechkurenkov, V. L., 164
Pedersen, T., 278
Pederson, T., 363
Pegg, A. E., 22, 25
Penrose, L. A., 337
Perez-Mosquera, G., 113
Perje, A. M., 112
Pero, R. W., 205
Perry, P., 31, 34, 58, 89, 92, 98, 171, 172, 191
Petersen, D. F., 333, 336, 339
Pettijohn, D., 22
Philippe, C., 118
Phillips, R. J. S., 263, 291, 298
Picciano, D. J., 386, 387, 389
Picken, D. I., 260
Pinkel, D., 329
Pitot, H. C., 162, 243
Podlasek, S., 113
Pokrovskaia, G. L., 163
Polani, P. E., 256
Pollard, C. E., 299, 314
Poste, G., 355
Povirk, L. F., 18
Prein, A. E., 169
PRESTON, R. J., 261–62, 263, 265, 267, 305
Prokof'yeva-Bel'govskaya, A. A., 163
Puck, T. T., 206, 336, 339, 363, 409
Pudney, M., 118
Purchase, I. F. H., 216, 389

Rackham, B. D., 176
Rai, K. S., 119

Raicu, P., 165
Ramaiya, L. K., 270
Rao, A. P., 373
Rao, P. N., 355, 357, 358, 359, 362, 363, 366, 367, 369, 372–375
Raposa, T., 335
Rasmussen, R. E., 12, 22, 238
Rathenberg, R., 257, 259
Ray, D. T., 112
Read, C. B., 119, 127, 128
Read, J., 82
Rees, E. D., 216
Regan, J. D., 15, 20
Reynolds, R. J., 25
Rhone, D., 367
Riazuddin, S., 20
Rieger, R., 82–83, 101, 104
Riles, L. S., 116, 119
Robbins, E., 363
Roberts, F. I., 165
Roberts, J. J., 12, 18
Roderick, T. H., 298
Rogers, A. W., 241
Röhrborn, G., 257
Roizman, B., 355
Romashov, D. D., 163
Rommelaere, J., 67
Rønne, M., 125
Ross, M. H., 108, 110, 111
Roth, L. M., 110
Rothenbuhler, W. C., 112
Rowley, J. D., 216
Rupert, C. S., 12
Rupp, W. D., 24
Russell, L. B., 285, 291, 292, 300
Russell, W. L., 284, 285, 290

Saez, F. A., 113
Samson, L., 22, 25
SAN, R. H. C., 238, 239, 240, 244
Sandberg, A. A., 65, 358
Sasaki, M. S., 205
Saura, A., 109, 123
Savage, J. R. K., 108, 298, 389
Savkovic, N. V., 261
Sawada, M., 217
Sawada, S., 19
Sax, K., 81, 261
Scarpelli, D. G., 165, 172
Schatz, S. A., 239
Schinzel, A., 391, 397
Schleiermacher, E., 250, 257
Schmid, E., 389
SCHMID, W., 228, 391, 397
Schnedl, W., 363
Schneider, A., 119
Schneider, I., 117, 118
Schneiderman, S., 363
Schor, S. L., 371

Name Index

Schordert-Slatkine, S., 357
SCHRECK, R. R., 29
Schubert, I., 83, 101, 104
Schwartzman, J. B., 96
Schwarz, 162
Scott, D., 83
Seabright, M., 123
Searle, A. G., 261, 263, 264, 282, 285, 290, 291, 294, 298, 300, 306, 307, 310, 313
Setlow, R. B., 12, 15, 20, 162, 238
Shafer, D. A., 62, 63
Sharma, A., 119
Sharma, A. K., 119
Shaw, D. D., 124
Shaw, M. W., 209
Shelby, M. D., 208
Shellenbarger, D. L., 118
Sheu, C. W., 302, 309, 311, 312
Shiriashi, Y., 65
Siciliano, M. J., 206
Singer, B., 16
Singh, K. R. P., 117
Sinnhuber, R. O., 165, 172
SHULER, C. F., 29
Skerfving, S., 396
Skogen-Hagensen, M. J., 329
Skowronek, W., 112
Slizynski, B. M., 292
Smith, B. D., 264
Smith, C. N., 110
Smith, K. C., 15, 16
Smith, S. G., 112, 113, 119
Snell, G. D., 260, 290, 292
Snodgrass, P. J., 243
Sognier, M. A., 374, 375
Sokoloff, A., 110, 112
Solberg, A. N., 163
Sorsa, V., 123
Sotomayor, R. E., 299, 314
Sperling, K., 355, 367
Spurr, A. R., 125
Sram, R. J., 265, 266, 299, 314
Srdic, Z., 118
Stadler, J. K., 363, 381, 383
Statham, C. N., 165
Steen, H. D., 346
Steinkamp, J. A., 327, 329
Stenman, S., 357
Stetka, D. G., 62, 208
Stevens, N. M., 119
STICH, H. F., 165, 172, 238, 239, 240, 243, 244
Stocker, E., 241
Stort, A. C., 112
Stott, W. T., 165, 172
Strong, L. C., 162
Stubblefield, E., 333, 341, 342, 355
Sturelid, S., 83, 91, 98, 99
Styles, J. A., 204

Sugatt, R. H., 169
Sugimura, T., 165
Summers, W. C., 25
Sumner, A. T., 123
Sun, C., 19
Sun, L., 16
Sunkara, P. S., 357
Surkova, N. I., 265
Sutton, W. S., 107
Svärdson, G., 162
Swartout, M. S., 299, 300

Taisescu, E., 165
Takagi, N., 355
Tarkowski, A. K., 279, 307
Tates, A. D., 257, 266
Taylor, J. H., 30, 190
Tazima, Y., 112
Tease, C., 126
Thangaard, G. H., 166
Thompson, J. N., Jr., 110
Thornton, J., 121
Tice, R., 65
Tjio, J. H., 206
Tobey, R. A., 332, 338, 339, 366
Tolmach, L. J., 12, 22
Tonomura, A., 205
Torvik-Greb, M., 112
Traut, W., 112
Trosko, J. E., 12
Tsoi, R. M., 164
Tsuchida, W. S., 269
Tsutsui, T., 214
Tsytsugina, V. G., 163, 164
Tyrrell, P. D., 278

Uchida, I. A., 269
Usehima, N., 113, 116
Unakul, W., 369

Vakhrameeva, N. A., 163
Van Buul, P. P. W., 65
Van Dilla, M. A., 327, 343
Varghese, A. J., 16
Vaughan, V. L., 363
Verma, R. A., 108
Vickery, V. R., 124
Vogel, W., 66
Vosa, C. G., 123

Waggie, K. S., 292
Wagoner, D. E., 112
Wainberg, M. A., 361
Waldeyer, W., 107
Waldren C. A., 370, 371, 373, 375
Walker, H. C., 269
Ward, D. C., 332
Wasserman, W. J., 357
Watson, W. A. F., 394

Webb, G. C., 123, 124, 125
Weigle, J. J., 24
Weinstein, D., 204, 207
Weinstein, I. B., 18
Welshons, W. J., 250
Wennström, J., 268, 269
Westra, J. G., 18
Wheeler, L. L., 123
Wheeler, M. R., 110
Wheless, L. L., 327
White, M. J. D., 108, 109, 112
Whitten, M. J., 110
Whorton, E. B., 387
Wickham, L., 285, 300
Wiemann, H., 311, 315
Wildrick, D. M., 358
Wilkes, A., 116
Williams, G. M., 238, 240, 243
Williams, R. W., 14, 19, 238
Wilson, B., 363
Wilson, E. B., 107, 108

Winking, H., 292
Witkin, E. M., 24, 25
Witkowski, R., 359
Wittman, 259, 262
Wogan, G. N., 243
WOLFF, S., 31, 34, 62, 89, 93, 171, 172, 191, 208
Woodhead, D. S., 175
Woollam, D. H. M., 120
Woyke, J., 112
Wray, W., 333, 341, 342
Wright, D. A., 357
Wyss, C., 117

Yaneshevsky, R., 367
Yoder, J., 396
Yosida, T. H., 339

Zakharov, A. F., 171
Zante, J., 335
Zirkle, C., 120

Subject Index

Subject Index

Acetyl-2-aminofluorene (AAF), 18
Acheta (house cricket), 124
Actinomycin D., 18
Adriamycin, 65, 374, 375
Aedes (mosquito), 111
Aedes aegypti, 117
Aedes albopictus, 117
Alkylating agents:
  DNA crosslinks with, 16-17, 22, 24
  DNA damaged by, 16-18
  mutagenic, 391
  in SCE, 37, 61, 63, 64, 66, 91, 92, 93
Ameca splendens, 175-76
American Type Culture Collection (ATCC), 117
Ames' bacterial mutation system, 204, 207, 208, 216
Ames *Salmonella* test, 150, 152, 153, 155
Androgenesis in fishes, 163
Aniline, 38
Anopheles mosquito, 111
Antheraea eucalypti, 117
Antibiotics, cytostatic, 391
Ants, 113, 124
Apis mellifera (honey bees), 112
Arginine deficient medium (ADM), 239, 240, 247-48
Ataxia telangiectasia, 64, 65, 162, 205
Australian sheep blowfly (*Lucilia cuprina*), 111-12
Autoradiography, 22, 190
Avian embryos (*see* Chick embryos)

BB-chromatid, 89, 93
BCNU (1, 3-bis-(2-chloroethyl)-3-nitrosourea), 266
Bean root tips (see *Vicia faba*)
Bedbug (*Cimex*), 113
Bees (*Apis mellifera*), 112
Benzo(a)pyrene, 18
Blatella germanica (German cockroach), 110-11, 117, 118
Bleomycin, 18, 88, 91, 371-74
Bloom's syndrome, 64, 65, 162, 190
Bone marrow in micronucleus test, 221-28
Brachystola magna (grasshopper), 107
BrdU techniques (bromodeoxyuridine):
  for chick embryo tests, 139, 142, 145
  for detection of SCE, 30-35, 37, 58, 59, 60, 66, 126, 139, 141, 142, 145, 171, *tables,* 32-34, 141-44
  for host-mediated assay, 234
  staining protocols, 31-35

substitution into DNA, 30-35
BrdUrd (5-bromodeoxyuridine), 89, 90, 93, 94, 191
BrUra (5-bromouracil), 89, 92
BUDR-labeling, 388-89, 397
Burkitt's lymphoma, 355
Busulphan, 397

Caffeine in SCE, 66-67
Captan, 314
Carcinogens:
  DNA repair synthesis and, 12, 237-48
  environmental, 325-26
  in fishes, cytogenetic studies of, 161-81
  mammalian cells in identification of, 203, 216-17
  SCE induction and, 37
  somatic mutation theory of, 161-62
Chemical mutagens:
  chemotherapeutic agents, 387-94, 397, 400, *tables,* 388-89, 392-94, 398-401
  in chick embryo test, 150-53
  chromosome aberrations in persons exposed to, 385-408
  deviations of chromosome number from, 396-97, *table,* 396
  industrial chemicals, 394, 396, *table,* 395
  investigations of, 385-87
  methodology of studies, 387-91
Chemotherapy, chromosomes affected by, 391, 394, 397, 400, *tables,* 392-94, 398-401
Chick embryo, in formation of SCEs, 39
Chick embryo cytogenetic test (CECT), 137-59, *tables,* 141-44, 146, 151
  application of, 154-55
  chemical mutagens in, 150-53
  chemicals evaluated in, 141-42, *tables,* 142-44
  chromosome aberrations analyzed, 149-50
  chromosome preparations for, 146-48
  procedure for, 140-42
  rate of, 150
  results of, 152-54
  solid-tissue preparations for, 147-48
  staining techniques for, 148-49
Chinese hamster cells, 5, 33, 58, 59, 61, 65, 66, 67, 91, 99, 124, 196
  in host-mediated assay, 233, 234
  in SCE, 59, 62, 65, 67
  in *Vicia faba* studies, 91
Chinese Hamster Ovary (CHO) cells:
  in chromosome aberration assays, 203,

425

Subject Index

205–6, 210, 211
  in premature chromosome condensation, 358, 361, 369, 372, 373
Chinese hamsters, 285, 355
  in flow cytometry, 334, 335, 341–47
  in male germ cell analysis, 250, 256, 257, 259, 260, 266
Chortoicetes (locust), 124
Chortophaga viridifasciata (grasshopper), 119, 124, 128
Chromatid break, 189–90
Chromatid gap, 5–6, 189
Chromatid interchange, 190
Chromosome aberrations, 1–9
  in chick embryo cytogenetic test, 149–50
  chromatid-type, 3–7, 188–90
  epidemiological approach to, 385–408
  of fishes, 163
  in human lymphocytes (*see* Human peripheral blood lymphocyte cultures)
  in male germ cell cytogenetics, 259, 260, tables, 258, 259, 260
  mammalian cells in assays for, 203–19
  in persons exposed to chemical mutagens, 385–408
  of *Vicia faba* root tips, 85–89, 94–95, 96
  (*see also* premature chromosome condensation)
Chromosome breakage, 2–3, 62, 205, 214–15
"Chromosome pulverization," 355
Chromosome-type aberrations (G or early S), 6–8 (*see also* Human peripheral blood lymphocyte cultures)
Cimex (bedbug), 113
Clastogens, 1, 7, 325
  DNA damage in living cells, 204
  ideal cells for study, 108–9
  insect cells for testing, 107–35
  in mammalian cell cultures, 208, 209, 215–16
  premature chromosome condensation in study of, 353, 359, 377
  in SCE, 37–39, 58, 59
CNU-ethanol, 257, 266
Cockayne's syndrome, 66
Cockroaches, 110–11, 118
Colcemid, 141, 147, 166, 208, 212, 213
  in premature chromosome condensation, 355, 361, 362, 363
Colchicine, 166, 204, 208
  in host-mediated assay, 233–34
  in male germ cell cytogenetics, 250
  Vicia faba root tips treated with, 85
Coregonus peled, 164
CP (cyclophosphamide), 39, 172, 174, 208, 256, 257, *table,* 209
Culex mosquito, 111
Cyclamate, 256

Cyclohexamine, 256
Cyprimus carpio, 164
Cytogenetic tests, 8–9
  short-term, 409–12
  training of investigators, 8–9

Decticus verrucivorus, 108
Dichroplus silveraguidoi, 113
Diepoxybutane, 64, 270
Diethystilbesterol, 37–38
Diol epoxide I, 18
Diphenyl, 38
DMBA (dimethylbenzanthracene), 18, 208
DNA:
  BrdU substitution into, 30–35
  BrUra and BrdUrd substitution in SCE, 89, 91
  in carcinogenesis of fishes, 162
  in chick embryos, 138
  chromosome breakage and, 204
DNA crosslinks, 16, 22, 24, 62
  with alkylating agents, 16–17, 22, 24
  with proteins, 16, 22, 24
DNA repair, 11–27
  base damage, 14–16, 20–22
  complex lesions, 14–15
  damage by alkylating agents, 16–18
  double-strand breaks, 14, 19–20
  excision repair, 22
  lesions caused by chemical agents, 16–19
  lesions from ionizing radiation, 13–15
  lesions from ultraviolet radiation, 15–16
  postreplication and induced repair, 24–25
  in premature chromosome condensation, 370, 374, 375
  pyrimidine dimers and, 15, 20, 162
  single-strand breaks, 11, 13–15, 17–18, 19
  in sister chromatid exchanges, 25, 61–63
DNA repair synthesis:
  advantages and limitations of, 242–43
  autoradiographic detection procedure, 239–42
  carcinogenic potential and, 244
  cell cultures in, 238–39
  in cultured human fibroblasts, 237–48
  DNA repair inhibition in, 244
  rat liver in, 240, 243–44
DNA synthesis:
  in human peripheral blood cultures, 185
  in male germ cell cytogenetics, 266
  in premature chromosome condensation, 357, 370–71
Dominant lethal test, 308–9, 315
Down's syndrome, 66, 205
Drosophila (fruit flies), 108, 110, 111, 116, 117, 118, 128
Drosophila melanogaster, 110, 111, 117, 118, 123, 126, 127, 167
Drug abuse, chromosome changes from, 394

EMS (ethyl methanesulfonate), 63, 65, 66, 139, 291, 299, 309, 310
ENU (ethylnitrosourea), 270
Environmental mutagens, 325-26
  premature chromosome condensation and, 377
  testing with flow cytometry (see Flow cytometry)
EOC (8-ethoxycaffeine), 85-86
Ephestia küniella (Mediterranean meal moth), 112
Epidemiological approach to chromosome aberrations, 385-408
Epstein-Barr virus, 38, 205
Escherichia coli, 24, 25
Ethylnitrosourea, 16

Fanconi's anemia, 162
  SCE in, 62, 64, 66
FdUrd (5-fluorodeoxyuridine), 93-94
Fibroblasts, human, DNA repair synthesis in, 237-48
Fishes:
  advantages and disadvantages as study material, 164-65, table, 165
  cytogenetic studies of genotoxic agents in, 161-81
  in vitro methods of studies, 175-76
  in vivo aquatic model system of studies, 168-78
  karyotypes of, 163, 167-70
  methodology of studies, 165-67
  mitotic cells in premature chromosome, condensation, 358
  model systems approach to cytogenetics of, 167-68, table, 167
  Soviet studies of radiation effects on, 163-64
Flour beetles (Tribolium), 112
Flow cytometers (FCM), 326-30, 334, 337, 338, 341-45, 347
  lasers in, 327, 328, 329
Flow cytometry:
  for cytogenic testing of environmental mutagens, 325-51
  instrumentation and technology, 326-31
  isolation and staining of single chromosomes, 333-35, 347-48
  preparation and staining of cells, 330-33, 347-48
  results, 335-45
  studies of isolated chromosomes, 340-45, 347-48
  whole-cell studies, 335-40
Flow microfluorometers (FMF), 326, 335 (see also Flow cytometers)
FPG (fluorescence plus Giemsa or harlequin) technique, 89, 91, 148, 191

Fruit flies (see Drosophila)
Fundulus embryos, 163

Giemsa staining, 2, 95 (see also FPG)
Grasshoppers, chromosome studies of, 107, 108, 113, 119, 124, 127, 128
Greater milkweed bug (Oncopeltus fasciatus), 113, 116
Guanine, 22, 25
Guinea pigs, 264, 266
Gynogenesis, radiation-induced, in fishes, 163

Habrobracon (parasitis wasp), 112
Halothane, 204-5
Hamsters:
  Armenian, 58
  Chinese (see Chinese hamsters)
  golden, 264, 266, 278
HeLa cells in premature chromosome condensation, 355, 357, 359, 361, 370, 373
Heritable translocation test in mice, 289-323
  application of, 310-12
  compared with other tests, 308-10
  conclusions and discussion, 314-17
  costing of, 312-14
  cytological features of, 295, 297
  dosage in treatment, 311
  importance of, 307-8
  phenotypes in, 291
  production of F_1 animals, 300-301
  properties of translocation heterozygotes, 291-97
  Robertsonian translocation, 292, 294, 297-300, 304, 306, 308
  screening of F_1 progeny, 307
  size of experiment, 311-12, table, 312
  spontaneous incidence of translocations, 297-98
  summary of results with chemical agents, 314-17, table, 316-17
  translocation by chemical mutagens, 299-300
  translocation by ionizing radiation, 298-99
  translocation in F_1 males, detecting, 301-6, tables, 301, 302, 306
  translocation in females, 300, 306-7, table, 307
  translocation male sterility, 294-95
  translocation semisterility, 291-94
Herpes simplex virus, 39
Herring gull (Larus argentatus), testing embryos for SCE, 145
Heteromorphism, 109
Hoechst staining, 30, 31, 34, 95
Host-mediated assay for cytogenic studies, 231-35
  cell preparation, 232-33
  diffusion chambers, 232, 233

host animals for, 233, 234
human cells in, 231, 232
House cricket (*Acheta*), 124
House fly (*Musca domestica*), 112, 124, 167
Howell-Jolly bodies, 223
Human chorionic gonadotropin (HCG), 279, 280
Human chromosomes, 108
  in flow cytometry, 337, 338, 341, 347
  premature condensation of (*see* premature chromosome condensation)
Human fibroblasts, DNA repair synthesis in, 237-48
Human peripheral blood lymphocyte cultures, 183-202, 205
  blood culture technique, 184-85
  bridges and fragments in anaphase cells, 191-92
  chromatid-type aberrations, 188-90
  chromosomal aberrations in metaphase cells, 185-91
  host-mediated assay of, 231, 232
  methods, problems, and results with assay system, 193-96, 206-11, *table*, 194-95
  micronuclei in interphase cells, 192-93
  mutagens in, 193, 196, *table*, 194-95
  SCE in, 190-91
  types of damage assayed, 185-86, 187
Hydroxylamine, 270

IMS, 300, 309
Industrial chemicals, chromosome damage from, 394, 396, *table*, 395
Insect cells, 107-35, 409
  chemicals tested with, 126-27
  chromosome banding, 123-26
  differential chromatid staining, 126-27
  genetic investigations of, 110-3, *tables*, 114, 115
  methods and techniques of study, 119-27
  organ cultures, 118-19
  tissue culture cells, 116-18
  tissues in situ, 113, 116
Interstitial deletion, 5
Intrachromal exchanges, 4
Isochromatid gap, 190

Leukemia, premature chromosome condensation in study of, 353, 367, 369, 375-79
Leukemia virus, 39
Locust (*Chortoicetes*), 124
Love Canal, 411
LSD (lysergic acid diethylamide), 394
Lucilia cuprina (Australian sheep blowfly), 111-12
Male germ cell cytogenetics, 249-76
  cell-killing effect, 256-57
  chemical mutagens, 265-68, 270-71, *tables*, 266, 268
  ionizing radiation (gamma rays) in, 259-60, *tables*, 260, 261, 262, 264
  propagation techniques, 250, 252
  radiation effects, 267-70
  RBE of gamma rays and X-rays, 264
  spermatogenesis of mice, 249-51
  spermatogonial mitoses, 252-60
  translocation multivalents at diakinesis, 260-67
  translocation test, 265-67
  X-rays in, 260-65
Maleic hydrazide, 86, 87, 91, 98, 99
Mammalian cells:
  abnormalities, recording of, 212-15
  anaphase abnormalities, 214, 215
  in assays for chromosome aberrations, 203-19
  data analysis, 215-16
  experimental procedure for, 211-16
  harvesting cultures of, 208-9
  identification of chromosome damage, 209-11
  metabolic activation system, 207-8
  methods of assays, 206-11
  test materials for studies of, 205-6
  treatment, continuous, 211-12
  treatment, pulse, 212
Mammalian oocytes in mutagenicity studies, 277-87
  chemical mutagens, 281-82, *table*, 283
  ionizing radiation, 282-85
  maturation of, 278-79
  techniques for, 280-81
Matthiola, 116
MC (mitomycin C), 37, 38, 39, 63, 64, 66, 67, 190, 257, 265, 266, 300, 309, 310, 314, *table*, 266
Mediterranean meal moth (*Ephestia küniella*), 112
8-Methoxypsoralen, 61-62
3-Methyladenine, 20
Methylnitrosourea, 16, 271
Mice:
  chromosomes, 108
  germ cell cytogenetics, 249-76 (*see also* Male germ cell cytogenetics)
  heritable translocation test (*see* Heritable translocation test in mice)
  in host-mediated assay, 233, 234
  in micronucleus test, 224-25
  oocytes of, 278
  SCE formation in, 39, 58, 59
  spermatogenesis of, 249-51
  spermatogonial mitoses of, 252-60, *table*, 257-60
  trophoblast cells of, 358-59
Micronuclei, formation and morphology of, 221-25

Subject Index

Micronucleus test, 221-29
  procedures for, 224-28
Minimal essential medium (MEM), 239
Misgurnus fossilis (loach), 163, 165
Mitomycin C (*see* MC)
Mitotic cells, premature chromosome condensation and, 354-59, 361-63
Mitotic index, 212-14
Mitotic poisons, 1, 204-5
  in mammalian cell cultures, 209, 210, 212-14, 215, 217
MMS (methyl methane sulfonate), 127, 139, 171-72, 174, 270, 299, 310
MNNG (N-methyl-N'-nitro-N-nitroso guanidine), 22, 25
MNU (methylnitrosourea), 16, 271
MOC (8-methoxycaffeine), 85-87
Mosquitoes, 111, 117
Mouse (*see* Mice)
6-MP, 309
MPNA (methylphenylnitrosamine), 85-86
MS, 310
Mudminnow (*see Umbra limi*)
Musca domestica (*see* House fly)
Mutagens (mutagenic agents):
  bean root tips treated with, 84-85
  chemical (*see* Chemical mutagens)
  in fishes, 161-78
  in heritable translocation test, 299-300, 314-17, *table*, 316-17
  in human peripheral blood lymphocyte cultures, 193, 196, *table,* 194-95
  human exposure to, 354
  mammalian cells in identification of, 203, 216-17
  in mammalian oocyte studies, 281-82, *table,* 283
  premature chromosome condensation for detection of, 353-84
  SCE induction and, 37, 84-85, 89-95, 150-53
Myleran, 270
Myrmecia (ants), 113, 124

N-acetoxy-acetylaminofluorene, 39
Natulan, 310
NTA (nitrilotriacetic acid), 312, 313
Nucleic acid antimetabolites, 391

Oncopeltus fasciatus (greater milkweed bug), 113, 116
Onion roots for chromosome studies, 82
Oocytes (*see* Mammalian oocytes)
Opossum chromosomes, 334
Osteichthes, 161

Periplaneta americana (cockroach), 117, 118
Pesticides, chromosome damage from, 396

Phosphotriesters, 16, 18
Plant chromosomes, 81-82, 409
Poecilia formosa, 162
Premature chromosome condensation (PCC), 353-84
  cell fusion technique in, 361-63
  as cytogenetic tool, 363, 366-67, 369
  discovery of, 355, 357
  measurement of chromosome aberrations, 369-79
  methodology of, 359-63
  mitotic cells in, 354-59, 361-63
  nature of, 355-59
  polyethylene glycol (PG) in, 362-63
  virus propagation in, 360-61
Procarbazine, 257, 260, *tables,* 258, 259, 268
Psychotropic agents, mutagenic, 391, 394
Pyrimidine dimers, 15, 20, 162

Rabbits, 264, 278
Rat liver in DNA repair synthesis, 240, 243-44
Rats:
  Long-Evans, 207
  oocytes of, 278
Rhesus monkeys, 264
RNA, alkylation of, 16, 17
RNA synthesis:
  in human peripheral blood cultures, 185
  in premature chromosome condensation, 357
RNase (ribonuclease) in slide preparation, 95

S9 (liver homogenate extract), 207
Salmo irideus, 164
Salmo salar (Atlantic salmon), 163, 164, 165
Samia cecropia, 116
SCD (*see* Sister chromatid differentiation)
SCE (*see* Sister chromatid exchange)
Schistocerca, 124
Scomber scombrus, 164
Scorpaena porcus, 164
Sendai virus in premature chromosome condensation, 353, 359-63
Sister chromatid differentiation (SCD), 31, 141, 142, 145, 146, 148, 149, 171
Sister chromatid exchange (SCE):
  agents capable of inducing (strongly positive), *table,* 40-52
  agents exhibiting mixed or weak induction, *table,* 53-57
  analysis, 2, 29-80
  avian species other than chick in testing of, 145
  basic information on, 35, 37
  biological significance of formation, 63
  BrdU techniques for detection of, 30-35, 37, 58, 59, 60, 126, 139, 142, 145, 171, *table,* 32-34

chemical mutagens and, 389, 391, 397, 398, 400, 401
chick embryos in detection of, 137–59
in chromosome fragility diseases, 64–66
clastogens in induction of, 37–39, 58, 59
DNA damage, repair, and synthesis in, 25, 61–63
DNA interchange in, 29–30, 67
in fish studies, 171, 172, 173, 176, 177
in human lymphocyte cultures, 190–91, 193, 196
interpretation of induction tests, 60–61
in vitro and in vivo studies of, 39, 58
mechanism of formation, 66–67
pesticide effects on, 396
in premature chromosome condensation, 381
Vicia faba studies of, 89–96, 98
viruses and, 38–39, 63
Specific locus mutation test, 309–10
Streptonigrin, 88
SV 40 virus, 63

TB-chromatid, 89, 93
TEM (triethylene melamine), 257, 270, 271, 291, 292, 299, 309, 310, 312, 313, 314, *tables*, 257, 258, 266
TEPA (tris (1-aziridinyl)-phosphine oxide), 86, 88, 265–66, 299, 314
THIOTEPA (tris (1-aziridinyl)-phosphine sulphide), 92, 93, 265
TMU (1,3,7,9-tetramethyluric acid), 86, 88
TPA (12-0-tetradecanoyl-phorbol-13-acetate), 63
Tradescantia species, chromosome studies, 81–82
Triatoma, 116, 117, 124
Triatoma infestans, 116, 118
Tribolium castaneum, Tribolium confusum (flour beetles), 112

Trimethylphosphate, 256
Trypanosoma cruzi, 116
TT-chromatid, 89, 93

UDS (unscheduled DNA synthesis), 237, 238
Ultraviolet radiation, 65, 66, 92, 162
Umbra limi (mudminnow), 169–75
Umbra pygmaea, 169

Vicia faba (broad bean), root tips, 81–105, 167
advantages and disadvantages in use of, 98–99
chromosome aberrations, 85–89, 94–95, 96
chromosomes, 82–83, 124, 125
early radiation experiments, 82
growing roots from seeds, 83–84
karyotypes, 82–83
mutagenic agents for treatment of, 84–85, 89–95
reconstructed karyotypes in study of mutagen sensitivity, 101–5
in SCE studies, 84, 89–96, 98
slide preparation and staining, 94–95
slide scoring, 96
treatment with mutagenic agents, 84–85
Vincristine, 204
Vinyl chloride, 394

Warramaba (grasshopper), 124
Wheat (*Triticum*) hybrids, chromosomes, 355
W-reaction of bacteriophage, 24–25

Xeroderma pigmentosum, 64, 65–66, 162
X-rays:
in fish studies, 163–64
in male germ cell cytogenetics, 260–65
in SCE, 65, 191–92
in *Vicia faba* studies, 85

About the Editor

T. C. Hsu holds the Olga Keith Wiess Chair for Cancer Research and is Professor of Biology and Chief of the Section of Cell Biology at the M. D. Anderson Hospital and Tumor Institute. In 1952, Dr. Hsu discovered the hypotonic method of spreading mitotic mammalian chromosomes so that they could be accurately counted and characterized for the first time. He has written more than 200 scientific articles and papers. He serves on the editorial boards of *Cytogenetics, Cancer Research,* and the *Journal of Experimental Zoology,* and on the advisory board of *Chromosoma.* He has also coedited 10 volumes of *An Atlas of Mammalian Chromosomes* and 3 volumes of *Chromosome Atlas: Fish, Amphibians, Reptiles and Birds.* His most recent book is *Human and Mammalian Cytogenetics: An Historical Perspective.*